MULTIVARIATE ANALYSIS — IV

Multivariate Analysis — IV

PROCEEDINGS OF THE FOURTH INTERNATIONAL SYMPOSIUM ON MULTIVARIATE ANALYSIS

Edited by

PARUCHURI R. KRISHNAIAH

Department of Mathematics and Statistics
University of Pittsburgh, Pittsburgh, Pa., U.S.A.

1977

NORTH-HOLLAND PUBLISHING COMPANY — AMSTERDAM · NEW YORK · OXFORD

Library of Congress Catalog Card Number 76 21717
North-Holland ISBN 0 7204 0520 3

Published by:
NORTH-HOLLAND PUBLISHING COMPANY
AMSTERDAM · NEW YORK · OXFORD

Distributors for the U.S.A. and Canada:
Elsevier North-Holland, Inc.
52 Vanderbilt Avenue
New York, N.Y. 10017

Library of Congress Cataloging in Publication Data

International Symposium on Multivariate Analysis,
4th, Wright State University, 1975.
Multivariate analysis — IV.

Includes index.
1. Multivariate analysis — Congresses.
I. Krishnaiah, Paruchuri R. II. Title.
QA278.I58 1975 519.5'3 76–21717
ISBN 0–7204–0520–3

PRINTED IN THE NETHERLANDS

H. Hotelling (1895–1973)

P. C. Mahalanobis (1893–1972)

Dedicated to the memory of the late
H. Hotelling and P. C. Mahalanobis

FOREWORD

From time to time I have been reviewing current trends of research work in theory and applications of statistical multivariate analysis. Each time, while going through the vast literature on multivariate analysis appearing in a wide variety of journals, I am specially struck by the following: While refinements of Fisherian methods continue to be made, relatively few new lines of investigations are started. New extensions of univariate methods to multiple measurements are being made, which are no doubt useful, but there has not been adequate discussion of the number or choice of variables. In spite of the enormous increase in multivariate methods they do not seem to be rich enough to meet all practical demands. Thus, the subject of multivariate analysis seems to be still in a developing stage, and there is a need for symposia of the kind which Dr. Krishnaiah is organizing to provide proper orientation to research in this area. We are all indebted to Dr. Krishnaiah for his foresight and thought and for giving his valuable time for a worthwhile endeavour.

The present volume contains the proceedings of the fourth Symposium on Multivariate Analysis. The first, the second and the third were held in the years 1965, 1968 and 1972 respectively. The papers presented cover a wide variety of topics including the early history of multivariate statistical analysis. The papers on distribution theory, decision procedures and pattern recognition contain new techniques in multivariate methodology. Of special interest and somewhat new to statisticians are papers on inference in stochastic processes and information and control.

Multivariate Analysis—IV is thus a welcome addition to the earlier three volumes edited by Dr. Krishnaiah. These volumes together with the *Journal of Multivariate Analysis*, which is also edited by Dr. Krishnaiah, provide a valuable record of the progress made in the past and of the current research work on multivariate analysis.

I hope the series of periodic symposia on multivariate analysis for assessing progress of research and providing stimulus for starting new lines of research will continue under the able leadership of Dr. Krishnaiah.

New Delhi
March 1, 1976

C. R. Rao

EDITOR'S PREFACE

Multivariate analysis deals with problems connected with correlated variables in the areas of statistics and probability. The techniques of multivariate analysis play a vital role in drawing inferences from the data that arise in atmospheric, biological, engineering, medical, physical, social and other sciences. The Fourth International Symposium on Multivariate Analysis was organized to stimulate research and disseminate knowledge in this important area. This symposium was sponsored by the Aerospace Research Laboratories and held at Wright State University, Dayton, Ohio, during the period 16–21 June 1975. While arranging the program, the field of multivariate analysis was interpreted in a broad sense as has been done in the case of the *Journal of Multivariate Analysis*. The present volume consists of the invited papers presented at the above symposium. In these papers, outstanding workers in the field discuss the present state-of-the-art on a broad spectrum of topics in the theory and applications of multivariate analysis. The areas covered in this volume include characterization problems, classification and pattern recognition, contingency tables, decision procedures, design and analysis of experiments, directional data analytic methods, distribution theory, growth curves, hydrology and meteorology, information and control theory, prediction and filtering, reliability, statistical physics, and time series and stochastic processes. This volume is of great value to mathematical statisticians and probabilists, as well as scientists in other disciplines who are interested in the applications and methodology of multivariate analysis techniques.

I wish to express my sincere gratitude to several persons for their valuable help. Dr. J. Martin, Principal Deputy Assistant Secretary of the Air Force for Research and Development, was kind enough to make the Opening Remarks. I am grateful to Professor R.C. Bose for delivering the Inaugural Address. Thanks are due to Professors R.R. Bahadur, F.J. Beutler, D.J. DeWaal, A. Dvoretzky, S. Geisser, S.G. Ghurye, S.S. Gupta, A.T. James, D.R. Jensen, G. Kallianpur, L. Kanal, S. Kotz, J.C. Lee, R.D. LePage, E. Lukacs, F.J. Schuurmann, J.N. Srivastava, B.E. Trumbo, G. Wahba, V.B. Waikar, and S. Zacks, and Drs. J. Chandra, H.L. Harter, H.M. Hughes, R. Lundegard, and W.F. Mikhail for presiding over different sessions. I wish to express my appreciation to Professors D.A. Dawson, Y. Fujikoshi, L. Gleser, T. Kailath, G. Kallianpur, C.G. Khatri, J.C. Lee, K.V. Mardia, D. Middleton, S.K. Mitra, H. Nagao, M. Neuts, M. Okamoto, T. Pitcher, B. Rajput, M.M. Rao, P.K. Sen, T.W.F. Stroud, and G. Wahba for reviewing the papers.

I am grateful to Professors J.C. Lee and C.C. Maneri for their valuable help in making the local arrangements. Thanks are due to my colleagues Dr. H.L. Harter and Dr. D.A. Lee for their constant encouragement and advice on various administrative matters. I wish to express my appreciation to Professor M.M. Rao for his valuable suggestions in the organization of the program. Special thanks are due to the contributors to this volume and to North-Holland Publishing Company for their excellent cooperation. Most of the editing of this volume was done during my stay at the Air Force Flight Dynamics Laboratory.

I am deeply indebted to the management of the Aerospace Research Laboratories for their enthusiastic support to the field of multivariate analysis for more than a decade by sponsoring the four international symposia on multivariate analysis and by supporting the research in this area in various ways. Last but not least, I wish to thank my wife, Indira, and my brother, Madhu R. Paruchuri, for their constant encouragement in the organization of the four symposia on multivariate analysis.

P.R. Krishnaiah

CONTENTS

FOREWORD vii

EDITOR'S PREFACE ix

PART I: MULTIVARIATE DISTRIBUTION THEORY

R.C. Bose: Early History of Multivariate Statistical Analysis* 3

Aryeh Dvoretzky: Asymptotic Normality of Sums of Dependent Random Vectors 23

D.A.S. Fraser and Kai W. Ng: Inference for the Multivariate Regression Model 35

Yasunori Fujikoshi: Asymptotic Expansions for the Distributions of Some Multivariate Tests 55

A.T. James: Tests for a Prescribed Subspace of Principal Components 73

C.G. Khatri: Quadratic Forms and Extension of Cochran's Theorem to Normal Vector Variables 79

P.R. Krishnaiah and Jack C. Lee: Inference on the Eigenvalues of the Covariance Matrices of Real and Complex Multivariate Normal Populations 95

J.C. Lee, T.C. Chang and P.R. Krishnaiah: Approximations to the Distributions of the Likelihood Ratio Statistics for Testing Certain Structures on the Covariance Matrices of Real Multivariate Normal Populations 105

E. Lukacs: A Characterization of a Bivariate Gamma Distribution .. 119

Claude McHenry and A.M. Kshirsagar: Use of Hotelling's Generalized T_0^2 in Multivariate Tests 129

* Inaugural address

PART II: ESTIMATION, DECISION PROCEDURES AND DESIGN OF EXPERIMENTS

J. Kiefer: Conditional Confidence and Estimated Confidence in Multidecision Problems (with Applications to Selection and Ranking) 143

P.A.W. Lewis, L.H. Liu, D.W. Robinson and M. Rosenblatt: Empirical Sampling Study of a Goodness of Fit Statistic for Density Function Estimation 159

I. Olkin and M. Sylvan: Correlational Analysis when Some Variances and Covariances are Known 175

C. Radhakrishna Rao: Prediction of Future Observations with Special Reference to Linear Models 193

S. Zacks: Problems and Approaches in Design of Experiments for Estimation and Testing in Non-Linear Models 209

PART III: TIME SERIES AND STOCHASTIC PROCESSES

Richard E. Barlow and Frank Proschan: Asymptotic Theory of Total Time on Test Processes with Applications to Life Testing 227

Takeyuki Hida: Topics on Nonlinear Filtering Theory 239

Masuyuki Hitsuda and Hisao Watanabe: On a Causal and Causally Invertible Representation of Equivalent Gaussian Processes 247

G. Kallianpur: A Stochastic Equation for the Optimal Non-Linear Filter 267

Emanuel Parzen: Multiple Time Series: Determining the Order of Approximating Autoregressive Schemes 283

Balram S. Rajput and N.N. Vakhania: On the Support of Gaussian Probability Measures on Locally Convex Topological Vector Spaces 297

M.M. Rao: Inference in Stochastic Processes–VI: Translates and Densities 311

V.A. Statulevičius: Application of Semi-Invariants to Asymptotic Analysis of Distributions of Random Processes 325

PART IV: INFORMATION AND CONTROL THEORY

A.V. Balakrishnan: Stochastic Control of Systems Governed by Partial Differential Equations 341

Frederick J. Beutler, Benjamin Melamed and Bernard P. Zeigler: Equilibrium Properties of Arbitrarily Interconnected Queueing Networks 351

Carl W. Helstrom: An Introduction to Quantum Estimation Theory 371

T. Kailath and L. Ljung: A Scattering Theory Framework for Fast Least-Squares Algorithms 387

David Middleton: A New Approach to Scattering Problems in Random Media 407

Yu. A. Rozanov: Some System Approaches to Water Resources Problems III: Optimal Control of Dam Storage 431

PART V: CONTINGENCY TABLES, DIRECTIONAL DATA STATISTICS, AND PATTERN RECOGNITION

Herman Chernoff: Some Applications of a Method of Identifying an Element of a Large Multidimensional Population 445

S. Das Gupta: Some Problems in Statistical Pattern Recognition 457

Richard H. Jones: Multivariate Statistical Problems in Meteorology 473

Maurice Kendall: Multivariate Contingency Tables and Some Further Problems in Multivariate Analysis 483

K.V. Mardia: Mahalanobis Distances and Angles 495

Madan L. Puri and J.S. Rao: Problems of Association for Bivariate Circular Data and a New Test of Independence 513

Minoru Siotani and Ruey-Hwa Wang: Asymptotic Expansions for Error Rates and Comparison of the *W*-Procedure and the *Z*-Procedure in Discriminant Analysis 523

LIST OF CONTRIBUTED PAPERS 547

AUTHOR INDEX 549

PART I

MULTIVARIATE DISTRIBUTION THEORY

P.R. Krishnaiah, ed., *Multivariate Analysis–IV*
© North-Holland Publishing Company (1977) 3–22

EARLY HISTORY OF MULTIVARIATE STATISTICAL ANALYSIS*

R.C. BOSE
Colorado State University, Fort Collins, Colo., U.S.A.

In this paper we give a brief history of Multivariate Statistical Analysis from 1908–1939 beginning with the work of Student on the distribution of the correlation coefficient in a bivariate normal population. We trace the development of (i) Fisher's work using his geometrical method in the discovery of the exact distributions of various types of correlation coefficients. (ii) Wishart's discovery of the product moment distribution and its elaboration in various directions by Bartlett, Wilks and Hsu. (iii) Hotelling's work on T^2 (the generalized Student ratio), principal components and cannonical correlations, and its extension by Girshick and Hsu. (iv) The work of the Indian School (Mahalanobis, Bose and Roy) on the rectangular coordinates and the exact distribution of the classical and Studentized form of the Mahalanobis distance, D^2, and its connection with Fisher's discriminant analysis. Our survey ends with the work of Roy, Fisher and Hsu on the distribution of latent roots of determinantal equations related to multivariate normal populations.

1. Introduction

I am glad to take this opportunity to give a brief history of early work on multivariate analysis at this international symposium held in honour of Professors Harold Hotelling and P.C. Mahalanobis. My survey covers the period 1908–1939 beginning with the work of Student on small sample theory and ending with the discovery of the distribution of the latent roots of determinantal equations related to the covariance matrices of multivariate normal distributions. Both Professors Hotelling and Mahalanobis did pioneering work in the field during the period under survey, and this, therefore, gives me an opportunity of presenting their work in its proper perspective. Towards the end of this period, i.e., from 1933–1939, I was myself a member of the Indian Statistical Institute and (with the collaboration of S.N. Roy) took an active part in the mathematical development of the ideas introduced by Mahalanobis. Professor Hotelling visited the

* This research was supported by The Aerospace Research Laboratories under Contract No. F33615–74–C–1198.

Institute in 1939–1940 and gave us a course of lectures on his recent multivariate work. A year earlier, 1938–1939, Professor R.A. Fisher was in Calcutta and discussed with us in detail the connections between his own and Hotelling's work and the work of the Indian school on the distributions related to Mahalanobis' D^2-statistic. However, though I told him about the work that Roy was doing on the latent roots he gave us no indication at all that he himself and P.L. Hsu were also investigating the same area. The papers by Roy, Hsu, and Fisher all appeared in the same year, 1939. Roy's discovery of the latent root distribution was, therefore, a completely independent piece of work. The cut off date of 1939 might seem somewhat arbitrary but, as a matter of fact, the discovery of the importance of latent roots led to an explosive development of the subject to which it would be impracticable to do any justice in the course of a short survey like the present one.

2. Beginnings of small sample theory

The beginnings of small sample theory can be traced back to the work of Student (1908–1909) in two important papers. In the first of these two papers he considers the problem of judging the significance of the mean of a small sample, taking into account the fact that the estimate of the standard deviation is itself subject to sampling errors. He, therefore sets out to find the distributions of s and t where

$$s^2 = \sum_{i=1}^{n} (x_i - \bar{x})^2/n, \qquad t = (\bar{x} - \mu)/s \tag{2.1}$$

where $x_1, x_2, \ldots, x_n$ is a sample of size n from a normal population with mean μ and variance σ^2. He obtained the first four moments of s^2 and guessed from these the correct form for the k-th moment. This enabled him to derive the distribution of s in the form

$$s^{n-2} e^{-ns^2/2\sigma^2} \mathrm{d}s. \tag{2.2}$$

Of course the distribution of $\Sigma(x_i - \mu)^2/n$ had been obtained by Helmert as early as (1875). Student showed that s and $\bar{x}$ are independently distributed. He obtained the probability integral of t and computed the necessary tables for samples of size n, for $4 \leq n \leq 10$. He illustrated the use of these results and checked their accuracy by Monte Carlo methods using Macdonell's data (1901–1902) on the height and left finger measurements of 3000 criminals, from which he obtained 50 samples of size 4.

In his second paper Student considers the problem of finding the distribution of the correlation coefficient r, from a sample of size n from a bivariate normal population. He obtains the correct distribution for samples of size 2, showing that the sample correlation is either $+1$ or -1 but the proportion depends on the population correlation ρ. If the proportions are A and B, $A+B=1$, then $B=\cos^{-1}\rho/\pi$. Again, using Monte Carlo methods he correctly concludes that when there is no correlation between two normally distributed variables

$$y = y_0(1-r^2)^{(n-4)/2}dr \tag{2.3}$$

gives fairly closely the distribution of r from samples of size n.

Soper (1913) obtained the approximate value

$$\sigma_r^2 = \frac{1-\rho^2}{(n-1)^{1/2}}\left(1+\frac{11\rho^2}{4n}\right) \tag{2.4}$$

as the standard deviation of r, for samples of size n from a normal bivariate population with correlation ρ. He also found that the mean value of r is

$$\bar{r} = \rho\left(1-\frac{1-\rho^2}{2n}\right) \tag{2.5}$$

and is therefore smaller than the true value ρ.

3. The geometrical method of Fisher, and exact small sample distributions

We have already noted that much of Student's work was based on empirical methods. Beginning with his important paper (1914–1915), R.A. Fisher, in a series of papers, laid down the ground work of exact small sample tests, and derived in a rigorous manner the necessary distributions. Here he introduced for the first time the brilliant technique of representing a sample $x_1, x_2, \ldots, x_n$ of size n, by a point P with coordinates $(x_1, x_2, \ldots, x_n)$ in the Euclidean space $\mathbf{R}_n$ of n-dimensions. The line $x_1 = x_2 = \ldots = x_n$, is the equiangular line which makes the same angle with each of the coordinate axes. Fisher notes that if M is the foot of the perpendicular from P to the equiangular line, and O is the origin of coordinates, then

$$OM = n^{1/2}\bar{x}, \qquad PM = n^{1/2}s \tag{3.1}$$

From this, Student's distribution of s follows at once. If we have a bivariate sample $(x_1, y_1), (x_2, y_2), \ldots, (x_n, y_n)$, and P' is the point

$(y_1, y_2, \ldots, y_n)$ then we may draw $P'M'$ perpindicular to the equiangular line. If OQ and OQ' are vectors from O parallel and equal to MP and $M'P'$, then the sample correlation r is the cosine of the angle between OQ and OQ'. Using this fact Fisher obtained the distribution of r in the form

$$\frac{(1-\rho^2)^{\frac{1}{2}(n-1)}}{\pi(n-3)!}(1-r^2)^{\frac{1}{2}(n-4)}\left(\frac{\partial}{\sin\theta\,\partial\theta}\right)^{n-2}\frac{\theta}{\sin\theta}\,\mathrm{d}r \tag{3.2}$$

where $\cos\theta = -\rho r$. This verifies that the distribution (2.3) given by Student for the case $\rho = 0$ is exact. Fisher also suggests in this paper the transformation

$$\zeta = \tanh^{-1}\rho = \tfrac{1}{2}\{\log(1+\rho) - \log(1-\rho)\},$$

$$z = \tanh^{-1}r = \tfrac{1}{2}\{\log(1+r) - \log(1-r)\}.$$

The distribution of z is not strictly normal, but it tends to normality rapidly as the sample size is increased, with mean ζ and standard deviation $1/(n-3)^{1/2}$.

The practical application of these results is given in another paper (1921) of Fisher which was primarily concerned to show that the sampling distribution was different for intraclass and interclass correlation coefficients, and to give an exact solution of the former comparable to that given for r in the earlier paper. The special simplicity of the solution in the intraclass case was one of the foundations of the recognition of the z-distribution and the analysis of variance.

In (1923) Fisher gave a rigorous proof of Student's result for the t-statistic and a little later (1926) showed how it could be used for testing various hypotheses, including the hypothesis of the equality of means of two normal populations with the same variance on the basis of a small sample from each. In another paper (1924a) Fisher went on to generalize Student's t to the well known z-statistic, which is half the logarithm of the ratio of two estimated variances based on different degrees of freedom, and was able with the help of this distribution, to give a unified treatment of practically all of the important distributions used in testing null hypotheses from univariate populations. Fisher (1924b) also obtained the distribution of the partial correlation coefficient, which turned out to have the same form as that of the correlation coefficient except for a change in the number of degrees of freedom. The effect of the elimination of variates by partial correlation is simply to reduce the effective size of the sample by unity for each independent variate eliminated.

4. The distribution of the multiple correlation coefficient and other associated distributions

Suppose there is a dependent variate y and n_1 independent variates $x_1, x_2, \ldots, x_{n_1}$. The multiple correlation R^2 is the correlation between y and that linear function of $x_1, x_2, \ldots, x_{n_1}$ with which its correlation is highest. Fisher (1928) showed that the distribution of R^2 depends only on a single parameter ρ^2. If the sample size is $n = n_1 + n_2 + 1$ then the distribution of R^2 is

$$\begin{aligned}&\frac{[\frac{1}{2}(n_1+n_2-2)]!}{[\frac{1}{2}(n_1-2)]![\frac{1}{2}(n_2-2)]!}(1-\rho^2)^{\frac{1}{2}(n_1+n_2)}(R^2)^{\frac{1}{2}(n_1-2)}(1-R^2)^{\frac{1}{2}(n_2-2)}\\ &\quad\times F[\tfrac{1}{2}(n_1+n_2), \tfrac{1}{2}(n_1+n_2), \tfrac{1}{2}n_1, \rho^2R^2]\,\mathrm{d}(R^2)\end{aligned} \tag{4.1}$$

where F is the hypergeometric function.

When $\rho^2 = 0$, the factors containing ρ^2 reduce to unity.

A little earlier than Fisher, P. Hall (1927) had obtained an approximate value of the mean $\bar{R}^2$ and the variance σ_R^2 of R^2 in the form

$$\bar{R}^2 = \rho^2 + \frac{a + (b-\frac{1}{2})\rho^2 + \rho^4}{a+b+\frac{1}{2}}, \qquad \sigma_R^2 = \frac{2\rho^2(1-\rho^2)^2}{a+b+\frac{1}{2}} \tag{4.2}$$

where

$$a = \tfrac{1}{2}n_1, \qquad b = \tfrac{1}{2}n_2 = \tfrac{1}{2}(n - n_1 - 1).$$

The exact value of $\bar{R}^2$ was obtained by Wishart (1931) as

$$\bar{R}^2 = 1 - \frac{b}{a+b}(1-\rho^2)F(1, 1, a+b+1, \rho^2). \tag{4.3}$$

In particular $R^2 = a/(a+b)$ when $\rho = 0$, and $R^2 = 1$, when $\rho^2 = 1$. Fisher calls the distribution (4.1), a type (A) distribution. He obtains a limiting form of this distribution when the parameter n_2 tends to infinity in such a way that $n_2\rho^2 \to \beta^2$, $n_2R^2 \to B^2$. The distribution now takes the limiting form

$$\frac{(\frac{1}{2}B^2)^{\frac{1}{2}(n_1-2)}}{[\frac{1}{2}(n_1-2)]!}\mathrm{e}^{-\frac{1}{2}(B^2-\beta^2)}\left\{1 + \frac{1}{n_1}\frac{B^2\beta^2}{2} + \frac{1}{n_1(n_1+2)}\frac{B^2\beta^2}{2.4} + \ldots\right\}\mathrm{d}(\tfrac{1}{2}B^2) \tag{4.4}$$

which may be written in terms of a Bessel function as

$$(B/\mathrm{i}\beta)^{\frac{1}{2}(n_1-2)}\mathrm{e}^{-\frac{1}{2}(B^2\beta^2)}I_{\frac{1}{2}(n_1-2)}(\mathrm{i}\beta B)\,\mathrm{d}(\tfrac{1}{2}B^2). \tag{4.5}$$

This distribution Fisher calls a type (B) distribution.

Fisher also shows in this paper that the non-central χ^2 distribution is a type (B) distribution. Thus, if $x_1, x_2, \ldots, x_{n_1}$ are normal variates with zero mean and common variance σ^2, and if we put

$$B^2 = \sum_{i=1}^{n_1} (x_i - a_i)^2/\sigma^2, \qquad \beta^2 = \sum_{i=1}^{n} a_i^2/\sigma^2 \tag{4.6}$$

the distribution (4.4) remains valid.

Fisher also obtains a third distribution the type (C) distribution when the multiple correlation is calculated for fixed $y_1, y_2, \ldots, y_n$, when $x_1, x_2, \ldots, x_n$ are normally correlated variates. This distribution is

$$\begin{aligned} &\frac{[\frac{1}{2}(n_1+n_2-2)]!}{[\frac{1}{2}(n_1-2)]![\frac{1}{2}(n_2-2)!]}(R^2)^{\frac{1}{2}(n_1-2)}(1-R^2)^{\frac{1}{2}(n_2-2)}e^{-\frac{1}{2}\beta^2} \\ &\times\left\{1+\frac{n_1+n_2}{n_1\cdot 1}\frac{R^2\beta^2}{2}+\frac{(n_1+n_2)(n_1+n_2+2)}{n_1(n_1+2)\cdot n_2!}\left(\frac{R^2\beta^2}{2}\right)+\cdots\right\}\mathrm{d}(R^2). \end{aligned} \tag{4.7}$$

5. Wishart's distribution of sample variances and covariances

Consider the p-variate normal distribution

$$\frac{|\sigma^{ij}|^{1/2}}{(2\pi)^{p/2}}\exp\left\{-\tfrac{1}{2}\sum_{i,j=1}^{p}\sigma^{ij}(x_i-\mu_i)(x_j-\mu_j)\right\}\prod_{i=1}^{p}\mathrm{d}x_i. \tag{5.1}$$

Then μ_i is the mean of the variate x_i. If σ_{ij} is the covariance of x_i and x_j, then

$$\Sigma = (\sigma_{ij}), \qquad \Sigma^{-1} = (\sigma^{ij}). \tag{5.2}$$

Suppose we have a sample of size n from this population, the observations of the i-th individual being $x_{1\alpha}, x_{2\alpha}, \ldots, x_{p\alpha}$. Then the sample mean and sample covariances may be defined as

$$\bar{x}_i = \sum_{\alpha=1}^{n} x_{i\alpha}/n, \qquad s_{ij} = \sum_{\alpha=1}^{n}(x_{i\alpha}-\bar{x}_i)(x_{j\alpha}-\bar{x}_j)/n. \tag{5.3}$$

Wishart (1928) using Fisher's geometrical representation and the method of quadratic coordinates found the joint distribution of the sample covariances s_{ij} which may be written as

$$\begin{aligned} &\frac{(n/2)^{p(n-1)/2}|\sigma^{ij}|^{(n-1)/2}}{\pi^{p(p-1)/4}\Gamma\left(\frac{n-1}{2}\right)\Gamma\left(\frac{n-2}{2}\right)\ldots\Gamma\left(\frac{n-p}{2}\right)}\exp\left\{-\tfrac{1}{2}n\sum\sigma^{ij}s_{ij}\right\} \\ &\times|s^{ij}|^{(n-p-2)/2}\prod_{i,j=1}^{p}\mathrm{d}s_{ij}. \end{aligned} \tag{5.4}$$

He then goes on to derive the moment coefficients of s_{ij}. This is a basic distribution and has formed the starting point of much of the subsequent work in multivariate analysis. Wishart and Bartlett (1933) gave another proof of the same distribution which is analytical and does not depend on geometrical methods. Yet another proof of the same distribution is due to Hsu (1939b).

Wilks (1932) used the Wishart distribution to give another proof of the distribution of multiple correlation R^2, obtained earlier by Fisher in (1928). He shows that there is a function $\phi(\theta)$, such that the k-th moment of $\phi(\theta)$ is the same as the k-th moment of R^2 for all positive values of k. The result follows from the theory of closure due to Stekloff (1914).

6. Hotelling's generalized T^2

Hotelling (1931) obtained the appropriate generalization of Student's t. To test the hypothesis that a p-variate sample of size n, has arisen from a p-variate normal population with mean vector $\mu_1, \mu_2, \ldots, \mu_p$ we can use the statistic

$$T^2 = \sum_{i,j=1}^{p} (\bar{x}_i - \mu_i)(\bar{x}_j - \mu_j) s^{ij}$$

keeping the notation of the last paragraph. Hotelling showed that T^2 is distributed as

$$\frac{\Gamma\left(\frac{n+1}{2}\right)}{n^{p/2}\Gamma\left(\frac{p}{2}\right)\Gamma\left(\frac{n-p+1}{2}\right)} \frac{(T^2)^{\frac{1}{2}(p-2)}}{\left(1+\frac{T^2}{n}\right)^{\frac{1}{2}(n+1)}} \mathrm{d}T^2. \tag{6.1}$$

Appropriate multivariate modifications can easily be found corresponding to other uses of Student's t, for example, we can test the equality of means of two p-variate normal populations on the basis of a random sample from each. The great advantage of using T^2 is the simplicity of its distribution, with its complete independence of any correlations that may exist in the populations.

7. Wilks generalized variance

Wilks (1932) introduced a new technique for the study of multivariate normal populations. He starts with a study of two basic integral equations.

His integral equation of type A is

$$\int_0^\infty z^k f(z)\mathrm{d}z = B^k \frac{\Gamma(a_1+k)\Gamma(a_2+k)\dots\Gamma(a_n+k)}{\Gamma(a_1)\Gamma(a_2)\dots\Gamma(a_n)} \tag{7.1}$$

where k and a's are real and positive and B and $f(z)$ are independent of k. He obtains $f(z)$ as a multiple integral

$$\begin{aligned} f(z) = {} & \frac{B^{-a_n} z^{a_n-1}}{\Gamma(a_1)\dots\Gamma(a_n)} \int_0^\alpha \int_0^\alpha \dots \int_0^\alpha v_1^{a_1-a_2-1} v_2^{a_2-a_3-1} \dots v_{n-1}^{a_{n-1}-a_n-1} \\ & \times \exp\left\{ -v_1 - \frac{v_2}{v_1} \dots - \frac{v_{n-1}}{v_{n-2}} - \frac{z}{Bv_{n-1}} \right\} \mathrm{d}v_1 \mathrm{d}v_2 \dots \mathrm{d}v_{n-1}. \end{aligned} \tag{7.2}$$

He gives a similar solution for the integral equation of type B

$$\int_0^B w^k g(w)\mathrm{d}w = CB^k \frac{\Gamma(b_1+k)\Gamma(b_2+k)\dots\Gamma(b_n+k)}{\Gamma(c_1+k)\Gamma(c_2+k)\dots\Gamma(c_n+k)} \tag{7.3}$$

where B and $g(w)$ are independent of k, $b_i < c_i$, $(i = 1, 2, \dots, n)$, the b's and c's are real positive integers, and

$$C = \frac{\Gamma(c_1)\Gamma(c_2)\dots\Gamma(c_n)}{\Gamma(b_1)\Gamma(b_2)\dots\Gamma(b_n)}. \tag{7.4}$$

He introduces the quantity $|\sigma_{ij}|$, the generalized variance as a measure of the dispersion of a p-variate normal population and obtains the distribution of the corresponding sample statistic

$$\xi = |s_{ij}|.$$

He shows that the k-th moment $M_k(\xi)$ of ξ is of the same form as the left hand side of (7.1), and then using the solution for the integral equation obtains the distribution of ξ, in which a $p-1$ fold multiple integral occurs. The distribution is explicitly obtained for the case $p = 1$ and 2. For $p = 2$ the distribution is

$$\frac{(n/2)^{n-2}\pi^{1/2}}{|\sigma_{ij}|^{(n-2)}\Gamma\left(\frac{n-1}{2}\right)\Gamma\left(\frac{n-2}{2}\right)} e^{-2A^{1/2}\xi^{1/2}} \xi^{(n-4)/2} \mathrm{d}\xi$$

where $A = (n/2)^p |\sigma^{ij}|$.

He also finds (i) the distribution and the moments of the ratio of two independent generalized variances, and (ii) the distribution and the moments of the ratio of the generalized variance to any one of its principal minors.

Wilks also obtains an alternative proof of the distribution of Hotelling's T^2, by his methods.

8. Principal components

Suppose there are p correlated variates $x_1, x_2, \dots, x_n$ which are standardized to $z_1, z_2, \dots, z_n$ so as to have zero means and unit standard deviations. Hotelling (1933) considers the problem of finding linear functions

$$\gamma_i = a_{i1}z_1 + a_{i2}z_2 + \dots + a_{in}z_{in}$$

which are independently distributed with unit variance and such that γ_1 is the component whose contribution to the variance of the x_i has as great a total as possible. Then γ_2 has the maximum contribution to the residual variance and so on. Then the functions $\gamma_1, \gamma_2, \dots, \gamma_n$ have been called the principal components by Hotelling.

Let r_{ij} be the correlation between z_i and z_j. Consider the set of equations

$$\begin{array}{ccccc} (1-k)a_1 + & r_{12}a_2 + \dots + & & r_{1n}a_n = 0, \\ r_{21}a_1 + & (1-k)a_2 + \dots + & & r_{2n}a_n = 0, \\ \dots & \dots & \dots & \dots & \dots \\ r_{n1}a_1 + & r_{n2}a_2 + \dots & + (1+k)a_n & - 0, \end{array} \tag{8.1}$$

$$f(k) = \begin{vmatrix} 1-k & r_{12} & \dots & r_{1n} \\ r_{21} & 1-k & \dots & r_{2n} \\ \dots & \dots & \dots & \dots \\ r_{n1} & r_{n2} & \dots & 1-k \end{vmatrix} = |I - kR|. \tag{8.2}$$

If k_1 is the maximum root of $f(k)$, we substitute k_1 for k in (8.1), and determine $a_1, a_2, \dots, a_n$ and normalize them to

$$\frac{a_i k_i^{1/2}}{(a_1^2 + a_2^2 + \dots + a_n^2)^{1/2}}, \qquad i = 1, 2, \dots, n$$

then the resulting values are the coefficients of the z_i in γ_1. If q largest roots are equal then the rank of $I - kR$ is $n - q$. We can choose q linearly independent orthogonal solutions and determine the coefficients of $\gamma_1, \gamma_2, \dots, \gamma_q$. We then proceed with the next highest root and so on.

If some of the principal components contribute negligible amounts to the variances, they may be neglected. Hotelling gives a geometrical interpretation of the principal components and illustrates their use in the analysis of psychometric data. In another paper Hotelling (1936a) gives a simplified method of calculating the principal components.

9. Cannonical correlations

The theory of cannonical correlations is due to Hotelling (1936b). Suppose there are two sets of correlated normal variates, the first set being x_α $(\alpha = 1, 2, \ldots, s)$, and the second set x_i $(i = 1, 2, \ldots, t)$, $s \leq t$. We shall use Greek subscripts α, β, for variates of the first set, and the subscripts i, j for the variates of the second set. Let $\Sigma_{\alpha\beta} = (\sigma_{\alpha\beta})$ be the covariance matrix of the variates of the first set, $\Sigma_{ij} = (\sigma_{ij})$ the covariance matrix for the variates of the second set and let $\Sigma_{\alpha i} = (\sigma_{\alpha i})$ be the covariance matrix giving the covariances of one variable of the first set and one variable of the second set. Consider the linear functions

$$u = \sum_\alpha a_\alpha x_\alpha, \qquad v = \sum_i b_i x_i \tag{9.1}$$

such that

$$\sum_{\alpha,\beta} \sigma_{\alpha\beta} a_\alpha a_\beta = 1, \qquad \sum_{i,j} \sigma_{ij} a_i a_j = 1 \tag{9.2}$$

i.e., u and v have variance unity. We want to make the correlation between u and v, maximum.

This is achieved by using the equations

$$\begin{aligned} &\sum_i \sigma_{\alpha i} b_i - \lambda \sum_\beta \sigma_{\alpha\beta} a_\beta = 0 \\ &\sum_\alpha \sigma_{\alpha i} a_\alpha - \mu \sum_j \sigma_{ij} b_j = 0 \end{aligned} \tag{9.3}$$

where

$$\lambda = \mu = \rho \tag{9.4}$$

and λ is a root of

$$\begin{bmatrix} -\lambda \Sigma_{\alpha\beta} & \Sigma_{\alpha i} \\ \Sigma_{i\alpha} & -\lambda \Sigma_{ij} \end{bmatrix} = 0. \tag{9.5}$$

Then there are $2s$ non-zero roots $\pm \rho_i$, $i = 1, 2, \ldots, s$. These are the cannonical correlations, and the corresponding variates are the cannonical variates. Hotelling (1936b) gives a geometrical interpretation of these correlations and variates, studies their properties, illustrates their use and derives a number of distributions associated with these. If we put

$$Q^2 = \rho_1^2\rho_2^2\ldots\rho_s^2, \qquad Z = (1-\rho_1^2)(1-\rho_2^2)\ldots(1-\rho_s^2)$$

then if $r_1, r_2, \ldots, r_s$, q and z are the corresponding sample quantities he shows that the joint distribution of s and z for the case $s = 2$, $t = 2$ is

$$\tfrac{1}{2}(n-2)(n-3)z^{\frac{1}{2}(n-3)}\mathrm{d}q\,\mathrm{d}z$$

and the joint distribution of r_1 and r_2 is

$$(n-2)(n-3)(r_1^2-r_2^2)(1-r_1^2)^{\frac{1}{2}(n-5)}(1-r_2^2)^{\frac{1}{2}(n-5)}\mathrm{d}r_1\,\mathrm{d}r_2$$

where

$$0 < z \leq 1, \qquad 0 \leq q \leq 1, \qquad z \leq (1-q^2),$$

on the assumption that the population cannonical correlations ρ_1 and ρ_2 are zero.

Girschick (1939) finds the joint moments of q and z for general s and t. He extends the joint distribution of q and z, and hence the joint distribution of the cannonical correlations to the case $s = 2$, $t > 2$.

10. Rectangular and normal coordinates

Consider the p-variate normal distribution (5.1). We shall keep to the notation introduced in paragraph 5. Let s_{ij} be the sample covariances and let $S = (s_{ij})$ be the sample covariance matrix and $\Sigma - (\sigma_{ij})$ be the population covariance matrix. Then since $s_{ij} = s_{ji}$ there are $p(p+1)/2$ distinct elements s_{ij}. Mahalanobis, Bose and Roy (1937) defined new sets of $p(p+1)/2$ variates which are functions of s_{ij}. Let $T = |t_{ij}|$ be an upper triangular matrix, i.e., $t_{ij} = 0$ if $i > t$, where

$$T'T = S \tag{10.1}$$

then the $p(p+1)/2$ variates t_{ij}, $i \leq j$, $i, j = 1, 2, \ldots, p$ were defined to be normal coordinates of the sample by them. We can take an n-dimensional Euclidean space $\mathbf{R}_n$, and using Fisher's representation given in paragraph 3 represent the sample by the points P_i $(i = 1, 2, \ldots, p)$, where the coordinates of P_i are the observed values of the i-th variate. Let M_i be the foot of the perpendicular from P_i to the equiangular line, and let OQ_i be equal and parallel to M_iP_i where O is the origin of coordinates. Let Z_i be the point on OQ_i such that $OZ_i/OQ_i = 1/n^{1/2}$. Then the points Z_i lie in an

$n-1$ dimensional space Σ_{n-1} perpendicular to the equiangular line. We now take a new system of rectangular axes immersed in the space $OZ_1Z_2\ldots Z_p$ such that OY_1 is identical with OZ_1, OY_2 lies in the plane OZ_1Z_2 and is perpendicular to OY_1, and in general OY_j $(j \leq p)$ is taken to lie in the subspace $OZ_1Z_2\ldots Z_j$ and is perpendicular to $OY_1, OY_2\ldots OY_{j-1}$. Then $t_{1j}, t_{2j}, \ldots, t_{ij}$ are the coordinates of Z_j with reference to this system of coordinates. Then the relation (10.1) determines s_{ij} as functions of t_{ij} and conversely the t_{ij}'s may be obtained as functions of s_{ij}'s. Mahalanobis, Bose and Roy obtain the distribution of the rectangular coordinates t_{ij} as

$$\frac{(n/2)^{p(n-1)/2}\left|\sigma^{ij}\right|^{(n-1)/2}}{\pi^{p(p-1)/4}\Gamma\left(\frac{n-1}{2}\right)\Gamma\left(\frac{n-2}{2}\right)\ldots\Gamma\left(\frac{n-p}{2}\right)} e^{-\frac{1}{2}n\,\mathrm{tr}(T\Sigma^{-1}T')}$$
$$\times\, t_{11}^{n-2}t_{22}^{n-3}\ldots t_{pp}^{n-p-1}\prod_{i\leq j} \mathrm{d}t_{ij}. \tag{10.2}$$

The normal coordinates l_{ij} are defined as certain linear functions of the t_{ij}'s. Let $\Delta = (\delta_{ij})$ be an upper triangular matrix such that

$$\Delta\Delta' = \Sigma^{-1} \tag{10.3}$$

Then the elements of Δ are uniquely determined. Let $L = (l_{ij})$ be an upper triangular matrix i.e., $l_{ij} = 0$ if $i > j$. Then the normal coordinates are defined by the linear transformation

$$L = T\Delta \tag{10.4}$$

The $p(p+1)/2$ normal coordinates are independently distributed. For $i \neq j$, $n^{1/2}l_{ij}$ is normally distributed with zero mean and unit variance, and nl_{ij}^2 has the χ^2 distribution with $n-i+1$ degrees of freedom.

The special simplicity of the distribution of the rectangular and normal coordinates makes them a suitable starting point for obtaining the distributions of many important multivariate statistics. In particular, Mahalanobis, Bose and Roy use them to get alternative proofs of the Wishart distribution and the distributions obtained by Wilks described in Section 7.

11. Mahalanobis D^2

The concept of measure of divergence between two p-variate normal populations is due to Mahalanobis (1925, 1927, 1928). There are two forms of the measure. In the first form the population variances and covariances are known. Suppose we have two p-variate normal populations with mean vectors

$$\underline{\mu}_1' = (\mu_{11}, \mu_{21}, \ldots, \mu_{p1}), \qquad \underline{\mu}_2' = (\mu_{12}, \mu_{22}, \ldots, \mu_{p2}) \tag{11.1}$$

and covariance matrices

$$\Sigma_1 = (\sigma_{ij1}), \qquad \Sigma_2 = (\sigma_{ij2}) \tag{11.2}$$

from which we have two samples of size n_1 and n_2 respectively.

Let the vector of mean differences be

$$\underline{\mu}' = (\mu_{11} - \mu_{12}, \mu_{21} - \mu_{22}, \ldots, \mu_{p1} - \mu_{p2}), \tag{11.3}$$

and

$$\sigma_{ij} = \frac{n_2\sigma_{ij1} + n_1\sigma_{ij2}}{n_1 + n_2} \tag{11.4}$$

Let $\underline{m}'$ be the corresponding vector of sample mean differences and let

$$s_{ij} = \frac{n_2 s_{ij1} + n_1 s_{ij2}}{n_1 + n_2} \tag{11.5}$$

where the s_{ij1} and s_{ij2} are sample covariances. Let

$$\Sigma = (\sigma_{ij}), \qquad S = (s_{ij}). \tag{11.6}$$

Then Mahalanobis defines the population distance Δ, and the sample distance D^2 by

$$\Delta^2 = \frac{1}{p}\underline{\mu}'\Sigma^{-1}\underline{\mu}, \qquad D_1^2 = \frac{1}{p}\underline{m}'\Sigma^{-1}\underline{m}, \qquad D^2 = D_1^2 - \frac{2}{\bar{n}} \tag{11.7}$$

where

$$\frac{1}{\bar{n}} = \frac{1}{2}\left(\frac{1}{n_1} + \frac{1}{n_2}\right)$$

i.e., $\bar{n}$ is the harmonic mean of the sample size. The correction $2/\bar{n}$ was introduced to make the expectation of D^2 equal to Δ^2. The D^2 here defined may be called the classical D^2. The first four moments of the classical D^2 were found by Mahalanobis (1930) using approximate methods. The distribution of D_1^2 was obtained by Bose (1936) in the form

$$\frac{\bar{n}p}{4}\left(\frac{D_1^2}{\Delta^2}\right)^{(p-2)/4} e^{-\bar{n}p(D_1^2+\Delta^2)/4} I_{1/2(p-2)}[\tfrac{1}{2}np(D_1^2\Delta^2)^{1/2}]\,dD_1^2. \tag{11.8}$$

Bose also calculated the k-th moment of D^2, and it turned out that the values already given by Mahalanobis for the first four moments were exact.

When the population variances and covariances are not available they must be estimated from the observed samples. One can then define a Studentized form of the D^2 statistic on the assumption that the covariances are the same for both populations. The Studentized D^2 is defined by

$$D^2 = \frac{1}{p} \underset{\sim}{m}' S^{-1} \underset{\sim}{m}. \tag{11.9}$$

The distribution of Studentized D^2 was obtained by Bose and Roy (1938) in the form

$$\text{const} \times \frac{(D^2)^{(p-2)/2}}{\left\{1 + \frac{p\bar{n}}{2N} D^2\right\}^{(N-1)/2}} \mathrm{d}D^2 \tag{11.10}$$

when $\Delta^2 = 0$, and $N = n_1 + n_2$. When Δ^2 is non-zero there is an additional factor

$$\mathrm{e}^{-\frac{1}{4}p\bar{n}\Delta^2} F_1\left(\frac{N-1}{2}, \frac{p}{2}, \frac{p^2\bar{n}^2\Delta^2 D^2}{8N + 4p\bar{n}D^2}\right) \tag{11.11}$$

where F_1 is the hypergeometric function defined by

$$F_1(\alpha, \rho, z) = 1 + \frac{\alpha}{1!\rho} + \frac{\alpha(\alpha+1)}{2!\rho(\rho+1)} z^2 + \frac{\alpha(\alpha+1)(\alpha+2)}{3!\rho(\rho+1)(\rho+2)} z^2 + \dots \tag{11.12}$$

and the constant is

$$\left(\frac{p\bar{N}}{2N}\right)^p \frac{\Gamma\left(\frac{N-1}{2}\right)}{\Gamma\left(\frac{p}{2}\right)\Gamma\left(\frac{N-p-1}{2}\right)}. \tag{11.13}$$

Instead of two populations we may have k p-variate populations from each of which a sample is available. We may wish to measure the $k(k-1)/2$ mutual distances between these populations, basing the estimates of covariances on all the k samples (and assuming that the population covariances are identical). The distribution of D^2 remains of the same form as (11.10) and (11.11) except for a change in the degrees of freedom. The actual distribution was obtained by Bose and Roy (1939).

12. Discriminant analysis

When two or more populations have been measured in several characters $x_1, x_2, \dots, x_p$ special interest attaches to certain linear functions, called discriminant functions by Fisher (1936), by which the populations are best discriminated. Fisher considers a linear compound

$$X = b^1x_1 + b^2x_2 + \dots + b^px_p \tag{12.1}$$

of the p characters and chooses $b^1, b^2, \dots, b^p$ so as to maximize the ratio of the variance of the chosen compound character to the within sample variance of the same character in the two (or more) populations. He illustrates the use of the discriminant functions for classification by considering an example of two closely related species of plants measured in respect of four characters. At Fisher's suggestion the same principle was used by Martin (1935) for sex differences in the measurements of the mandible and by Barnard (1935) in showing how to obtain from a set of dated series the particular compound of cranial measurements showing most distinctly a progressive or secular trend. Fairfield Smith (1936) also used the discriminant function for plant selection.

The discriminant function is closely connected to Mahalanobis D^2 and Hotelling's T^2. It should be noted that the compounding coefficients obtained after maximization are really functions of the sample variances and covariances and the ratio of the variance of the compound character to its within sample variance is proportional to Hotelling's T^2 or Mahalanobis D^2. Thus, the distribution of the Studentized D^2 is in effect the non-central T^2 distribution.

Fisher (1938) shows how the distributions of the classical and Studentized D^2 are connected with his type (B) and type (C) distributions discussed in Section 4.

If we have samples of n_1 and n_2 objects and measure them with respect to p characters $x_1, x_2, \dots, x_p$ the analogy between the discriminant function (12.1) and the procedure of multiple regression is brought out by introducing a formal dependent variate y which is given the value $n_2/(n_1+n_2)$ for objects of the first class and $-n_1/(n_1+n_2)$ for objects of the second class. Then

$$\sum y = 0, \qquad \sum y^2 = n_1n_2/(n_1+n_2) = \lambda^2 \text{ (say)} \tag{12.3}$$

The multiple regression equation for predicting the value of Y from the observed values of $x_1, x_2, \dots, x_p$ is now of the form

$$Y - \sum_{i=1}^{p} b^i(x_1 - \bar{x}_i). \tag{12.4}$$

Fisher shows that if we now calculate the regression coefficients in the usual manner we obtain the values of the coefficients in the discriminant function (12.1). From this he shows that the translation of the distribution (11.10) and (11.11) given by Bose and Roy (1938) into the distribution (4.7) of type (C) is simply given by the substitution

$$\lambda p \Delta^2 = \beta^2, \qquad \lambda p D^2 = nR^2/(1-R^2) = T^2 \tag{12.5}$$

where R^2 is the multiple correlation coefficient, T^2 is Hotelling's T^2, and $n = n_1 + n_2 - 2$.

He also shows that the sampling distribution of the classical D^2 statistic found by Bose (1936) is equivalent to the limiting distribution (4.4) of type (B) obtained by him in (1928). The translation is given by

$$\beta^2 = \lambda^2 p \Delta^2, \qquad B^2 = \lambda^2 p D^2 + p$$

where $\lambda^2 = n_1 n_2/(n_1 + n_2)$.

13. Latent roots of determinantal equations

S. N. Roy (1939) showed how Fisher's technique of forming a linear compound can be used for obtaining statistics appropriate for testing the hypothesis of the equality of the covariance matrices of two p-variate normal populations on the basis of a random sample from each. The coefficients of the linear compound must now be so chosen as to make maximum the ratio of the variance of the first sample (for the compound character) and the variance of the second sample for the same character. Let the compounding coefficients be $\lambda_1, \lambda_2, \ldots, \lambda_p$ then one obtains the equations

$$\sum_{j=1}^{p} \lambda_j (s_{ij1} - K^2 s_{ij2}) = 0, \qquad i = 1, 2, \ldots, p \tag{13.1}$$

for determining the compounding coefficients, where K^2 is the ratio of the variances of the compounding character for the two samples. Eliminating λ_j between these p equations we get the determinantal equation

$$| s_{ij1} - K^2 s_{ij2} | = 0$$

of the p-th degree. One of these K roots corresponds to the maximum K, one to the minimum K and the rest to stationary values of K. Thus, the simultaneous distribution of the roots $K_1^2, K_2^2, \ldots, K_p^2$ is the appropriate generalization of the distribution of the statistic $F = s_1^2/s_2^2$ for testing the hypothesis of the equality of the variances of two univariate normal populations. This simultaneous distribution is obtained by Roy as

$$\text{const} \times \prod_{i=1}^{p} \frac{K_i^{n_1-p-1}}{\left(1+\frac{n_1}{n_2}K_i^2\right)^{(n_1+n_2-2)/2}} \prod_{i<j} (K_i^2 - K_j^2) \prod_{i=1}^{p} \mathrm{d}K_i \tag{13.2}$$

where $K_1^2 \geq K_2^2 \geq \ldots K_p^2$. Roy then goes on to consider the application of this distribution to test the hypothesis of the equality of covariance matrices of two p-variate normal populations.

Hsu (1939a) obtained the joint distribution of the latent roots of a slightly different but related determinantal equation. He starts with a set of $p(n_1 + n_2)$ variables

$$y_{ir}, z_{it}, \qquad (i = 1, 2, \ldots, p; \quad r = 1, 2, \ldots, n_1; \quad t = 1, 2, \ldots, n_2)$$

following the distribution

$$\text{const} \times \exp\left\{-\tfrac{1}{2} \sum_{i,j=1}^{p} \alpha_{ij}(a_{ij} + b_{ij})\right\} \prod \mathrm{d}y \, \mathrm{d}z, \tag{13.3}$$

where

$$a_{ij} = \sum_{r=1}^{n} y_{ir} y_{jr}, \qquad b_{ij} = \sum_{t=1}^{n} z_{it} z_{jt}. \tag{13.4}$$

He considers the determinantal equation

$$|a_{ij} - \theta(a_{ij} + b_{ij})| = 0 \tag{13.5}$$

and obtains the joint distribution of its roots. He assumes that $n_2 > p$, but n_1 may be greater or less than p.

Case I. If $n_1 \geq p$, then the distribution is given by

$$c_1 \times \left\{\prod_{i=1}^{p} \theta_i\right\}^{(n_1-p-1)/2} \left\{\prod_{i=1}^{p} (1-\theta_i)\right\}^{\frac{1}{2}(n_2-p-1)} \prod_{i<j}^{p} (\theta_i - \theta_j) \prod_{i=1}^{p} \mathrm{d}\theta_i. \tag{13.6}$$

Case II. If $n_1 \leq p$, then the distribution is given by

$$c_2 \times \left\{\prod_{i=1}^{n_1} \theta_i\right\}^{(p-n_1-1)/2} \left\{\prod_{i=1}^{n_1} (1-\theta_i)\right\}^{\frac{1}{2}(n_2-p-1)} \prod_{i<j}^{n_1} (\theta_i - \theta_j) \prod_{i=1}^{n} \mathrm{d}\theta_i. \tag{13.7}$$

He gives explicit values of the constants c_1 and c_2. He then goes on to prove the following:

Theorem. *If the $p(p+1)/2$ variables s_{ij} $(i \leq j = 1, 2, \ldots, p)$ have such a domain of existence that the symmetric matrix (s_{ij}) is always non-singular and their joint distribution is*

$$df = g(\lambda_1, \lambda_2, \ldots, \lambda_p) \prod ds_{ij} \tag{13.8}$$

where $\lambda_1, \lambda_2, \ldots, \lambda_p$ *are the latent roots of* (s_{ij}) *in descending order, then their distribution is*

$$\frac{\pi^{\frac{1}{2}p(p+1)}}{\prod_{i=1}^{p} \Gamma\frac{1}{2}(p-i+1)} \prod_{i<j}^{p} (\lambda_i - \lambda_j) g(\lambda_1, \lambda_2, \ldots, \lambda_p) \prod_{i=1}^{p} d\lambda_i. \tag{13.9}$$

As an application of this theorem he obtains the joint distribution of Hotelling's cannonical correlations: If $(x_1, x_2, \ldots, x_p)$, $(x_{p+1}, x_{p+2}, \ldots, x_{p+q})$ be two mutually independent sets of variates, $p \leq q$ such that the first set is normally distributed, then if $\theta_1, \theta_2, \ldots, \theta_p$ are the cannonical correlations arranged in descending order of magnitude, their joint distribution is given by

$$K \cdot \left\{ \prod_{i=1}^{p} \theta_i \right\}^{\frac{1}{2}(q-p-1)} \left\{ \prod_{i=1}^{p} (1-\theta_i) \right\}^{\frac{1}{2}(n-p-q-2)} \prod_{i<j}^{p} (\theta_i - \theta_j) \prod_{i=1}^{p} d\theta_i \tag{13.10}$$

where

$$K = \frac{\pi^{p/2} \prod_{i=1}^{p} \Gamma\frac{1}{2}(n-i)}{\prod_{i=1}^{p} \{\Gamma\frac{1}{2}(n-q-i)\Gamma\frac{1}{2}(p-i+1)\Gamma\frac{1}{2}(q-i+1)\}}. \tag{13.11}$$

This is a generalization of Hotelling's result for $p = q = 2$.

Fisher (1939) arrives at the equation (13.5) from the point of view of discriminant analysis. He considers in detail the case $p = 2$, and obtains the distribution (13.6). He points out various applications of the distributions of statistics obtained from non-linear equations.

As pointed out in the introduction, the discovery of the use and distribution of the latent roots in (1939) lead to an explosive development of multivariate analysis. It is therefore appropriate to bring to a close this short review of the early history of multivariate analysis at this point.

References

Barnard, M.M. (1935). The secular variations of skull characters in four series of Egyptian skulls. *Ann. Eugen.*, **6**, 352–371.

Bose, R.C. (1936). On the exact distribution and moment coefficients of the D^2-statistic. *Sankhyā*, **2**, 379–384.

Bose, R.C. and Roy, S.N. (1938). Distribution of the Studentized D^2-statistic. *Sankhyā*, **4**, 19–38.
Bose, R.C. and Roy, S.N. (1939). The use and distribution of the Studentized D^2-statistic when the variances and covariances are based on k samples. *Sankhyā*, **4**, 535–542.
Fisher, R.A. (1914–15). Frequency distribution of the values of the correlation coefficient in samples from an indefinitely large population. *Biometrika*, **10**, 507–521.
Fisher, R.A. (1921). On the "probable error" of a coefficient of correlation deduced from a small sample. *Metron*, **1**, pt. 4, 1–32.
Fisher, R.A. (1923). Note on Dr. Burnside's recent paper on errors of observation. *Proc. Camb. Phil. Soc.*, **21**, 655–658.
Fisher, R.A. (1924a). On a distribution yielding the error functions of several well known statistics. *Proc. Int. Math. Congress*, Toronto, 805–813.
Fisher, R.A. (1924b). The distribution of the partial correlation coefficient. *Metron*, **3**, 329–332.
Fisher, R.A. (1926). Applications of "Students" distribution. *Metron*, **5**, pt. 3, 90–104.
Fisher, R.A. (1928). The general sampling distribution of the multiple correlation coefficient. *Proc. Royal. Soc. Lond.*, A, **121**, 654–673.
Fisher, R.A. (1936). The use of multiple measurements in taxonomic problems. *Ann. Eugen.*, **7**, 179–188.
Fisher, R.A. (1938). The statistical utilization of multiple measurements. *Ann. Eugen.*, **8**, 376–386.
Fisher, R.A. (1939). The sampling distribution of some statistics obtained from non-linear equations. *Ann. Eugen.*, **9**, 238–249.
Girschik, M.A. (1939). On the sampling theory of the roots of determinantal equations. *Ann. Math. Statist.*, **10**, 203–224.
Hall, P. (1927). Multiple and partial correlation coefficients in the case of an m-fold variate system. *Biometrika*, **19**, 100–109.
Helmert, C.F. (1875). Über die Berechnung des wahrscheinlichen Fehlers aus einer endlichen Anzahl wahrer Beobachtungsfehler. *Zeitschrift für Math. und Physik.*, **20**, 300–303.
Hotelling, H. (1931). The generalization of "Student's ratio". *Ann. Math. Statist.*, **2**, 360–378.
Hotelling, H. (1933). Analysis of a complex of statistical variables. *J. Ed. Psych.*, **24**, 417–441 and 498–520.
Hotelling, H. (1936a). Simplified calculation of principal components. *Psychometrika*, **1**, 27–35.
Hotelling, H. (1936b). Relations between two sets of variates. *Biometrika*, **28**, 321–377.
Hsu, P.L. (1939a). On the distribution of the roots of certain determinantal equations. *Ann. Eugen.*, **9**, 250–258.
Hsu, P.L. (1939b). A new proof of the joint product moment distribution. *Proc. Camb. Phil. Soc.*, **35**, 336–338.
Macdonell, W.R. (1901–2). On criminal anthropology and the identification of criminals. *Biometrika*, **1**, 177–227.
Mahalanobis, P.C. (1925, 1927). Analysis of race-mixture in Bengal. Presidential Address Anthropological Section, Indian Science Congress. Also *Jour. and Proc. Asiat. Soc. Bengal*, **23**, 301–333.
Mahalanobis, P.C. (1928). Statistical study of the Chinese head. *Man in India*, **8**, 107–122.
Mahalanobis, P.C. (1930). On tests and measures of group divergence, Part 1: Theoretical formulae. *Jour. and Proc. Asiat. Soc. Bengal*, New series, **26**, 541–588.
Mahalanobis, P.C., Bose, R.C. and Roy, S.N. (1937). Normalization of statistical variates and the use of rectangular coordinates in the theory of sampling distributions. *Sankhyā*, **3**, 1–40.

Martin, E.A. (1936). A study of the Egyptian series of mandibles with special reference to mathematical methods of sexing. *Biometrika*, **28**, 149–178.
Roy, S.N. (1939). p-statistics or some generalizations in analysis of variance appropriate to multivariate problems. *Sankhyā*, **4**, 381–396.
Smith, Fairfield (1936). A discriminant function for plant selection. *Ann. Eugen.*, **7**, 240–250.
Soper, H.E. (1913). On the probable error of the correlation coefficient to a second approximation. *Biometrika*, **9**, 91–115.

P. R. Krishnaiah, ed., *Multivariate Analysis–IV*
© North-Holland Publishing Company (1977) 23–34

ASYMPTOTIC NORMALITY OF SUMS OF DEPENDENT RANDOM VECTORS*

Aryeh DVORETZKY
The Hebrew University of Jerusalem, Jerusalem, Israel
and *Columbia University, New York, U.S.A.*

1. Introduction

1.1. Limiting distributions of sums of independent (numerical) random variables have been exhaustively studied and there exists a satisfactory theory of the subject. The corresponding theory for dependent random variables is, naturally, more fragmentary. A rather general prescription for obtaining results on limit laws of sums of dependent random variables is the following [2, 3]:

If the contribution of each individual summand is negligible then conditions implying convergence to any specified law in the independent case, imply the same conclusion also in the most general dependent case, provided quantities such as means etc. are replaced by conditional means etc., the conditioning being relative to the preceding sum.

Similar results can be established for convergence of sums of dependent random vectors to any infinitely divisible law. Here we confine ourselves to asymptotic normality. This is done in order to simplify the exposition and in view of the great importance of normal distributions in multivariate analysis.

The distributions of a sequence of random vectors converge to a limiting distribution if the one-dimensional distributions of every linear combination of the components of these vectors converge to the corresponding one-dimensional distribution. This observation can be utilized to derive the results about sums of random vectors from those about sums of random variables. We do not follow this course here and present a self-contained treatment. Though following the pattern of [3] we simplify some of the arguments and elaborate others.

* Research supported by National Science Foundation grant no. NSF–MPS–7518471.

To keep the paper within bounds we confine ourselves to random vectors with finite second moments (see, however, 4.4). Specialized to independent summands our results reduce to the well-known fact that the Lindeberg condition implies asymptotic normality (Cramér [1]).

1.2. We deal with vectors in d-dimensional space E^d (d a fixed positive integer). The vectors are considered as column vectors and are denoted by bold type. The transposed row vectors are indicated by primes.

By $|\cdot|$ we designate the norm (length) of the vector. The same notation is used to indicate the absolute value of a real or complex number.

$1\{\cdot\}$ denotes the indicator of the set within the braces.

$\mathscr{B}(\cdot)$ stands for the σ-field generated by the indicated quantity (-ies).

$\mathscr{F}_n$ and $\mathscr{F}_{n,k}$ are σ-fields in the appropriate probability spaces. Conditional expectations relative to these σ-fields are denoted by $\mathbf{E}_n$ and $\mathbf{E}_{n,k}$.

Σ, with or without subscripts, is a $d \times d$ matrix, usually the covariance matrix of the corresponding random vector.

$\mathrm{tr}(\cdot)$ stands for the trace (sum of diagonal elements) of the indicated matrix.

$\mathscr{L}(\cdot)$ denotes the law (distribution) of the indicated random quantity. By $\mathscr{L}_1 * \mathscr{L}_2$ we denote the convolution of $\mathscr{L}_1$ and $\mathscr{L}_2$, i.e. the law corresponding to the sum of two independent vectors, one distributed according to $\mathscr{L}_1$ and the other according to $\mathscr{L}_2$.

$\mathscr{N}(\cdot,\cdot)$ denotes the normal law with the indicated parameters.

$\rightarrow$ stands for $\lim_{n\to\infty}$.

$\xrightarrow{\mathrm{P}}$ denotes convergence in probability.

$\sum_k$ is an abbreviation for $\sum_{k=1}^{k_n}$.

$\max_k$ is an abbreviation for $\max_{k=1,\ldots,k_n}$.

All equalities between random quantities are a.s. (holding with probability 1) equalities. Similarly for inequalities between random variables.

1.3. The main results are stated and discussed in the next Section. Section 3 establishes some lemmas which are used in the following one to prove the theorems of Section 2. The last section contains some remarks and generalizations.

2. Main results

2.1. To state the results in convenient generality we consider double arrays of random vectors

$$\boldsymbol{X}_{n,1}, \boldsymbol{X}_{n,2}, \ldots, \boldsymbol{X}_{n,k_n} \quad (n = 1, 2, \ldots). \tag{2.1}$$

The random vectors in the same row (same n) are on the same probability space. No relation is assumed between the probability spaces corresponding to different rows (in particular they may be identical).

We assume throughout (except in 4.4) that the random vectors in (2.1) are square integrable.

We put

$$\boldsymbol{S}_{n,k} = \boldsymbol{S}_{n,1} + \ldots + \boldsymbol{S}_{n,k} \quad (k = 0, 1, \ldots, k_n), \tag{2.2}$$

$$\boldsymbol{S}_n = \boldsymbol{S}_{n,k_n} \quad (\boldsymbol{S}_{n,0} = 0),$$

$$\mathcal{F}_{n,k} = \mathcal{B}(\boldsymbol{S}_{n,k}) \quad \text{(in particular } \mathcal{F}_{n,0} \text{ is the trivial field).}$$

We use the notations

$$\boldsymbol{\mu}_{n,k} = \mathbf{E}_{n,k-1} \boldsymbol{X}_{n,k} \tag{2.3}$$

and

$$\Sigma_{n,k} = \mathbf{E}_{n,k-1}(\boldsymbol{X}_{n,k} - \boldsymbol{\mu}_{n,k})(\mathbf{X}'_{n,k} - \boldsymbol{\mu}'_{n,k}). \tag{2.4}$$

$\boldsymbol{\mu}_{n,k}$ and $\Sigma_{n,k}$ exist a.s. and are, in general, random vectors and matrices, respectively.

We put

$$\boldsymbol{\mu}_n = \sum_k \boldsymbol{\mu}_{n,k}, \qquad \Sigma_n = \sum_k \Sigma_{n,k}. \tag{2.5}$$

2.2. We start with a special result patterned after a classical version of Lindeberg's theorem for sums of independent random vectors.

Theorem 1. *Let $\boldsymbol{\mu}_n$ and Σ_n $(n = 1, 2, \ldots)$ be non-random (a.s. constant) vectors and matrices respectively. Then the conditions*

$$\boldsymbol{\mu}_n \to \boldsymbol{\mu}, \qquad \Sigma_n \to \Sigma, \tag{2.6}$$

and

$$\sum_k \mathbf{E}\,|\boldsymbol{X}_{n,k} - \boldsymbol{\mu}_{n,k}|^2\, 1\{|\boldsymbol{X}_{n,k} - \boldsymbol{\mu}_{n,k}| > \varepsilon\} \to 0 \text{ for every } \varepsilon > 0, \tag{2.7}$$

imply

$$\mathscr{L}(\boldsymbol{S}_n) \to \mathscr{N}(\boldsymbol{\mu}, \Sigma). \tag{2.8}$$

We emphasize that the conditioning is relative to the preceding sum. Thus even in the special case $\boldsymbol{\mu}_{n,k} = \boldsymbol{0}$, for all n and k, Theorem 1 is more general than the corresponding result for martingale differences (which would have entailed $\mathbf{E}(\boldsymbol{X}_{n,k} \mid \mathscr{B}(\boldsymbol{S}_{n,1}, \ldots, \boldsymbol{S}_{n,k-1})) = \boldsymbol{0}$).

Condition (2.7) is, of course, the Lindeberg condition. If all $\boldsymbol{\mu}_{n,k} = \boldsymbol{0}$ it reduces to

$$\sum_k \mathbf{E} \mid \boldsymbol{X}_{n,k} \mid^2 1\{\mid \boldsymbol{X}_{n,k} \mid > \varepsilon\} \to 0 \quad \text{for every} \quad \varepsilon > 0. \tag{2.9}$$

Actually (2.9) implies (2.8) whatever the $\boldsymbol{\mu}_{n,k}$ (by Lemma 3.3 of [3]).

2.3. A more general result is the following.

Theorem 2. *Let* $\boldsymbol{\mu}$ *be a non-random vector and let* Σ *be a non-random matrix. Then*

$$\boldsymbol{\mu}_n \xrightarrow{P} \boldsymbol{\mu}, \qquad \Sigma_n \xrightarrow{P} \Sigma, \tag{2.10}$$

and (2.7) *imply* (2.8).

Here we no longer assume that $\boldsymbol{\mu}_n$ and Σ_n are constant. We do, however, retain the essential assumption that $\boldsymbol{\mu}$ and Σ are constant.

We note the following immediate consequence of this theorem.

Corollary 1. *Let* $\boldsymbol{a}_{n,k}$ *and* $B_{n,k}$, $(n = 1, 2, \ldots;\ k = 1, \ldots, k_n)$, *be constant vectors and positive semi-definite matrices, respectively. Then*

$$\sum_k \boldsymbol{a}_{n,k} \to \boldsymbol{\mu}, \qquad \sum_k B_{n,k} \to \Sigma, \tag{2.11}$$

$$\mathbf{E} \sum_k \mid \boldsymbol{\mu}_{n,k} - \boldsymbol{a}_{n,k} \mid \to 0, \qquad \mathbf{E} \sum_k \mid \operatorname{tr}(\Sigma_{n,k} - B_{n,k}) \mid \to 0, \tag{2.12}$$

and (2.7) *imply* (2.8).

To deduce this corollary it is enough to observe that (2.11) and (2.12) imply (2.10).

3. Some lemmas

3.1. The first lemma is very important technically, its purpose is to substitute the conditioning by $\mathcal{B}(S_{n,k-1})$ in the preceding theorems by a finer conditioning realtive to an increasing sequence of σ-fields.

Lemma 1. *Let* $S_1, S_2, \ldots, S_n$ *be random vectors defined on a probability space* $(\Omega, \mathcal{F}, P)$. *Then there exists a probability space* $(\tilde{\Omega}, \tilde{\mathcal{F}}, \tilde{P})$ *and random vectors* $\tilde{S}_1, \tilde{S}_2, \ldots, \tilde{S}_n$ *defined on it such that*

$$\mathcal{L}(\tilde{S}_{k-1}, \tilde{S}_k) = \mathcal{L}(S_{k-1}, S_k), \qquad (k = 2, \ldots, n), \tag{3.1}$$

$$\mathcal{L}(\tilde{S}_k \mid \mathcal{B}(\tilde{S}_1, \ldots, \tilde{S}_{k-1})) = \mathcal{L}(\tilde{S}_k \mid \mathcal{B}(\tilde{S}_{k-1})), \quad (k = 2, \ldots, k_n). \tag{3.2}$$

Here $\mathcal{L}(\cdot,\cdot)$ denotes the joint law and $\mathcal{L}(\cdot \mid \cdot)$ the conditional law. The conclusion asserts the Markovian character of $\tilde{S}_1, \tilde{S}_2, \ldots, \tilde{S}_n$. (It follows, of course, from (3.1) that $\mathcal{L}(\tilde{S}_k) = \mathcal{L}(S_k)$.)

Proof. For completeness we give the proof though it is similar to that of Lemma 3.1 in [3] which states the same result for random variables.

There is nothing to prove for $n = 1, 2$. Let $n > 2$ and assume the lemma for $n - 1$. Denote by $\hat{S}_1, \ldots, \hat{S}_{n-1}$ and $(\hat{\Omega}, \hat{\mathcal{F}}, \hat{P})$ the random vectors and probability space whose existence is asserted by the lemma for $n - 1$. Let $\mathcal{B}$ be the Borel σ-field on E^d and put $\tilde{\Omega} = \hat{\Omega} \times E^d$, $\tilde{\mathcal{F}} = \hat{\mathcal{F}} \times \mathcal{B}$. Let $f(v, B)$, for $v \in E^d$ and $B \in \mathcal{B}$, be a regular version of the conditional probability $\mathbf{P}(S_n \in B \mid S_{n-1} = v)$.

Define $\tilde{S}_k(\hat{\omega}, v) = \hat{S}_k(\hat{\omega})$ for $k = 1, \ldots, n-1$ and $\tilde{S}_n(\hat{\omega}, v) = v$. Furthermore let $\tilde{P}$ be defined by

$$\tilde{P}(F \times E^d) = \hat{P}(F) \tag{3.3}$$

for every $F \in \hat{\mathcal{F}}$, and

$$\tilde{\mathbf{P}}(\tilde{S}_n \in B \mid \tilde{S}_1 = v_1, \ldots, \tilde{S}_{n-1} = v_{n-1}) = f(v_{n-1}, B) \tag{3.4}$$

for all $B \in \mathcal{B}$ and $v_1, \ldots, v_{n-1} \in E^d$.

To complete the proof we have to verify that (3.1) and (3.2) are satisfied for $k = n$. (3.2) follows immediately from the fact that $v_1, \ldots, v_{n-2}$ do not appear on the right side of (3.4). By the induction assumption $\mathcal{L}(\tilde{S}_{n-1}) = \mathcal{L}(S_{n-1})$ and, by (3.4), $\mathcal{L}(\tilde{S}_n \mid \tilde{S}_{n-1} = v) = \mathcal{L}(S_n \mid S_{n-1} = v)$ for all v. These two statements imply (3.1) for $k = n$ and the proof is achieved.

3.2. The next lemma is tailored to normal distributions.

Lemma 2. *Let* $\mathcal{F}_{-1} = \mathcal{F}_0 \subset \mathcal{F}_1 \subset \ldots \subset \mathcal{F}_{n-1} \subset \mathcal{F}$ *be* σ*-fields in a probability space. Let* $\boldsymbol{\mu}_1, \ldots, \boldsymbol{\mu}_n$ *and* $\Sigma_1, \ldots, \Sigma_n$ *be random vectors and random positive semi-definite matrices with* $\boldsymbol{\mu}_k$ *and* Σ_k *being* $\mathcal{F}_{k-1}$*-measurable* $(k = 1, \ldots, n)$ *and put* $\boldsymbol{\mu} = \boldsymbol{\mu}_1 + \ldots + \boldsymbol{\mu}_n$, $\Sigma = \Sigma_1 + \ldots + \Sigma_n$. *If* $Y_1, \ldots, Y_n$ *are random vectors such that* $\mathcal{L}(Y_k \mid \mathcal{F}) = \mathcal{N}(\boldsymbol{\mu}_k, \Sigma_k)$ *and* $\mathbf{Y}_1 \mid \mathcal{F}, \ldots, \mathbf{Y}_n \mid \mathcal{F}$ *are independent, then*

$$\mathcal{L}(\mathbf{Y}_k + \ldots + \mathbf{Y}_n \mid \mathcal{F}) \\ = \mathcal{N}(\boldsymbol{\mu}_k + \ldots + \boldsymbol{\mu}_n, \Sigma_k + \ldots + \Sigma_n) \quad (k = 1, \ldots, n). \tag{3.5}$$

If $\boldsymbol{\mu}$ *and* Σ *are* $\mathcal{F}_0$*-measurable then* (3.5) *holds with* $\mathcal{F}$ *replaced by* $\mathcal{F}_{k-2}$ *and, moreover,* $\mathbf{Y}_1 \mid \mathcal{F}_{k-2}, \ldots, \mathbf{Y}_{k-1} \mid \mathcal{F}_{k-2}$ *and* $\mathbf{Y}_k + \ldots + \mathbf{Y}_n \mid \mathcal{F}_{k-2}$ *are independent.*

If, in particular, $\boldsymbol{\mu}$ *and* Σ *are constant, then* $\mathcal{L}(\mathbf{Y}_1 + \ldots + \mathbf{Y}_n) = \mathcal{N}(\boldsymbol{\mu}, \Sigma)$.

Proof. (3.5) follows immediately from the assumptions. (It is easy to write a formal proof, but it suffices to understand the case when $\mathcal{F}$ is atomic.) If $\boldsymbol{\mu}$ and Σ are $\mathcal{F}_0$-measurable then $\boldsymbol{\mu}_k + \ldots + \boldsymbol{\mu}_n = \boldsymbol{\mu} - (\boldsymbol{\mu}_1 + \ldots + \boldsymbol{\mu}_{k-1})$ and $\Sigma_1 + \ldots + \Sigma_k = \Sigma - (\Sigma_1 + \ldots + \Sigma_{k-1})$ are $\mathcal{F}_{k-2}$-measurable. Therefore the conditioning by $\mathcal{F}$ in (3.5) can be replaced by the coarser conditioning by $\mathcal{F}_{k-2}$. The final statement is simply (3.5), with $\mathcal{F}$ replaced by $\mathcal{F}_{k-2}$, for $k = 1$ and $\mathcal{F}_0$ the trivial field.

3.3. Our next lemma is technical. For the case of random variables it can be traced, in essence, to [4].

Lemma 3. *Let* $\mathcal{F}_0 \subset \mathcal{F}_1 \subset \ldots \subset \mathcal{F}_n$ *be* σ*-fields in a probability space. Let* $\mathbf{X}_k$ $(k = 1, \ldots, n)$ *be* $\mathcal{F}_k$*-measurable random vectors and let* $\mathbf{Y}_k$ $(k = 1, \ldots, n)$ *be random vectors such that* $\mathbf{X}_k \mid \mathcal{F}_{k-1}$, $\mathbf{Y}_k \mid \mathcal{F}_{k-1}$ *and* $\mathbf{Y}_{k+1} + \ldots + \mathbf{Y}_n \mid \mathcal{F}_{k-1}$ *are independent. Then*

$$|\mathbf{E}\, e^{it'(\mathbf{X}_1 + \ldots + \mathbf{X}_n)} - \mathbf{E}\, e^{it'(\mathbf{Y}_1 + \ldots + \mathbf{Y}_n)}| \le \sum_{k=1}^{n} \mathbf{E} |\mathbf{E}_{k-1} e^{it'\mathbf{X}_k} - \mathbf{E}_{k-1} e^{it'\mathbf{Y}_k}| \tag{3.6}$$

for every vector $\boldsymbol{t}$.

Proof. Putting

$$\mathbf{S}_k = \mathbf{X}_1 + \ldots + \mathbf{X}_k, \qquad \mathbf{R}_k = \mathbf{Y}_{k+1} + \ldots + \mathbf{Y}_n \quad (k = 0, 1, \ldots, n), \tag{3.7}$$

we have

$$\begin{aligned} e^{it'S_n} - e^{it'R_0} &= \sum_{k=1}^{n} (e^{it'(S_k+R_k)} - e^{it'(S_{k-1}+R_{k-1})}) \\ &= \sum_{k=1}^{n} e^{it'(S_{k-1}+R_k)}(e^{it'X_k} - e^{it'Y_k}). \end{aligned} \tag{3.8}$$

Since S_{k-1} is $\mathcal{F}_{k-1}$-measurable and $X_k \mid \mathcal{F}_{k-1}$, $Y_k \mid \mathcal{F}_{k-1}$ and $R_k \mid \mathcal{F}_{k-1}$ are independent

$$\begin{aligned} &\mathbf{E}_{k-1}\, e^{it'(S_{k-1}+R_k)}(e^{it'X_k} - e^{it'Y_k}) \\ &= \mathbf{E}_{k-1}\, e^{it'(S_{k-1}+R_k)}\mathbf{E}_{k-1}(e^{it'X_k} - e^{it'Y_k}). \end{aligned} \tag{3.9}$$

Therefore,

$$|\mathbf{E}^{it'(S_k+R_n)}(e^{it'X_k} - e^{it'Y_k})| \le \mathbf{E}\,|\mathbf{E}_{k-1}(e^{it'X_k} - e^{it'Y_k})|. \tag{3.10}$$

Taking expectations in (3.8) and applying (3.10) we obtain (3.6).

3.4. The last lemma is simple and, quite likely, it has been often applied. However, we are not aware of it being explicitly stated — let alone proved — even in the one-dimensional case.

Let $\mathcal{L}_1$ and $\mathcal{L}_2$ be d-dimensional laws and denote by $G_1(\boldsymbol{x})$ and $G_2(\boldsymbol{x})$ the corresponding distribution functions. Then

$$\begin{aligned} \Delta(\mathcal{L}_1, \mathcal{L}_2) = \inf_{h>0} \{G_1(\boldsymbol{x} - h\mathbf{1}) \le G_2(\boldsymbol{x}) \le G_1(\boldsymbol{x} + h\mathbf{1}) \\ \text{for all} \quad \boldsymbol{x} \in \mathrm{E}^d\} \end{aligned} \tag{3.11}$$

defines a metric on the space of laws (here $\mathbf{1}$ stands for the vector all of whose components = 1). A sequence $\mathcal{L}_n$ of random laws is said to converge in probability to the non-random law $\mathcal{L}$ iff $\Delta(\mathcal{L}_n, \mathcal{L}) \xrightarrow{P} 0$.

Lemma 4. *Let* S_n, Z_n $(n = 1, 2, \dots)$ *be random vectors satisfying*

$$\mathcal{L}(S_n) \to \mathcal{L}_1, \qquad \mathcal{L}(Z_n \mid \mathcal{B}(S_n)) \xrightarrow{P} \mathcal{L}_2. \tag{3.12}$$

Then

$$\mathcal{L}(S_n + Z_n) \to \mathcal{L}_1 * \mathcal{L}_2. \tag{3.13}$$

Proof. We prove this result through the fundamental theorem relating convergence of laws and of the corresponding characteristic functions. It follows from the definition (3.11) that, for any random vector Z, we have

$$|\mathbf{E}(e^{it'Z_n} \mid \mathcal{B}(S_n)) - \mathbf{E}e^{it'Z}| \le |t|\,\Delta(\mathcal{L}(Z_n \mid \mathcal{B}(S_n)), \mathcal{L}(Z)). \tag{3.14}$$

Therefore if $\mathscr{L}(\boldsymbol{Z})=\mathscr{L}_2$ it follows from (3.12) that the left-hand side of (3.14) tends in probability to zero for every $\boldsymbol{t}\in \mathrm{E}^d$. Since $\mathrm{E}e^{it'\boldsymbol{Z}_n}=\mathbf{E}\,\mathrm{E}(e^{it'\boldsymbol{Z}_n}|\mathscr{B}(\boldsymbol{S}_n))$ it follows that

$$\mathscr{L}(\boldsymbol{Z}_n)\rightarrow \mathscr{L}_2. \tag{3.15}$$

By the same argument

$$\begin{aligned}&\mathrm{E}e^{it'(\boldsymbol{S}_n+\boldsymbol{Z}_n)}-\mathrm{E}e^{it'\boldsymbol{S}_n}\mathrm{E}e^{it'\boldsymbol{Z}_n}\\ &\quad=\mathbf{E}(e^{it'\boldsymbol{S}_n}(\mathbf{E}(e^{it'\boldsymbol{Z}_n}|\mathscr{B}(\boldsymbol{S}_n))-\mathrm{E}e^{it'\boldsymbol{Z}_n}))\rightarrow \mathbf{0}.\end{aligned} \tag{3.16}$$

Hence the limit laws of $\boldsymbol{S}_n+\boldsymbol{Z}_n$ are the same as those of $\tilde{\boldsymbol{S}}_n+\tilde{\boldsymbol{Z}}_n$, where $\tilde{\boldsymbol{S}}_n, \tilde{\boldsymbol{Z}}_n$ are independent and $\mathscr{L}(\tilde{\boldsymbol{S}}_n)=\mathscr{L}(\boldsymbol{S}_n)$, $\mathscr{L}(\tilde{\boldsymbol{Z}}_n)=\mathscr{L}(\boldsymbol{Z})$. But by (3.12) and (3.15) $\mathscr{L}(\tilde{\boldsymbol{S}}_n+\tilde{\boldsymbol{Z}}_n)\rightarrow \mathscr{L}_1 * \mathscr{L}_2$.

Remark. A variation of this lemma is obtained on substituting $\mathscr{L}(\boldsymbol{S}_n+\boldsymbol{Z}_n)\rightarrow \mathscr{L}_1 * \mathscr{L}_2$ in place of the first assumption in (3.12) and concluding that $\mathscr{L}(\boldsymbol{S}_n)\rightarrow \mathscr{L}_1$. (This again follows from (3.15) and (3.16).)

4. Proof of the main results

4.1. By Lemma 1 we may assume the conditioning in the theorems of Section 2 being relative not to $\mathscr{B}(\boldsymbol{S}_{n,k})$ but to finer σ-fields $\mathscr{F}_{n,k}$ satisfying

$$\mathscr{F}_{n,0}\subset \mathscr{F}_{n,1}\subset \ldots \subset \mathscr{F}_{n,k-1}\subset \mathscr{F}_{n,k}\subset \ldots \subset \mathscr{F}_{n,k_n}. \tag{4.1}$$

Though Theorems 1 and 2 were stated for arbitrary $\boldsymbol{\mu}_{n,k}$ it is clearly enough to consider the case $\boldsymbol{\mu}_{n,k}=\mathbf{0}$ for all n and k. To simplify the writing we make this assumption throughout the present section.

4.2. Proof of Theorem 1. Let $\boldsymbol{Y}_{n,k}$ $(k=1,\ldots,k_n)$ be random vectors such that $\boldsymbol{Y}_{n,1}|\mathscr{F}_{n,k_n},\ldots,\boldsymbol{Y}_{n,k_n}|\mathscr{F}_{n,k_n}$ are independent and distributed according to $\mathscr{N}(\mathbf{0},\Sigma_{n,1}),\ldots,\mathscr{N}(\mathbf{0},\Sigma_{n,k_n})$ respectively. (No loss of generality is involved in assuming the existence of such $\boldsymbol{Y}_{n,k}$ since it is always possible to imbed the given probability space in a larger one which admits such $\boldsymbol{Y}_{n,k}$. A similar remark applies whenever necessary in the sequel.)

By Lemma 2, $\boldsymbol{Y}_{n,k+1}+\ldots+\boldsymbol{Y}_{n,k_n}|\mathscr{F}_{n,k-1}$ is distributed according to $\mathscr{N}(\mathbf{0},\Sigma_{k+1}+\ldots+\Sigma_n)$ and is independent of $\boldsymbol{Y}_{n,k}|\mathscr{F}_{n,k-1}$ which is distributed according to $\mathscr{N}(\mathbf{0},\Sigma_k)$. By Lemma 3 we have

$$|\mathbf{E}\,e^{it'\boldsymbol{S}_n}-e^{-\frac{1}{2}t'\Sigma_n t}|\leq \sum_k \mathbf{E}|\mathbf{E}_{n,k-1}e^{it'\boldsymbol{X}_{n,k}}-e^{-\frac{1}{2}t'\Sigma_{n,k}t}|. \tag{4.2}$$

By the fundamental theorem on characteristic functions and (2.6) it remains to show that the left-hand side of (4.2)$\to 0$ for every $t \in E^d$. Using $\mu_{n,k} = \mathbf{0}$ and (2.6) the expression whose absolute value is indicated on the right in (4.2) may be rewritten as

$$\begin{aligned}&\mathbf{E}_{n,k-1}(e^{it'X_{n,k}} - 1 - it'X_{n,k} + \tfrac{1}{2}(t'X_{n,k})^2)\\&\quad - (e^{-\frac{1}{2}t'\Sigma_{n,k}t} - 1 + \tfrac{1}{2}t'\Sigma_{n,k}t).\end{aligned} \tag{4.3}$$

From $|e^{iu} - (1 + iu - \frac{1}{2}u^2)| \le \min(\frac{1}{6}|u|^3, u^2)$ for real u it follows that the expression under $\mathbf{E}_{n,k-1}$ in (4.3) is dominated by

$$\tfrac{1}{6}|t'X_{n,k}|^3 1\{|X_{n,k}| \le \varepsilon\} + (t'X_{n,k})^2 1\{|X_{n,k}| > \varepsilon\} \tag{4.4}$$

for every $\varepsilon > 0$; from $|e^{-u} - 1 + u| \le \frac{1}{2}u^2$ for $u \ge 0$ it follows that the expression within the last parentheses in (4.3) is dominated by

$$\tfrac{1}{2}(t'\Sigma_{n,k}t)^2. \tag{4.5}$$

Since $\mathbf{E}_{n,k-1}|X_{n,k}|^2 = \mathrm{tr}(\Sigma_{n,k})$, it follows from (2.5) and the estimates (4.4), (4.5) that the right-hand side of (4.2) cannot exceed

$$\begin{aligned}&\frac{\varepsilon}{6}|t|^3 \mathrm{tr}(\Sigma_n) + |t|^2 \mathbf{E} \sum_k |X_{n,k}|^2 1\{|X_{n,k}| > \varepsilon\}\\&\quad + \tfrac{1}{2}|t|^2 \mathbf{E} \sum_k (\mathrm{tr}(\Sigma_{n,k}))^2.\end{aligned} \tag{4.6}$$

The second summand in (4.6) tends to 0 by (2.7); the last summand also tends to 0 since $\Sigma_k(\mathrm{tr}(\Sigma_{n,k}))^2 \le \mathrm{tr}(\Sigma_n)\ \max_k \mathrm{tr}(\Sigma_{n,k})$ and $\mathbf{E}\max\mathrm{tr}(\Sigma_{n,k}) \to 0$, again by (2.7). Since $\varepsilon > 0$ is arbitrary this completes the proof.

4.3. *Proof of Theorem* 2. We shall derive this theorem from the preceding one.

First we prove Theorem 2 under the additional assumption that the random variables $\mathrm{tr}(\Sigma_n)$ are uniformly bounded.

$$\mathrm{tr}(\Sigma_n) < c \quad (n = 1, 2, \ldots). \tag{4.8}$$

Then the random matrices cI-Σ_n are positive definite (I denotes the unit $d \times d$ matrix). We adjoin to the n-th row of (2.1) a further random vector $X_{n,k_n+1} = Z_n$ such that $\mathscr{L}(Z_n \mid \mathscr{F}_{n,k_n}) = \mathscr{N}(\mathbf{0}, cI - \Sigma_n)$. Then, by Theorem 1, $\mathscr{L}(S_n + Z_n) \to \mathscr{N}(\mathbf{0}, cI)$. The conclusion (2.8) follows from Lemma 4 (or, rather, from its variant stated in the remark at the end of Section 3).

In the general case let $c > \mathrm{tr}(\Sigma)$ and put $\tilde{X}_{n,k} = X_{n,k}$ $1\{\mathrm{tr}(\Sigma_{n,1} + \ldots + \Sigma_{n,k}) \le c\}$. By (2.10), $\mathbf{P}\{\Sigma_k \tilde{X}_{n,k} \ne S_n\} \to 0$, hence $\mathscr{L}(\Sigma_k \tilde{X}_{n,k})$

and $\mathscr{L}(\boldsymbol{S}_n)$ have the same limit laws. Since $\Sigma_{n,1}+\ldots+\Sigma_{n,k}$ is $\mathscr{F}_{n,k-1}$-measurable we have $\mathbf{E}_{k,n-1}\tilde{\boldsymbol{X}}_{n,k}=0$,

$$\mathbf{E}_{k,n-1}\tilde{\boldsymbol{X}}'_{n,k}\tilde{\boldsymbol{X}}_{n,k}=\Sigma_{n,k}\,1\{\operatorname{tr}(\Sigma_{n,1}+\ldots+\Sigma_{n,k})<c\}.$$

Therefore, $\tilde{\boldsymbol{X}}_{n,k}$ $(n=1,2,\ldots,k=1,\ldots,k_n)$ satisfy the conditions of Theorem 2 and also (4.8). This completes the proof.

5. Remarks

5.1. Results about asymptotic normality of cumulative sums of random vectors are easily obtained from those on double arrays. We say that a sequence of random vectors $\boldsymbol{S}_n$ is asymptotically $\mathcal{N}(\boldsymbol{a}_n, B_n)$ — and express this by $\boldsymbol{S}_n\sim\mathcal{N}(\boldsymbol{a}_n, B_n)$ — iff

$$\Delta\left(\mathscr{L}\left(\frac{\boldsymbol{S}_n}{(\operatorname{tr}B_n)^{1/2}}\right),\mathcal{N}\left(\frac{\boldsymbol{a}_n}{(\operatorname{tr}B_n)^{1/2}},\frac{B_n}{(\operatorname{tr}B_n)^{1/2}}\right)\right)\to 0.$$

(Here $\boldsymbol{a}_n$ and B_n are constant vectors and matrices.)

We give one example of a result of this nature.

Corollary 2. *Let* $\boldsymbol{X}_1,\boldsymbol{X}_2,\ldots,\boldsymbol{X}_n,\ldots$ *be random vectors and put* $\boldsymbol{S}_n=\boldsymbol{X}_1+\ldots+\boldsymbol{X}_n$. *Let* $\boldsymbol{a}_n, B_n$ $(n=1,2,\ldots)$ *be constant vectors and matrices. Put* $\boldsymbol{\mu}_n=\mathbf{E}(\boldsymbol{X}_n\mid\mathscr{B}(\boldsymbol{S}_n))$ *and* $\Sigma_n=\mathbf{E}(\boldsymbol{X}_n-\boldsymbol{\mu}_n)(\boldsymbol{X}'_n-\boldsymbol{\mu}'_n)$, *and* $\sigma_n^2=\operatorname{tr}(B_1)+\ldots+\operatorname{tr}(B_n)$, *then the conditions*

$$\frac{1}{\sigma_n}\sum_{k=1}^{n}(\boldsymbol{\mu}_k-\boldsymbol{a}_k)\xrightarrow{\mathrm{P}}\mathbf{0},\qquad \frac{1}{\sigma_n}\sum_{k=1}^{n}(\Sigma_k-B_k)\xrightarrow{\mathrm{P}}0, \tag{5.1}$$

and

$$\sum_{k=1}^{n}\mathbf{E}|\boldsymbol{X}_k-\boldsymbol{\mu}_k|^2\,1\{\boldsymbol{X}_k-\boldsymbol{\mu}_k|>\varepsilon\}\to 0\quad\textit{for every}\quad\varepsilon>0, \tag{5.2}$$

imply

$$\mathscr{L}(\boldsymbol{S}_n)\sim\mathcal{N}\left(\frac{1}{\sigma_n}\sum_{k=1}^{n}\boldsymbol{a}_k,\frac{1}{\sigma_n}\sum_{k=1}^{n}B_k\right).$$

The last 0 in (5.1) is, of course, the 0-matrix. (5.2) is the Lindeberg condition. As in the case of (2.8), the assumption obtained from (5.2) upon deleting the $\boldsymbol{\mu}_n$, implies (5.2).

5.2. Since our proofs are based on showing that the characteristic functions approach the characteristic function corresponding to the limit normal law it is possible, in specific cases, to obtain information on the rate of convergence from results linking the distance between laws and the difference of the corresponding characteristic functions. This is one of the reasons we preferred not to reduce the results about sums of random vectors to those about random variables in the way mentioned in 1.1. In some cases the estimates obtained directly for the vectors may be better than those obtained from estimating first the linear combinations of the components.

5.3. Results about arrays with infinite rows follow easily from those proved here. With obvious modifications of notation we have

Corollary 3. *Let the rows in* (2.1) *be infinite and let* $\mathscr{L}_n$ *be any limit law of the sequence* $\mathscr{L}(\boldsymbol{S}_{n,1}), \mathscr{L}(\boldsymbol{S}_{n,2}), \ldots, \mathscr{L}(\boldsymbol{S}_{n,k}), \ldots$. *Then* (2.10) *and* (2.7) *imply* $\mathscr{L}_n \to \mathscr{N}(\boldsymbol{\mu}, \Sigma)$.

Here condition (2.10) may be interpreted in a sense which does not imply the convergence of the series defining $\boldsymbol{\mu}_n$ and $\boldsymbol{S}_n$. Thus the first part of (2.10) may be understood as $\limsup_{k=\infty} |\boldsymbol{\mu}_{n,1} + \cdots + \boldsymbol{\mu}_{n,k} - \boldsymbol{\mu}| \xrightarrow{P} 0$, and similarly for the second part.

5.4. An easy way of getting rid of the moment assumptions is through truncation. Thus the following result is an immediate consequence of Theorem 2.

Corollary 4. *Let* $H_{n,k}$ $(n = 1, 2, \ldots, k = 1, \ldots, k_n)$ *be positive numbers and put* $\bar{\boldsymbol{X}}_{n,k} = \boldsymbol{X}_{n,k} 1\{|\boldsymbol{X}_{n,k}| \leq H_{n,k}\}$. *If*

$$\mathbf{P}\left(\bigcup_{k=1}^{k_n} \{|\boldsymbol{X}_{n,k}| > H_{n,k}\}\right) \to 0 \tag{5.3}$$

holds and the $\bar{\boldsymbol{X}}_{n,k}$ *satisfy the conditions imposed on* $\boldsymbol{X}_{n,k}$ *in Theorem* 2, *then* (2.8) *holds.*

Note that if $H_{n,k} = H$ for all n and k, then the Lindeberg condition for $\bar{\boldsymbol{X}}_{n,k}$ is equivalent to

$$\sum_k \mathbf{P}(|\bar{\boldsymbol{X}}_{n,k} - \bar{\boldsymbol{\mu}}_{n,k}| > \varepsilon) \to 0 \quad \text{for every} \quad \varepsilon > 0, \tag{5.4}$$

where $\bar{\boldsymbol{\mu}}_{n,k} = E_{n,k-1} \bar{\boldsymbol{X}}_{n,k}$.

Let $|\boldsymbol{X}| \le H$ and $\mathscr{F}$ be any σ-field in the probability space. Then $|\mathbf{E}(\boldsymbol{X} \mid \mathscr{F})| \le \varepsilon + H\,\mathbf{P}(|\boldsymbol{X}| > \varepsilon \mid \mathscr{F})$. Hence $\mathbf{P}(|\mathbf{E}(\boldsymbol{X} \mid \mathscr{F})| > \varepsilon + \varepsilon H) \le \mathbf{P}(\mathbf{P}(|\boldsymbol{X}| > \varepsilon \mid \mathscr{F}) > \varepsilon) = \mathbf{P}(|\boldsymbol{X}| > \varepsilon)$. This shows that (5.4) is implied by

$$\sum_k \mathbf{P}(|\boldsymbol{X}_{n,k}| > \varepsilon) \to 0 \quad \text{for every} \quad \varepsilon > 0, \tag{5.5}$$

if all $H_{n,k} \le H$.

Specializing Corollary 4 to $H_{n,k} = H$ and noticing that (5.5) implies (5.3) we obtain

Corollary 5. *Let the random vectors* (2.1) *satisfy* (5.5). *Let H be a positive constant and put* $\bar{\boldsymbol{X}}_{n,k} = \boldsymbol{X}_{n,k} 1\{|\boldsymbol{X}_{n,k}| \le H\}$, $\bar{\boldsymbol{\mu}}_{n,k} = \mathbf{E}\,\bar{\boldsymbol{X}}_{n,k}$, $\bar{\Sigma}_{n,k} = \mathbf{E}(\bar{\boldsymbol{X}}_{n,k} - \bar{\boldsymbol{\mu}}_{n,k})(\bar{\boldsymbol{X}}'_{n,k} - \bar{\boldsymbol{\mu}}'_{n,k})$. *Then the conditions*

$$\sum_k \bar{\boldsymbol{\mu}}_{n,k} \xrightarrow{\mathrm{P}} \boldsymbol{\mu}, \qquad \sum_k \bar{\Sigma}_{n,k} \to \Sigma, \tag{5.6}$$

where $\boldsymbol{\mu}$ and Σ are non-random, imply (2.8).

We remark that, under (5.5), if (5.6) holds for truncation with one H then it holds, with the same $\boldsymbol{\mu}$ and Σ, for truncation with any H (or, indeed, for other truncations, for example truncating the i-th component of $\boldsymbol{X}_{n,k}$ by H_i where H_i are arbitrary positive numbers).

We also note that in the independent case (5.5) is equivalent to $\mathbf{P}(\max_k |\boldsymbol{X}_{n,k}| > \varepsilon) \to 0$ for every $\varepsilon > 0$.

5.5. Finally we remark that if all random vectors (2.1) are defined on the same probability space, then our proofs yield $\mathscr{L}(\boldsymbol{S}_n \mid \mathscr{F}_{n,0}) \to \mathscr{N}(\boldsymbol{\mu}, \Sigma)$, where $\boldsymbol{\mu}$ and Σ are $\mathscr{F}_{n,0}$-measurable and $\mathscr{F}_{n,0}$ is any σ-field such that all $\boldsymbol{X}_{n,k}$ are $\mathscr{F}_0$-measurable. This fact can be used to study vector-valued stochastic processes.

References

[1] Cramér, H. (1937). *Random Variables and Probability Distributions.* Cambridge Tracts No. 36. Cambridge.

[2] Dvoretzky, A. (1970). "The Central limit problem for dependent random variables", *Proceedings of the International Congress of Mathematicians.* Nice.

[3] Dvoretzky, A. (1970). "Asymptotic normality for sums of dependent random variables", *Proceedings of the Sixth Berkeley Symposium on Mathematical Statistics and Probability.* Vol. 2, 513–535.

[4] Lévy, P. (1937). *Théorie de l'Addition des Variables Aléatoires.* Paris, Gauthier-Villars.

P.R. Krishnaiah, ed., *Multivariate Analysis–IV*

INFERENCE FOR THE MULTIVARIATE REGRESSION MODEL

D.A.S. FRASER
and
Kai W. NG
University of Toronto, Toronto, Canada

The recent sequential methods of inference examined in D.A.S. Fraser and Jock MacKay (Parameter factorization and inference based on significance, likelihood, and objective posterior, Annals Statistics 3 (1975)) are applied to the multivariate normal regression model. The parameters are separated: location as opposed to scale; parameters for a first group of variables as opposed to those for the remaining group of variables. Two orders of examining the parameters in sequence are investigated. A new disguised matrix *t* distribution is obtained (related to the disguised Wishart distribution). The factorization methods cast light on anomalies in the Bayesian methods for the multivariate regression model.

0. Introduction

Some of the larger challenges for statistical inference arise in the context of many parameters. Inference methods for a sequence of parameter components has been examined by Fraser and MacKay [4]: for a model satisfying certain regularity conditions it is shown that significance, likelihood, and objective posteriors are essentially equivalent in the inference results they provide. More recently it has been shown that the equivalence extends to confidence methods and that a sequence of confidence regions for component parameters combine to give an overall confidence region with the product confidence level.

This paper examines statistical inference for component parameters of the multivariate regression model, examining in detail the case with normal error. Inference for the regression coefficients or for the variance matrix in the normal case is relatively straightforward. Some attention has been given to further separation of the parameters; for example, Dawid, Stone, and Zidek [2] consider the first q response variables in contrast to last $p-q$ variables.

In this paper we apply the parameter separation methods discussed in Fraser and MacKay [4] to the multivariate regression model. We examine two proper factorization sequences based on the regression coefficients and variance matrix division and the first q variables, last $p-q$ variables division. The resulting inference procedures are examined in detail for the normal; some new distribution theory is obtained for the disguised Wishart and for a new disguised matrix t distribution.

We also examine a third and improper factorization which casts some new light on certain Bayesian anomalies.

1. The multivariate regression model

Consider a response Y recording p response variables on n repetitions:

$$Y = \begin{pmatrix} y_{11} & \cdots & y_{1n} \\ \vdots & & \vdots \\ y_{p1} & \cdots & y_{pn} \end{pmatrix}. \tag{1.1}$$

And suppose initially that Y is multivariate normal $\mathrm{MN}_p(\mathscr{B}X, I_n \otimes \Sigma)$ where X is a given design matrix and $(\mathscr{B}, \Sigma)$ is the parameter giving regression coefficients and the variance matrix. For purposes of interrelating component parameters we structure the model as

$$Y = \mathscr{B}X + \Delta Z \tag{1.2}$$

where Z has normal(0, 1) entries and $\mathscr{B} \in R^{pr}$ and Δ is positive lower triangular (PLT). Equation (1.2) can be rewritten as

$$\begin{pmatrix} X \\ Y \end{pmatrix} = \begin{pmatrix} I & 0 \\ \mathscr{B} & \Delta \end{pmatrix} \begin{pmatrix} X \\ Z \end{pmatrix} = \theta \begin{pmatrix} X \\ Z \end{pmatrix} \tag{1.3}$$

in which the response is obtained as a matrix transformation

$$\theta = \begin{pmatrix} I & 0 \\ \mathscr{B} & \Delta \end{pmatrix} \tag{1.4}$$

of the variation Z; θ belongs to a Lie group G and X is recorded with the variation and with the response to permit the matrix multiplication. We can now generalize the model and let Z have a distribution say with density f_λ involving possibly a parameter λ. The model either in the usual normal case or in its generalized form is now a structural model as examined in Fraser and MacKay [4].

The analysis of a structural model has been recorded in detail [4] as part of the comparison of inference procedures; certain additional details for the multivariate regression model may be found in Fraser [3]; we summarize here. On the sample space for Z delete the set of measure zero for which the rank of $(X'Z')$ is less than $p+r$ (assume $n \geq p+r$). View

$$\begin{pmatrix} X \\ Z \end{pmatrix} = \begin{pmatrix} \boldsymbol{x}_1' \\ \vdots \\ \boldsymbol{x}_r' \\ \boldsymbol{z}_1' \\ \vdots \\ \boldsymbol{z}_p' \end{pmatrix} \tag{1.5}$$

as a matrix of $r+p$ row vectors; then successively orthonormalize $\boldsymbol{z}_1', \ldots, \boldsymbol{z}_p'$ to preceding row vectors in the matrix array obtaining

$$\begin{pmatrix} X \\ D(Z) \end{pmatrix} = \begin{pmatrix} \boldsymbol{x}_1' \\ \vdots \\ \boldsymbol{x}_r' \\ \boldsymbol{d}_1'(Z) \\ \vdots \\ \boldsymbol{d}_p'(Z) \end{pmatrix} \tag{1.6}$$

where the $\boldsymbol{d}$'s are orthonormal and for example $\boldsymbol{d}_2$ is the unit residual after linear regression on $\boldsymbol{x}_1, \ldots, \boldsymbol{x}_r, \boldsymbol{z}_1$. Let $B = B(Z)$ and $T = T(Z)$ be matrices of linear regression coefficients on the row vectors in (1.6); then

$$Z = BX + TD(Z),$$

$$\begin{pmatrix} X \\ Z \end{pmatrix} = \begin{pmatrix} I & 0 \\ B & T \end{pmatrix} \begin{pmatrix} X \\ D(Z) \end{pmatrix} = g \begin{pmatrix} X \\ D(Z) \end{pmatrix} \tag{1.7}$$

where

$$B(Z) = \begin{pmatrix} b_{11}(Z) & \ldots & b_{1r}(Z) \\ \vdots & & \vdots \\ b_{p1}(Z) & \ldots & b_{pr}(Z) \end{pmatrix} \tag{1.8}$$

$$T(Z) = \begin{pmatrix} s_1(Z) & & \dots & 0 \\ t_{21}(Z) s_2(Z) & & & \vdots \\ & & & \\ \vdots & & \ddots & \\ t_{p1}(Z) \dots t_{pp-1}(Z) s_p(Z) \end{pmatrix}. \tag{1.9}$$

The preceding provides suitable notation.

Now let Z have a distribution with density f_λ and let Y be an observed response obtained from the model (1.2). Then the function $D(Z) = D(Y) = D$ is observed concerning the realization Z; the likelihood function for λ from D is

$$k_\lambda(D) = \int f_\lambda(BX + TD) \frac{|T|^{n-r}}{|T|_\Delta} \mathrm{d}B \, \mathrm{d}T \tag{1.10}$$

where the integration is over the range of B and T and

$$|T|_\Delta = s_1^1 \dots s_p^p$$

is the ascending determinant. The conditional distribution for the unobserved B, T given the observed D is

$$k_\lambda^{-1}(D) f_\lambda(BX + TD) s_1^{n-r-1} \dots s_p^{n-r-p} \mathrm{d}B \, \mathrm{d}T. \tag{1.11}$$

For the case of standard normal Z this distribution becomes

$$k \operatorname{etr}\{-\tfrac{1}{2}BXX'B' - \tfrac{1}{2}TT'\} s_1^{n-r-1} \dots s_p^{n-r-p} \mathrm{d}B \, \mathrm{d}T \tag{1.12}$$

where

$$k = \frac{\Pi_1^p A_{n-r-j+1}}{(2\pi)^{pn/2}}, \qquad A_n = \frac{2\pi^{n/2}}{\Gamma(n/2)}$$

and the distribution involves $\mathrm{MN}_r(0; (XX')^{-1})$ variables, normal(0, 1) variables, and chi variables. For reference note that $B(Y)$ is the maximum-likelihood least-squares estimate of β and the inner product matrix

$$S(Y) = (Y - B(Y)X)(Y - B(Y)X)' = T(Y)T'(Y)$$

is the sample Wishart matrix.

2. First parameter separation

Consider a parameter separation in which we examine inference for scale parameters first and location second and then within each of these examine the last $p-q$ variables first and the first q variables second. Specifically we factor θ as

$$\theta = \begin{pmatrix} I & 0 \\ \mathcal{B} & \Delta \end{pmatrix} = \begin{pmatrix} I & 0 \\ 0 & \Delta \end{pmatrix} \begin{pmatrix} I & 0 \\ M & I \end{pmatrix}$$

$$= \left(\begin{array}{c:cc} I & 0 & 0 \\ \hdashline 0 & I & 0 \\ 0 & 0 & \Delta_{22} \end{array}\right) \left(\begin{array}{c:cc} I & 0 & 0 \\ \hdashline 0 & I & 0 \\ 0 & \Lambda_{21} & I \end{array}\right) \left(\begin{array}{c:cc} I & 0 & 0 \\ \hdashline 0 & \Delta_{11} & 0 \\ 0 & 0 & I \end{array}\right) \left(\begin{array}{c:cc} I & 0 & 0 \\ \hdashline 0 & I & 0 \\ M_2 & 0 & I \end{array}\right) \left(\begin{array}{c:cc} I & 0 & 0 \\ \hdashline M_1 & I & 0 \\ 0 & 0 & I \end{array}\right) \quad (2.1)$$

$$= \theta_5\theta_4\theta_3\theta_2\theta_1$$

where $\theta_i \in H_i$ which is a subgroup of G. Also note that the factorization is a proper inference factorization in the sense that $G_i = H_i \ldots H_1$ is for each $i = 1, \ldots, 5$ a subgroup of G.

The component parameters Δ_{22}, Λ_{21}, Δ_{11}, M_2, M_1 are easily related to components of the regression and variance matrices:

$$\mathcal{B} = \begin{pmatrix} \mathcal{B}_1 \\ \mathcal{B}_2 \end{pmatrix}, \qquad \Sigma = \begin{pmatrix} \Sigma_{11} & \Sigma_{12} \\ \Sigma_{21} & \Sigma_{22} \end{pmatrix} \quad (2.2)$$

where the partition of p is q to $p-q$;

$$H_{21} = \Sigma_{21}\Sigma_{11}^{-1}, \qquad \Sigma_{22\cdot1} = \Sigma_{22} - \Sigma_{21}\Sigma_{11}^{-1}\Sigma_{12} \quad (2.3)$$

where H_{21} is the regression matrix of the last $p-q$ on the first q variables and $\Sigma_{22\cdot1}$ is the residual variance matrix;

$$\mathcal{B}_{2\cdot1} = \mathcal{B}_2 - H_{21}\mathcal{B}_1 \quad (2.4)$$

where $\mathcal{B}_{2\cdot1}$ is the residual regression coefficient matrix;

$$\Delta = \begin{pmatrix} \Delta_{11} & 0 \\ \Delta_{21} & \Delta_{22} \end{pmatrix}, \qquad \Delta_{21} = \Delta_{22}\Lambda_{21}\Delta_{11}$$

where the diagonal components are PLT. Then

$$\begin{aligned} &\Sigma_{22\cdot1} = \Delta_{22}\Delta_{22}', \qquad H_{21} = \Delta_{22}\Lambda_{21}, \\ &\Sigma_{11} = \Delta_{11}\Delta_{11}', \qquad \mathcal{B}_{2\cdot1} = \Delta_{22}M_2, \qquad \mathcal{B}_1 = \Delta_{11}M_1 \end{aligned} \quad (2.5)$$

gives the connection between the regression and variance matrix components and the components $\theta_5, \theta_4, \ldots, \theta_1$.

The factorization (2.1) of θ leads [4] to a reverse factorization of g:

$$g = \begin{pmatrix} I & 0 \\ B & T \end{pmatrix} = \begin{pmatrix} I & 0 \\ B & I \end{pmatrix}\begin{pmatrix} I & 0 \\ 0 & T \end{pmatrix}$$

$$= \begin{pmatrix} I & 0 & 0 \\ B_1 & I & 0 \\ 0 & 0 & I \end{pmatrix}\begin{pmatrix} I & 0 & 0 \\ 0 & I & 0 \\ B_2 & 0 & I \end{pmatrix}\begin{pmatrix} I & 0 & 0 \\ 0 & T_{11} & 0 \\ 0 & 0 & I \end{pmatrix}\begin{pmatrix} I & 0 & 0 \\ 0 & I & 0 \\ 0 & T_{21} & I \end{pmatrix}\begin{pmatrix} I & 0 & 0 \\ 0 & I & 0 \\ 0 & 0 & T_{22} \end{pmatrix} \tag{2.6}$$

$$= h_1 h_2 \dots h_5$$

where $h_i \in H_i$ and

$$T = \begin{pmatrix} T_{11} & 0 \\ T_{21} & T_{22} \end{pmatrix}$$

is partitioned q to $p-q$. The distribution (1.11) for g can be expressed in terms of these components

$$k_\lambda^{-1}(D) f_\lambda(BX + TD) \,|T_{11}|^{n-r}\, |T_{22}|^{n-r-q} \mathrm{d}B_1 \mathrm{d}B_2 \frac{\mathrm{d}T_{11}}{|T_{11}|_\Delta} \mathrm{d}T_{21} \frac{\mathrm{d}T_{22}}{|T_{22}|_\Delta}. \tag{2.7}$$

Let f_5 be the conditional density for h_5 given D; f_4 the conditional density for h_4 given h_5 and D; and so on. By integrating and dividing successively we can complete the factorization of the density for g:

$$f_5(T_{22})\mathrm{d}T_{22} f_4(T_{21})\mathrm{d}T_{21} \dots f_1(B_1)\mathrm{d}B_1. \tag{2.8}$$

For tests of significance and confidence regions for the components of θ we need the preceding factorization but used for coordinates taken about the observed Y. Thus $Y = \theta g D$ is written as $Y = \theta g^* Y$ or as

$$Y = \theta_5 \theta_4 \theta_3 \theta_2 \theta_1 h_1^* h_2^* h_3^* h_4^* h_5^* Y \tag{2.9}$$

in terms of the components; for details see [4]. The transformation from g to g^* is given by $g = g^* g(Y)$ which for the components gives

$$\begin{aligned}
T_{22} &= T_{22}^* T_{22}(Y), & \mathrm{d}T_{22} &= |T_{22}(Y)|_\nabla \mathrm{d}T_{22}^*, \\
T_{21} &= T_{21}^* T_{11}(Y) + T_{22}^* T_{21}(Y), & \mathrm{d}T_{21} &= |T_{11}(Y)|^{p-q} \mathrm{d}T_{21}^*, \\
T_{11} &= T_{11}^* T_{11}(Y), & \mathrm{d}T_{11} &= |T_{11}(Y)|_\nabla \mathrm{d}T_{11}^*, \\
B_2 &= B_2^* + T_{21}^* B_1(Y) + T_{22}^* B_2(Y), & \mathrm{d}B_2 &= \mathrm{d}B_2^*, \\
B_1 &= B_1^* + T_{11}^* B_1(Y), & \mathrm{d}B_1 &= \mathrm{d}B_1^*,
\end{aligned} \tag{2.10}$$

where $|T|_\nabla = s_1^p \dots s_p^1$ is the descending determinant. Let f^5 be the conditional density for h_5^* given D, f^4 be the conditional density for h_4^* given h_5^*

and D and so on. We avoid cumbersome notation now by deleting the * but emphasizing *that coordinates are now with respect to* Y. From (2.8) and (2.10) we then obtain

$$
\begin{aligned}
f^5(T_{22})\mathrm{d}T_{22} &= f_5(T_{22}T_{22}(Y))\,|T_{22}(Y)|_{\nabla}\,\mathrm{d}T_{22},\\
f^4(T_{21})\mathrm{d}T_{21} &= f_4(T_{21}T_{11}(Y)+T_{22}T_{21}(Y))\,|T_{11}(Y)|^{p-q}\,\mathrm{d}T_{21},\\
f^3(T_{11})\mathrm{d}T_{11} &= f_3(T_{11}T_{11}(Y))\,|T_{11}(Y)|_{\nabla}\,\mathrm{d}T_{11},\\
f^2(B_2)\mathrm{d}B_2 &= f_2(B_2+T_{21}B_1(Y)+T_{22}B_2(Y))\,\mathrm{d}B_2,\\
f^1(B_1)\mathrm{d}B_1 &= f_1(B_1+T_{11}B_1(Y))\,\mathrm{d}B_1.
\end{aligned}
\tag{2.11}
$$

For tests of significance and for confidence regions the relation (2.9) becomes

$$
\theta_5 = h_5^{-1},\qquad \theta_4 = h_4^{-1},\qquad \theta_3 = h_3^{-1},\qquad \theta_2 = h_2^{-1},\qquad \theta_1 = h_1^{-1} \tag{2.12}
$$

where as noted earlier we are deleting the * for easier notation. The relations (2.12) become

$$
\begin{aligned}
&\Sigma_{22\cdot 1} = T_{22}^{-1}T_{22}'^{-1}, \qquad H_{21} = -T_{22}^{-1}T_{21},\\
&\Sigma_{11} = T_{11}^{-1}T_{11}'^{-1}, \qquad \mathcal{B}_{2\cdot 1} = -T_{22}^{-1}B_2, \qquad \mathcal{B}_1 = -T_{11}^{-1}B_1.
\end{aligned}
\tag{2.13}
$$

The distributions (2.11) can then be used to form successive tests or confidence regions for $\Sigma_{22\cdot 1}$, H_{21}, Σ_{11}, $\mathcal{B}_{2\cdot 1}$, $\mathcal{B}_1$.

3. First parameter separation; normal error

Now consider the preceding parameter separation for the case of standard normal Z with density

$$
f(Z) = (2\pi)^{-pn/2}\operatorname{etr}\{-\tfrac{1}{2}ZZ'\}.
$$

Using the orthonormal properties of D we obtain

$$
\begin{aligned}
\operatorname{tr} ZZ' &= \operatorname{tr}\{BXX'B'\} + \operatorname{tr} TT'\\
&= \operatorname{tr} B_1XX'B_1' + \operatorname{tr} B_2XX'B_2' + \operatorname{tr} T_{11}T_{11}' + \operatorname{tr} T_{21}T_{21}' + \operatorname{tr} T_{22}T_{22}'.
\end{aligned}
$$

The conditional density (1.12) for g given D can be easily factored using normal and chi integration constants; it is the product of the following

$$f_5(T_{22})\mathrm{d}T_{22} = \frac{\Pi_1^{p-q} A_{n-r-q-j+1}}{(2\pi)^{(p-q)(n-r-q)/2}} \operatorname{etr}\{-\tfrac{1}{2}T_{22}T'_{22}\} \, |T_{22}|^{n-r-q} \frac{\mathrm{d}T_{22}}{|T_{22}|_\Delta},$$

$$f_4(T_{21})\mathrm{d}T_{21} = \frac{1}{(2\pi)^{(p-q)q/2}} \operatorname{etr}\{-\tfrac{1}{2}T_{21}T'_{21}\}\,\mathrm{d}T_{21},$$

$$f_3(T_{11})\mathrm{d}T_{11} = \frac{\Pi_1^{q} A_{n-r-j+1}}{(2\pi)^{q(n-r)/2}} \operatorname{etr}\{-\tfrac{1}{2}T_{11}T'_{11}\} \, |T_{11}|^{n-r} \frac{\mathrm{d}T_{11}}{|T_{11}|_\Delta}, \tag{3.1}$$

$$f_2(B_2)\mathrm{d}B_2 = \frac{|XX'|^{(p-q)/2}}{(2\pi)^{(p-q)r/2}} \operatorname{etr}\{-\tfrac{1}{2}B_2XX'B'_2\}\,\mathrm{d}B_2,$$

$$f_1(B_1)\mathrm{d}B_1 = \frac{|XX'|^{q/2}}{(2\pi)^{qr/2}} \operatorname{etr}\{-\tfrac{1}{2}B_1XX'B_1\}\,\mathrm{d}B_1.$$

For distributions relative to the observed Y we need some additional notation:

$$S(Y) = T(Y)T'(Y) = \begin{pmatrix} S_{11}(Y) & S_{12}(Y) \\ S_{21}(Y) & S_{22}(Y) \end{pmatrix}$$

where the partition is q to $p-q$;

$$S_{22\cdot 1}(Y) = S_{22}(Y) - S_{21}(Y)S_{11}^{-1}(Y)S_{12}(Y),$$

$$H_{21}(Y) = S_{21}(Y)S_{11}^{-1}(Y), \qquad S_{11}(Y) = T_{11}(Y)T'_{11}(Y), \tag{3.2}$$

$$H_{21}(Y) = T_{21}(Y)T_{11}^{-1}(Y), \qquad S_{22\cdot 1}(Y) = T_{22}(Y)T'_{22}(Y). \tag{3.3}$$

We now substitute the normal expressions (3.1) into the general expressions (2.11) and obtain the following component distributions f^5, f^4, f^3, f^2, f^1 relative to the observed Y:

$$\frac{\Pi_1^{p-q} A_{n-r-q-j+1}}{(2\pi)^{(p-q)(n-r-q)/2}} \operatorname{etr}\{-\tfrac{1}{2}T'_{22}T_{22}S_{22\cdot 1}(Y)\}$$

$$\times \frac{|T_{22}(Y)|_\nabla}{|T_{22}(Y)|_\Delta} |S_{22\cdot 1}(Y)|^{(n-r-q)/2} |T_{22}|^{n-r-q} \frac{\mathrm{d}T_{22}}{|T_{22}|_\Delta},$$

$$\frac{|S_{11}(Y)|^{(p-q)/2}}{(2\pi)^{(p-q)q/2}} \operatorname{etr}\{-\tfrac{1}{2}[T_{21} + T_{22}H_{21}(Y)]S_{11}(Y)[\cdot\cdot]'\}\,\mathrm{d}T_{21},$$

$$\frac{\Pi_1^{q} A_{n-r-j+1}}{(2\pi)^{q(n-r)/2}} \operatorname{etr}\{-\tfrac{1}{2}T'_{11}T_{11}S_{11}(Y)\} \tag{3.4}$$

$$\times \frac{|T_{11}(Y)|_\nabla}{|T_{11}(Y)|_\Delta} |S_{11}(Y)|^{(n-r)/2} |T_{11}|^{n-r} \frac{\mathrm{d}T_{11}}{|T_{11}|_\Delta},$$

$$\frac{|XX'|^{(p-q)/2}}{(2\pi)^{(p-q)r/2}} \operatorname{etr}\{-\tfrac{1}{2}[B_2 + T_{21}B_1(Y) + T_{22}B_2(Y)]XX'[\cdot\cdot]'\}\,\mathrm{d}B_2,$$

$$\frac{|XX'|^{q/2}}{(2\pi)^{qr/2}} \operatorname{etr}\{-\tfrac{1}{2}[B_1 + T_{11}B_1(Y)XX'[\cdot\cdot]'\}\,\mathrm{d}B_1.$$

The distribution for T_{22} can be inverted to give the confidence or posterior distribution for $\Sigma_{22\cdot 1}$. We have

$$\Sigma_{22\cdot 1} = (T'_{22}T_{22})^{-1}, \qquad \mathrm{d}T_{22} = \frac{\mathrm{d}\Sigma_{22\cdot 1}}{2^{p-q}\,|\Delta_{22}|_{\nabla}\,|\Sigma_{22\cdot 1}|^{(p-q+1)/2}}$$

where the differential comes easily from differential properties of measures on the group. Thus the inverted distribution for $\Sigma_{22\cdot 1}$ is

$$\begin{aligned} &\frac{\Pi_1^{p-q} A_{n-r-q-j+1}}{(2\pi)^{(p-q)(n-r-q)/2}} \operatorname{etr}\{-\tfrac{1}{2}\Sigma_{22\cdot 1}^{-1} S_{22\cdot 1}(Y)\} \\ &\times \frac{|S_{22\cdot 1}(Y)|^{(n-r-q)/2}}{|\Sigma_{22\cdot 1}|^{(n-r-q)/2}} \frac{|T_{22}(Y)|_{\nabla}}{|T_{22}(Y)|_{\Delta}} \frac{|\Delta_{22}|_{\Delta}}{|\Delta_{22}|_{\nabla}} \\ &\times \frac{\mathrm{d}\Sigma_{22\cdot 1}}{2^{p-q}\,|\Sigma_{22\cdot 1}|^{(p-q+1)/2}}. \end{aligned} \tag{3.5}$$

The distribution for T_{21} can then be inverted to give the confidence or posterior distribution for H_{21}:

$$H_{21} = -T_{22}^{-1}T_{21}, \qquad \mathrm{d}T_{21} = \frac{\mathrm{d}H_{21}}{|\Sigma_{22\cdot 1}|^{q/2}};$$

$$\frac{|S_{11}(Y)|^{(p-q)/2}}{(2\pi)^{(p-q)q/2}\,|\Sigma_{22\cdot 1}|^{q/2}} \operatorname{etr}\{-\tfrac{1}{2}\Sigma_{22\cdot 1}^{-1}[H_{21} - H_{21}(Y)]S_{11}(Y)[\cdot\cdot]'\}\,\mathrm{d}H_{21}. \tag{3.6}$$

The distribution for T_{11} can be inverted to give the confidence or posterior distribution for Σ_{11}:

$$\Sigma_{11} = (T'_{11}T_{11})^{-1}, \qquad \mathrm{d}T_{11} = \frac{\mathrm{d}\Sigma_{11}}{2^{q}\,|\Delta_{11}|_{\nabla}\,|\Sigma_{11}|^{(q+1)/2}};$$

$$\begin{aligned} &\frac{\Pi_1^{q} A_{n-r-j+1}}{(2\pi)^{q(n-r)/2}} \operatorname{etr}\{-\tfrac{1}{2}\Sigma_{11}^{-1} S_{11}(Y)\} \\ &\times \frac{|S_{11}(Y)|^{(n-r)/2}}{|\Sigma_{11}|^{(n-r)/2}} \frac{|T_{11}(Y)|_{\nabla}}{|T_{11}(Y)|_{\Delta}} \frac{|\Delta_{11}|_{\Delta}}{|\Delta_{11}|_{\nabla}} \frac{\mathrm{d}\Sigma_{11}}{2^{q}\,|\Sigma_{11}|^{(q+1)/2}}. \end{aligned} \tag{3.7}$$

The distribution for B_2 can be inverted to give the confidence distribution for $\mathcal{B}_{2\cdot 1}$:

$$\begin{aligned} &\frac{|XX'|^{(p-q)/2}}{(2\pi)^{(p-q)r/2}\,|\Sigma_{22\cdot 1}|^{r/2}} \\ &\times \operatorname{etr}\{-\tfrac{1}{2}\Sigma_{22\cdot 1}^{-1}[\mathcal{B}_{2\cdot 1} + H_{21}B_1(Y) - B_2(Y)]XX'[\cdot\cdot]'\}\,\mathrm{d}\mathcal{B}_{2\cdot 1}. \end{aligned} \tag{3.8}$$

And finally the distribution for B_1 can be inverted to give the confidence distribution for $\mathscr{B}_1$:

$$\frac{|XX'|^{q/2}}{(2\pi)^{qr/2}|\Sigma_{11}|^{r/2}} \operatorname{etr}\{-\tfrac{1}{2}\Sigma_{11}^{-1}[\mathscr{B}_1 - B_1(Y)]XX'[\cdot\cdot]'\}\,d\mathscr{B}_1. \tag{3.9}$$

Note that the distribution for $\mathscr{B}_{2\cdot 1}$ is

$$\mathrm{MN}(B_2(Y) - H_{21}B_1(Y), (XX')^{-1} \otimes \Sigma_{22\cdot 1})$$

and for $\mathscr{B}_1$ is $\mathrm{MN}(B_1(Y), (XX')^{-1} \otimes \Sigma_{11})$.

Now let us consider the distribution form for the confidence distributions for $\Sigma_{22\cdot 1}$ and Σ_{11}. A positive definite matrix W is said to have a disguised Wishart distribution $\mathrm{DW}_p(m, V)$ if the probability element is

$$\frac{\Pi_1^p A_{m-j+1}}{(2\pi)^{pm/2}} \operatorname{etr}\{-\tfrac{1}{2}WV^{-1}\} \frac{|W|^{m/2}}{|V|^{m/2}} \frac{|T_W|_\nabla}{|T_W|_\Delta} \frac{|T_V|_\Delta}{|T_V|_\nabla} \frac{dW}{2^p|W|^{(p+1)/2}} \tag{3.10}$$

where V is a positive definite matrix of order p and T_W and T_V are positive lower triangular matrices such that

$$W = T_W'T_W, \qquad V = T_V'T_V.$$

Note that for $p = 1$ it becomes a scaled chi-square distribution with m degrees of freedom. Also note that the disguised Wishart density is the ordinary Wishart density multiplied by a quantity

$$\frac{|T_W|_\nabla}{|T_W|_\Delta} \cdot \frac{|T_V|_\Delta}{|T_V|_\nabla}.$$

The standard disguised Wishart $\mathrm{DW}_p(m, I)$ has been obtained by Tan and Guttman (1971). Observe that $T_W V^{-1} T_W'$ where T_W is as described above is distributed as the standard Wishart; thus the name for the distribution.

If W is distributed as $\mathrm{DW}_p(m, V)$ we say that W^{-1} is distributed as $\mathrm{IDW}_p(m, V^{-1})$. Thus we see that the inverted distribution for Σ_{11} is $\mathrm{IDW}_q(n - r, S_{11}(Y))$, and the inverted distribution for $\Sigma_{22\cdot 1}$ is $\mathrm{IDW}_{p-q}(n - r - q, S_{22\cdot 1})$. In particular if $q = 1$ the confidence distribution for $\Sigma_{11}^{-1} = 1/\sigma_1^2$ is a chi-square $(n - r)$ distribution scaled by the residual sum of squares $S_{11}(Y)$ for the first response variable; this is of course in accord with confidence theory based on the first response variable alone.

It is of interest to compare the inverted distributions for Σ_{11} and $\Sigma_{22\cdot 1}$ with those obtained by ordinary Bayesian procedures. For multivariate normal regression the Jeffrey's prior is commonly chosen; Box and Tiao [1] and Zellner [6]. In Appendix 1 of their paper Dawid, Stone, and Zidek [2] renew the support for this prior by arguing that this prior would not give rise to marginalization paradoxes. The Jeffreys' prior lead to an inverted Wishart posterior $\mathrm{IW}_p(n-r, S(Y))$ for Σ and then leads by the Bayesian argument to an inverted Wishart $\mathrm{IW}_q(n-r-p+q, S_{11}(Y))$ for Σ_{11}. For the case $q=1$ the Bayesian posterior for $\Sigma_{11}^{-1} = 1/\sigma_1^2$ is a chi-square $(n-r-p+1)$ distribution scaled by the residual sum of squares $S_{11}(Y)$ for the first response variable. This Bayesian analysis is supposed to apply to any model having a response Y that is $\mathrm{MN}(\mathscr{B}X, I_n \otimes \Sigma)$ and it disagrees with results obtained from the present model (1.3). But even leaving our present model (1.3) aside, the Bayesian results remain counter-intuitive in that the degrees of freedom $n-r-p+1$ are inappropriate to a residual sum of squares with a chi-square $(n-r)$ distribution. We will return to this Bayesian question in Sections 4 and 5.

4. Second parameter separation

As a second illustration of the methods for separating parameter components suppose we examine the first q variables first and the last $p-q$ variables second and then within each of these examine location first and scale second. Specifically we factor θ as

$$\theta = \begin{pmatrix} I & 0 & 0 \\ \mathscr{B}_1 & I & 0 \\ 0 & 0 & I \end{pmatrix} \begin{pmatrix} I & 0 & 0 \\ 0 & \Delta_{11} & 0 \\ 0 & 0 & I \end{pmatrix} \begin{pmatrix} I & 0 & 0 \\ 0 & I & 0 \\ 0 & \Delta_{21} & I \end{pmatrix} \begin{pmatrix} I & 0 & 0 \\ 0 & I & 0 \\ \mathscr{B}_2 & 0 & I \end{pmatrix} \begin{pmatrix} I & 0 & 0 \\ 0 & I & 0 \\ 0 & 0 & \Delta_{22} \end{pmatrix} \quad (4.1)$$

$$= \theta_5\theta_4\theta_3\theta_2\theta_1$$

where $\theta_i \in K_i$ which is a subgroup of G and where $K_i \cdots K_1$ is for each i a subgroup of G.

The components $\mathscr{B}_1, \Delta_{11}, \Delta_{21}, \mathscr{B}_2, \Delta_{22}$ are easily related to components of the regression and variance matrices:

$$\Sigma_{11} = \Delta_{11}\Delta_{11}', \qquad H_{21} = \Delta_{21}\Delta_{11}^{-1},$$
$$\mathscr{B}_{2\cdot1} = \mathscr{B}_2 - H_{21}\mathscr{B}_1, \qquad \Sigma_{22\cdot1} = \Delta_{22}\Delta_{22}'. \quad (4.2)$$

The reverse factorization of g is

$$g = \begin{pmatrix} I & 0 & 0 \\ 0 & I & 0 \\ 0 & 0 & T_{22} \end{pmatrix} \begin{pmatrix} I & 0 & 0 \\ 0 & I & 0 \\ E_2 & 0 & I \end{pmatrix} \begin{pmatrix} I & 0 & 0 \\ 0 & I & 0 \\ 0 & U_{21} & I \end{pmatrix} \begin{pmatrix} I & 0 & 0 \\ 0 & T_{11} & 0 \\ 0 & 0 & I \end{pmatrix} \begin{pmatrix} I & 0 & 0 \\ E_1 & I & 0 \\ 0 & 0 & I \end{pmatrix}$$

$$= \begin{pmatrix} I & 0 & 0 \\ T_{11}E_1 & T_{11} & 0 \\ T_{22}E_2 + T_{22}U_{21}T_{11}E_1 & T_{22}U_{21}T_{11} & T_{22} \end{pmatrix}$$

$$= h_1h_2h_3h_4h_5; \quad (4.3)$$

$$\begin{aligned} dg &= dB\, dT \\ &= |T_{22}|^{r+q} dT_{22} dE_2 dU_{21} dE_2 |T_{11}|^{r+p-q} dT_{11}. \end{aligned} \tag{4.3}$$

The distribution for g given D is then available by simple adjustment of (2.7):

$$k_\lambda^{-1}(D) f_\lambda(BX + TD) dE_1 \frac{|T_{11}|^{n+p-q}}{|T_{11}|_\Delta} dT_{11} dU_{21} dE_2 \frac{|T_{22}|^n}{|T_{22}|_\Delta} dT_{22}. \tag{4.4}$$

We now examine in detail the preceding distribution for the special case of normal error. The exponential term in the normal density has the form

$$-\tfrac{1}{2}\operatorname{tr} T_{11}' T_{11}(I + E_1 XX' E_1')$$
$$-\tfrac{1}{2}\operatorname{tr} T_{22}' T_{22}(I + U_{21} T_{11} T_{11}' U_{21}' + [E_2 + U_{21} T_{11} E_1] XX' [\cdot\cdot]').$$

For the factorization of (4.4) in accord with the five components we need some integral identities. We first derive these identities.

The distributions given by $f^5(T_{22})$ and $f^3(T_{11})$ in the array (3.4) imply the following:

$$\int_T \operatorname{etr}\{-\tfrac{1}{2}T'TP\} \frac{|L_p|_\nabla}{|L_p|_\Delta} |P|^{m/2} \frac{|T|^m}{|T|_\Delta} dT = \frac{(2\pi)^{pm/2}}{\Pi_1^p A_{m-j+1}} \tag{4.5}$$

where T is a $p \times p$ ($p < m$) PLT matrix and P is a $p \times p$ positive definite matrix with PLT square root L_p ($L_p L_p' = P$). Now suppose that T has distribution as indicated by (4.5) and also suppose that a $p \times q$ matrix M has an independent multivariate normal distribution

$$\frac{|Q|^{p/2}}{(2\pi)^{pq/2}} \operatorname{etr}\{-\tfrac{1}{2}MQM'\} dM.$$

We transform $(T, M) \mapsto (T, B)$ where $M = TB$ and obtain the joint distribution

$$\frac{\Pi_1^p A_{m-j+1}}{(2\pi)^{p(m+q)/2}} \operatorname{etr}\{-\tfrac{1}{2}T'T(P + BQB')\} \frac{|L_p|_\nabla}{|L_p|_\Delta} |P|^{m/2} |Q|^{p/2} \frac{|T|^{m+q}}{|T|_\Delta} dT dB$$

We integrate out T using the identity (4.5) and obtain the marginal for B:

$$\frac{\Pi_1^p A_{m-j+1}}{\Pi_1^p A_{m+q-j+1}} \frac{|P|^{m/2} |Q|^{p/2}}{|P + BQB'|^{(m+q)/2}} \frac{|L_p|_\nabla}{|L_p|_\Delta} \frac{|L|_\Delta}{|L|_\nabla} dB \tag{4.6}$$

where L_p and L are the PLT square roots of P and $P + BQB'$ respectively. This distribution is a modification of the matrix $-t$ distribution and its density has the familiar adjustment factor

$$\frac{|L_p|_\nabla}{|L_p|_\Delta} \frac{|L|_\Delta}{|L|_\nabla}.$$

We call it, by association, the disguised matrix t, $\mathrm{DM}t(p \times q, m, P, Q)$.

Let W be distributed as $\mathrm{DW}_p(m, P^{-1})$ and T a PLT matrix such that $W = T'T$. And let an independent $p \times q$ matrix M be distributed as $\mathrm{MN}(0; Q^{-1} \otimes I_p)$. Then the $p \times q$ matrix $B = T^{-1}M$ is distributed as $\mathrm{DM}t(p \times q, m, P, Q)$.

We now use the relation (4.5) and the distribution (4.6) to factor the normal case distribution (4.4):

$$f_5(E_1)\mathrm{d}E_1 = \frac{\prod_{j=1}^{q} A_{(n-r)-j+1}}{\prod_{j=1}^{q} A_{n-j+1}} \frac{|XX'|^{q/2}}{|I + E_1XX'E_1'|^{n/2}} \frac{|L_Q|_\Delta}{|L_Q|_\nabla} \mathrm{d}E_1$$

(where L_Q is the PLT square root of $Q = I + E_1XX'E_1'$);

$$f_4(T_{11})\mathrm{d}T_{11} = \frac{\prod_{j=1}^{q} A_{n-j+1}}{(2\pi)^{qn/2}} \operatorname{etr}(-\tfrac{1}{2}T_{11}'T_{11}Q)|Q|^{n/2} \cdot \frac{|L_Q|_\nabla}{|L_Q|_\Delta} |T_{11}|^n \frac{\mathrm{d}T_{11}}{|T_{11}|_\Delta};$$

$$f_3(U_{21})\mathrm{d}U_{21} = \frac{\prod_{j=1}^{p-q} A_{(n-r-q)-j+1}}{\prod_{j=1}^{p-q} A_{(n-r)-j+1}} \frac{|T_{11}|^{p-q}}{|I + U_{21}T_{11}T_{11}'U_{21}'|^{(n-r)/2}} \frac{|L_V|_\Delta}{|L_V|_\nabla} \mathrm{d}U_{21};$$

(where L_V is the PLT square root of $V = I + U_{21}T_{11}T_{11}'U_{21}'$);

$$f_2(E_2)\mathrm{d}E_2 = \frac{\prod_{1}^{p-q} A_{(n-r)-j+1}}{\prod_{1}^{p-q} A_{n-j+1}} \frac{|V|^{(n-r)/2}|XX'|^{(p-q)/2}}{|V + [E_2 + U_{21}T_{11}E_1]XX'[\cdot\cdot]'|^{n/2}} \cdot \frac{|L_V|_\nabla}{|L_V|_\Delta} \frac{|L_W|_\Delta}{|L_W|_\nabla} \mathrm{d}E_2$$

(where L_W is the PLT square root of $W = V + [E_2 + U_{21}T_{11}E_1]XX'[\cdot\cdot]'$);

$$f_1(T_{22})\mathrm{d}T_{22} = \frac{\prod_{1}^{p-q} A_{n-j+1}}{(2\pi)^{(p-q)n/2}} \operatorname{etr}(-\tfrac{1}{2}T_{22}'T_{22}W) \cdot \frac{|L_W|_\nabla}{|L_W|_\Delta} |W|^{n/2} |T_{22}|^n \frac{\mathrm{d}T_{22}}{|T_{22}|_\Delta}. \tag{4.7}$$

For tests and confidence regions we need coordinates relative to the observed response; thus we change from g to g^* by means of $g = g^* g(Y)$. Then if we are particularly interested in confidence distributions we can invert the distribution for g^* by means of $\theta = g^{*-1}$; or we can proceed directly from the distribution of g by means of the composite relation $\theta g = g(Y)$ with fixed $g(Y)$. For brevity here and for the advantage of some interesting distribution theory we omit the distribution for g^* and record the results of the composite transformation $g(Y) = \theta g$,

$$\begin{aligned}
B_1(Y) &= \mathcal{B}_1 + T_{11}(Y)E_1,\\
T_{11}(Y) &= \Delta_{11}T_{11},\\
T_{21}(Y) &= (\Delta_{21} + T_{22}(Y)U_{21})T_{11},\\
B_2(Y) &= \mathcal{B}_2 + T_{21}(Y)E_1 + T_{22}(Y)E_2,\\
T_{22}(Y) &= \Delta_{22}T_{22},
\end{aligned} \tag{4.8}$$

or equivalently

$$\begin{aligned}
E_1 &= T_{11}^{-1}(Y)(B_1(Y) - \mathcal{B}_1),\\
T'_{11}T_{11} &= T'_{11}(Y)\Sigma_{11}^{-1}T_{11}(Y),\\
U_{21}T_{11} &= T_{22}^{-1}(Y)(T_{21}(Y) - H_{21}T_{11}(Y)),\\
E_2 + U_{21}T_{11}E_1 &= T_{22}^{-1}(Y)(B_2(Y) - H_{21}B_1(Y) - \mathcal{B}_{2\cdot 1}),\\
T'_{22}T_{22} &= T'_{22}(Y)\Sigma_{22\cdot 1}^{-1}T_{22}(Y).
\end{aligned} \tag{4.9}$$

Note that we can use these relations with the distributions f_5, f_4, f_3, f_2, f_1 for tests and confidence regions to assess the parameter components $\mathcal{B}_1$, Σ_{11}, H_{21}, $\mathcal{B}_{2\cdot 1}$ and $\Sigma_{22\cdot 1}$ in this order. The differentials corresponding to (4.9) are readily calculated.

$$\begin{aligned}
dE_1 &= \frac{d\mathcal{B}_1}{|T_{11}(Y)|^r},\\
dT_{11} &= \frac{|T_{11}(Y)|_\nabla \, d\Sigma_{11}}{2^q |\Delta_{11}|_\nabla |\Sigma_{11}|^{(q+1)/2}},\\
dU_{21} &= \frac{|T_{11}(Y)|^{p-q}}{|T_{22}(Y)|^q} dH_{21},\\
dE_2 &= \frac{d\mathcal{B}_2}{|T_{22}(Y)|^r},\\
dT_{22} &= \frac{|T_{22}(Y)|_\nabla \, d\Sigma_{22\cdot 1}}{2^{p-q} |\Delta_{22}|_\nabla |\Sigma_{22\cdot 1}|^{(p-q+1)/2}}.
\end{aligned} \tag{4.10}$$

From (4.8) and (3.3) we have that

$$
\begin{aligned}
Q &= T_{11}^{-1}(Y)(S_{11}(Y)+[\mathcal{B}_1 - B_1(Y)]XX'[\cdots]')T_{11}'^{-1}(Y)\\
&= T_{11}^{-1}(Y)\tilde{Q}T_{11}'^{-1}(Y),\\
L_Q &= T_{11}^{-1}(Y)L_{\tilde{Q}}, \qquad |Q| = |S_{11}(Y)|^{-1}|\tilde{Q}|,\\
V &= T_{22}^{-1}(Y)(S_{22\cdot 1}(Y)+[H_{21} - H_{21}(Y)]S_{11}(Y)[\cdots]')T_{22}'^{-1}(Y)\\
&= T_{22}^{-1}(Y)\tilde{V}T_{22}'^{-1}(Y),\\
L_V &= T_{22}^{-1}(Y)L_{\tilde{V}}, \qquad |V| = |S_{22\cdot 1}(Y)|^{-1}|\tilde{V}|,\\
W &= T_{22}^{-1}(Y)(\tilde{V}+[\mathcal{B}_{2\cdot 1} + H_{21}B_1(Y) - B_2(Y)]XX'[\cdots])T_{22}'^{-1}(Y)\\
&= T_{22}^{-1}(Y)\tilde{W}T_{22}'^{-1}(Y),\\
L_W &= T_{22}^{-1}(Y)L_{\tilde{W}}, \qquad |W| = |S_{22\cdot 1}(Y)|^{-1}|\tilde{W}|.
\end{aligned}
$$

We now substitute these identities in the densities $f_5, \ldots, f_1$ obtaining the following confidence distributions:

$$
p_5(\mathcal{B}_1) = \frac{\prod_{j=1}^{q} A_{n-r-j+1}}{\prod_{j=1}^{q} A_{n-j+1}} \frac{|S_{11}(Y)|^{(n-r)/2}|XX'|^{q/2}}{|\tilde{Q}|^{n/2}} \cdot \frac{|L_{\tilde{Q}}|_\Delta |T_{11}(Y)|_\nabla}{|L_{\tilde{Q}}|_\nabla |T_{11}(Y)|_\Delta},
$$

$$
p_4(\Sigma_{11}) = \frac{\prod_{1}^{q} A_{n-j+1}}{(2\pi)^{qn/2}} \operatorname{etr}(-\tfrac{1}{2}\Sigma^{-1}\tilde{Q})\frac{|\tilde{Q}|^{n/2}}{|\Sigma_{11}|^{n/2}} \cdot \frac{|L_{\tilde{Q}}|_\nabla |\Delta_{11}|_\Delta}{|L_{\tilde{Q}}|_\Delta |\Delta_{11}|_\nabla} \frac{1}{2^q |\Sigma_{11}|^{(q+1)/2}},
$$

$$
p_3(H_{21}) = \frac{\prod_{j=1}^{p-q} A_{n-r-q-j+1}}{\prod_{j=1}^{p-q} A_{n-r-j+1}} \frac{|S_{22\cdot 1}(Y)|^{(n-r-q)/2}|S_{11}(Y)|^{(p-q)/2}}{|\tilde{V}|^{(n-r)/2}} \cdot \frac{|L_{\tilde{V}}|_\Delta |T_{22}(Y)|_\nabla}{|L_{\tilde{V}}|_\nabla |T_{22}(Y)|_\Delta},
$$

$$p_2(\mathcal{B}_{2\cdot1}) = \frac{\prod_{j=1}^{p-q} A_{n-r-j+1}}{\prod_{j=1}^{p-q} A_{n-j+1}} \frac{|\tilde{V}|^{(n-r)/2}|XX'|^{(p-q)/2}}{|\tilde{W}|^{n/2}} \frac{|L_{\tilde{W}}|_\Delta}{|L_{\tilde{W}}|_\nabla} \frac{|L_{\tilde{V}}|_\nabla}{|L_{\tilde{V}}|_\Delta},$$

$$p_1(\Sigma_{22\cdot1}) = \frac{\prod_{j=1}^{p-q} A_{n-j+1}}{(2\pi)^{(p-q)n/2}} \operatorname{etr}(-\tfrac{1}{2}\Sigma_{22\cdot1}^{-1}\tilde{W}) \frac{|\tilde{W}|^{n/2}}{|\Sigma_{22\cdot1}|^{n/2}}$$

$$\cdot \frac{|L_{\tilde{W}}|_\nabla}{|L_{\tilde{W}}|_\Delta} \frac{|\Delta_{22}|_\Delta}{|\Delta_{22}|_\nabla} \frac{1}{2^{p-q}|\Sigma_{22\cdot1}|^{(p-q+1)/2}}.$$

For the particular case $q = 1$ the confidence distribution for $\boldsymbol{\beta}_1'(\beta_{11}, \ldots, \beta_{1r})$ is multivariate-t with $n - r$ degrees of freedom, located at $\boldsymbol{b}_1(Y) = B_1(Y)$ and scaled by the residual sum-of-squares $S_{11}(Y)$ for the first response variable; this is the same as from the univariate model for the first response ($p = q = 1$) and of course is in agreement with standard confidence theory.

Now if Jeffrey's prior is appended to the multivariate model the marginal posterior for $\boldsymbol{\beta}_1'$ becomes multivariate-t with $n - r - p + 1$ degrees-of-freedom, but located and scaled as before; this has $p - 1$ fewer degrees-of-freedom which suggests less precise inference. As this posterior depends only on the first response we could perhaps expect it to agree with that from the univariate model; but this is *not* so, contrary to assertions in Dawid, Stone, and Zidek [2].

5. Invalid parameter separations and Bayesian difficulties

In the concluding paragraph of Section 3 we noted the inconsistency of the degrees of freedom in a Bayesian posterior distribution. We now examine some further aspects of this inconsistency.

Suppose that Y has the $\mathrm{MN}(0, I \otimes \Sigma)$ distribution; this is a special case of the Section 3 analysis with $r = 0$. A possible structural model is

$$Y = \Gamma Z \tag{5.1}$$

where Γ is a $p \times p$ matrix $|\Gamma| > 0$ and Z is standard normal. The transformation Γ can be factored as

$$\begin{aligned}\Gamma &= \Delta\Omega \\ &= \begin{pmatrix} I & 0 \\ 0 & \Delta_{22} \end{pmatrix} \begin{pmatrix} I & 0 \\ \Lambda_{21} & I \end{pmatrix} \begin{pmatrix} \Delta_{11} & 0 \\ 0 & I \end{pmatrix} \Omega \\ &= \theta_4\theta_3\theta_2\theta_1\end{aligned} \tag{5.2}$$

where Δ is PLT and Ω is a positive orthogonal matrix; we have

$$\Sigma_{22\cdot 1} = \Delta_{22}\Delta'_{22}, \qquad \Sigma_{11} = \Delta_{11}\Delta'_{11}, \qquad H_{21} = \Delta_{22}\Lambda_{21}.$$

The factorization (5.2) is a semi-direct product for the group $G = H_4 \ldots H_1$. Each H_i is a subgroup *but* $G_i = H_i \ldots H_1$ is not a subgroup of G for $i = 3, 2$. The conditions in [4] for a proper factored analysis are thus not fulfilled.

Consider however a *formal* analysis following the methods earlier. The reverse factorization for $g = h_1 \ldots h_4$ is

$$g = R\begin{pmatrix} T_{11} & 0 \\ 0 & I \end{pmatrix} \begin{pmatrix} I & 0 \\ T_{21} & I \end{pmatrix} \begin{pmatrix} I & 0 \\ 0 & T_{22} \end{pmatrix} \tag{5.3}$$

where R is positive orthogonal and both T_{11} and T_{22} are PLT. The left invariant measure (standardized) on G can be factored as

$$\mathrm{d}\mu(g) = \frac{\mathrm{d}g}{|g|^p} = \mathrm{d}R \frac{\mathrm{d}T_{11}}{|T_{11}|^{p-q}|T_{11}|_\nabla} \mathrm{d}T_{21} \frac{\mathrm{d}T_{22}}{|T_{22}|_\nabla}. \tag{5.4}$$

The matrix $Y = T(Y)D(Y)$ can be factored into PLT and semi-orthogonal components. Suppose that g is used as coordinates relative to $D(Y)$. Then the conditional distribution for g given D is

$$\begin{aligned} &k^{-1}(D)f(gD)|g|^n \mathrm{d}\mu(g) \\ &\quad = k^{-1}(D)(2\pi)^{-pn/2}\mathrm{etr}\{-\tfrac{1}{2}gg'\} \\ &\qquad \cdot |T_{11}|^n |T_{22}|^n \mathrm{d}R \frac{\mathrm{d}T_{11}}{|T_{11}|^{p-q}|T_{11}|_\nabla} \mathrm{d}T_{21} \frac{\mathrm{d}T_{22}}{|T_{22}|_\nabla}. \end{aligned}$$

The exponential term can be written as

$$-\tfrac{1}{2}\{\mathrm{tr}\, T_{11}T'_{11} + \mathrm{tr}\, T_{21}T'_{21} + \mathrm{tr}\, T_{22}T'_{22}\}$$

and the distribution factored following the pattern in Sections 2 and 3:

$$
\begin{aligned}
f_4(T_{22}) &= \frac{\prod_1^{p-q} A_{n-j+1}}{(2\pi)^{(p-q)n/2}} \operatorname{etr}(-\tfrac{1}{2}T_{22}T'_{22}) \frac{|T_{22}|^n}{|T_{22}|_\nabla}, \\
f_3(T_{21}) &= (2\pi)^{-(p-q)q/2} \operatorname{etr}(-\tfrac{1}{2}T_{21}T'_{21}), \\
f_2(T_{11}) &= \frac{\prod_1^{q} A_{n-p+q-j+1}}{(2\pi)^{q(n-p+q)/2}} \operatorname{etr}(-\tfrac{1}{2}T_{11}T'_{11}) |T_{11}|^{n-p+q} \frac{1}{|T_{11}|_\nabla}, \\
f_1(R) &= \frac{1}{A_p \dots A_2}.
\end{aligned}
\tag{5.5}
$$

We now relocate the coordinates relative to the observed Y which gives the following changes

$$
\begin{aligned}
T_{22} &\to T_{22}T_{22}(Y) & \mathrm{d}T_{22} &\to |T_{22}(Y)|_\nabla \mathrm{d}T_{22} \\
T_{21} &\to T_{21}T_{11}(Y) + T_{22}T_{21}(Y) & \mathrm{d}T_{21} &\to |T_{11}(Y)|^{p-q} \mathrm{d}T_{21} \\
T_{11} &\to T_{11}T_{11}(Y) & \mathrm{d}T_{11} &\to |T_{11}(Y)|_\nabla \mathrm{d}T_{11}.
\end{aligned}
$$

The distributions relocated to Y are

$$
\begin{aligned}
f^4(T_{22}) &= \frac{\prod_1^{p-q} A_{n-j+1}}{(2\pi)^{(p-q)n/2}} \operatorname{etr}[-\tfrac{1}{2}T'_{22}T_{22}S_{22\cdot 1}(Y)] |S_{22\cdot 1}(Y)|^{n/2} \frac{|T_{22}|^n}{|T_{22}|_\nabla}, \\
f^3(T_{21}) &= \frac{|S_{11}(Y)|^{(p-q)/2}}{(2\pi)^{(p-q)q/2}} \operatorname{etr}\{-\tfrac{1}{2}[T_{21} + T_{22}H_{21}(Y)]S_{11}(Y) \\
&\quad \cdot [T_{21} + T_{22}H_{21}(Y)]'\}, \\
f^2(T_{11}) &= \frac{\prod_1^{q} A_{n-p+q-j+1}}{(2\pi)^{q(n-p+q)/2}} \operatorname{etr}[-\tfrac{1}{2}T'_{11}T_{11}S_{11}(Y)] |S_{11}(Y)|^{(n-p+q)/2} \\
&\quad \cdot \frac{|T_{11}|^{n-p+q}}{|T_{11}|_\nabla}, \\
f^1(R) &= f_1(R) = \frac{1}{A_p \dots A_2}.
\end{aligned}
\tag{5.6}
$$

The relations between the parameter components and the sample space components are

$$\Sigma_{22\cdot 1} = \Delta_{22}\Delta'_{22} = (T'_{22}T_{22})^{-1}, \qquad \mathrm{d}T_{22} = \frac{\mathrm{d}\Sigma_{22\cdot 1}}{2^{p-q}\,|\Delta_{22}|_{\nabla}\,|\Sigma_{22\cdot 1}|^{(p-q+1)/2}};$$

$$H_{21} = \Delta_{22}\Lambda_{21} = -T_{22}^{-1}T_{21}, \qquad \mathrm{d}T_{21} = \frac{\mathrm{d}H_{21}}{|\Sigma_{22\cdot 1}|^{q/2}};$$

$$\Sigma_{11} = \Delta_{11}\Delta'_{11} = (T'_{11}T_{11})^{-1}, \qquad \mathrm{d}T_{11} = \frac{\mathrm{d}\Sigma_{11}}{2^{q}\,|\Delta_{11}|_{\nabla}\,|\Sigma_{11}|^{(q+1)/2}}.$$

These relations can be used with (5.6) to obtain confidence distributions; we obtain

(a) The posterior for $\Sigma_{22\cdot 1}$ is $\mathrm{IW}_{p-q}(n, S_{22\cdot 1}(Y))$.

(b) The posterior for H_{21} given $\Sigma_{22\cdot 1}$ is

$$\mathrm{MN}(H_{21}(Y), S_{11}^{-1}(Y) \otimes \Sigma_{22\cdot 1}).$$

(c) The posterior for Σ_{11} is $\mathrm{IW}_q(n-p+q, S_{11}(Y))$.

These posteriors coincide with those obtained by Bayesian procedures using Jeffreys' invariant prior. They are obtained here by *formal* analysis ignoring the needed factorization criteria from [4]. This gives some additional background for the Bayesian inconsistency on the degrees of freedom for a component variance matrix; see the concluding remarks to Sections 3 and 4.

References

[1] Box, G.E.P. and Tiao, G.C. (1973). *Bayesian Inference in Statistical Analysis*, Reading, Mass.: Addison-Wesley.

[2] Dawid, A.P., Stone, M., and Zidek, J.W. (1973). Marginalization paradoxes in Bayesian and structural inference, *Jour. Roy. Statist. Soc.* B **35**, 189–233.

[3] Fraser, D.A.S. (1968). *The Structure of Inference*, New York: John Wiley and Sons.

[4] Fraser, D.A.S. and MacKay, Jock (1975). Parameter factorization and inference based on significance, likelihood and objective posterior, *Annals Statistics* **3**, 559–572.

[5] Tan, W.Y. and Guttman, I. (1971). A disguised Wishart Variable and a Related Theorem, *Jour. Roy. Statist. Soc.* B **33**, 147–152.

[6] Zellner, A. (1971). *An Introduction to Bayesian Inference in Econometrics*, New York: John Wiley and Sons.

P.R. Krishnaiah, ed., *Multivariate Analysis–IV*

ASYMPTOTIC EXPANSIONS FOR THE DISTRIBUTIONS OF SOME MULTIVARIATE TESTS

Yasunori FUJIKOSHI

Kobe University, Nada, Kobe, Japan

In this paper, asymptotic expansions of the null and nonnull distributions of the three criteria based on the likelihood ratio (= LR) test and the sums of the last few roots of the multivariate F and beta matrices are obtained for testing the hypothesis on the dimensionality of regression coefficients in a MANOVA model and the hypothesis that the smallest $p-k$ canonical correlations are all zero. We also obtain asymptotic expansions of the null distributions of LR statistics for testing the equality of several latent roots of a covariance matrix. The asymptotic expansions are derived by using perturbation expansions of sample latent roots with multiple population roots. The expansions in the null and the nonnull cases are given in terms of central χ^2-distributions and the standard normal distribution function and its derivatives respectively.

1. Introduction

The exact null and nonnull distributions of some test statistics in multivariate analysis have been obtained in a closed form. However, it seems that in most cases the exact distributions cannot be obtained in a form which is of practical use except for some particular values of parameters. Therefore, the study of approximations to the distributions are very important. Asymptotic expansions play an important part in approaches to the distribution approximations. Many asymptotic expansions have been obtained, based on the method of Box [7] and other methods (cf. Siotani [23]). These expansions have been used for approximate evaluation of the significant points of tests for large or moderately large samples, and also for investigations of power and robustness of various tests. Approximations based on Pearson Type distributions have also been used. Recently Lee, Chang and Krishnaiah [18], and Lee, Krishnaiah and Chang [19] approximated the distributions of certain powers of a class of likelihood ratio (= LR) statistics with Pearson Type I distribution.

Department of Mathematics, Faculty of General Education, Kobe University Nada, Kobe, Japan.

In this paper we consider the distributions of certain statistics associated with testing the hypotheses on the reduction of dimensionality under the MANOVA, Canonical Correlation and Principal Components models. We derive perturbation expansions for the latent roots of a sample matrix occurring in each of the above models. These results are applied to obtain asymptotic expansions of the null and nonnull distributions of the three criteria based on the likelihood ratio test and the sums of the last few roots of the multivariate F and beta matrices for testing (I) the hypothesis on the dimensionality in a MANOVA model and (II) the hypothesis that the smallest $p-k$ canonical correlations are all zero, and LR statistics for testing (III) the equality of several latent roots of a covariance matrix. The expansions in the null case are given in terms of central χ^2-distributions with the errors of $O(N^{-2})$, where N denotes the sample size. The expansions in the nonnull case are given in terms of the standard normal distribution function and its derivatives with the errors of $O(N^{-1})$. It may be noted that Sugiura [26] has obtained asymptotic expansions of the nonnull distributions of LR statistics for testing (III) when the population roots are all simple. Krishnaiah and Waiker [14], [15] treated the problems of reduction of dimensionality under the above three models within the framework of the simultaneous test procedures.

The problems of testing linear hypothesis about regression coefficients in MANOVA, test of independence of two sets of variables and testing sphericity are special cases of the problems treated in this paper. Asymptotic expansions of the null and nonnull distributions of the tests statistics associated with the above special cases have been obtained in various papers in the literature.

2. Outline of the asymptotic expansion method

The testing problems (I), (II) and (III) mentioned in Section 1 can be reduced as follows: Let $\Theta = \mathrm{diag}(\theta_1, \ldots, \theta_p)$ be a diagonal matrix of unknown parameters $\theta_1 \geq \ldots \geq \theta_p$ and let $d_1 \geq \ldots \geq d_p$ be the latent roots of a random matrix M, which can be expressed as

$$M = \Theta + \varepsilon V^{(1)} + \varepsilon^2 V^{(2)} + \ldots, \tag{2.1}$$

where $\varepsilon = N^{-1/2}$, N denotes the sample size and the elements of $V^{(j)}$ are $O_p(1)$. The problems can be expressed as ones of testing $H_0: \theta_{a+1} = \ldots = \theta_{a+b} = \lambda$ against H_1: not H_0, where for (I) and (II) $\lambda = 0$ and $a + b = p$, and for (III) λ is a given positive number or unknown. The test statistics for these hypotheses take the form of $T = T(d_{a+1}, \dots, d_{a+b})$ which is a symmetric function of $d_{a+1}, \dots, d_{a+b}$. To find asymptotic expansions for the distribution of T, we expand T under each of H_0 and H_1 as a power-series in ε:

$$\begin{aligned}\tilde{T} &= c(\varepsilon)\{T(d_{a+1}, \dots, d_{a+b}) - T(\theta_{a+1}, \dots, \theta_{a+b})\} \\ &= S_0 + \varepsilon S_1 + \varepsilon^2 S_2 + \dots,\end{aligned} \tag{2.2}$$

where $c(\varepsilon)$ is a power in ε. Then we consider the characteristic function (= ch. f.) of $\tilde{T}$ which can be expressed as

$$\boldsymbol{E}\,[e^{\phi S_0}\{1 + \varepsilon\phi S_1 + \varepsilon^2\phi(S_2 + \tfrac{1}{2}\phi S_1^2) + \dots\}], \tag{2.3}$$

where $\phi = it$. After we evaluate the expectations in (2.3), the desired asymptotic expansions for the distribution of T may be obtained by the inversion formula.

To obtain (2.2), we need to get the perturbation expansions for $d_{a+1}, \dots, d_{a+b}$ in terms of Θ and $V^{(j)}$. The expansion of d_α when θ_α is simple is well known, see, e.g., Bellman [6], p. 63 and, more recently, Sugiura [25]. However, the formula cannot be used for the null case, $\theta_{a+1} = \dots = \theta_{a+b}$ and for the nonnull case with multiple population roots. Therefore, we proceed to the general case in which some of the θ_i are equal. Let the multiplicity of the latent roots of Θ be $b_1, \dots, b_h$. Let

$$\begin{aligned}&\theta_1 = \dots = \theta_{b_1} = \lambda_1, \qquad \theta_{b_1+1} = \dots = \theta_{b_1+b_2} = \lambda_2, \\ &\dots, \theta_{p-b_h+1} = \dots = \theta_p = \lambda_h,\end{aligned} \tag{2.4}$$

where $\lambda_1 > \dots > \lambda_h$. We partition the matrices into submatrices with $b_1, \dots, b_h$ rows and colums

$$\Theta = \begin{pmatrix} \lambda_1 I & 0 \dots 0 \\ 0 & \lambda_2 I \dots 0 \\ \vdots & \vdots \ \ddots \ \vdots \\ 0 & 0 \dots \lambda_h I \end{pmatrix}, \qquad V^{(j)} = \begin{pmatrix} V_{11}^{(j)} & V_{12}^{(j)} \dots V_{1h}^{(j)} \\ V_{21}^{(j)} & V_{22}^{(j)} \dots V_{2h}^{(j)} \\ \vdots & \vdots \ \ddots \ \vdots \\ V_{h1}^{(j)} & V_{h2}^{(j)} \dots V_{hh}^{(j)} \end{pmatrix} \tag{2.5}$$

Let J_α be the set of integers $b_1 + \dots + b_{\alpha-1} + 1, \dots, b_1 + \dots + b_\alpha$. The following Lemma is fundamental for our asymptotic expansion.

Lemma 1. *Let $d_1 \geq \dots \geq d_p$ be the latent roots of M defined by* (2.1). *Assume that $\Theta = \operatorname{diag}(\theta_1, \dots, \theta_p)$ has the structure defined by* (2.4). *Then* *the set of d_j, $j \in J_\alpha$ is equal to the latent roots of Z_α for small positive number ε:*

$$Z_\alpha = \lambda_\alpha I_{b_\alpha} + \varepsilon Z_\alpha^{(1)} + \varepsilon^2 Z_\alpha^{(2)} + \varepsilon^3 Z_\alpha^{(3)} + \varepsilon^4 Z_\alpha^{(4)} + \varepsilon^5 Z_\alpha^{(5)} + O(\varepsilon^6), \quad (2.6)$$

where

$$Z_\alpha^{(1)} = V_{\alpha\alpha}^{(1)}, \qquad Z_\alpha^{(2)} = V_{\alpha\alpha}^{(2)} + \sum_{\beta\neq\alpha} \lambda_{\alpha\beta} V_{\alpha\beta}^{(1)} V_{\beta\alpha}^{(1)},$$

$$Z_\alpha^{(3)} = V_{\alpha\alpha}^{(3)} + \sum_{\beta\neq\alpha} \lambda_{\alpha\beta} \{ V_{\alpha\beta}^{(1)} V_{\beta\alpha}^{(2)} + V_{\alpha\beta}^{(2)} V_{\beta\alpha}^{(1)} - \lambda_{\alpha\beta} V_{\alpha\alpha}^{(1)} V_{\alpha\beta}^{(1)} V_{\beta\alpha}^{(1)} \}$$
$$+ \sum_{\beta\neq\alpha} \sum_{\gamma\neq\alpha} \lambda_{\alpha\beta}\lambda_{\alpha\gamma} V_{\alpha\beta}^{(1)} V_{\beta\gamma}^{(1)} V_{\gamma\alpha}^{(1)},$$

$$\begin{aligned} Z_\alpha^{(4)} = V_{\alpha\alpha}^{(4)} &+ \sum_{\beta\neq\alpha} \lambda_{\alpha\beta} [V_{\alpha\beta}^{(1)} V_{\beta\alpha}^{(3)} + V_{\alpha\beta}^{(2)} V_{\beta\alpha}^{(2)} + V_{\alpha\beta}^{(3)} V_{\beta\alpha}^{(1)} \\ &- \lambda_{\alpha\beta} \{ V_{\alpha\alpha}^{(1)} V_{\alpha\beta}^{(1)} V_{\beta\alpha}^{(2)} + V_{\alpha\alpha}^{(1)} V_{\alpha\beta}^{(2)} V_{\beta\alpha}^{(1)} + V_{\alpha\alpha}^{(2)} V_{\alpha\beta}^{(1)} V_{\beta\alpha}^{(1)} \} \\ &+ \lambda_{\alpha\beta}^2 V_{\alpha\alpha}^{(1)^2} V_{\alpha\beta}^{(1)} V_{\beta\alpha}^{(1)}] + \sum_{\beta\neq\alpha} \sum_{\gamma\neq\alpha} \lambda_{\alpha\beta}\lambda_{\alpha\gamma} [V_{\alpha\beta}^{(1)} V_{\beta\gamma}^{(1)} V_{\gamma\alpha}^{(2)} \\ &+ V_{\alpha\beta}^{(1)} V_{\beta\gamma}^{(2)} V_{\gamma\alpha}^{(1)} + V_{\alpha\beta}^{(2)} V_{\beta\gamma}^{(1)} V_{\gamma\alpha}^{(1)} - (\lambda_{\alpha\beta} + \lambda_{\alpha\gamma}) V_{\alpha\alpha}^{(1)} V_{\alpha\beta}^{(1)} V_{\beta\gamma}^{(1)} V_{\gamma\alpha}^{(1)} \\ &- \lambda_{\alpha\beta} V_{\alpha\beta}^{(1)} V_{\beta\alpha}^{(1)} V_{\alpha\gamma}^{(1)} V_{\gamma\alpha}^{(1)}] + \sum_{\beta\neq\alpha} \sum_{\gamma\neq\alpha} \sum_{\delta\neq\alpha} \lambda_{\alpha\beta}\lambda_{\alpha\gamma}\lambda_{\alpha\delta} V_{\alpha\beta}^{(1)} V_{\beta\gamma}^{(1)} V_{\gamma\delta}^{(1)} V_{\delta\alpha}^{(1)}, \end{aligned} \quad (2.7)$$

and the elements of $Z^{(5)}$ are homogeneous polynomials of degree 5 in the elements of $V^{(j)}$ satisfying $\nu_1 + 2\nu_2 + \ldots = 5$, and $\lambda_{\alpha\beta} = (\lambda_\alpha - \lambda_\beta)^{-1}$.

The proof can be obtained on the same lines as given in Lawley [16]. It may be noted that Lawley [16], [17] gave similar results in terms of $Y = V^{(1)} + \varepsilon V^{(2)} + \ldots$ when $b_1 = \ldots = b_{h-1} = 1$.

3. Tests for dimensionality in MANOVA

3.1. Introduction. In a MANOVA model we observe a random matrix Y whose rows are independently distributed as a p-variate normal distribution with unknown covariance matrix Σ and $\boldsymbol{E}[Y] = A\Xi$, where A is a known $N \times b$ matrix of rank $b \leq N$ and Ξ is an unknown $b \times p$ matrix of regression coefficients. Consider the test of H_{01}: rank $(C\Xi) = k$ against H_{11}: rank $(C\Xi) > k$, where C is a known $q \times b$ matrix of rank q. We consider three statistics for this test:

(1) LR statistic,

$$T_1 = \log \prod_{j=k+1}^{p} (1 + d_j),$$

(2) The sum of the last few roots of the multivariate F matrix,

$$T_2 = \sum_{j=k+1}^{p} d_j,$$

(3) The sum of the last few roots of the multivariate beta matrix,

$$T_3 = \sum_{j=k+1}^{p} d_j/(1 + d_j),$$

where $d_1 \geq \ldots \geq d_p$ are the latent roots of $S_h S_e^{-1}$,

$$S_e = Y'(I_N - A(A'A)^{-1}A')Y$$

and

$$S_h = Y'A(A'A)^{-1}C'[C(A'A)^{-1}C']^{-1}C(A'A)^{-1}A'Y.$$

Anderson [2] showed that T_1 is a LR statistic and under H_{01}, the limiting distribution of T_1 is a LR statistic and under H_{01}, the limiting distribution of T_1 is a χ^2-distribution with $f = (p-k)(q-k)$ degrees of freedom. Lawley [17] gave a correction factor for T_1 by considering the asymptotic mean of T_1 under H_{01}. Asymptotic normality of T_1 under H_{11} was obtained by Fujikoshi [10]. For testing a different type hypothesis on the dimensionality, see Rao [22] and Fujikoshi [10]. When $k = 0$, the test statistics T_j are reduced to ones for testing linear hypotheses about regression coefficients and the asymptotic expansions have been investigated by many authors in the literature.

3.2. Reduction. In dealing with the distribution of $d_1, \ldots, d_p$ we may assume that S_e and S_h are independent central and noncentral Wishart matrices having $W_p(n, I_p)$ and $W_p(q, I_p; \Omega)$ respectively, where $n = N - b$, $\Omega = \mathrm{diag}(\omega_1, \ldots, \omega_p)$ and $\omega_1 \geq \ldots \geq \omega_p$ are the latent roots of $(C\Xi)'[C(A'A)^{-1}C']^{-1}C\Xi\Sigma^{-1}$. To obtain asymptotic ($n \to \infty$) distributions of T_j, we note that Ω depends on n through A. We make the following assumptions for Ω as in Hsu [12], Anderson [1] and Sugiura [24], etc.:

A1 : $\Omega = O(n) = m\Theta = m\,\mathrm{diag}(\theta_1, \ldots, \theta_p)$,

A2 : the multiplicity of the latent roots of Ω does not depend on n, where $m = n - 2\eta$ and η is a correction factor. We may assume that Θ has the structure (2.4). In the following we treat the case of $q \geq p$. Then we have

$$\frac{1}{m} S_h = \Theta + \frac{1}{\sqrt{m}} (X\Theta^{1/2} + \Theta^{1/2}X') + \frac{1}{m} W, \tag{3.1}$$

where $W = XX' + \tilde{X}\tilde{X}'$ and $X : p \times p$ and $\tilde{X} : p \times (q-p)$ are matrices with independent standard normal variates as their elements. Let

$$\frac{1}{m} S_e = I_p + \frac{1}{\sqrt{m}} V. \tag{3.2}$$

The limiting distribution of $V = V'$ is normal with mean 0 and functionally independent elements are uncorrelated. From (3.1) and (3.2) we have that $d_1 \geq \ldots \geq d_p$ are the latent roots of

$$S_h S_e^{-1} = \Theta + \frac{1}{\sqrt{m}} D^{(1)} + \frac{1}{m} D^{(2)} + \frac{1}{m\sqrt{m}} D^{(3)} + \ldots, \tag{3.3}$$

where

$$\begin{aligned} &D^{(1)} = X\Theta^{1/2} + \Theta^{1/2}X' - \Theta V, \qquad D^{(2)} = W - D^{(1)}V, \\ &D^{(j)} = -D^{(j-1)}V, \qquad j = 3, 4, \ldots . \end{aligned} \tag{3.4}$$

The testing problem is equivalent to one of testing $H_{01} : \theta_k > \theta_{k+1} = \ldots = \theta_p = 0$ against $H_{11} : \theta_{k+1} \geq \ldots \geq \theta_p \geq 0$ and $\theta_{k+1} > 0$.

3.3. *Approximate null distributions.* Under H_{01}, without loss of generality, we may assume $\lambda_h = 0$ and $b_h = p - k$. Let $\Theta_1 = \operatorname{diag}(\theta_1, \ldots, \theta_k)$ and partition V and $[X, \tilde{X}]$ into

$$V = \begin{pmatrix} V_{[11]} & V_{[12]} \\ V_{[21]} & V_{[22]} \end{pmatrix}\begin{matrix} k \\ p-k \end{matrix}, \qquad Y = [X, \tilde{X}] = \begin{pmatrix} Y_{11} & Y_{12} \\ Y_{21} & Y_{22} \end{pmatrix}\begin{matrix} k \\ q-k \end{matrix}. \tag{3.5}$$
$$\qquad\quad k \quad p-k \qquad\qquad\qquad\qquad\qquad k \quad p-k$$

Applying Lemma 1 to (3.3), we have that $d_{k+1}, \ldots, d_p$ are equal to the latent roots of

$$Z = \frac{1}{m}\left\{Z^{(2)} + \frac{1}{\sqrt{m}} Z^{(3)} + \frac{1}{m} Z^{(4)} + \frac{1}{m\sqrt{m}} Z^{(5)} + O_p(m^{-2})\right\}, \tag{3.6}$$

where

$$\begin{aligned} Z^{(2)} &= Y_{22}Y_{22}', \quad Z^{(3)} = -Z^{(2)}V_{[22]} - Y_{22}Y_{12}'\Theta_1^{-1/2}Y_{21}' - Y_{21}\Theta_1^{-1/2}Y_{12}Y_{22}', \\ Z^{(4)} &= Y_{21}\Theta_1^{-1/2}Y_{12}(Y_{21}\Theta_1^{-1/2}Y_{12})' + Y_{21}\Theta_1^{-1/2}Y_{11}\Theta_1^{-1/2}Y_{12}Y_{22}' \\ &\quad + (Y_{21}\Theta_1^{-1/2}Y_{11}\Theta_1^{-1/2}Y_{12}Y_{22}')' - Y_{22}Y_{12}'\Theta_1^{-1}Y_{12}Y_{22}' \\ &\quad - Z^{(2)}\{Y_{21}\Theta_1^{-1}Y_{21}' - V_{[21]}\Theta_1^{-1/2}Y_{21}' - (V_{[21]}\Theta_1^{-1/2}Y_{21}')'\} - Z^{(3)}V_{[22]} \end{aligned} \tag{3.7}$$

and the elements of $Z^{(5)}$ are homogeneous polynomials of degree 5 in the elements of Y and V. Then LR statistic T_1 can be expressed as

$$\begin{aligned} mT_1 &= m \log |I_{p-k} + Z| \\ &= \operatorname{tr} Z^{(2)} + \frac{1}{\sqrt{m}} \operatorname{tr} Z^{(3)} + \frac{1}{m} R^{(4)} + \frac{1}{m\sqrt{m}} R^{(5)} + O_p(m^{-2}), \end{aligned} \tag{3.8}$$

where

$$R^{(4)} - \operatorname{tr} Z^{(4)} - \tfrac{1}{2}\operatorname{tr}(Z^{(2)})^2, \qquad R^{(5)} = \operatorname{tr} Z^{(5)} - \operatorname{tr} Z^{(2)}Z^{(3)}. \tag{3.9}$$

Hence the ch. f. of mT_1 is given by

$$\begin{aligned} \boldsymbol{E}\Big[e^{\phi \operatorname{tr} Z^{(2)}} \Big\{ 1 + \frac{\phi}{\sqrt{m}} \operatorname{tr} Z^{(3)} + \frac{\phi}{m} \Big(R^{(4)} + \frac{\phi}{2} (\operatorname{tr} Z^{(3)})^2 \Big) \\ + \frac{\phi}{m\sqrt{m}} \Big(R^{(5)} + \phi R^{(4)} \operatorname{tr} Z^{(3)} + \frac{\phi^2}{6} (\operatorname{tr} Z^{(3)})^3 \Big) \Big\} \Big] + O(m^{-2}). \end{aligned} \tag{3.10}$$

The expectations in (3.10) are computed in a straightforward manner. Finally we can get ch. f. of mT_1 as

$$\begin{aligned} (1-2\phi)^{-f/2} \Big[1 + \frac{f}{2m} \Big\{ \tfrac{1}{2}(q-p-1) + \sum_{j=1}^{k} \theta_j^{-1} + 2\eta \Big\} \{ 1 - (1-2\phi)^{-1} \} \\ + O(m^{-2}) \Big]. \end{aligned} \tag{3.11}$$

Inversion of the ch. f. yields:

Theorem 1. *Under assumptions* A1 *and* A2, *the null distribution of the* LR *statistic* T_1 *for testing* H_{01} *against* H_{11} *can be asymptotically approximated by*

$$\begin{aligned} \boldsymbol{P}(mT_1 \le x) = \boldsymbol{P}(\chi_f^2 \le x) + \frac{f}{2m} \Big\{ \tfrac{1}{2}(q-p-1) + \sum_{j=1}^{k} \theta_j^{-1} + 2\eta \Big\} \\ \times \{ \boldsymbol{P}(\chi_f^2 \le x) - \boldsymbol{P}(\chi_{f+2}^2 \le x) \} + O(m^{-2}) \end{aligned} \tag{3.12}$$

where $f = (p-k)(q-k)$. *Further letting* $m_1 = n + \tfrac{1}{2}(q-p-1) + \Sigma_{j=1}^{k} \theta_j^{-1}$,

$$\boldsymbol{P}(m_1 T_1 \le x) = \boldsymbol{P}(\chi_f^2 \le x) + O(m_1^{-2}). \tag{3.13}$$

Lawley [17] suggested $m_1 T_1$ as a better approximation to a χ^2-distribution with f degrees of freedom by showing $\boldsymbol{E}[m_1 T_1] = f + O(n^{-2})$. Our result (3.13) confirms this fact.

Similarly the ch. f.'s of mT_2 and mT_3 can be expressed by (3.10) for $R^{(4)} = \operatorname{tr} Z^{(4)}$, $R^{(5)} = \operatorname{tr} Z^{(5)}$ and $R^{(4)} = \operatorname{tr} Z^{(4)} - \operatorname{tr}(Z^{(2)})^2$, $R^{(5)} = \operatorname{tr} Z^{(5)} - 2 \operatorname{tr} Z^{(2)} Z^{(3)}$ respectively. Hence we have

Theorem 2. *Under assumptions* A1 *and* A2, *the null distributions of* T_2 *and* T_3 *can be approximated by*

$$\begin{aligned} \boldsymbol{P}(m_j T_j \leq x) = \boldsymbol{P}(\chi_f^2 \leq x) + \frac{f}{4m_j} l_j \{\boldsymbol{P}(\chi_f^2 \leq x) \\ - 2\boldsymbol{P}(\chi_{f+2}^2 \leq x) + \boldsymbol{P}(\chi_{f+4}^2 \leq x)\} + \mathrm{O}(m_j^{-2}), \qquad j = 2, 3, \end{aligned} \tag{3.14}$$

where

$$\begin{aligned} l_2 &= p + q - 2k + 1, \qquad l_3 = -l_2, \\ m_2 &= n - p - 1 + k + \sum_{j=1}^{k} \theta_j^{-1}, \qquad m_3 = n + q - k + \sum_{j=1}^{k} \theta_j^{-1}. \end{aligned} \tag{3.15}$$

In Theorem 2 the correction factors m_2 and m_3 are chosen so that $\boldsymbol{E}[m_j T_j] = f + \mathrm{O}(m_j^{-2})$.

3.4. Approximate nonnull distributions. From Lemma 1 it follows that the set of d_j, $j \in J_\alpha$ is equal to the latent roots of

$$Z_\alpha = \lambda_\alpha I_{b_\alpha} + \frac{1}{\sqrt{m}} D_{\alpha\alpha}^{(1)} + \frac{1}{m}\left\{D_{\alpha\alpha}^{(2)} + \sum_{\beta \neq \alpha} \lambda_{\alpha\beta} D_{\alpha\beta}^{(1)} D_{\beta\alpha}^{(1)}\right\} + \mathrm{O}_p(m^{-3/2}), \tag{3.16}$$

where $D_{\alpha\beta}^{(j)}$'s mean the submatrices of D partitioned in the manner of $V^{(j)}$ in (2.5). Under H_{11}, we make the following assumption:

A3 : $\theta_k > \theta_{k+1}$, i.e., $b_1 + \ldots + b_a = k$ for some a.

Then we have

$$\begin{aligned} \tilde{T}_1 &= \sqrt{m}\left\{T_1 - \log \prod_{j=k+1}^{p} (1 + \theta_j)\right\} \\ &= \sqrt{m} \sum_{\alpha=a+1}^{h} \{\log |I_{b_\alpha} + Z_\alpha| - b_\alpha \log(1 + \lambda_\alpha)\}. \end{aligned} \tag{3.17}$$

Expanding the right-hand side of (3.17), the ch. f. of $\tilde{T}_1$ can be expressed by

$$\begin{aligned} \boldsymbol{E}\Bigg[\mathrm{e}^{\phi\, \mathrm{tr}(AX+BV)}\Bigg\{1 + \frac{\phi}{\sqrt{m}}\Bigg[\mathrm{tr}\, \Gamma_{[2]} D^{(2)} \\ + \sum_{\alpha=a+1}^{h} (\mathrm{tr}\, \Gamma_{(\alpha)} D^{(1)} R_\alpha D^{(1)} - \tfrac{1}{2} \mathrm{tr}(\Gamma_{(\alpha)} D^{(1)})^2)\Bigg]\Bigg\}\Bigg] + \mathrm{O}(m^{-1}) \end{aligned} \tag{3.18}$$

where $A = 2\Gamma_{[2]}\Theta^{1/2}$, $B = -\Gamma_{[2]}\Theta$, $\Gamma_{(\alpha)} = (1 + \lambda_\alpha)^{-1} I_{(\alpha)}$, $R_\alpha = \Sigma_{\beta \neq \alpha} \lambda_{\alpha\beta} I_{(\alpha)}$, $\Gamma_{[2]} = \Sigma_{\alpha=a+1}^{h} \Gamma_{(\alpha)}$ and $I_{(\alpha)}$ is a block-diagonal matrix given by $I_{(\alpha)} = \mathrm{diag}(0, \ldots, 0, I_{b_\alpha}, 0, \ldots, 0)$. The expectations with respect to V, i.e., S_e are computed, based on Lemma 5 in Sugiura [25]. After some computations, we obtain the following form for ch. f. of $\tilde{T}_1$:

$$e^{-\tau^2 t^2/2}\left[1+\frac{1}{\sqrt{m}}\{h_1\phi+h_3\phi^3\}+O(m^{-1})\right] \tag{3.19}$$

where for $s_i=\Sigma_{j=k+1}^{p}\{\theta_j/(1+\theta_j)\}^i$, $\tau^2=4s_1-2s_2$ and

$$\begin{aligned} h_1 &= q(p-k)+(k-q-2\eta)s_1+\tfrac{1}{2}s_2+\tfrac{1}{2}s_1^2 \\ &\quad+\sum_{l=k+1}^{p}\sum_{j=1}^{k}(\theta_l+\theta_j+\theta_l\theta_j)/\{(1+\theta_l)(\theta_l-\theta_j)\}, \\ h_3 &= 4s_1-8s_2+\tfrac{20}{3}s_3-2s_4. \end{aligned} \tag{3.20}$$

Inverting (3.20), we have

Theorem 3. *Under assumptions* A1, A2 *and* A3, *the nonnull distribution of* $\tilde{T}=\sqrt{m}\{T_1-\log\Pi_{j=k+1}^{p}(1+\theta_j)\}$ *can be asymptotically approximated by*

$$P(\tilde{T}/\tau\le x)=\Phi(x)-\frac{1}{\sqrt{m}}\{h_1\Phi^{(1)}(x)/\tau+h_3\Phi^{(3)}(x)/\tau^3\}+O(m^{-1}), \tag{3.21}$$

where $\Phi^{(j)}(x)$ *means the j-th order derivative of the standard normal distribution function* $\Phi(x)$ *and* τ^2 *and* h_j *are given by* (3.20).

Similarly we have

Theorem 4. *Under assumptions* A1, A2 *and* A3, *the nonnull distributions of* $\sqrt{m}\{T_2-\Sigma_{j=k+1}^{p}\theta_j\}/\tau$ *and* $\sqrt{m}(T_3-\Sigma_{j=k+1}^{p}\theta_j(1+\theta_j)^{-1}\}/\tau$ *are given by the right-hand side of* (3.21) *with* τ^2 *and* h_j *given below. For* T_2:

$$\tau^2=2(2t_1+t_2),\qquad h_1=b(p-k)+(p+1-2\eta)t_1$$
$$+\sum_{l=k+1}^{p}\sum_{j=1}^{k}(\theta_l+\theta_j+\theta_l\theta_j)(\theta_l-\theta_j)^{-1},\qquad h_3=4t_1+8t_2+\tfrac{8}{3}t_3$$

and for T_3:

$$\tau^2=2(c_2-c_4),$$
$$h_1=kc_1-2\eta(c_1-c_2)+(q-p-1)c_2+c_1c_2+c_3$$
$$+\sum_{l=k+1}^{p}\sum_{j=1}^{k}(\theta_l+\theta_j+\theta_l\theta_j)(\theta_l-\theta_j)^{-1}(1+\theta_l)^{-2},$$

$$h_3 = 4 \sum_{l=k+1}^{p} \theta_l(1+\theta_l)^{-4}[1 - 4\theta_l(1+\theta_l)^{-1} + \tfrac{11}{3}\{\theta_l(1+\theta_l)^{-1}\}^2 - \{\theta_l(1+\theta_l)^{-1}\}^3],$$

where $t_i = \sum_{j=k+1}^{p} \theta_j^i$ *and* $c_i = \sum_{j=k+1}^{p}(1+\theta_j)^{-i}$.

Recently Fujikoshi [11] obtained asymptotic expansions of the nonnull distributions of T_j in terms of linear combinations of noncentral χ^2-distributions under the assumptions of $\Omega_1 = \text{diag}(\omega_1, \ldots, \omega_k) = O(n)$ and $\Omega_2 = \text{diag}(\omega_{k+1}, \ldots, \omega_p) = O(1)$.

4. Tests for canonical correlations

4.1. Introduction. Let $r_1^2 \geq \ldots \geq r_p^2$ and $\rho_1^2 \geq \ldots \geq \rho_p^2$ be the squares of the sample and the population canonical correlations between p-components and q-components ($p \leq q$) of a $p+q$ variate normal population, based on a sample of size $n+1$. For testing $H_{02}: \rho_k^2 > \rho_{k+1}^2 = \ldots = \rho_p^2 = 0$ against $H_{12}: \rho_{k+1}^2 \geq \ldots \geq \rho_p^2 \geq 0$ and $\rho_{k+1}^2 > 0$, we consider the following three statistics analogous to the statistics proposed for MANOVA:

(1) LR statistic, $Q_1 = -\log \prod_{j=k+1}^{p}(1 - r_j^2)$,

(2) $Q_2 = \sum_{j=k+1}^{p} r_j^2/(1 - r_j^2)$,

(3) $Q_3 = \sum_{j=k+1}^{p} r_j^2$.

The test statistic Q_1, which is in fact a LR statistic (for a proof, see Fujikoshi [10]), was given by Bartlett [4]. Under H_{02}, Bartlett [4] has given an approximation as $\chi_f^2 = \{n - \frac{1}{2}(p+q+1)\}Q_1$ with $f = (p-k)(p-q)$. Lawley [17] has shown that, under H_{02},

$$\boldsymbol{E}\left[\left\{n - k - \tfrac{1}{2}(p+q+1) + \sum_{j=1}^{k} \rho_j^{-2}\right\} Q_1\right] = f + O(n^{-2}). \qquad (4.1)$$

However such a correction factor for other two statistics Q_2 and Q_3 has not been demonstrated so far. In this section we obtain asymptotic expansions of the null and nonnull distributions of Q_j. When $k = 0$, the testing problems become to test the hypothesis between two sets of variates. Asymptotic expansions of the nonnull distributions of the corresponding test statistics have been obtained under each of local and fixed alternatives in the literature.

4.2. Reduction. Let $g_1 \geq \ldots \geq g_p$ and $\theta_1 \geq \ldots \geq \theta_p$ be the transformed canonical correlations defined by $g_j = r_j^2/(1 - r_j^2)$ and $\theta_j = \rho_j^2/(1 - \rho_j^2)$.

From the result of Constantine [8], Lee [20] it follows that $g_1 \geq \ldots \geq g_p$ are distributed like the roots of

$$(\Theta^{1/2} U \Theta^{1/2} + X U^{1/2} \Theta^{1/2} + \Theta^{1/2} U^{1/2} X' + W) S^{-1}, \tag{4.2}$$

where $\Theta = \text{diag}\,(\theta_1, \ldots, \theta_p)$, S and U are Wishart matrices having $W_p(n - q, I_p)$ and $W_p(n, I_p)$ respectively, W and X are the same matrices as in (3.1). Furthermore, S, U and $[X, \tilde{X}]$ are independently distributed. Let

$$\frac{1}{m} S = I_p + \frac{1}{\sqrt{m}} V, \qquad \frac{1}{m} U = I_p + \frac{1}{\sqrt{m}} L, \tag{4.3}$$

where $m = n - 2\eta$ and η is a correction factor. Expanding S and U in terms of V and L, we have that $g_1 \geq \ldots \geq g_p$ are the latent roots of

$$G = \Theta + \frac{1}{\sqrt{m}} G^{(1)} + \frac{1}{m} G^{(2)} + \frac{1}{m\sqrt{m}} G^{(3)} + \ldots, \tag{4.4}$$

where $G^{(j)} = D^{(j)} + H^{(j)}$, $D^{(j)}$ are given by (3.4) and

$$H^{(1)} = \Theta^{1/2} L \Theta^{1/2}, \qquad H^{(2)} = \tfrac{1}{2}(X L \Theta^{1/2} + \Theta^{1/2} L X') - H^{(1)} V,$$

$$H^{(j+1)} = (-1)^{j+1} \frac{1 \cdot 3 \ldots (2j-3)}{2 \cdot 4 \ldots 2j} (X L^j \Theta^{1/2} + \Theta^{1/2} L^j X') - H^{(j)} V, \tag{4.5}$$

$$j = 2, 3, \ldots.$$

The test statistics are expressed in terms of g_i as $Q_1 = \log \prod_{i=k+1}^{p} (1 + g_i)$, $Q_2 = \sum_{j=k+1}^{p} g_j$ and $Q_3 = \sum_{j=k+1}^{p} g_j/(1 + g_j)$. The testing problem is equivalent to one of testing $H_{02}: \theta_k > \theta_{k+1} = \ldots = \theta_p = 0$ against $H_{12}: \theta_{k+1} \geq \ldots \geq \theta_p$ and $\theta_{k+1} > 0$.

4.3. Approximate null and nonnull distributions. From the above reduction it is clear that asymptotic expansions of the null and nonnull distributions of Q_j can be treated by following similar lines as in Sections 3.3 and 3.4. We note that the expectations with respect to L, i.e. U are similarly computed as ones with respect to V, i.e. S_e. The final results are given in the following theorems.

Theorem 5. *Under H_{02}, we have*

$$P(m_1 Q_1 \leq x) = P(\chi_f^2 \leq x) + O(m_1^{-2}), \tag{4.6}$$

and for $j = 2, 3$,

$$P(m_jQ_j \le x) = P(\chi_f^2 \le x) + \frac{f}{4m_j} l_j\{P(\chi_f^2 \le x) - 2P(\chi_{f+2}^2 \le x) + P(\chi_{f+4}^2 \le x)\} + O(m_j^{-2}), \quad (4.7)$$

where $f = (p-k)(q-k)$, $l_2 = p+q-2k+1$, $l_3 = -l_2$ *and the correction factors are given by*

$$m_1 = n - k - \tfrac{1}{2}(p+q+1) + \sum_{j=1}^{k} \rho_j^{-2},$$
$$m_2 = n - (p+q+1) + \sum_{j=1}^{k} \rho_j^{-2}, \quad (4.8)$$
$$m_3 = n - 2k + \sum_{j=1}^{k} \rho_j^{-2}.$$

Theorem 6. *Assume that* $\rho_k^2 > \rho_{k+1}^2$. *Let* $\tilde{Q}_1 = \sqrt{m}\{Q_1 + \log \Pi_{j=k+1}^p (1-\rho_j^2)\}$, $\tilde{Q}_2 = \sqrt{m}\{Q_2 - \Sigma_{j=k+1}^p \rho_j^2/(1-\rho_j^2)\}$ *and* $\tilde{Q}_3 = \sqrt{m}\{Q_3 - \Sigma_{j=k+1}^p \rho_j^2\}$, *where* $m = n - 2\eta$ *and* η *is a fixed constant. Then, under* H_{12}, *we have*

$$P(\tilde{Q}_j/\tau_j \le x) = \Phi(x) - \frac{1}{\sqrt{m}}\{h_{1j}\Phi^{(1)}(x)/\tau_j + h_{3j}\Phi^{(3)}(x)/\tau_j^3\} + O(m^{-1}), \qquad j = 1,2,3, \quad (4.9)$$

where $\tau_1^2 = 4s_1$, $\tau_2^2 = 4(t_1 + t_2)$, $\tau_3^2 = 4(s_1 - 2s_2 + s_3)$ *for* $s_i = \Sigma_{j=k+1}^p \rho_j^{2i}$ *and* $t_i = \Sigma_{j=k+1}^p \{\rho_j^2/(1-\rho_j^2)\}^i$, *and the coefficients* h_{ij} *are given by*

$$h_{11} = q(p-k) + \sum_{l=k+1}^{p}\sum_{j=1}^{k} (\rho_l^2 + \rho_j^2 - 2\rho_j^2\rho_l^2)/(\rho_l^2 - \rho_j^2)$$
$$h_{12} = q(p-k) + (p+q+1)t_1 + \sum_{l=k+1}^{p}\sum_{j=1}^{k} (\rho_l^2 + \rho_j^2)/(\rho_j^2 - \rho_l^2),$$
$$h_{13} = q(p-k) - (p+q-k+1)s_1 + s_2 + s_1^2 + \sum_{l=k+1}^{p}\sum_{j=1}^{k} (1-\rho_l^2)(\rho_l^2 + \rho_j^2 - 2\rho_l^2\rho_j^2)/(\rho_l^2 - \rho_j^2) \quad (4.10)$$
$$h_{31} = 4(s_1 - s_2), \qquad h_{32} = 4(t_1 + 3t_2 + 2t_3),$$
$$h_{33} = 4(s_1 - 6s_2 + 12s_3 - 10s_4 + 3s_5).$$

It may be noted that the correction factors m_2 and m_3 are also chosen so that, under H_{02}, $E[m_jQ_j] = f + O(m_j^{-2})$ and the nonnull distributions (4.9) up to the order $n^{-1/2}$ do not depend on η. Asymptotic expansions of the distributions of Q_j under local alternatives were obtained by Fujikoshi [11].

5. Tests of equality of latent roots of a covariance matrix

5.1. Introduction. Let S be a sample covariance matrix calculated from a sample of size $n+1$ drawn from a p-variate normal population with unknown covariance matrix Σ; then nS is distributed as Wishart $W_p(n,\Sigma)$. Let $d_1 \geq \ldots \geq d_p$ and $\delta_1 \geq \ldots \geq \delta_p > 0$ be the latent roots of S and Σ respectively. For testing $H_{03}: \delta_{a+1} = \ldots = \delta_{a+b} = \delta$ $(0 \leq a < a+b \leq p)$ against H_{13}: not H_{03}, we consider the following test statistics:

(1) When δ is known,

$$L_1 = -\log \prod_{j=a+1}^{a+b} \{d_j/\delta\} + \sum_{j=a+1}^{a+b} \{d_j/\delta\} - b.$$

(2) When δ is unknown

$$L_2 = -\log \prod_{j=a+1}^{a+b} d_j + b \log\left\{\frac{1}{b}\sum_{j=a+1}^{a+b} d_j\right\}.$$

Anderson [3] proved that L_2 is a LR statistic. However it may be noted that L_1 is not precisely a LR statistic (cf. Fujikoshi [9]). The limiting distributions of nL_1 and nL_2 are shown by Anderson [3] to be χ^2 distributions with degrees of freedom $b(b+1)/2$ and $b(b+1)/2-1$ respectively. Bartlett [5] and Lawley [16] have investigated χ^2-approximations to L_1 and L_2 for tests of equality of the smallest $p-b$ latent roots of Σ, i.e., the case of $a+b=p$. Lawley [16] has suggested correction factors to L_1 and L_2, based on their asmptotic means. James [13] obtained Lawley's correction factor to L_2, by using a different method. Recently Sugiura [26] has derived asymptotic expansions of the distributions of L_1 and L_2 under H_{13} when $\delta_{a+1},\ldots,\delta_{a+b}$ are all simple roots. In this section we will give asymptotic expansions of the distributions of L_1 and L_2 under H_{03}. When $a=0$ and $b=p$, L_2 is a LR statistic for testing sphericity and the distribution of L_1 may be regarded as one of a LR statistic for testing the equality of Σ to a given Σ_0. Asymptotic expansions of the distributions of these statistics have been obtained in the literature.

5.2. Approximate null distributions. Without loss of generality, under H_{03}, we may start by putting $\delta = 1$ and $\Sigma = \Lambda = \mathrm{diag}\,(\lambda_1,\ldots,\lambda_a, 1,\ldots,1, \lambda_{a+b+1},\ldots,\lambda_p)$, where $\lambda_j = \delta_j/\delta$. Here we assume that $\lambda_a > 1$ and $\lambda_{a+b+1} < 1$. Let $S = \Lambda + (1/\sqrt{n})V$ and let V and Λ partition as

$$V = \begin{pmatrix} V_{11} & V_{12} & V_{13} \\ V_{21} & V_{22} & V_{23} \\ V_{31} & V_{32} & V_{33} \end{pmatrix}, \qquad \Lambda = \begin{pmatrix} \Lambda_1 & 0 & 0 \\ 0 & I_b & 0 \\ 0 & 0 & \Lambda_3 \end{pmatrix},$$

where $V_{11}: a \times a$, $V_{22}: b \times b$ and $\Lambda_1: a \times a$. Then, from Lemma 1, it follows that $d_{a+1} \geq \ldots \geq d_{a+b}$ are equal to the latent roots of

$$B = I_b + \frac{1}{\sqrt{n}} V_{22} - \frac{1}{n} U_{21}\Omega U_{12} + \frac{1}{n\sqrt{n}} \{U_{21}\Omega U_{11}\Omega U_{12} - V_{22}U_{21}\Omega^2 U_{12}\}$$
$$+ \frac{1}{n^2} B^{(4)} + \frac{1}{n^2\sqrt{n}} B^{(5)} + O_p(n^{-3}), \tag{5.2}$$

where the elements of $B^{(j)}$, $j = 4, 5$ are homogeneous polynomials of j degree in the elements of V and

$$U_{11} = \begin{pmatrix} V_{11} & V_{13} \\ V_{31} & V_{33} \end{pmatrix}, \qquad U_{12} = \begin{pmatrix} V_{12} \\ V_{32} \end{pmatrix} = U'_{21},$$
$$\Omega^{-1} = \begin{pmatrix} \Lambda_1 - I_a & 0 \\ 0 & \Lambda_3 - I_{p-a-b} \end{pmatrix}. \tag{5.3}$$

Expressing L_1 as $-\log|B| + \operatorname{tr} B - b$ and then expanding each term, the ch. f. of nL_1 can be expressed by

$$\boldsymbol{E}\left[e^{\frac{1}{2}\phi \operatorname{tr} V_{22}^2}\left\{1 + \frac{\phi}{\sqrt{n}} q_1 + \frac{\phi}{n}\left(q_2 + \frac{\phi}{2} q_1^2\right) + \frac{1}{n\sqrt{n}} r\right\}\right] + O(n^{-2}), \tag{5.4}$$

where

$$q_1 = -\operatorname{tr} V_{22}U_{21}\Omega U_{12} - \tfrac{1}{3}\operatorname{tr} V_{22}^3,$$
$$q_2 = \operatorname{tr} V_{22}U_{21}\Omega U_{11} U_{12} - \operatorname{tr} V_{22}^2 U_{21}\Omega^2 U_{12} + \tfrac{1}{2}\operatorname{tr}(U_{21}\Omega U_{12})^2$$
$$+ \operatorname{tr} V_{22}^2 U_{21}\Omega U_{12} + \tfrac{1}{4}\operatorname{tr} V_{22}^4, \tag{5.5}$$

and r is a homogeneous polynomial of degree 5 in the elements of V. We have to carry out the expectations in (5.4). The expectations can be regarded as ones with respect to W defined by

$$W = \begin{pmatrix} W_{11} & W_{12} \\ W_{21} & W_{22} \end{pmatrix}, \qquad W_{12} = \begin{pmatrix} \Lambda_1 & 0 \\ 0 & \Lambda_3 \end{pmatrix}^{-1/2} \begin{pmatrix} S_{12} \\ S_{32} \end{pmatrix} = W'_{21},$$
$$W_{11} = \begin{pmatrix} \Lambda_1 & 0 \\ 0 & \Lambda_3 \end{pmatrix}^{-1/2} \begin{pmatrix} S_{11} & S_{13} \\ S_{31} & S_{33} \end{pmatrix} \begin{pmatrix} \Lambda_1 & 0 \\ 0 & \Lambda_3 \end{pmatrix}^{-1/2}, \qquad W_{22} = S_{22} \tag{5.6}$$

and S_{ij} are the submatrices of S partitioned in the manner as in (5.1). nW has a Wishart distribution $W_p(n, I_p)$. Consider the following transformation from W to $(W_{11\cdot 2}, R, Y)$:

$$W_{11\cdot 2} = W_{11} - W_{12} W_{22}^{-1} W_{21},$$
$$R = \sqrt{n}\, W_{12} W_{22}^{-1}, \qquad Y = \sqrt{\frac{2}{n}} \log W_{22}. \tag{5.7}$$

The last transformation $Y = \sqrt{2/n} \log W_{22}$ has been used in Nagao [21]. Then it is well known that $nW_{11\cdot 2}$ and nW_{22} are distributed as $W_{p-b}(n - b, I_{p-b})$ and $W_b(n, I_b)$ respectively, and the $(p-b)b$ elements of R are independently distributed as $N[0, 1]$. Furthermore, these three matrices are mutually independent. Hence the expectations with respect to $W_{11\cdot 2}$ and R are easily computed. The expectations with respect to Y can be computed by following similar lines as in Nagao [21]. After the calculations, we can write (5.4) as

$$(1-2\phi)^{-f/2}\left[1 + \frac{f}{2n} l\{1-(1-2\phi)^{-1}\} + O(n^{-2})\right], \tag{5.8}$$

where $f = b(b+1)/2$ and

$$l = -(p-b) - \tfrac{1}{6}\left(2b + 1 - \frac{2}{b+1}\right) - \frac{1}{b+1}\{\Sigma_{(\beta)} \delta_\beta/(\delta_\beta - \delta)\}^2$$
$$+ \Sigma_{(\beta)}\{\delta/(\delta_\beta - \delta)\}^2. \tag{5.9}$$

Here the symbol $\Sigma_{(\beta)}$ means the sum of $\beta = 1, \ldots, a, a+b+1, \ldots, p$. From (5.8) we have

Theorem 7. *Assume that $\delta_a > \delta > \delta_{a+b+1}$. Let $m = n - 2\eta$ and η is a fixed constant. Then, the null distribution of L_1 can be asymptotically approximated by*

$$P(mL_1 \le x) = P(\chi_f^2 \le x) + \frac{f}{2m}(l + 2\eta)\{P(\chi_f^2 \le x) - P(\chi_{f+2}^2 \le x)\}$$
$$+ O(m^{-2}), \tag{5.10}$$

where $f = b(b+1)/2$ and l is given by (5.9). Further letting $\tilde{m} = n + l$,

$$P(\tilde{m}L_1 \le x) = P(\chi_f^2 \le x) + O(\tilde{m}^{-2}). \tag{5.11}$$

Next consider the ch. f. of $nL_2 = n\{-\log|B| + b \log(1/b \operatorname{tr} B)\}$, which can be expressed by

$$E\left[\exp\left\{\tfrac{1}{2}\phi\left(\operatorname{tr} V_{22}^2 - \frac{1}{b}(\operatorname{tr} V_{22})^2\right)\right\}\left\{1 + \frac{\phi}{\sqrt{n}}\tilde{q}_1 + \frac{\phi}{n}\left(\tilde{q}_2 + \frac{\phi}{2}\tilde{q}_1^2\right) + \frac{1}{n\sqrt{n}}\tilde{r}\right]$$
$$+ O(n^{-2}), \tag{5.12}$$

where

$$\tilde{q}_1 = q_1 + \frac{1}{b}\,\mathrm{tr}\,V_{22}\left\{\mathrm{tr}\,U_{21}\Omega U_{12} + \frac{1}{3b}(\mathrm{tr}\,V_{22})^2\right\},$$

$$\tilde{q}_2 = q_2 - \frac{1}{2b}(\mathrm{tr}\,U_{21}\Omega U_{12})^2 - \frac{1}{b}\,\mathrm{tr}\,V_{22}\{\mathrm{tr}\,U_{21}\Omega U_{11}\Omega U_{12} - \mathrm{tr}\,V_{22}U_{21}\Omega^2 U_{12}\}$$

$$- \left(\frac{1}{b}\,\mathrm{tr}\,V_{22}\right)^2\left\{\mathrm{tr}\,U_{21}\Omega U_{12} + \frac{1}{4b}(\mathrm{tr}\,V_{22})^2\right\}, \tag{5.13}$$

with q_1 and q_2 defined by (5.5), and $\tilde{r}$ is a homogeneous polynomial of degree 5 in the elements of V. The expectations in (5.12) can be evaluated by following similar lines as ones in (5.4). After much simplification and inversion, we obtain the following Theorem:

Theorem 8. *Under the same assumption as in Theorem* 7, *the null distributions of* mL_2 *and* $\tilde{m}L_2$ *can be expressed as* (5.10) *and* (5.11) *respectively for* $f = b(b+1)/2 - 1$ *and*

$$l = -(p-b) - \frac{1}{6b}(2b^2 + b + 2) + \Sigma_{(\beta)}\{\delta/(\delta_\beta - \delta)\}^2. \tag{5.14}$$

It may be noted that the correction factors $\tilde{m} = n + l$ to L_1 and L_2 in a special case of $a + b = p$ have been obtained by Lawley [16].

References

[1] Anderson, T.W. (1951). The asymptotic distribution of certain characteristic roots and vectors. *Proc. Second Berkeley Symp. on Math. Statist. and Prob.* pp. 103–130. Univ. of California Press.

[2] Anderson, T.W. (1951). Estimating linear restrictions on regression coefficients for multivariate normal distributions. *Ann. Math. Statist.* **22**, 327–351.

[3] Anderson, T.W. (1963). Asymptotic theory for principal component analysis. *Ann. Math. Statist.* **34**, 122–148.

[4] Bartlett, M.S. (1938). Further aspects of the theory of multiple regression. *Proc. Camb. Phil. Soc.* **34**, 33–40.

[5] Bartlett, M.S. (1954). A note on the multiplying factors for various χ^2 approximations. *J. Roy. Statist. Soc. Ser.* B **16**, 296–298.

[6] Bellman, R. (1960). *Introduction to Matrix Analysis.* McGraw-Hill, New York.

[7] Box, G.E.P. (1949). A general distribution theory for a class of likelihood criteria. *Biometrika* **36**, 317–346.

[8] Constantine, A.G. (1963). Some non-central distribution problems in multivariate analysis. *Ann. Math. Statist.* **34**, 1270–1285.

[9] Fujikoshi, Y. (1973). Likelihood ratio tests of certain hypotheses about principal components. *J. Japan Statist. Soc.* **4**, 5–9.
[10] Fujikoshi, Y. (1974). The likelihood ratio tests for the dimensionality of regression coefficients. *J. Multivariate Anal.* **4**, 327–340.
[11] Fujikoshi, Y. (1975). The nonnull distributions of three test statistics for the dimensionality of mean vectors and the last $p - k$ canonical correlations. Technical Report No. 300, Dept. of Math. and Statist., University of Calgary.
[12] Hsu, P.L. (1941). On the limiting distribution of the roots of a determinantal equation. *J. London Math. Soc.* **16**, 183–194.
[13] James, A.T. (1969). Tests of equality of latent roots of the covariance matrix. In *Multivariate Analysis–II* (P.R. Krishnaiah, Ed.), pp. 205–218. Academic Press, New York.
[14] Krishnaiah, P.R. and Waikar, V.B. (1971). Simultaneous tests for the equality of latent roots against certain alternatives–I. *Ann. Inst. Statist. Math.* **23**, 451–468.
[15] Krishnaiah, P.R. and Waiker, V.B. (1972). Simultaneous tests for equality of latent roots against certain alternatives–II. *Ann. Inst. Statist. Math.* **24**, 81–85.
[16] Lawley, D.N. (1956). Tests of significance for the latent roots of covariance and correlation matrices. *Biometrika* **55**, 171–178.
[17] Lawley, D.N. (1959). Tests of significance in canonical analysis. *Biometrika* **46**, 59–66.
[18] Lee, J.C., Chang, T.C. and Krishnaiah, P.R. (1975). Approximations to the distributions of the likelihood ratio statistics for testing certain structures on the covariance matrices of real multivariate normal populations. ARL. 75–0167, Wright-Patterson AFB, Ohio.
[19] Lee, J.C., Krishnaiah, P.R. and Chang, T.C. (1975). Approximations to the distributions of the determinants real and complex multivariate beta matrices. ARL. 75–0168 Wright-Patterson AFB, Ohio.
[20] Lee, Y.S. (1971). Distribution of the canonical correlations and asymptotic expansions for the distributions of certain independence test statistics. *Ann. Math. Statist.* **42**, 526–537.
[21] Nagao, H. (1973). On some test criteria for covariance matrix. *Ann. Statist.* **1**, 700–709.
[22] Rao, C.R. (1965). *Linear Statistical Inference and Its Application.* Wiley, New York.
[23] Siotani, M. (1974). Recent development in asymptotic expansions for the nonnull distributions of multivariate test statistics. Technical Report No. 32, Dept. of Statistics, Kansas State Univ..
[24] Sugiura, N. (1973). Further asymptotic formulas for the non-null distributions of three statistics for multivariate linear hypothesis. *Ann. Inst. Statist. Math.* **25**, 153–163.
[25] Sugiura, N. (1973). Derivatives of the characteristic roots of a symmetric or a Hermitian matrix with two applications in multivariate analysis. *Communications in Statistics* **1**, 393–417.
[26] Sugiura, N. (1975). Asymptotic non-null distributions of the likelihood ratio criteria for the equality of several characteristic roots of a Wishart matrix. To appear in Statistical Papers in Honor of Junjiro Ogawa.

P.R. Krishnaiah, ed., *Multivariate Analysis–IV*
© North-Holland Publishing Company (1977) 73–77

TESTS FOR A PRESCRIBED SUBSPACE OF PRINCIPAL COMPONENTS

A.T. JAMES
The University of Adelaide, Adelaide, Australia

Let $S(m \times m)$ be a sample estimate of a variance matrix Σ, derived from a normal population, and $H_1(m \times p)$ a matrix whose columns, as vectors, form an orthonormal basis for a subspace $\mathcal{R}(H_1)$ of $\mathbf{R}^m$. The hypothesis that $\mathcal{R}(H_1)$ is spanned by principal components of Σ can be tested by using the fact that if f_i, $i = 1, \ldots p$ are the latent roots of the matrix $H_1'SH_1H_1'S^{-1}H_1 - I_p$ then $r_i^2 = f_i/(1+f_i)$ are squares of canonical correlations, with a null distribution if the hypothesis is true.

1. Introduction

Prescribed linear functions of a vector variate are principal components if and only if they are uncorrelated with each other and with all orthogonal linear functions, as Mallows [6] on p. 138 in the last paragraph has pointed out. He went on to derive the likelihood ratio criterion and use its asymptotic distribution. T.W. Anderson's [1] use of the test as an asymptotic test for a prescribed single principal component is often given in textbooks.

This paper develops two further points which follow from Mallows' idea. Firstly, one can test whether a prescribed subspace is spanned by principal components by calculating canonical correlation coefficients of linear functions whose coefficient vectors span the space, with linear functions spanning the orthogonal complement. Secondly, canonical correlations have exact tests for finite samples.

M. Vaughton [7] has derived the test for a single vector.

2. Test of prescribed subspaces

Let Σ be the variance matrix of a random vector $y \in \mathbf{R}^m$. A p-dimensional subspace $\mathcal{S}$ of $\mathbf{R}^m$ is spanned by eigenvectors of Σ if and only if $H_1'y$ is canonically uncorrelated with $H_2'y$ where H_1 and H_2 are $p \times m$ and $(m-p) \times m$ matrices of orthonormal column vectors which span $\mathcal{S}$

and its orthogonal complement $\mathscr{S}^{\perp}$ respectively; $\mathscr{R}(H_1) = \mathscr{S}$, $\mathscr{R}(H_2) = \mathscr{S}^{\perp}$,· $H = [H_1 : H_2] \in O(m)$.

If $\mathbf{y}$ is normally distributed and S is an estimate of Σ on n degrees of freedom, the sample canonical correlations are the roots of the equation

$$\det(B - r^2 T) = 0$$

where $B = H_1' S H_2 (H_2' S H_2)^{-1} H_2' S H_1$ and $T = H_1' S H_1$.

In place of r_i^2, one can use the statistics $f_i = r_i^2/(1 - r_i^2)$ which are roots of the equation

$$\det(B - fW) = 0$$

where $W = T - B$.

The matrices B and W appear in the analysis of variance for the regression associated with the canonical correlation.

Multivariate analysis of variance

	d.f.	S.S. and P.
Regression	$m - p$	B
Deviations	$n - m + p$	W
Total	n	T

As the roots f_i depend only upon $\mathscr{S}$ and S, it must be possible to eliminate H_2 and express them as the roots of a matrix depending only upon H_1 and S. The matrix W can be written as the inverse of the top left $p \times p$ submatrix of the inverse of $H'SH$, viz.

$$\begin{aligned} W &= H_1' S H_1 - H_1' S H_2 (H_2' S H_2)^{-1} H_2' S H_1 \\ &= ([(H'SH)^{-1}]_{11})^{-1} \\ &= ([H' S^{-1} H]_{11})^{-1} \\ &= (H_1' S^{-1} H_1)^{-1}. \end{aligned}$$

The f_i are then the latent roots of the matrix

$$\begin{aligned} BW^{-1} = (T - W)W^{-1} &= TW^{-1} - I_p \\ &= H_1' S H_1 H_1' S^{-1} H_1 - I_p \end{aligned}$$

As a test that $\mathscr{S}$ is spanned by eigenvectors, the usual multivariate statistics such as

$$w = \Pi(1+f_i)^{-1} = \det(H_1'SH_1H_1'S^{-1}H_1),$$

$$t = \Sigma f_i = \operatorname{tr}(H_1'SH_1H_1'S^{-1}H_1 - I_p)$$

or largest f_i can be used.

3. Non orthogonal basis of subspace

Alternatively the matrix

$$SH_1H_1'S^{-1}H_1H_1' - H_1H_1' = SES^{-1}E - E, \quad \text{where } E = H_1H_1'$$

has the same roots together with a further $m-p$ zero roots.

If A is a $n \times p$ matrix of non orthogonal column vectors which span $\mathcal{S}$, then

$$E = A(A'A)^{-1}A'$$

because both sides are uniquely characterized as symmetric idempotent matrices of range $\mathcal{S}$.

Then

$$SES^{-1}E - E = SA(A'A)^{-1}A'S^{-1}A(A'A)^{-1}A' - A(A'A)^{-1}A'$$

and apart from $m-p$ zero roots, its roots will be the same as those of

$$(A'SA)(A'A)^{-1}(A'S^{-1}A)(A'A)^{-1} - I_p.$$

4. Distribution under the alternative hypothesis

As Dr. W.N. Venables has pointed out to me, the distribution of $F = \operatorname{diag}(f_i)$ on the alternative hypothesis is the distribution of f variables derived from Constantine's [2] distribution of the canonical correlation coefficients:

$$\det(I_p+\Phi)^{-\frac{1}{2}n}{}_2F_1^{(m)}(\tfrac{1}{2}n, \tfrac{1}{2}n; \tfrac{1}{2}(m-p); \Phi(I_p+\Phi)^{-1}, F(I_p+F)^{-1})$$

$$\frac{\Gamma_p(\frac{1}{2}n)\pi^{\frac{1}{2}p^2}}{\Gamma_p(\frac{1}{2}(n-m+p))\Gamma_p(\frac{1}{2}(m-p))\Gamma_p(\frac{1}{2}p)}$$

$$\det F^{\frac{1}{2}(m-2p-1)}\det(I_p+F)^{-\frac{1}{2}n}\prod_{i<j}^{p}(f_i - f_j)\,\mathrm{d}f_i \ldots \mathrm{d}f_p$$

where $\Phi = \mathrm{diag}(\phi_i)$ and the ϕ_i are the latent roots of the matrix

$$H_1'\Sigma H_1 H_1'\Sigma^{-1} H_1 - I_p.$$

The power of the criteria is then given by their power for testing canonical correlations.

5. Confidence intervals based on significance tests

Although the significance test for a principal component or a subspace of them, is a natural and valid one, a serious objection can be raised to a say 95% confidence interval consisting of the set complementary to those rejected at the 5% level by the test.

The objection is that the statistic used to test the hypothesis changes with the hypothesis and a sort of multiple comparisons problem arises. Associated with the objection is the anomaly that the confidence procedure can set up a 95% confidence interval for a principal component in some cases when sphericity is not rejected at the 5% level i.e. one purports to set up a 95% confidence interval for a principal component whose actual existence is not established with comparable evidence or certainty.

James, Wilkinson & Venables [5] have discussed a similar problem in estimating the ratio of means and have given a third alternative to the solutions of Creasy [3] and Fieller [4], which overcomes the objection. For the major principal axis of a bivariate normal distribution, a similar solution is possible. A paper on it is in preparation.

6. Subspace of multivariate means

The Creasy Fieller problem generalizes to the confidence interval for the direction of a mean vector in $\mathbf{R}^m$ with normal spherically distributed estimate or the confidence interval for a plane spanned by the columns of a matrix of means.

The test for the latter is as follows. Let

$$\left[\frac{Y_1}{Y_2}\right]$$

be a $n \times m$ matrix distributed as

$$N\left(\left[\frac{M}{O}\right], I_n \otimes \Sigma\right)$$

where M is a $p \times m$ matrix with $p > m$. The hypothesis $\mathcal{R}(M) \subset \mathcal{S}$ where $\mathcal{S}$ is a m dimensional subspace of $\mathbf{R}^p$ can be tested by formulating a Multivariate Analysis of Variance:

	d.f.	S.S. and P.
$\mathcal{S}$	m	
$\mathcal{S}^\perp$	$p - m$	$Y_1' P(\mathcal{S}^\perp) Y_1$
Residual	$n - p$	$Y_2' Y_2$

where $P(\mathcal{S}^\perp)$ is a $p \times p$ matrix which projects orthogonally on $\mathcal{S}^\perp$. The hypothesis can be tested with the usual tests; but the inversion of a test does not produce a valid confidence interval for $\mathcal{S}$.

7. Discriminant functions

There are tests for prescribed simple discriminant functions or subspaces of multiple discriminant functions. Confidence intervals based on such tests are subject to the same objection.

References

[1] Anderson, T.W. (1963). Principal component analysis. *Ann. Math. Statist.* **34**, 122–148.
[2] Constantine, A.G. (1963). Some noncentral distribution problems in multivariate analysis. *Ann. Math. Statist.* **34**, 1270–1285.
[3] Creasy, M.A. (1954). Limits for the ratio of means. *J. Roy. Statist. Soc.*, (B), **16**, 186–192.
[4] Fieller, E.C. (1954). Some problems in interval estimation *J. Roy. Statist. Soc.*, (B), **16**, 175–186.
[5] James, A.T., Wilkinson, G.N. & Venables, W.N. (1974). Interval estimates for a ratio of means. *Sankhya* **36**.
[6] Mallows, C.L. (1961). Latent vectors of random symmetric matrices. *Biometrika* **48**, 133–149.
[7] Vaughton, Margaret G. (1970). Analysis of examination results. M.Sc. thesis. The University of Adelaide.

P.R. Krishnaiah, ed. *Multivariate Analysis–IV*
© North-Holland Publishing Company (1977) 79–94

QUADRATIC FORMS AND EXTENSION OF COCHRAN'S THEOREM TO NORMAL VECTOR VARIABLES

C.G. KHATRI

Indian Statistical Institute
and
Gujarat University, Ahmedabad, India

This note investigates the necessary and the sufficient conditions for the quadratic forms $Q_1, \ldots, Q_m$ to be distributed independently and/or as a linear function of independent chi-squares (central or non-central) variables where $Q = \Sigma_{i=1}^m Q_i$ is distributed as the linear function of independent chi-square variables. Thus, this result becomes a generalisation of Cochran's theorem in normal random variables. The results are generalised to complex and real normal vector variables introducing the idea of pseudo-Wishart (central or non-central) variables.

1. Introduction

Let $\boldsymbol{x}$ be a $p \times 1$ random vector distributed as normal with mean vector $\boldsymbol{\mu}$ and covariance matrix $\boldsymbol{V}$. This will be denoted by $\boldsymbol{x} \sim N_p(\boldsymbol{\mu}, \boldsymbol{V})$. Let $\boldsymbol{x}'\boldsymbol{A}\boldsymbol{x} = q$ be a quadratic form in $\boldsymbol{x}$. Let y be distributed as (non-central) chi-square with r degrees of freedom (d.f.) and non-central parameter v. This will be denoted by $y \sim \chi^2(r, v)$. When $\boldsymbol{V} > \boldsymbol{0}$ (i.e. $\boldsymbol{V}$ is positive definite), Baldessari (1967) has established necessary and sufficient conditions for q to be distributed as $\Sigma_{i=1}^k b_i\chi_i^2(r_i, v_i)$ where $b_i \neq b_{i'}$ for $i \neq i'$ and $\chi_1^2, \ldots, \chi_k^2$ are independent non-central chi-squares. In this note, we extend this result to singular $\boldsymbol{V}$ and to a set of quadratic forms q_j ($j = 1, 2, \ldots, m$) to be distributed as $\Sigma_{i=1}^k b_i\chi_{ij}^2(r_{ij}, v_{ij})$ where χ_{ij}^2's are independent (non-central) chi-squares. Here some of the r_{ij}'s may be zero. This leads to the extension of the Cochran's theorem to the more general situation and the modifications of the results due to Graybill and Marsaglia (1957). All these results are extended to normal vector variables giving the extensions of the results of Khatri (1962, 1963) and further, they are extended to complex normal variates.

2. Notation

We have mentioned the meaning of the notations, $\boldsymbol{x} \sim N_p(\boldsymbol{\mu}, \boldsymbol{V})$ and $y \sim \chi^2(r, v)$. The notation, $\boldsymbol{X} \sim N_{p,n}(\boldsymbol{\mu}, \boldsymbol{V}_1, \boldsymbol{V}_2)$ means that if the column vectors of $\boldsymbol{X}$ and $\boldsymbol{\mu}$ are $\boldsymbol{x}_i$ and $\boldsymbol{\mu}_i$, $i = 1, 2, \ldots, n$ respectively, then $\boldsymbol{x}_0 = (\boldsymbol{x}'_1, \ldots, \boldsymbol{x}'_n)' \sim N_{pn}(\boldsymbol{\mu}_0, \boldsymbol{V})$ where $\boldsymbol{\mu}_0 = (\boldsymbol{\mu}'_1, \ldots, \boldsymbol{\mu}'_n)'$, $\boldsymbol{V} = \boldsymbol{V}_2 \otimes \boldsymbol{V}_1$ and $\boldsymbol{A} \otimes \boldsymbol{B} = (a_{ij}\boldsymbol{B})$ is a Kronecker product of $\boldsymbol{A}$ and $\boldsymbol{B}$. Let $\boldsymbol{X}$ be distributed as $N_{p,n}(\boldsymbol{\mu}, \boldsymbol{V}, \boldsymbol{I}_n)$. Then, the distribution of $\boldsymbol{X}\boldsymbol{X}' = \boldsymbol{S}$ will be called (non-central) Wishart and will be denoted by $\boldsymbol{S} \sim W_p(n, \boldsymbol{V}, \boldsymbol{\mu}\boldsymbol{\mu}')$. Here, notice that n is the degrees of freedom (d.f.), $\boldsymbol{V}$ is the scale factor and $\boldsymbol{\mu}\boldsymbol{\mu}'$ will be non-central matrix. If $n < p$, then $\boldsymbol{S}$ has a pseudo-Wishart and if rank of $\boldsymbol{V} < p$, then $\boldsymbol{S}$ has a singular Wishart.

$\boldsymbol{A}^-$, $\boldsymbol{A}'$ and $\zeta(\boldsymbol{A})$ indicate respectively a g-inverse of $\boldsymbol{A}$ (i.e. $\boldsymbol{A}\boldsymbol{A}^-\boldsymbol{A} = \boldsymbol{A}$), the transpose of $\boldsymbol{A}$ and the rank of $\boldsymbol{A}$. $\boldsymbol{A}^+$ will be denoted as the Moore–Penrose inverse of $\boldsymbol{A}$ satisfying (i) $\boldsymbol{A}\boldsymbol{A}^+\boldsymbol{A} = \boldsymbol{A}$, (ii) $\boldsymbol{A}^+\boldsymbol{A}\boldsymbol{A}^+ = \boldsymbol{A}^+$, (iii) $\boldsymbol{A}^+\boldsymbol{A}$ and $\boldsymbol{A}\boldsymbol{A}^+$ are symmetric (or Hermitian) if $\boldsymbol{A}$ is real (or complex). Here, we shall restrict the matrices to real space only. A square matrix $\boldsymbol{A}$ will be said to be semi-simple if there exists a nonsingular matrix $\boldsymbol{P}$ such that $\boldsymbol{P}\boldsymbol{A}\boldsymbol{P}^{-1}$ is a diagonal matrix. Note that every semi-simple matrix $\boldsymbol{A}$ can be represented in the spectral decomposition, namely,

$$\boldsymbol{A} = \sum_{i=1}^{k} w_i \boldsymbol{P}_i \tag{2.1}$$

where w_i's are the nonzero distinct eigen values of $\boldsymbol{A}$, $\boldsymbol{P}_i^2 = \boldsymbol{P}_i$ and $\boldsymbol{P}_i\boldsymbol{P}_{i'} = \boldsymbol{0}$ for $i \neq i' = 1, 2, \ldots, k$. (Here, w_i's may be complex and $\boldsymbol{P}_i$'s can be complex matrices, even when $\boldsymbol{A}$ is a real matrix). A matrix can be represented in the spectral decomposition (2.1) iff it is semi-simple. Let $\boldsymbol{A}$ be a symmetric matrix and $\boldsymbol{V}$ be a positive semi-definite (p.s.d.) matrix (i.e. $\boldsymbol{V} = \boldsymbol{B}\boldsymbol{B}'$). Then, will $\boldsymbol{V}\boldsymbol{A}$ be always semi-simple? The answer to this question is "no"; for example see Mitra and Rao (1968) and more clearly by the following illustration: Take

$$\boldsymbol{V} = \begin{pmatrix} 1 \\ -1 \end{pmatrix} \begin{pmatrix} 1 & -1 \end{pmatrix} \quad \text{and} \quad \boldsymbol{A} = \begin{pmatrix} a & b \\ b & c \end{pmatrix}$$

with $a - b = b - c = 1$.

$$\boldsymbol{V}\boldsymbol{A} = \begin{pmatrix} 1 & 1 \\ -1 & -1 \end{pmatrix} \quad \text{while} \quad \boldsymbol{V}\boldsymbol{A}\boldsymbol{V} = \boldsymbol{0}.$$

A quadratic form q in $\boldsymbol{x}$ will be denoted by $q = \boldsymbol{x}'\boldsymbol{A}\boldsymbol{x}$ where $\boldsymbol{A}$ is a symmetric matrix.

3. Some results in matrix algebra

We have noted in Section 2 that if $\boldsymbol{V}$ is p.s.d. and $\boldsymbol{A}$ is symmetric, then $\boldsymbol{VA}$ may not have a spectral decomposition. The following lemma gives the uniqueness of the spectral decomposition and its proof is omitted.

Lemma 1. *If there exists a spectral decomposition of a matrix, then its spectral representation is unique.*

Lemma 2. *Let* $\boldsymbol{A}$ *be a symmetric matrix and* $\boldsymbol{V} = \boldsymbol{BB}'$ *be a p.s.d. matrix. Then, let the spectral decomposition of* $\boldsymbol{B}'\boldsymbol{AB}$ *be* $\boldsymbol{B}'\boldsymbol{AB} = \Sigma_{j=1}^{k} w_j \boldsymbol{F}_j$. *Then the spectral decomposition of* $\boldsymbol{VAVV}^-$ *is* $\boldsymbol{VAVV}^- = \Sigma_{j=1}^{k} w_j \boldsymbol{E}_j$ *and* $\boldsymbol{E}_j = \boldsymbol{BF}_j\boldsymbol{B}'\boldsymbol{V}^-$, $\boldsymbol{F}_j = \boldsymbol{F}_j\boldsymbol{B}'\boldsymbol{V}^-\boldsymbol{B} = (\boldsymbol{B}'\boldsymbol{V}^-\boldsymbol{B})\boldsymbol{F}_j$ *for each* $j = 1, 2, \ldots, k$.

Proof. Since $\boldsymbol{B}'\boldsymbol{AB} = \Sigma w_j \boldsymbol{F}_j$, so $w_j \boldsymbol{F}_j = \boldsymbol{B}'\boldsymbol{ABF}_j$ i.e.,

$$w_j \boldsymbol{B}'\boldsymbol{V}^-\boldsymbol{BF}_j = \boldsymbol{B}'\boldsymbol{V}^-\boldsymbol{BB}'\boldsymbol{ABF}_j = \boldsymbol{B}'\boldsymbol{ABF}_j = w_j \boldsymbol{F}_j$$

for each j. Similarly, we can have $\boldsymbol{F}_j = \boldsymbol{F}_j(\boldsymbol{B}'\boldsymbol{V}^-\boldsymbol{B})$. This proves Lemma 2, on account of $\boldsymbol{VAVV}^- = \boldsymbol{B}(\boldsymbol{B}'\boldsymbol{AB})\boldsymbol{B}'\boldsymbol{V}^- = \Sigma_{j=1}^{k} w_j \boldsymbol{E}_j$ with $\boldsymbol{E}_j = \boldsymbol{BF}_j\boldsymbol{B}'\boldsymbol{V}^-$, $\boldsymbol{E}_j^2 = \boldsymbol{BF}_j\boldsymbol{B}'\boldsymbol{V}^-\boldsymbol{BF}_j\boldsymbol{B}'\boldsymbol{V}^- = \boldsymbol{E}_j$ for each j and $\boldsymbol{E}_j\boldsymbol{E}_{j'} = 0$ for $i \neq j'$.

Note 1. $\boldsymbol{E}_j\boldsymbol{V} = \boldsymbol{BF}_j\boldsymbol{B}'$ is symmetric and

$$\boldsymbol{B}(\boldsymbol{B}'\boldsymbol{AB} - w_j\boldsymbol{I})^+\boldsymbol{B}' = (\boldsymbol{VAVV}^- - w_j\boldsymbol{I})^+\boldsymbol{V}$$

for $j = 0, 1, \ldots, k$ with $w_0 = 0$, $\boldsymbol{F}_0 = \boldsymbol{I} - \Sigma_{j=1}^{k} \boldsymbol{F}_j$ and $\boldsymbol{E}_0 = \boldsymbol{I} - \Sigma_{j=1}^{k} \boldsymbol{E}_j$.

Note 2. If $k = 2$ in Lemma 2, then $\boldsymbol{E}_i = (\boldsymbol{VA} - w_j\boldsymbol{I})\boldsymbol{VAVV}^-/w_i(w_i - w_j)$ for $(i, j) = (1, 2)$ and $(2, 1)$, and

$$[(\boldsymbol{VA})^2 - (w_1 + w_2)\boldsymbol{VA} + w_1 w_2 \boldsymbol{I}]\boldsymbol{VAV} = \boldsymbol{0}. \tag{3.1}$$

In this case, (3.1) and $\operatorname{tr} \boldsymbol{E}_i = r_i$ $(i = 1, 2)$ imply and are implied by $\zeta(\boldsymbol{VAV}) = r_1 + r_2$ and $\operatorname{tr}(\boldsymbol{VA})^i = w_1^i r_1 + w_2^i r_2$ for $i = 1, 2, 3, 4$.

Note 3. $w_1, \ldots, w_k$, $w_0 = 0$ are distinct eigenvalues of $\boldsymbol{VA}$ with multiplicities $r_1, r_2, \ldots, r_0$ respectively iff

$$\left[\prod_{j=0}^{k} (\boldsymbol{VA} - w_j\boldsymbol{I})\right]\boldsymbol{V} = \boldsymbol{0} \quad \text{and} \quad \zeta\left\{\left[\prod_{\substack{j=0 \\ j \neq j'}}^{k} (\boldsymbol{VA} - w_j\boldsymbol{I})\right]\boldsymbol{V}\right\} = r_j \tag{3.2}$$

for $j \neq j' = 0, 1, 2, \ldots, k$.

Lemma 3. *Let* $A_1, A_2, \ldots, A_m$ *and* A *be square matrices such that* $A = \Sigma_{i=1}^m A_i$. *Consider the following conditions*:

(a) $A_i^2 = A_i$ *for each* $i = 1, 2, \ldots, m$,
(b) $A_i A_{i'} = 0$ *for each* $i \neq i'$ *and* $\zeta(A_i^2) = \zeta(A_i)$ *for each* i,
(c) $A^2 = A$,
(d) $\zeta(A) = \Sigma_{i=1}^m \zeta(A_i)$.

Then, (i) (c) *and* (d) $\Rightarrow$ (a) *and* (*b*) *and* (ii) *any two of* (a), (b) *and* (c) *imply all conditions.*

For a proof, see Khatri (1968) and Rao (1973).

Note 4. If $A_1, \ldots, A_m$ and A are square matrices such that $A = \Sigma_{i=1}^m A_i$ and $\zeta(A) = \Sigma_{i=1}^m \zeta(A_i)$, then $A_i A^- A_i A^- = A_i A^-$, $A^- A_i A^- A_i = A^- A_i$ and $A_i A^- A_{i'} = 0$ for $i \neq i'$, $i, i' = 1, 2, \ldots, m$, and for any g-inverse A^- of A.

Lemma 4. *Let* $D_\lambda^{(1)}$ *and* $D_\lambda^{(2)}$ *be* $\text{diag.}(\lambda_1 I_{r_1}, \ldots, \lambda_k I_{r_k})$ *and* $\text{diag.}(\lambda_1 I_{s_1}, \ldots, \lambda_k I_{s_k})$ *respectively*, $(\lambda_1, \ldots, \lambda_k)$ *being a set of distinct nonzero elements*, $k > 1$ *and* $\Sigma_{i=1}^k r_i = \Sigma_{i=1}^k s_i$. *Let* $X = (X_{ij})$ *be a matrix such that* $X D_\lambda^{(1)} X' = D_\lambda^{(2)}$, X_{ij} *is an* $s_i \times r_j$ *sub-matrix and* $X'X$ *has unit elements in the diagonal places. Then*, $X_{ij} = 0$ *for* $i \neq j$, $r_i = s_i$ *and* X_{ii} *is an orthogonal matrix for each* i *provided either* $k = 2$ *and* λ_j*'s are of the same sign, or* $k = 2, 3$ *and* λ_j*'s are of different signs.*

The following example indicates that Lemma 4 is not true for any k: Let $k = 3$, $D_\lambda^{(1)} = \text{diag}(1, 2, 2, 2, 3)$, $D_\lambda^{(2)} = \text{diag}(1, 1, 2, 3, 3)$, and

$$X = \begin{pmatrix} 2 & 1 & -1 & \cdot & \cdot \\ 2 & -1 & 1 & \cdot & \cdot \\ \cdot & \cdot & \cdot & 2\sqrt{2} & \cdot \\ \cdot & \sqrt{3} & \sqrt{3} & \cdot & 2 \\ \cdot & \sqrt{3} & \sqrt{3} & \cdot & -2 \end{pmatrix} \Big/ 2\sqrt{2}.$$

Proof. Let $k = 2$ and λ_1, λ_2 both have the same sign. Without loss of generality, assume that $\lambda_1 > \lambda_2 > 0$. Let $Y = (D_\lambda^{(2)})^{-1/2} X (D_\lambda^{(1)})^{1/2}$. Then

$$YY' = I \text{ or } Y'Y = I. \tag{3.4}$$

Partitioning $Y = (Y_{ij})$ according as $X = (X_{ij})$, and the diagonal elements of $X'X = (D_\lambda^{(1)})^{-1/2}(Y'D^{(2)}Y)(D_\lambda^{(1)})^{-1/2}$ are unities, we get for $(i,j) = (1,2)$ and $(2,1)$,

$$\lambda_i r_i = \lambda_i \operatorname{tr}(X'_{ii}X_{ii} + X'_{ij}X_{ij}) = \lambda_i \operatorname{tr}(Y'_{ii}Y_{ii}) + \lambda_j \operatorname{tr}(Y'_{ij}Y_{ij}) \tag{3.5}$$

and

$$Y'_{ii}Y_{ii} + Y'_{ij}Y_{ij} = I_{r_i}. \tag{3.6}$$

Using (3.6) in (3.5), we get

$$\operatorname{tr}(Y'_{ij}Y_{ij}) = 0 \quad \text{or} \quad Y_{ij} = 0 \quad \text{for} \quad (i,j) = (1,2) \text{ and } (2,1). \tag{3.7}$$

(3.7) shows that for $(i,j) = (1,2)$ and $(2,1)$

$$r_i = s_i, \qquad X_{ij} = 0, \qquad X_{ii} \text{ is orthogonal.} \tag{3.8}$$

This proves the first part. For the second part, let us assume without loss of generality that there is only one negative value of λ and others are positive. Suppose λ_k is negative and $\lambda_1 > \lambda_2 > \ldots > \lambda_{k-1} > 0$. Then, using the Sylvester's law of inertia for quadratic forms connected with matrices $XD_\lambda^{(1)}X'$ and $D_\lambda^{(2)}$, we must have $r_1 + \ldots + r_{k-1} = s_1 + \cdots + s_{k-1} = n_1$ (say) and $r_k = s_k = n$ (say). Write

$$D_\lambda^{(i)} = \operatorname{diag}(D_1^{(i)}, \lambda_k I_n), \qquad D^{(i)} = \operatorname{diag}((D_1^{(i)})^{1/2}, \sqrt{-\lambda_k I_n})$$

and

$$B = \operatorname{diag}(I_{n_1}, -I_n).$$

Then, $Y = (D^{(2)})^{-1}X(D^{(1)})$ in $XD_\lambda^{(1)}X' = D_\lambda^{(2)}$ gives

$$YBY' = B \text{ or } BYBY' = I \text{ or } Y'BYB = I. \tag{3.9}$$

Writing

$$Y = \left(\begin{array}{c|c} Y_1 & Y_2 \\ \hline Y_3 & Y_4 \end{array}\right)\begin{matrix} n_1 \\ n \end{matrix} \quad \text{and} \quad X = \left(\begin{array}{c|c} X_1 & X_2 \\ \hline X_3 & X_4 \end{array}\right)\begin{matrix} n_1 \\ n \end{matrix}$$
$$\qquad n_1 \quad n \qquad\qquad\qquad\qquad n_1 \quad n$$

in (3.9) and $X'X = (D^{(1)})^{-1}(Y'(D^{(2)})^2Y)(D^{(1)})^{-1}$, we get

$$(-\lambda_k)(X'_4X_4 + X'_2X_2) = (-\lambda_k)Y'_4Y_4 + Y'_2D_1^{(2)}Y_2 \tag{3.10a}$$

and

$$Y'_4Y_4 = I_n + Y'_2Y_2. \tag{3.10b}$$

Taking traces in (3.10) and then using (3.10b) in (3.10a), we get

$$(-\lambda_k)\operatorname{tr} Y'_2Y_2 + \operatorname{tr}(Y'_2D_1^{(2)}Y_2) = 0 \text{ or } Y_2 = 0 \tag{3.11}$$

because $-\lambda_k > 0$ and $\boldsymbol{Y}_2'\boldsymbol{D}_1^{(2)}\boldsymbol{Y}_2$ is positive semi-definite. Using (3.11) in $\boldsymbol{YBY}' = \boldsymbol{B}$ gives $\boldsymbol{Y}_3 = \boldsymbol{0}$ because $\boldsymbol{Y}_4'\boldsymbol{Y}_4 = \boldsymbol{I}_n$. Thus, (3.9) reduces to $\boldsymbol{Y}_1'\boldsymbol{Y}_1 = \boldsymbol{I}_{n_1}$ and $\boldsymbol{X}_1\boldsymbol{D}_1^{(1)}\boldsymbol{X}' = \boldsymbol{D}_1^{(2)}$ which is the equation for $k-1$. Now, using the proof for the first part, we get the required Lemma 4.

Lemma 5. *Let $\boldsymbol{A}_1, \boldsymbol{A}_2, \ldots, \boldsymbol{A}_m$, $\boldsymbol{A}$ be $n \times n$ symmetric matrices such that $\boldsymbol{A} = \Sigma_{i=1}^m \boldsymbol{A}_i$ and their spectral decompositions are*

$$\boldsymbol{A}_i = \sum_{j=1}^{k} \lambda_j \boldsymbol{E}_{ij}, \qquad \boldsymbol{E}_{ij}^2 = \boldsymbol{E}_{ij}, \qquad \boldsymbol{E}_{ij}\boldsymbol{E}_{ij'} = \boldsymbol{0} \quad \text{for } j \neq j'$$

and

$$\boldsymbol{A} = \sum_{j=1}^{k} \lambda_j \boldsymbol{E}_j, \qquad \boldsymbol{E}_j^2 = \boldsymbol{E}_j, \qquad \boldsymbol{E}_j\boldsymbol{E}_{j'} = \boldsymbol{0} \quad \text{for } j \neq j'.$$

(Here, some of the $\boldsymbol{E}_{ij}$'s may be null). If $\zeta(\boldsymbol{A}) = \Sigma_{i=1}^m \zeta(\boldsymbol{A}_i)$ and either $k = 2$ and λ_j's have the same sign, or $k = 2, 3$ and λ_j's have different signs, then $\boldsymbol{E}_{ij}\boldsymbol{E}_{i'j'} = \boldsymbol{0}$ for either ($i \neq i'$ and $j, j' = 1, 2, \ldots, k$) or ($i, i' = 1, 2, \ldots, m$ and $j \neq j'$), and $\boldsymbol{E}_{ij} = \boldsymbol{A}_i\boldsymbol{E}_j/\lambda_j = \boldsymbol{E}_j\boldsymbol{A}_i/\lambda_j$ for all i, j. Note that $\boldsymbol{A}_i\boldsymbol{A}_{i'} = \boldsymbol{0}$ for $i \neq i' = 1, 2, \ldots, m$.

If $k \geq 3$ and λ_j's have the same sign, the Lemma 5 may not be true. This can be illustrated by the example given in Lemma 4. Let j-th column of $\boldsymbol{X}$ (given below Lemma 4) be $\boldsymbol{x}_j$ ($j = 1, 2, 3, 4, 5$).

Let $\boldsymbol{A}_1 = \boldsymbol{x}_1\boldsymbol{x}_1'$, $\boldsymbol{A}_i = 2\boldsymbol{x}_i\boldsymbol{x}_i'$ ($i = 2, 3, 4$) and $\boldsymbol{A}_5 = 3\boldsymbol{x}_5\boldsymbol{x}_5'$. $\boldsymbol{A} = \Sigma_{i=1}^5 \boldsymbol{A}_i$. Then $\zeta(A) = \Sigma_{i=1}^5 \zeta(A_i) = 5$ and Lemma 5 does not hold if the condition on k is omitted.

Proof. Since $\boldsymbol{E}_{ij}$ and $\boldsymbol{E}_j$ are idempotent symmetric matrices, we can write $\boldsymbol{E}_{ij} = \boldsymbol{F}_{ij}\boldsymbol{F}_{ij}'$ and $\boldsymbol{E}_j = \boldsymbol{F}_j\boldsymbol{F}_j'$ where $\boldsymbol{F}_{ij}'\boldsymbol{F}_{ij} = \boldsymbol{I}_{r_{ij}}$ and $\boldsymbol{F}_j'\boldsymbol{F}_j = \boldsymbol{I}_{r_j}$ for each i, j, $r_{ij} = \zeta(\boldsymbol{E}_{ij})$ and $r_j = \zeta(\boldsymbol{E}_j)$. Let $\boldsymbol{D}_\lambda^{(2)} = \operatorname{diag}(\lambda_1 \boldsymbol{I}_{r_1}, \ldots, \lambda_k \boldsymbol{I}_{r_k})$, $\boldsymbol{F} = (\boldsymbol{F}_1, \ldots, \boldsymbol{F}_k)$, $\boldsymbol{Q} = (\boldsymbol{Q}_1, \ldots, \boldsymbol{Q}_k)$, $\boldsymbol{Q}_j = (\boldsymbol{F}_{1j}, \boldsymbol{F}_{2j}, \ldots, \boldsymbol{F}_{mj})$, $\boldsymbol{P}_j = \Sigma_{i=1}^m \boldsymbol{E}_{ij} = \boldsymbol{Q}_j\boldsymbol{Q}_j'$ and $s_j = \Sigma_{i=1}^m r_{ij}$ for $j = 1, 2, \ldots, k$. Then, it is easy to see that

$$\boldsymbol{A} = \boldsymbol{Q}\boldsymbol{D}_\lambda^{(2)}\boldsymbol{Q}' = \boldsymbol{F}\boldsymbol{D}_\lambda^{(1)}\boldsymbol{F}', \qquad \boldsymbol{F}'\boldsymbol{F} = \boldsymbol{I}_r \tag{3.12}$$

and

$$\text{diagonal elements of } \boldsymbol{Q}'\boldsymbol{Q} \text{ are unities.} \tag{3.13}$$

Now, on account of rank condition,

$$r = \zeta(\boldsymbol{A}) = \zeta(\boldsymbol{Q}) \leq \sum_{j=1}^{k} \zeta(\boldsymbol{Q}_j) \leq \sum_{j=1}^{k} \sum_{i=1}^{m} r_{ij}$$

$$= \sum_{i=1}^{m} \left(\sum_{j=1}^{k} r_{ij} \right) = \sum_{i=1}^{m} \zeta(\boldsymbol{A}_i) = r,$$

we get

$$r = \sum_{j=1}^{k} r_j = \sum_{j=1}^{k} s_j, \qquad \zeta(\boldsymbol{P}_j) = \sum_{i=1}^{m} \zeta(\boldsymbol{E}_{ij}) \quad \text{for each } j. \tag{3.14}$$

Using (3.14) in (3.12), we get

$$\boldsymbol{X}\boldsymbol{D}_\lambda^{(2)}\boldsymbol{X}' = \boldsymbol{D}_\lambda^{(1)}, \ \boldsymbol{X} = \boldsymbol{F}'\boldsymbol{Q} \text{ is nonsingular and diagonal elements of } \boldsymbol{X}'\boldsymbol{X} \text{ are unities.} \tag{3.15}$$

Using Lemma 4 in (3.15), we get for $j, j' = 1, 2, \ldots, k$,

$$r_j = s_j, \ \boldsymbol{F}_j'\boldsymbol{Q}_j \text{ is orthogonal and } \boldsymbol{F}_j'\boldsymbol{Q}_{j'} = \boldsymbol{0} \text{ for } j \neq j'. \tag{3.16}$$

Using (3.16) in (3.12), we get after some simplifications

$$\boldsymbol{E}_j = \boldsymbol{E}_j\boldsymbol{P}_j = \boldsymbol{P}_j\boldsymbol{E}_j = \boldsymbol{P}_j \ \text{ for } j = 1, 2, \ldots, k. \tag{3.17}$$

Hence idempotency of

$$\sum_{j=1}^{k} \boldsymbol{E}_j = \sum_{i=1}^{m}\sum_{j=1}^{k} \boldsymbol{E}_{ij} \quad \text{and} \quad \zeta\Big(\sum_{j=1}^{k} \boldsymbol{E}_j\Big) = \sum_{i=1}^{m}\sum_{j=1}^{m} \zeta(\boldsymbol{E}_{ij})$$

give the required result with the help of Lemma 3(i).

Note 5. In Lemma 5, $\zeta(\boldsymbol{A}) = \Sigma_{i=1}^m \zeta(\boldsymbol{A}_i)$ is an essential condition for $k > 1$. The following two sets of matrices

$$2\boldsymbol{A}_1 = \begin{pmatrix} 1 & \sqrt{3} \\ \sqrt{3} & -1 \end{pmatrix}, \quad 2\boldsymbol{A}_2 = \begin{pmatrix} 1 & -\sqrt{3} \\ -\sqrt{3} & -1 \end{pmatrix}, \quad \boldsymbol{A} = \begin{pmatrix} 1 & \cdot \\ \cdot & -1 \end{pmatrix} = \boldsymbol{A}_1 + \boldsymbol{A}_2$$

or

$$\boldsymbol{A}_1 = \operatorname{diag}(2, 0, 2), \qquad \boldsymbol{A}_2 = \operatorname{diag}(0, 2, 2), \qquad \boldsymbol{A} = \operatorname{diag}(2, 2, 4) = \boldsymbol{A}_1 + \boldsymbol{A}_2$$

do not satisfy this rank condition and $\boldsymbol{A}_1\boldsymbol{A}_2 \neq \boldsymbol{0}$.

Note 6. When $k = 1$, we do not need the condition $\zeta(\boldsymbol{A}) = \Sigma_{i=1}^m \zeta(\boldsymbol{A}_i)$, because it is automatically satisfied.

Note 7. When $\boldsymbol{A}_i$'s are not symmetric in Lemma 5, then results of Lemma 5 are no longer true. Consider,

$$\boldsymbol{A}_1 = \begin{pmatrix} 1 & 1 \\ 0 & 0 \end{pmatrix}, \quad \boldsymbol{A}_2 = \begin{pmatrix} 0 & -1 \\ 0 & -1 \end{pmatrix}, \quad \boldsymbol{A}_1 + \boldsymbol{A}_2 = \boldsymbol{A} = \begin{pmatrix} 1 & 0 \\ 0 & -1 \end{pmatrix}.$$

Lemma 6. (a) $\zeta(\boldsymbol{BC}) = \rho(\boldsymbol{C})$ *iff* $\boldsymbol{BCx} = \boldsymbol{0}$ *for all possible non-null vectors* $\boldsymbol{x} \Rightarrow \boldsymbol{Cx} = \boldsymbol{0}$,

(b) $\zeta(\boldsymbol{BC})=\zeta(\boldsymbol{B})$ *iff* $\boldsymbol{C}'\boldsymbol{B}'\boldsymbol{y}=\boldsymbol{0}$ *for all possible non-null vectors* $\boldsymbol{y} \Rightarrow \boldsymbol{B}'\boldsymbol{y}=\boldsymbol{0}$.

This follows immediately from the Frobenius theorem on ranks, namely

$$\zeta(\boldsymbol{BCP}) \geq \zeta(\boldsymbol{BC})+\zeta(\boldsymbol{CP})-\zeta(\boldsymbol{C}).$$

Corollary 1. *Let* $\boldsymbol{A}=\Sigma_{i=1}^{m}\boldsymbol{A}_i$. *Then* $\zeta(\boldsymbol{A})=\zeta(\boldsymbol{A}_1,\ldots,\boldsymbol{A}_m)$ *iff* $\boldsymbol{Ax}=\boldsymbol{0}$ *for all possible non-null vectors* $\boldsymbol{x} \Rightarrow \boldsymbol{A}_i'\boldsymbol{x}=\boldsymbol{0}$ *for all i.*

Lemma 7. *Let* $\boldsymbol{A}, \boldsymbol{A}_1, \boldsymbol{A}_2, \ldots, \boldsymbol{A}_m$ *be* $n\times n$ *square matrices such that* $\boldsymbol{A}=\Sigma_{i=1}^{m}\boldsymbol{A}_i$ *can be represented in the spectral decomposition by* $\boldsymbol{A}=\Sigma_{j=1}^{k}\lambda_j\boldsymbol{E}_j$ *for* $k>1$. *Then, there exist idempotent matrices* $\boldsymbol{E}_{ij}$ *such that* $\boldsymbol{A}_i=\Sigma_{j=1}^{k}\lambda_j\boldsymbol{E}_{ij}$ *and* $\boldsymbol{E}_{ij}\boldsymbol{E}_{i'j'}=\boldsymbol{0}$ *for either* $i\neq i'$ *or* $j\neq j'$ *(some* $\boldsymbol{E}_{ij}$*'s may be zero) iff* $\zeta(\boldsymbol{A})=\zeta(\boldsymbol{A}_1',\ldots,\boldsymbol{A}_m')'=\Sigma_{i=1}^{m}\Sigma_{j=1}^{k}\zeta(\boldsymbol{A}_i\boldsymbol{E}_j)$.

Proof. If part: Follows from Lemmas 3(i) and 6 after noting the idempotency of $\boldsymbol{AA}^+=\Sigma_{i=1}^{m}\Sigma_{j=1}^{k}\boldsymbol{A}_i\boldsymbol{E}_j/\lambda_j$ and by defining $\boldsymbol{E}_{ij}=\boldsymbol{A}_i\boldsymbol{E}_j/\lambda_j$.

Only if: Follows immediately by noting the idempotency and orthogonality of matrices, $\Sigma_{i=1}^{m}\boldsymbol{E}_{ij}$ for $j=1,2,\ldots,k$.

Lemma 8. *Let* $\boldsymbol{A}_1, \boldsymbol{A}_2, \ldots, \boldsymbol{A}_m, \boldsymbol{A}$ *be* $n\times n$ *matrices such that* $\Sigma_{i=1}^{m}\boldsymbol{A}_i=\boldsymbol{A}$. *Consider the following statements: for* $k>1$,

(a) $\boldsymbol{A}_i=\Sigma_{j=1}^{k}\lambda_j\boldsymbol{E}_{ij}$ *for* $i=1,2,\ldots,m$, $\boldsymbol{E}_{ij}^2=\boldsymbol{E}_{ij}$ *and* $\boldsymbol{E}_{ij}\boldsymbol{E}_{ij'}=\boldsymbol{0}$ *for* $j\neq j'$, *(some* $\boldsymbol{E}_{ij}$*'s may be null).*

(b) $\boldsymbol{A}_i\boldsymbol{A}_{i'}=\boldsymbol{0}$ *for* $i\neq i'=1,2,\ldots,$ *and* $\zeta(\boldsymbol{A}_i^2)=\zeta(\boldsymbol{A}_i)$ *for each i.*

(c) $\boldsymbol{A}=\Sigma_{j=1}^{k}\lambda_j\boldsymbol{E}_j$, $\boldsymbol{E}_j^2=\boldsymbol{E}_j$ *and* $\boldsymbol{E}_j\boldsymbol{E}_{j'}=0$ *for* $j\neq j'$, *and*

(d) $\rho(\boldsymbol{A})=\Sigma_{i=1}^{m}\Sigma_{j=1}^{k}\rho(\boldsymbol{A}_i\boldsymbol{E}_j)=\rho(\boldsymbol{A}_1':\boldsymbol{A}_2':\ldots:\boldsymbol{A}_m')'$, *where* $\lambda_1,\lambda_2,\ldots,\lambda_k$ *are distinct and nonzero elements.*

Then (i) (a) *and* (b) $\Rightarrow$ (c) *and* (d), (ii) (b) *and* (c) $\Rightarrow$ (a) *and* (d), *and* (iii) (c) *and* (d) $\Rightarrow$ (a) *and* (b).

Proof. (i) Let (a) and (b) be given. Take $\boldsymbol{P}_j=\Sigma_{i=1}^{m}\boldsymbol{E}_{ij}$ and $\boldsymbol{A}=\Sigma_{j=1}^{k}\lambda_j\boldsymbol{P}_j$. Further, it is easy to see that (a) and (b) implies $\boldsymbol{P}_j^2=\boldsymbol{P}_j$ and $\boldsymbol{P}_j\boldsymbol{P}_{j'}=\boldsymbol{0}$ for $j\neq j'$, i.e. (a) and (b) $\Rightarrow$ (c) and $\zeta(\boldsymbol{P}_j)=\Sigma_{i=1}^{m}\zeta(\boldsymbol{E}_{ij})$. Noting $\boldsymbol{E}_{ij}=\boldsymbol{A}_i\boldsymbol{P}_j/\lambda_j$ and $\zeta(\boldsymbol{A})=\Sigma_{j=1}^{k}\zeta(\boldsymbol{P}_j)$, we get (d). This proves (i).

(ii) Let (b) and (c) be given. Then it is easy to see that

$$\boldsymbol{A}_i^2\boldsymbol{E}_j=\lambda_j\boldsymbol{A}_i\boldsymbol{E}_j \quad \text{and} \quad \boldsymbol{E}_j\boldsymbol{A}_i^2=\lambda_j\boldsymbol{E}_j\boldsymbol{A}_i. \tag{3.18}$$

This gives

$$\boldsymbol{E}_j\boldsymbol{A}_i\boldsymbol{E}_{j'}=\boldsymbol{0} \text{ for } j\neq j',$$

and

$$E_jA_iE_j = E_jA_i = A_iE_j.$$

Hence, defining

$$A_iE_j = \lambda_j E_{ij}, \tag{3.19}$$

it is easy to see that

$$A_iE_{ij} = \lambda_i E_{ij}, \qquad E_{ij}^2 = E_{ij}, \qquad E_{ij}E_{ij'} = 0 \quad \text{for } j \neq j'.$$

(3.19) gives

$$A_i = \sum_{j=1}^{k} \lambda_j E_{ij} \quad \text{provided } A_i = A_iE_0, \quad E_0 = \sum_{j=1}^{k} E_j.$$

To see this, (3.18) gives $A_i^2E_0 = A_iA = A_i^2$ and $\zeta(A_i^2) = \zeta(A_i)$, gives the required result. Thus, (b) and (c) $\Rightarrow$ (a) and then we get (d). This proves (ii).

(iii) is established by Lemma 7. Thus Lemma 8 is established.

Lemma 9. *Let $A_1, A_2, \ldots, A_m$ and A be symmetric matrices such that $A = \Sigma_{i=1}^m A_i$. Consider the following statements with the distinct nonzero elements $\lambda_1, \ldots, \lambda_k$:*

(a) $A_i = \Sigma_{j=1}^k \lambda_j E_{ij}$, $E_{ij}^2 = E_{ij}$ *and* $E_{ij}E_{ij'} = 0$ *for* $j \neq j'$ *for* $k > 1$ *and all i,*

(b) $A_iA_{i'} = 0$ *for every* $i \neq i'$,

(c) $A = \Sigma_{j=1}^k \lambda_j E_j$, $E_j^2 = E_j$ *and* $E_jE_{j'} = 0$ *for* $j \neq j'$, $k > 1$,

(d) $\zeta(A) = \Sigma_{i=1}^m \zeta(A_i)$, *and*

(e) $\zeta(A) = \Sigma_{i=1}^m \Sigma_{j=1}^k \zeta(A_iE_j)$, $k > 1$.

Then, (i) (a) and (b) $\Rightarrow$ all conditions, (ii) (b) and (c) $\Rightarrow$ all conditions, (iii) (c), (d), (e) $\Rightarrow$ all conditions and (iv) (a), (c) and (d) $\Rightarrow$ all conditions either for $k = 2$ if λ_j's are of the same sign, or for $k = 2, 3$ if λ_j's are of different signs.

This follows from Lemmas 8 and 5. When $k = 1$, the result has been established by Graybill and Marsaglia (1957) and Khatri (1968) (see Lemma 3).

Lemma 10. *Let $A_1, A_2, \ldots, A_m$ and A be symmetric matrices such that $A = \Sigma_{i=1}^m A_i$. Consider the following statements:*

(a) $A_i^3 = A_i$ *for each i,*

(b) $A_iA_{i'} = 0$ *for* $i \neq i'$,

(c) $A^3 = A$,

(d) $\zeta(A) = \Sigma_{i=1}^m \zeta(A_i)$, *and*

(e) $\zeta(A) = \Sigma_{i=1}^m \zeta[A_iA(A + I)] + \Sigma_{i=1}^m \zeta[A_iA(A - I)]$.

Then, (i) (a) *and* (b) $\Rightarrow$ *all conditions,* (ii) (b) *and* (c) $\Rightarrow$ *all conditions,* (iii) (c), (d) *and* (e) $\Rightarrow$ *all conditions and* (iv) (a), (c) *and* (d) $\Rightarrow$ *all conditions.*

This follows from Lemma 9 by noting $\boldsymbol{E}_1 = (\boldsymbol{A}^2 + \boldsymbol{A})/2$, $\boldsymbol{E}_2 = (\boldsymbol{A}^2 - \boldsymbol{A})/2$ and $\boldsymbol{A} = \boldsymbol{E}_1 - \boldsymbol{E}_2$.

Note 8. In tripotent matrices, condition like (e) is important. The conditions (c) and (d) are satisfied for the following matrices:

$$\boldsymbol{A}_1 = \frac{1}{1-a^2}\begin{pmatrix} 1 & a \\ a & a^2 \end{pmatrix}, \quad \boldsymbol{A}_2 = \frac{1}{1-a^2}\begin{pmatrix} -a^2 & -a \\ -a & -1 \end{pmatrix}, \quad \boldsymbol{A}_1 + \boldsymbol{A}_2 = \begin{pmatrix} 1 & \cdot \\ \cdot & -1 \end{pmatrix}$$

but the condition (e) is not satisfied. This is true for Lemma 9.

Note 9. In Lemmas 9 and 10, if $\boldsymbol{A}_i$'s are not symmetric matrices with real elements, then results (i), (ii) and (iii) are valid, but we cannot say about the result (iv), (see also note 7).

4. Quadratic forms and extension of Cochran's theorem

The following two lemmas will be useful in the development of the results on quadratic forms.

Lemma 11. *Let* $y_1, \ldots, y_k$ *and* $\boldsymbol{x}$ *be independently distributed,* $\boldsymbol{x} \sim N(\boldsymbol{0}, \boldsymbol{I})$ *and* $y_j \sim \chi^2(r_j, v_j)$. *Then* $\Sigma_{j=1}^{k} w_j y_j + \boldsymbol{l}'\boldsymbol{x} + c$ *is distributed as* $\Sigma_{i=1}^{m} \lambda_i \chi_i^2(s_i, \alpha_i)$, *where* $(w_1, \ldots, w_k)$ *and* $(\lambda_1, \ldots, \lambda_m)$ *are sets of distinct nonzero elements and* $\chi_i^2(s_i, \alpha_i)$, $i = 1, 2, \ldots, m$, *are independent non-central chi-squares, iff* $k = m$, $\boldsymbol{l} = \boldsymbol{0}$, $c = 0$, *and by renaming (if necessary),* $s_j = r_j$, $\lambda_j = w_j$ *and* $\alpha_j = v_j$ *for* $j = 1, 2, \ldots, k$.

Proof. Using the characteristic function arguments, the necessary and sufficient condition is, for all real t,

$$\left(\prod_{j=1}^{k} (1 - 2itw_j)^{-r_j/2}\right) \exp\left[-\sum_{j=1}^{k} itv_j(1 - 2it^2 w_j)^{-1} + itc - t^2\boldsymbol{l}'\boldsymbol{l}/2\right] =$$

$$= \left(\prod_{i=1}^{m} (1 - 2it\lambda_i)^{-s_i/2}\right) \exp\left[-\sum_{i=1}^{m} it\alpha_i(1 - 2it\lambda_i)^{-1}\right]. \tag{4.1}$$

Then, arguing as in the previous papers (see Khatri (1962, 1963) and Laha (1956)), we have, for all real t,

$$\prod_{j=1}^{k}(1-2itw_j)^{r_j}=\prod_{i=1}^{m}(1-2it\lambda_i)^{s_i} \text{ and} \tag{4.2a}$$

$$\sum_{j=1}^{k} v_j(1-2itw_j)^{-1}-c-itl'l/2=\sum_{i=1}^{m}\alpha_i(1-2it\lambda_i)^{-1}. \tag{4.2b}$$

From (4.2), we get the required result.

Lemma 12. *Let* $\boldsymbol{x}$ *be* $N(\boldsymbol{\mu}, \boldsymbol{V})$ *and let* $q_i = \boldsymbol{x}'\boldsymbol{A}_i\boldsymbol{x}$, $i = 1, 2$, *be two quadratic forms. Then,* q_1 *and* q_2 *are independently distributed iff* (i) $\boldsymbol{VA}_1\boldsymbol{VA}_2\boldsymbol{V} = \boldsymbol{VA}_2\boldsymbol{VA}_1\boldsymbol{V} = \boldsymbol{0}$, (ii) $\boldsymbol{VA}_2\boldsymbol{VA}_1\boldsymbol{\mu} = \boldsymbol{VA}_1\boldsymbol{VA}_2\boldsymbol{\mu} = \boldsymbol{0}$ *and* (iii) $\boldsymbol{\mu}'\boldsymbol{A}_1\boldsymbol{VA}_2\boldsymbol{\mu} = 0$.

This has been established by Khatri (1963) and also see Rao (1973), and Rao and Mitra (1971).

Note 10. If $\boldsymbol{\mu} = \boldsymbol{Vd}$ for some $\boldsymbol{d}$, then (ii) and (iii) are satisfied and we have only one condition (i).

Note 11. If $\boldsymbol{VA}_1\boldsymbol{\mu} = \boldsymbol{VA}_1\boldsymbol{Vd}$ for some $\boldsymbol{d}$, then we have only one condition (i).

Note 12. If $\boldsymbol{V}$ is nonsingular, then we have only one condition $\boldsymbol{A}_1\boldsymbol{VA}_2 = \boldsymbol{A}_2\boldsymbol{VA}_1 = \boldsymbol{0}$.

Corollary 2. *Let* $\boldsymbol{x}$ *be* $N(\boldsymbol{\mu}, \boldsymbol{V})$ *and let* $q_i = \boldsymbol{x}'\boldsymbol{A}_i\boldsymbol{x}$ *be quadratic forms* $i = 1, 2, \dots, k$. *Then* $q_1, \dots, q_k$ *are independently distributed iff any two of* $q_1, q_2, \dots, q_k$ *are independently distributed.*

Theorem 1. *Let* $\boldsymbol{x}$ *be* $N(\boldsymbol{\mu}, \boldsymbol{V})$ *and let* $q = \boldsymbol{x}'\boldsymbol{Ax} + 2\boldsymbol{l}'\boldsymbol{x} + c$ *be a quadratic form. Then,* $q \sim \sum_{j=1}^{k} \lambda_j \chi_j^2(r_j, v_j)$, *where* $\lambda_1, \dots, \lambda_k$ *are distinct nonzero elements and* $\chi_j^2(r_j, v_j)$, $j = 1, 2, \dots, k$ *are independent non-central chi-squares, iff* (i) $\lambda_1, \lambda_2, \dots, \lambda_k$ *are distinct nonzero eigen values of* $\boldsymbol{VA}$ *with multiplicities* $r_1, r_2, \dots, r_k$, (ii) $\boldsymbol{V}(\boldsymbol{A\mu} + \boldsymbol{l}) = \boldsymbol{VAV}(\boldsymbol{VAV})^-\boldsymbol{V}(\boldsymbol{A\mu} + \boldsymbol{l})$ *and* (iii) $v_j = (\boldsymbol{l}' + \boldsymbol{\mu}'\boldsymbol{A})[\boldsymbol{V} - (\boldsymbol{VAVV}^- - \lambda_j\boldsymbol{I})^+(\boldsymbol{VAV} - \lambda_j\boldsymbol{V})](\boldsymbol{A\mu} + \boldsymbol{l})/\lambda_j^2$ *for* $j = 1, 2, \dots, k$ *with*

$$\textstyle\sum_{j=1}^{k} \lambda_j v_j = \boldsymbol{\mu}'\boldsymbol{A\mu} + 2\boldsymbol{l}'\boldsymbol{\mu} + c = (\boldsymbol{l} + \boldsymbol{A\mu})'(\boldsymbol{VAVV}^-)^+\boldsymbol{V}(\boldsymbol{l} + \boldsymbol{A\mu}).$$

Proof. Let $\boldsymbol{V}$ be of rank r. Then, we can write $\boldsymbol{V} = \boldsymbol{BB}'$ where $\boldsymbol{B}$ is a $p \times r$ matrix of rank r. Then with probability one,

$$x = By + \mu, \qquad y \sim N(0, I). \tag{4.3}$$

Since $B'AB$ is a symmetric matrix, there exists an orthogonal matrix $C = (C_1, \ldots, C_m, C_0)$ such that

$$B'AB = \sum_{i=1}^{m} w_i C_i C_i', \; w_i\text{'s being the distinct eigenvalues of } VA. \tag{4.4}$$

Define $y'C = z' = (z_1', \ldots, z_m', z_0')$. Then, using (4.3) and (4.4) in q, we can write

$$q = \sum_{i=1}^{m} w_i y_i + l_0' z_0 + c_0 \tag{4.5}$$

where $y_1, \ldots, y_m$, z_0 are independent, $z_0 \sim N(0, I)$, $y_i \sim \chi^2(s_i, \alpha_i)$, $l_0 = 2C_0'B'(A\mu + l)$, s_i = multiplicity of w_i, $\alpha_i = (l' + \mu'A)BC_iC_i'B'(A\mu + l)/w_i^2$ and $c_0 = \mu'A\mu + 2\mu'l + c - \Sigma_{i=1}^m w_i\alpha_i$. Now, applying Lemma 11, we get the required result by noting

$$l_0 = 0 \Leftrightarrow V(A\mu + l) = VAV(VAV)^- V(A\mu + l)$$

$$\sum_{i=1}^{m} w_i\alpha_i = (l + A\mu)'(VAVV^-)^+ V(A\mu + l) \text{ and}$$

$$\alpha_i = (l + A\mu)'[V - (VAVV^- - w_iI)^+(VAV - w_iV)](A\mu + l)/w_i^2.$$

Note 13. Using note 3, condition (i) of Theorem 1 can be written in an alternative form given by (3.2).

Note 14. When $c = 0$, and $l = 0$ and $\mu = Vd$ for some d, then conditions (ii) and (iii) are satisfied and we have only one condition (i). In this case, v_j can be rewritten as $v_j = \mu'V^-E_j\mu$ where the spectral decomposition of $VAVV^-$ is

$$VAVV^- = \sum_{j=1}^{k} \lambda_j E_j, \; E_j^2 = E_j \quad \text{and} \quad E_jE_{j'} = 0 \quad \text{for } j \neq j'.$$

This is true when V is nonsingular, (see Baldessari (1967)).

Note 15. In Theorem 1, if A is positive semi-definite $c = l_0'Al_0$ and $l = Al_0$ some l_0, then the conditions (ii) and (iii) are automatically satisfied and we require only condition (i), because $V(VAV)^-$ and $V(VAV)^-V$ are respectively g-inverses of VA and A respectively giving $VA\,V(VAV)^-VA = VA$ and $A\,V(VAV)^-VA = A$.

Note 16. When $k = 1$, then the conditions reduce to (i) $\boldsymbol{VAVAV} = \lambda \boldsymbol{VAV}$, (ii) $\boldsymbol{V}(\boldsymbol{A\mu} + \boldsymbol{l}) = \boldsymbol{VAV}(\boldsymbol{A\mu} + \boldsymbol{l})$ and (iii) $v\lambda = \boldsymbol{\mu}'\boldsymbol{A\mu} + 2\boldsymbol{l}'\boldsymbol{\mu} + c = (\boldsymbol{l} + \boldsymbol{A\mu})'V(\boldsymbol{l} + \boldsymbol{A\mu})/\lambda$. Further, if $\boldsymbol{l} = \boldsymbol{0}$, $c = 0$ and $\boldsymbol{\mu} = \boldsymbol{Vd}$, then there is only one condition (i). This result was established by Khatri (1963).

Note 17. Let $k = 2$ in Theorem 1. Then, the three conditions can be rewritten as

$$\text{(i)}\ (\boldsymbol{VA})^3\boldsymbol{V} - (\lambda_1 + \lambda_2)(\boldsymbol{VA})^2\boldsymbol{V} + \lambda_1\lambda_2\boldsymbol{VAV} = \boldsymbol{0},$$

$$r_1 = (\mathrm{tr}(\boldsymbol{VA})^2 - \lambda_2\,\mathrm{tr}(\boldsymbol{VA}))/\lambda_1(\lambda_1 - \lambda_2) \text{ and}$$

$$r_2 = (\mathrm{tr}(\boldsymbol{VA})^2 - \lambda_1\,\mathrm{tr}(\boldsymbol{VA}))/\lambda_2(\lambda_2 - \lambda_1),$$

$$\text{(ii)}\ [(\boldsymbol{VA})^2 - (\lambda_1 + \lambda_2)(\boldsymbol{VA}) + \lambda_1\lambda_2\boldsymbol{I}]\boldsymbol{V}(\boldsymbol{A\mu} + \boldsymbol{l}) = \boldsymbol{0} \text{ and}$$

$$\text{(iii)}\ \lambda_1 v_1 = (\boldsymbol{l} + \boldsymbol{A\mu})'(\boldsymbol{AVA} - \lambda_2\boldsymbol{V})(\boldsymbol{l} + \boldsymbol{A\mu})/\lambda_1(\lambda_1 - \lambda_2),$$

$$\lambda_2 v_2 = (\boldsymbol{l} + \boldsymbol{A\mu})'(\boldsymbol{AVA} - \lambda_1\boldsymbol{V})(\boldsymbol{l} + \boldsymbol{A\mu})/\lambda_2(\lambda_2 - \lambda_1) \text{ and}$$

$$\lambda_1 v_1 + \lambda_2 v_2 = \boldsymbol{\mu}'\boldsymbol{A\mu} + 2\boldsymbol{l}'\boldsymbol{\mu} + c$$

$$= (\boldsymbol{l} + \boldsymbol{A\mu})\{(\lambda_1 + \lambda_2)\boldsymbol{V} - \boldsymbol{VAV}\}(\boldsymbol{l} + \boldsymbol{A\mu})/\lambda_1\lambda_2.$$

Here, by note 2, the condition (i) can be written in terms of traces and rank, namely $\zeta(\boldsymbol{VAV}) = r_1 + r_2$ and $\mathrm{tr}(\boldsymbol{VA})^i = \lambda_1^i r_1 + \lambda_2^i r_2$ for $i = 1, 2, 3, 4$. This generalises the results of Shanbhag (1968), (1970) and the result of Tan (1974) is obtained by taking $\lambda_1 = 1$ and $\lambda_2 = -1$ with $\boldsymbol{l} = \boldsymbol{0}$, $c = 0$.

Note 18. In Theorem 1, q is distributed as $\sum_{j=1}^{k} \lambda_j \chi_j^2(r_j)$, where $\chi_j^2(r_j)$ $j = 1, 2, \ldots, k$ are independent central chi-squares, iff (i) $\lambda_1, \ldots, \lambda_k$ are the distinct nonzero eigenvalues of $\boldsymbol{VA}$ with multiplicities $r_1, \ldots, r_k$ and (ii) $\boldsymbol{\mu}'\boldsymbol{A\mu} + 2\boldsymbol{l}'\boldsymbol{\mu} + c = 0$, and $\boldsymbol{V}(\boldsymbol{A\mu} + \boldsymbol{l}) = \boldsymbol{0}$.

Theorem 2. *Let $\boldsymbol{x}$ be $N(\boldsymbol{\mu}, \boldsymbol{V})$ and let $q_i = \boldsymbol{x}'\boldsymbol{A}_i\boldsymbol{x}$ for $i = 1, 2$ be two quadratic forms. Then q_1 and q_2 are distributed as independent*

$$\sum_{j=1}^{k_i} \lambda_{ij}\chi_{ij}^2(r_{ij}, v_{ij}) \quad \text{for } i = 1, 2,$$

(where χ_{ij}^2's are independent non-central chi-squares and $(\lambda_{11}, \ldots, \lambda_{1k_1})$ and $(\lambda_{21}, \ldots, \lambda_{2k_2})$ have distinct nonzero elements) iff (i) $\boldsymbol{VA}_1\boldsymbol{VA}_2\boldsymbol{V} = \boldsymbol{0} = \boldsymbol{VA}_2\boldsymbol{VA}_1\boldsymbol{V}$, *and the conditions of Theorem* 1 *for $\boldsymbol{x}'\boldsymbol{A}_i\boldsymbol{x}$ to be distributed as $\sum \lambda_{ij}\chi_{ij}^2(r_{ij}, v_{ij})$ for $i = 1, 2$.*

The extension of the Cochran's theorem can be given by the following theorem and it generalises the results of Styan (1970).

Theorem 3. *Let* $\boldsymbol{x}$ *be* $N(\boldsymbol{\mu}, \boldsymbol{V})$ *and* $\boldsymbol{\mu} = \boldsymbol{V}\boldsymbol{d}$ *for some* $\boldsymbol{d}$. *Let* $q_1, \ldots, q_m$ *and* q *be quadratic forms in* $\boldsymbol{x}$ *such that* $q = \sum_{i=1}^m q_i = \boldsymbol{x}'\boldsymbol{A}\boldsymbol{x}$ *and* $q_i = \boldsymbol{x}'\boldsymbol{A}_i\boldsymbol{x}$. *Then, consider the following conditions with distinct nonzero* $\lambda_1, \lambda_2, \ldots, \lambda_k$, $(k > 1)$;

(a) q_i *is distributed as* $\sum_{j=1}^k \lambda_j \chi^2_{ij}(r_{ij}, v_{ij})$ *with* $\chi^2_{i1}, \ldots, \chi^2_{ik}$ *being independent non-central chi-squares, for each* $i = 1, 2, \ldots, m$ *(some* r_{ij}*'s can be zero)*;

(b) $q_1, q_2, \ldots, q_m$ *are independently distributed*;

(c) q *is distributed as* $\sum_{j=1}^k \lambda_j \chi^2_j(r_j, v_j)$ *with* $\chi^2_1, \ldots, \chi^2_k$ *being independent non-central chi-squares,*

(d) $\zeta(\boldsymbol{VAV}) = \sum_{i=1}^m \zeta(\boldsymbol{VA}_i\boldsymbol{V})$, *and*

(e) $\zeta(\boldsymbol{VAV}) = \sum_{i=1}^m \sum_{j=1}^k \zeta[\boldsymbol{VA}_i\{\boldsymbol{V} - (\boldsymbol{VAVV}^- - \lambda_j \boldsymbol{I})^+(\boldsymbol{VAV} - \lambda_j \boldsymbol{V})\}]$

Then, (i) (a) *and* (b) $\Rightarrow$ *all conditions,* (ii) (a), (c) *and* (d) $\Rightarrow$ *all conditions, either for* $k = 2$ *if* λ_j*'s are of the same sign, or for* $k = 2$ *and* 3 *if* λ_j*'s are of different signs,* (iii) (b) *and* (c) $\Rightarrow$ *all conditions and* (iv) (c), (d) *and* (e) $\Rightarrow$ *all conditions.*

Result (iv) is known as the extension of Cochran's theorem. Note that if $\boldsymbol{V} = \boldsymbol{BB}'$ and $\boldsymbol{\mu} = \boldsymbol{B\nu}$ where $\boldsymbol{B}$ is a $p \times r$ matrix of rank $r = \zeta(\boldsymbol{V})$, then with probability one $\boldsymbol{x} = \boldsymbol{By}$ where $y \sim N(\boldsymbol{\nu}, \boldsymbol{I})$. Using this, the above five conditions are equivalent to the five conditions of Lemma 8 with matrices $\boldsymbol{B}'\boldsymbol{AB}$ and $\boldsymbol{B}'\boldsymbol{A}_i\boldsymbol{B}$ for $i = 1, 2, \ldots, m$ by applying Theorems 1, 2 and Lemma 1. This proves Theorem 3.

Note 19. Suppose in Theorem 3, $\boldsymbol{\mu} \neq \boldsymbol{Vd}$ for some $\boldsymbol{d}$. Then, we have to modify the conditions so as to include the conditions (ii) and (iii) of Theorem 1. Thus, over and above the five conditions of Theorem 3, we need the following two conditions:

(f) $\boldsymbol{VA}_i\boldsymbol{\mu} = \boldsymbol{VA}_i\boldsymbol{V}(\boldsymbol{VA}_i\boldsymbol{V})^-\boldsymbol{VA}_i\boldsymbol{\mu}$ and $\boldsymbol{\mu}'\boldsymbol{A}_i\boldsymbol{\mu} = \boldsymbol{\mu}'\boldsymbol{A}_i\boldsymbol{V}(\boldsymbol{VA}_i\boldsymbol{V})^-\boldsymbol{VA}_i\boldsymbol{\mu}$, for $i = 1, 2, \ldots, m$ and

(g) $\boldsymbol{VA\mu} = \boldsymbol{VAV}(\boldsymbol{VAV})^-\boldsymbol{VA\mu}$ and $\boldsymbol{\mu}'\boldsymbol{A\mu} = \boldsymbol{\mu}'\boldsymbol{AV}(\boldsymbol{VAV})^-\boldsymbol{VA\mu}$.

Then, we shall have the following results:

(i) (a), (b) and (g) $\Rightarrow$ all conditions, (ii) (a), (c) and (d) $\Rightarrow$ all conditions, either for $k = 2$ if λ_j's are of the same sign, or for $k = 2$ and 3 if λ_j's are of different signs, (iii) (b), (c), (f) $\Rightarrow$ all conditions and (iv) (c), (d), (e) and (f) $\Rightarrow$ all conditions.

Note 20. In the light of note 15, if A_i's are positive semi-definite, the conditions (f) and (g) of note 19 are satisfied. Hence, Theorem 3 becomes valid even when $\boldsymbol{\mu} \neq \boldsymbol{V}\boldsymbol{d}$ provided A_i's are positive semi-definite.

Note 21. When $k = 1$, we do not require condition (d) in (ii) and condition (e) is redundant in Theorem 3.

5. Wishartness and independence

Let $\boldsymbol{X}$ be an $n \times p$ random matrix distributed as $N_{n,p}(\boldsymbol{\mu}, \boldsymbol{V}, \boldsymbol{W})$ and let $\boldsymbol{Q} = \boldsymbol{X}'\boldsymbol{A}\boldsymbol{X} + \boldsymbol{L}'\boldsymbol{X} + \boldsymbol{X}'\boldsymbol{L} + \boldsymbol{C}$. Suppose that $\boldsymbol{Q}$ is distributed as $\Sigma_{j=1}^{k} \lambda_j \boldsymbol{S}_j$ where $\boldsymbol{S}_1, \ldots, \boldsymbol{S}_k$ are independent and $\boldsymbol{S}_j \sim W_p(r_j, \boldsymbol{W}, \boldsymbol{\Delta}_j)$ for $j = 1, 2, \ldots, k$. The results corresponding to Theorem 1 can be obtained by replacing $\boldsymbol{\mu}'\boldsymbol{A}\boldsymbol{\mu} + 2\boldsymbol{l}'\boldsymbol{\mu} + c$ by $\boldsymbol{\mu}'\boldsymbol{A}\boldsymbol{\mu} + \boldsymbol{L}'\boldsymbol{\mu} + \boldsymbol{\mu}'\boldsymbol{L} + \boldsymbol{C}$, v_j by $\boldsymbol{\Delta}_j$, $\boldsymbol{A}\boldsymbol{\mu} + \boldsymbol{l}$ by $\boldsymbol{A}\boldsymbol{\mu} + \boldsymbol{L}$, because $\boldsymbol{b}'\boldsymbol{Q}\boldsymbol{b}/\boldsymbol{b}'\boldsymbol{W}\boldsymbol{b}$ will be distributed as $\Sigma_{j=1}^{k} \lambda_j \chi_j^2$, $\chi_1^2, \ldots, \chi_k^2$ being independent non-central chi-squares given by $\chi_j^2(r_j, v_j)$, $v_j = \boldsymbol{b}'\boldsymbol{\Delta}_j\boldsymbol{b}$; for every non-null vector $\boldsymbol{b}$ such that $\boldsymbol{b}'\boldsymbol{W}\boldsymbol{b} \neq 0$. Thus, there is no need of rewriting the same Theorem 1 again. The modifications required are mentioned above. All the notes connected with Theorem 1 can be modified with respect to Wishart variates.

The extension of Cochran's Theorem can be written down parallel to Theorem 3 by replacing chi-squares by Wishart variates, q_i by $\boldsymbol{Q}_i = \boldsymbol{X}'\boldsymbol{A}_i\boldsymbol{X}$ for $i = 1, 2, \ldots, m$, and q by $\boldsymbol{Q} = \boldsymbol{X}'\boldsymbol{A}\boldsymbol{X} = \Sigma \boldsymbol{Q}_i$. All the notes connected with Theorem 3 can be expressed in the similar way. All these results are not expressed explicitly on account of shortage of space.

6. Remarks

Complex multivariate normal distribution is defined by Goodman (1963) and studied by various authors for various problems. Let $\boldsymbol{X}_1$ and $\boldsymbol{X}_2$ be $p \times n$ random matrices such that

$$(\boldsymbol{X}_1, \boldsymbol{X}_2) \sim N_{n,2p}\left((\mu_1, \mu_2), I_n, \begin{pmatrix} V_1 & V_2 \\ -V_2 & V_1 \end{pmatrix}\right),$$

where $\boldsymbol{V}_2$ is skew symmetric. Then $\boldsymbol{Z} = \boldsymbol{X}_1 + \mathrm{i}\boldsymbol{X}_2$ will have complex multivariate normal and will be denoted by $CN_{n,p}(\boldsymbol{\mu}, \boldsymbol{I}_n, \boldsymbol{V})$, $\boldsymbol{\mu} = \boldsymbol{\mu}_1 + \mathrm{i}\boldsymbol{\mu}_2$ and $\boldsymbol{V} = \boldsymbol{V}_1 + \mathrm{i}\boldsymbol{V}_2$. Note that $\boldsymbol{V}$ is Hermitian positive semi-definite (h.p.s.d) matrix. The distribution of $\boldsymbol{S} = \bar{\boldsymbol{Z}}'\boldsymbol{Z}$ is non-central complex Wishart and can be denoted by $CW_p(n, \boldsymbol{V}, \bar{\boldsymbol{\mu}}'\boldsymbol{\mu})$ where $\bar{\boldsymbol{Z}}'$ is the complex conjugate transpose of $\boldsymbol{Z}$. As in the real Wishart distribution, we have singular and/or pseudo complex Wishart distribution.

All the matrix results given in Section 2 are valid for all complex matrices by changing symmetric matrices by Hermitian matrices and p.s.d. by Hermitian p.s.d. matrices. Thus, there is no need to restate them. In the same way, all the results of Sections 4 and 5 are valid for the complex normal case by the above appropriate notations in place of real variates. Thus, there is no need to restate all the results of Sections 4 and 5.

References

Baldessari, B. (1967). The distribution of a quadratic form of normal random variables. *Ann. Math. Statist.* **38**, 1700–1704.

Goodman, N.R. (1963). Statistical analysis based on a certain multivariate complex Gaussian distribution (an introduction). *Ann. Math. Statist.* **34**, 152–177.

Graybill, F.A. and Marsaglia, G. (1957). Idempotent matrices and quadratic forms in general linear hypothesis. *Ann. Math. Statist.* **28**, 678–686.

Khatri, C.G. (1962). Conditions for Wishartness and independence of second degree polynomials in normal vectors. *Ann. Math. Statist.*, **33**, 1002–1007.

Khatri, C.G. (1963). Further contribution to Wishartness and independence of second degree polynomials in normal vectors. *J. Indian Statist. Assoc.*, **1**, 61–70.

Khatri, C.G. (1968). Some results for the singular normal mutlivariate regression models. *Sankhyā Ser. A*, **30**, 267–280.

Laha, R.G. (1956). On the stochastic independence of two second degree polynomial statistics in normally distributed variates. *Ann. Math. Statist.*, **27**, 790–796.

Mitra, S.K. and Rao, C.R. (1968). Simultaneous reduction of a pair of quadratic forms. *Sankhyā Ser. A*, **30**, 313–322.

Rao, C.R. (1973). *Linear Statistical Inferences and its Applications.* John Wiley & Sons, N.Y. (Second edition).

Rao, C.R. and Mitra, S.K. (1971). *Generalized Inverse of Matrices and its Applications.* John Willey, N.Y.

Shanbhag, D.N. (1968). Some remarks concerning Khatri's result on quadratic forms. *Biometrika*, **55**, 593–595.

Shanbhag, D.N. (1970). On the distribution of a quadratic form. *Biometrika*, **57**, 222–223.

Styan, G.P.H. (1970). Notes on the distribution of quadratic forms in singular normal variables. *Biometrika*, **57**, 567–572.

Tan, W.Y. (1975). Some matrix results and extensions of Cochran's theorem. *SIAM J. Appl. Math.* **28**, No. 3, 1975.

P.R. Krishnaiah, ed., *Multivariate Analysis–IV*

INFERENCE ON THE EIGENVALUES OF THE COVARIANCE MATRICES OF REAL AND COMPLEX MULTIVARIATE NORMAL POPULATIONS

P.R. KRISHNAIAH*
University of Pittsburgh, Pittsburgh, PA, U.S.A.

and

Jack C. LEE**
Wright State University, Dayton, Ohio, U.S.A.

In this paper, the authors derived asymptotic expressions for the joint distributions of the linear combinations as well as the ratios of the roots of the real and complex Wishart matrices. These distributions are useful in drawing inference on the eigenvalues of the covariance matrices.

1. Introduction

There are several physical situations where one would be interested in drawing inference on the eigenvalues of the covariance matrices of the real and complex multivariate normal populations. For example, in the area of the principal component analysis, it would be of interest to find out the number of eigenvalues which are significant. One can test the hypotheses on the eigenvalues of the covariance matrices by using various functions of the roots of the sample covariance matrix. Krishnaiah and Waikar (1971, 1972) proposed procedures for testing the hypothesis of the equality of the eigenvalues and its various subhypotheses simultaneously. These procedures are based upon certain ratios of the roots. Krishnaiah and Schuurmann (1974) computed the tables for the exact distributions of the ratios of the extreme roots as well as the ratios of the individual roots to the sum of the roots for the real and complex Wishart matrices; these tables are useful in the application of the simultaneous test procedures of Krishnaiah and

* The work of this author was performed at the Aerospace Research Laboratories and the Air Force Flight Dynamics Laboratory.

** Part of this work was performed at the Aerospace Research Laboratories in the capacity of Technology Incorporated Visiting Research Associate under Contract F33615–73–C–4155.

Waikar (1971, 1972). Recently, Sugiyama and Tong (1975) gave an asymptotic expression for the distribution function of the ratio of the sum of the first few roots to the sum of all the roots for the real and complex Wishart matrices; this expression is based upon the distribution function of the normal variate and the derivatives of this distribution function.

In Section 2 of this paper, we discuss certain simultaneous procedures for inference on the eigenvalues of the covariance matrices. An asymptotic expression is given in Section 3 for the joint density of several linear functions of the roots of the real Wishart matrix. In Section 4, we give an asymptotic expression for the joint distribution of certain ratios of the linear combinations of the roots of the real Wishart matrix. The results of Sections 3 and 4 are restricted to the situations when the population roots are distinct. Finally, in Section 5, we derive asymptotic expressions for the joint distributions of the linear combinations as well as the ratios of the linear combinations of the roots of the complex Wishart matrix when the population roots are distinct.

2. Simultaneous tests for the equality of the eigenvalues

Let the columns of $X: p \times n$ be distributed independently and identically with zero mean vector and covariance matrix Σ. Also, let $l_1 \geq \ldots \geq l_p$ be the eigenvalues of $S = (s_{jh}) = XX'$ whereas $\lambda_1 \geq \ldots \geq \lambda_p$ are the latent roots of Σ. In addition, let $\mu_j = \boldsymbol{c}_j'\boldsymbol{\lambda}$, $\nu_j = \boldsymbol{d}_j'\boldsymbol{\lambda}$, $\boldsymbol{\lambda}' = (\lambda_1, \ldots, \lambda_p)$, $\hat{\mu}_j = \boldsymbol{c}_j'\boldsymbol{l}$, $\hat{\nu}_j = \boldsymbol{d}_j'\boldsymbol{l}$, and $\boldsymbol{l}' = (l_1, \ldots, l_p)$. Also, we assume that $\bigcap_{j=1}^q H_j$ and $\bigcap_{j=1}^q H_j^*$ are equivalent to the hypothesis H_0 where $H_0: \lambda_1 = \ldots = \lambda_p$, $H_j: \mu_j = 0$, $H_j^*: \mu_i = \nu_j$, $A_j: \mu_j > 0$, $A_j^*: \mu_j > \nu_j$, $A_0 = \bigcup_{j=1}^q A_j$ and $A_0^* = \bigcup_{j=1}^q A_j^*$.

We will first assume that the common value (λ) of λ_i's under H_0 is known. In this case, we accept or reject H_j against A_j accordingly as

$$\hat{\mu}_j \lesseqgtr c_\alpha$$

where

$$\mathbf{P}[\hat{\mu}_j \leq c_\alpha;\, j = 1, \ldots, q \mid H_0] = (1 - \alpha). \tag{2.1}$$

The total hypothesis H_0 is accepted against A_0 if all the component hypotheses H_j are accepted.

Next, let us assume that λ is unknown. Then we accept or reject H_j^* against A_j^* accordingly as

$$(\hat{\mu}_j / \hat{\nu}_j) \lesseqgtr d_\alpha$$

where

$$\mathbf{P}[(\hat{\mu}_j/\hat{\nu}_j) \leq d_\alpha;\, j = 1, \ldots, q \mid H_0] = (1-\alpha). \tag{2.2}$$

The total hypothesis H_0 when tested against A_0^* is accepted if all the component hypotheses $H_1^*, \ldots, H_q^*$ are accepted. We can similarly consider procedures against two-sided alternatives by making obvious modifications. The simultaneous test procedures proposed by Krishnaiah and Waikar to test H_0^* against

$$\bigcup_{i=1}^{p-1} [\lambda_i > \lambda_{i+1}], \qquad \bigcup_{i=1}^{p-1} \left[p\lambda_i > \sum_{j=1}^{p} \lambda_j\right] \quad \text{and} \quad \bigcup_{i=1}^{p-1} [\lambda_i > \lambda_p]$$

are special cases of the above test procedures. Next, let $H_{1j} : \boldsymbol{c}_j'\boldsymbol{\lambda} \geq a\boldsymbol{d}_j'\boldsymbol{\lambda}$ and $A_{1j} : \boldsymbol{c}_j'\boldsymbol{\lambda} < a\boldsymbol{d}_j'\boldsymbol{\lambda}$ for $j = 1, \ldots, q$ where a is known. Then, we accept or reject H_{1j} $(j = 1, \ldots, q)$ accordingly as

$$\boldsymbol{c}_j'\boldsymbol{l} \gtrless ab_\alpha \boldsymbol{d}_j'\boldsymbol{l}$$

where

$$\mathbf{P}\left[\frac{\boldsymbol{c}_j'\boldsymbol{l}}{\boldsymbol{d}_j'\boldsymbol{l}} \geq ab_\alpha;\, j = 1, \ldots, q \,\middle|\, \bigcap_{j=1}^{q} H_{1j}\right] = (1-\alpha). \tag{2.3}$$

Since the probability integral in eq. (2.3) involves nuisance parameters, it would be of interest to construct bounds (free from nuisance parameters) on this probability integral. A special case of this problem that is of interest is to test the hypothesis that $\Sigma_{j=1}^{r} \lambda_j / \Sigma_{j=1}^{p} \lambda_j > a$. It has been proposed in the literature (e.g., see Rao (1965)) to use the statistic $\Sigma_{j=1}^{r} l_j / \Sigma_{j=1}^{p} l_j$ as a measure to find out as to whether the first r principal components are adequate to explain the variation among experimental units.

In this paper, we consider asymptotic distribution problems associated with the above tests when $l_1, \ldots, l_p$ are the roots of the real or complex Wishart matrix.

3. Joint distribution of the linear combinations of the roots

In this section, we derive an asymptotic expression for the joint density of the linear combinations of the roots of the real Wishart matrix when the population roots are distinct.

Let

$$L_g = \sqrt{n}\left\{n^{-1} \sum_{j=1}^{p} c_{gj} l_j - \sum_{j=1}^{p} c_{gj} \lambda_j\right\} \tag{3.1}$$

for $g = 1, \ldots, q$. Lawley (1956) obtained the following asymptotic expression for l_j when the population roots are distinct:

$$\frac{l_h}{n} = \lambda_h + Y_{1h} + Y_{2h} + Y_{3h} + \text{higher order terms} \tag{3.2}$$

where

$$Y_{1h} = (n^{-1}s_{hh} - \lambda_h), \qquad Y_{2h} = \sum_{k \neq h} \lambda_{hk}^{-1} s_{hk}^2 / n^2$$

$$Y_{3h} = -\sum_{j \neq h} \lambda_{hj}^{-2}(n^{-1}s_{hh} - \lambda_h)\frac{s_{hj}^2}{n^2} + \sum_{j \neq h} \lambda_{hj}^{-2}\left(\frac{s_{jj}}{n} - \lambda_j\right)\frac{s_{hj}^2}{n^2}$$

and $\lambda_{hk} = \lambda_h - \lambda_k$. Using (3.2) in (3.1), we obtain the following asymptotic expression for the joint characteristic function of $L_1, \ldots, L_q$:

$$\phi_1(t_1, \ldots, t_q) = C_{11}(t) + C_{12}(t) + O(n^{-1}) \tag{3.3}$$

where

$$C_{11}(t) = \mathbf{E}\left[\exp\left\{\mathrm{i}\sqrt{n}\sum_{g=1}^{q}\sum_{h=1}^{p} c_{gh} Y_{1h} t_g\right\}\right], \tag{3.4}$$

$$C_{12}(t) = \frac{1}{\sqrt{n}}\mathbf{E}\left[\eta_1(Y)\exp\left\{\mathrm{i}\sqrt{n}\sum_{g=1}^{q}\sum_{h=1}^{p} c_{gh} Y_{1h} t_g\right\}\right], \tag{3.5}$$

$$\eta_1(Y) = n\sum_{g=1}^{q}\sum_{h=1}^{p} (\mathrm{i}\, t_g) c_{gh} Y_{2h}. \tag{3.6}$$

After some algebraic manipulations, it is seen that

$$C_{11}(t) = \operatorname{etr}[-i\sqrt{n}B_1\Sigma]\left|I - 2\mathrm{i}B_1\frac{\Sigma}{\sqrt{n}}\right|^{-n/2} \tag{3.7}$$

$$C_{12}(t) = C_{11}(t)\left[\frac{\mathrm{i}}{\sqrt{n}}\sum_{g=1}^{q}\sum_{h=1}^{p}\sum_{j \neq h} c_{gh}\lambda_{hj}^{-1}\lambda_j\lambda_h t_g \delta_{hh}^*(t)\delta_{jj}^*(t)\right] \tag{3.8}$$

where etr denotes the exponential of the trace and

$$\delta_{hh}^*(t) = \left(1 - 2\mathrm{i}\sum_{g=1}^{q} t_g c_{g_h}\lambda_h / n\right)^{-1}$$

and

$$B_1 = \sum_{g=1}^{q} t_g \operatorname{diag}(c_{g1}, \ldots, c_{gp}) = \operatorname{diag}(b_1, \ldots, b_p).$$

But we know that

$$\begin{aligned} |I - A|^{-\beta} &= \exp[-\beta \log|I - A|] \\ &= 1 + \beta\sum_{r=1}^{\infty}\frac{\operatorname{tr} A^r}{r} + \frac{\beta^2}{2!}\left(\sum_{r=1}^{\infty}\operatorname{tr} A^r\right)^2 + \cdots. \end{aligned} \tag{3.9}$$

Using (3.9) in (3.7) and (3.8), we obtain the following asymptotic expression for $\phi_1(t_1, \ldots, t_q)$:

$$\phi_1(t_1, \ldots, t_q) = \exp[-\tfrac{1}{2}t'B_2t]\Bigg[1 + \frac{1}{\sqrt{n}} \sum_{h=1}^{p} \sum_{k \neq h}^{p} \sum_{j=1}^{q} c_{jh}\lambda_{hk}^{-1}\lambda_h\lambda_k(\mathrm{i}t_j) + \frac{4}{3\sqrt{n}} \sum_{h=1}^{p} \sum_{j_1=1}^{q} \sum_{j_2=1}^{q} \sum_{j_3=1}^{q} c_{j_1h}c_{j_2h}c_{j_3h}\mathrm{i}^3 t_{j_1}t_{j_2}t_{j_3}\lambda_h^3 + O(n^{-1})\Bigg] \tag{3.10}$$

where $B_2 = 2\Sigma_{h=1}^{p} \boldsymbol{b}_h^* \boldsymbol{b}_h^{*\prime}$ and $\boldsymbol{b}_h^{*\prime} = (\lambda_h c_{1h}, \ldots, \lambda_h c_{qh})$. We assume that B_2 is nonsingular.

Next, let

$$N(\boldsymbol{x}; \Omega) = \frac{1}{(2\pi)^{p/2}|\Omega|^{1/2}} \exp(-\tfrac{1}{2}\boldsymbol{x}'\Omega^{-1}\boldsymbol{x}) \tag{3.11}$$

where $\boldsymbol{x}' = (x_1, \ldots, x_p)$ and $\Omega = (\omega_{jk})$. Also, let $j_1, \ldots, j_s$ be s integers not necessarily all different such that $1 \leq j_i \leq p$. Then, the multivariate Hermite polynomials $H_{j_1 \ldots j_s}(\boldsymbol{x})$ studied by Hermite (see Appel and Kampe de Feriet (1926)) were defined as follows:

$$H_{j_1 \ldots j_s}(\boldsymbol{x}) = \frac{(-1)^s}{N(\boldsymbol{x}; \Omega)} \frac{\partial^s}{\partial x_{j_1} \ldots \partial x_{j_s}} N(\boldsymbol{x}; \Omega). \tag{3.12}$$

It is also known (e.g., see Khatri and Mitra (1969)) that the characteristic function of $H_{j_1 \ldots j_s}(\boldsymbol{x})N(\boldsymbol{x}; \Omega)$ is $\mathrm{i}^s\theta_{j_1} \ldots \theta_{j_s} \exp(-\tfrac{1}{2}\Sigma\, \theta_j\theta_k\omega_{jk})$.

Now, inverting (3.10), we obtain the following expression for the joint density of $L_1, \ldots, L_q$:

$$f_1(L_1, \ldots, L_q) = N(\boldsymbol{L}; B_2^{-1})\Bigg[1 + \frac{1}{\sqrt{n}}\Bigg\{ \sum_{h=1}^{p} \sum_{k \neq h}^{p} \sum_{j=1}^{q} c_{jh}\lambda_{hk}^{-1}\lambda_h\lambda_k H_j(\boldsymbol{L}) + \frac{4}{3} \sum_{h=1}^{p} \sum_{j_1=1}^{q} \sum_{j_2=1}^{q} \sum_{j_3=1}^{q} c_{j_1h}c_{j_2h}c_{j_3h} H_{j_1j_2j_3}(\boldsymbol{L})\lambda_h^3\Bigg\} + O(n^{-1})\Bigg]. \tag{3.13}$$

where $\boldsymbol{L}' = (L_1, L_2, \ldots, L_q)$.

4. Joint distribution of the ratio of linear combinations of roots

In this section we derive an asympotic expression for the joint density of the ratio of linear combinations of the roots of the real Wishart matrix when the population roots are distinct.

Let

$$T_g = \sqrt{n}\left[\sum_{h=1}^{p} c_{gh}l_h\left(\sum_{h=1}^{p} d_{gh}l_h\right)^{-1} - \sum_{h=1}^{p} c_{gh}\lambda_h\left(\sum_{h=1}^{p} d_{gh}\lambda_h\right)^{-1}\right] \tag{4.1}$$

for $g = 1, 2, \ldots, q$. The characteristic function of $T_1, T_2, \ldots, T_q$ is

$$\phi_2(\boldsymbol{t}) = C_{21}(\boldsymbol{t}) + C_{22}(\boldsymbol{t}) + O(n^{-1}) \tag{4.2}$$

where

$$C_{21}(\boldsymbol{t}) = \mathbf{E}\left[\exp\left\{\mathrm{i}\sqrt{n}\sum_{g=1}^{q}\sum_{h=1}^{p} t_g\,\Lambda_g^2 a_{gh} Y_{1h}\right\}\right], \tag{4.3}$$

$$C_{22}(\boldsymbol{t})\frac{1}{\sqrt{n}}\mathbf{E}\left[\eta_2(Y)\exp\left\{\mathrm{i}\sqrt{n}\sum_{g=1}^{q}\sum_{h=1}^{p} t_g\,\Lambda_g^2 a_{gh} Y_{1h}\right\}\right] \tag{4.4}$$

$$\eta_2(Y) = n\left[\sum_{g=1}^{q}\sum_{h=1}^{p}(\mathrm{i}t_g)\Lambda_g^2 a_{gh} Y_{2h} - \sum_{g=1}^{q}(\mathrm{i}t_g\Lambda_g^3)\left(\sum_{h=1}^{p} a_{gh} Y_{1h}\right)\left(\sum_{h=1}^{p} d_{gh} Y_{1h}\right)\right], \tag{4.5}$$

$$\Lambda_g^{-1} = \sum_{h=1}^{p} d_{gh}\lambda_h, \tag{4.6}$$

$$a_{gh} = \Lambda_g^{-1}c_{gh} - \tilde{\Lambda}_g d_{gh} \tag{4.7}$$

$$\tilde{\Lambda}_g = \sum_{h=1}^{p} c_{gh}\lambda_h. \tag{4.8}$$

By using arguments similar to those used in Section 3, we obtain the following asymptotic expression for the joint characteristic function of $T_1, \ldots, T_q$:

$$\begin{aligned}\phi_2(\boldsymbol{t}) = \exp(-\tfrac{1}{2}\boldsymbol{t}'B_3\boldsymbol{t})\Bigg[1 &+ \frac{1}{\sqrt{n}}\sum_{h=1}^{p}\sum_{k\neq h}^{p}\sum_{j=1}^{q} a^*_{jh}\lambda^{-1}_{hk}\,\lambda_h\lambda_k(\mathrm{i}t_j)\\ &- \frac{2}{\sqrt{n}}\sum_{h=1}^{p}\sum_{j=1}^{q} b^*_{jh}\lambda_h^2(\mathrm{i}t_j)\\ &+ \frac{4}{3\sqrt{n}}\sum_{h=1}^{p}\sum_{j_1=1}^{q}\sum_{j_2=1}^{q}\sum_{j_3=1}^{q}\lambda_h^3 a^*_{j_1h}a^*_{j_2h}a^*_{j_3h}\,\mathrm{i}^3 t_{j_1}t_{j_2}t_{j_3}\\ &- \frac{4}{\sqrt{n}}\sum_{h=1}^{p}\sum_{h'=1}^{p}\sum_{j_1=1}^{q}\sum_{j_2=1}^{q}\sum_{j_3=1}^{q}\lambda_h^2\lambda_{h'}^2 a^*_{j_1h}a^*_{j_2h'}b^*_{hh',j_3}\,\mathrm{i}^3 t_{j_1}t_{j_2}t_{j_3}\\ &+ O(n^{-1})\Bigg]\end{aligned} \tag{4.9}$$

where

$$B_3 = 2\sum_{h=1}^{p}\lambda_h^2\boldsymbol{a}_h^*\boldsymbol{a}_h^{*\prime}, \qquad \boldsymbol{a}_h^* = (\Lambda_1^2 a_{1h}, \ldots, \Lambda_q^2 a_{qh}),$$

$$a^*_{gh} = \Lambda_g^2 a_{gh}, \qquad b^*_{gj} = \Lambda_j^3 a_{gj}d_{gj} \quad \text{and} \quad b^*_{gf,j} = \Lambda_j^3 a_{jg}d_{jf}$$

and B_3 is assumed to be non-singular.

By inverting the above characteristic function, we obtain the following expression for the joint density of $T_1, \ldots, T_q$:

$$
\begin{aligned}
f_2(T_1, \ldots, T_q) = N(\boldsymbol{T}; B_3^{-1})\Bigg[1 + \frac{1}{\sqrt{n}}\Bigg\{ & \sum_{h=1}^{p} \sum_{k\neq h} \sum_{j=1}^{q} a^*_{jh}\lambda_{hk}^{-1}\lambda_h\lambda_k H_j(\boldsymbol{T}) \\
& - 2\sum_{h=1}^{p}\sum_{j=1}^{q} b^*_{jh}\lambda_h^2 H_j(\boldsymbol{T}) \\
& + \frac{4}{3}\sum_{h=1}^{p}\sum_{j_1=1}^{q}\sum_{j_2=1}^{q}\sum_{j_3=1}^{q} \lambda_h^3 a^*_{j_1h}a^*_{j_2h}a^*_{j_3h}H_{j_1j_2j_3}(\boldsymbol{T}) \\
& - 4\sum_{h=1}^{p}\sum_{h'=1}^{p}\sum_{j_1=1}^{q}\sum_{j_2=1}^{q}\sum_{j_3=1}^{q} \lambda_h^2\lambda_{h'}^2 a^*_{j_1h}a^*_{j_2h'} \\
& \cdot b^*_{hh',j_3}H_{j_1j_2j_3}(\boldsymbol{T}) \\
& + O(n^{-1})\Bigg]
\end{aligned}
\tag{4.10}
$$

5. Asymptotic distributions in the complex case

Let $Z = Z_1 + \mathrm{i}Z_2$ be a $p \times n$ matrix and let the rows of $(Z_1' : Z_2')$ be distributed independently as a multivariate normal with zero mean vector and covariance matrix

$$
\begin{pmatrix} \Sigma_1 & \Sigma_2 \\ -\Sigma_2 & \Sigma_1 \end{pmatrix}
$$

where Σ_1 and Σ_2 are of order $p \times p$. Then, the columns of Z are distributed independently as complex multivariate normal (in the sense of Wooding (1958)) with the covariance matrix $2(\Sigma_1 + \mathrm{i}\Sigma_2)$. Also, the distribution of $\tilde{S} = Z\bar{Z}' = (w_{ab})$, $w_{ab} = u_{ab} + \mathrm{i}v_{ab}$, is known to be a central complex Wishart distribution. In this section, we assume that $\Sigma_2 = 0$ and $\Sigma_1 = \frac{1}{2}\mathrm{diag.}(\delta_1, \ldots, \delta_p)$ where $\delta_1 \geq \ldots \geq \delta_p$. In this section, we give asymptotic expressions for the joint density of $(\tilde{L}_1, \ldots, \tilde{L}_q)$ as well as the joint density of $(\tilde{T}_1, \ldots, \tilde{T}_q)$ where

$$
\tilde{L}_g = \sqrt{n}\sum_{j=1}^{p} c_{gj}(n^{-1}\theta_j - \delta_j) \tag{5.1}
$$

$$
\tilde{T}_g = \sqrt{n}\left[\left(\sum_{j=1}^{p} c_{gj}\theta_j\right)\left(\sum_{h=1}^{p} d_{gh}\theta_h\right)^{-1} \left(\sum_{j=1}^{p} c_{gj}\delta_j\right)\left(\sum_{h=1}^{p} d_{gh}\delta_h\right)^{-1}\right] \tag{5.2}
$$

and $\theta_1 \geq \ldots \geq \theta_p$ are the roots of $\tilde{S}$.

It is known (e.g., see Sugiura (1973)) that

$$\frac{\theta_h}{n} = \delta_h + \tilde{Y}_{1h} + \tilde{Y}_{2h} + \text{higher order terms} \tag{5.3}$$

where

$$\tilde{Y}_{1h} = \left(\frac{u_{hh}}{n} - \delta_h\right), \qquad \tilde{Y}_{2h} = \sum_{j \neq h} \frac{1}{n^2} \delta_{hj}^{-1} \{u_{hj}^2 + v_{hj}^2\}$$

and $\delta_{hk} = \delta_h - \delta_k$. Using (5.3) in (5.1) and following similar lines as in Section 3, we obtain the following asymptotic expression for the joint characteristic function of $\tilde{L}_1, \ldots, \tilde{L}_q$:

$$\begin{aligned}\phi_3(t) = \exp[-\tfrac{1}{2}t'\tilde{B}_2 t]\Bigg[1 &+ \frac{1}{\sqrt{n}} \sum_{\alpha=1}^{p} \sum_{j \neq \alpha}^{p} \sum_{k=1}^{q} \delta_{\alpha j}^{-1} \delta_\alpha \delta_j c_{k\alpha} \mathrm{i} t_k \\ &+ \frac{1}{3\sqrt{n}} \sum_{\alpha=1}^{p} \sum_{j_1=1}^{q} \sum_{j_2=1}^{q} \sum_{j_3=1}^{q} c_{j_1\alpha} c_{j_2\alpha} c_{j_3\alpha} \mathrm{i}^3 t_{j_1} t_{j_2} t_{j_3} \delta_\alpha^3 \\ &+ O(n^{-1})\Bigg]\end{aligned} \tag{5.4}$$

where $2\tilde{B}_2 = B_2$. Inverting the right-hand side of (5.4) we obtain the following expression for the joint density of $\tilde{L}_1, \ldots, \tilde{L}_q$:

$$\begin{aligned}f_3(\tilde{L}_1, \ldots, \tilde{L}_q) = N(\tilde{\boldsymbol{L}}; \tilde{B}_2^{-1})\Bigg[1 &+ \frac{1}{\sqrt{n}} \sum_{\alpha=1}^{p} \sum_{j \neq \alpha}^{p} \sum_{k=1}^{q} \delta_{\alpha j}^{-1} \delta_\alpha \delta_j c_{k\alpha} H_k(\tilde{\boldsymbol{L}}) \\ &+ \frac{1}{3\sqrt{n}} \sum_{\alpha=1}^{p} \sum_{j_1=1}^{q} \sum_{j_2=1}^{q} \sum_{j_3=1}^{q} c_{j_1\alpha} c_{j_2\alpha} c_{j_3\alpha} H_{j_1 j_2 j_3}(\tilde{\boldsymbol{L}}) \delta_\alpha^3 \\ &+ O(n^{-1})\Bigg].\end{aligned} \tag{5.5}$$

Next, following the same lines as in Section 4, we obtain the following expression for the joint density of $\tilde{T}_1, \ldots, \tilde{T}_q$:

$$\begin{aligned}f_4(\tilde{T}_1, \ldots, \tilde{T}_q) = N(\tilde{\boldsymbol{T}}; \tilde{B}_3^{-1})\Bigg[1 &+ \frac{1}{\sqrt{n}}\Bigg\{ \sum_{\alpha=1}^{p} \sum_{j \neq \alpha} \sum_{k=1}^{q} \delta_{\alpha j}^{-1} \delta_\alpha \delta_j a_{k\alpha}^* H_k(\tilde{\boldsymbol{T}}) \\ &- \sum_{\alpha=1}^{p} \sum_{g=1}^{q} \delta_\alpha^2 b_{g\alpha}^* H_g(\tilde{\boldsymbol{T}}) \\ &+ \frac{1}{3} \sum_{\alpha=1}^{p} \sum_{j_1=1}^{q} \sum_{j_2=1}^{q} \sum_{j_3=1}^{q} \delta_\alpha^3 a_{j_1\alpha}^* a_{j_2\alpha}^* a_{j_3\alpha}^* H_{j_1 j_2 j_3}(\tilde{\boldsymbol{T}}) \\ &- \sum_{\alpha=1}^{p} \sum_{\beta=1}^{p} \sum_{j_1=1}^{q} \sum_{j_2=1}^{q} \sum_{j_3=1}^{q} \delta_\alpha^2 \delta_\beta^2 a_{j_1\alpha}^* a_{j_2\beta}^* b_{\alpha\beta, j_3}^* H_{j_1 j_2 j_3}(\tilde{\boldsymbol{T}})\Bigg\} \\ &+ O(n^{-1})\Bigg]\end{aligned} \tag{5.6}$$

where $\tilde{B}_3 = \frac{1}{2} B_3$.

Acknowledgment

The authors wish to thank Professor Y. Fujikoshi for his helpful comments on an earlier version of the manuscript.

References

Appel, P. and Kampe De Feriet, J. (1926). *Functions Hypergéométriques et Hyperspheriques*. Gauthier-Villars, Paris.

Khatri, C.G. and Mitra, S.K. (1969). Some identities and approximations concerning positive and negative multinomial distributions. In *Multivariate Analysis-II* (P.R. Krishnaiah, editor), Academic Press, Inc., New York.

Krishnaiah, P.R. and Waikar, V.B. (1971). Simultaneous tests for equality of latent roots against certain alternatives-I. *Ann. Inst. Statist. Math.* **23**, 451–468.

Krishnaiah, P.R. and Waikar V.B. (1972). Simultaneous tests for equality of latent roots against certain alternatives-II. *Ann. Inst. Statist. Math.* **24**, 81–85.

Krishnaiah, P.R. and Schuurmann, F.J. (1974). On the evaluation of some distributions that arise in simultaneous tests for the equality of the latent roots of the covariance matrix. *J. Multivariate Anal.* **4**, 265–282.

Lawley, D.N. (1956). Tests of significance for the latent roots of covariance and correlation matrices, *Biometrika* **43**, 128–136.

Rao, C.R. (1965). The use and interpretation of principal component analysis in applied research. *Sankhyā Ser. A* **26**, 329–358.

Sigiura, N. (1973). Derivatives of the characteristic root of a symmetric or a Hermitian matrix with two applications in multivariate analysis. *Comm. Statist.* **1**, 397–417.

Sugiyama, T. and Tong, H. (1975). On a statistic useful in dimensionality reduction in multivariate linear stochastic systems. Tech. Rept. No. 62, Department of Mathematics, University of Manchester.

Wooding, R.A. (1956). The multivariate distribution of complex normal variables. *Biometrika* **43**, 212–215

P.R. Krishnaiah, ed., *Multivariate Analysis–IV*
© North-Holland Publishing Company (1977) 105–118

APPROXIMATIONS TO THE DISTRIBUTIONS OF THE LIKELIHOOD RATIO STATISTICS FOR TESTING CERTAIN STRUCTURES ON THE COVARIANCE MATRICES OF REAL MULTIVARIATE NORMAL POPULATIONS

Jack C. LEE*
Wright State University, Dayton, Ohio, U.S.A.

T.C. CHANG*
University of Cincinnati, Cincinnati, Ohio, U.S.A.

P.R. KRISHNAIAH**
University of Pittsburgh, Pittsburgh, PA, U.S.A.

In this paper, the authors consider approximations to the distributions of the likelihood ratio statistics for testing the hypotheses on certain structures of the covariance matrices of the real multivariate normal populations.

1. Introduction

The problems of testing the hypotheses on the structures of the covariance matrix of the multivariate normal population have received considerable attention in the literature since these problems have applications in various disciplines. Wilks [21, 22] is one of the earliest workers in this area. In this paper, we investigate approximations to the distributions of the likelihood ratio statistics for testing the hypotheses of (1) multiple independence of several sets of variables, (2) sphericity, (3) equality of the covariance matrix to a given matrix, and (4) equality of the covariance matrices of independent sets of variables. These approximations are based upon fitting suitable Pearson type distributions by using the first four moments of the above statistics. Using the above approximations, we construct percentage points of these statistics. The accuracy of these approximations is sufficient for practical purposes.

* The work of Lee and Chang was performed at the Aerospace Research Laboratories when they are Visiting Research Associates of the Technology Incorporated under contract F33615–73–C–4155.

** The work of this author was performed at the Aerospace Research Laboratories and the Airforce Flight Dynamics Laboratory.

Exact distributions of the test statistics considered here are very difficult to compute. One can, of course, use Box's asymptotic expression by taking a sufficient number of terms to compute the tables and this may be preferable (at least in some cases) to exact expressions from a computational point of view. However, this asymptotic expression is also difficult to compute when we need to take several terms in the series and this may be the case when the sample size is not large. Since Pearson type approximations are quite simple from a computational point of view and since their accuracy is sufficient for practical purposes, they are definitely preferable to either exact expressions or Box's asymptotic series if we are interested in computing the percentage points of the statistics considered in this paper.

2. Approximations to the distributions of the likelihood ratio tests

Let $\boldsymbol{X}' = (\boldsymbol{X}_1', \ldots, \boldsymbol{X}_q')$ be distributed as a multivariate normal with mean vector $\boldsymbol{\mu}'$ and covariance matrix Σ. Also, let $\mathbf{E}(\boldsymbol{X}_i) = \boldsymbol{\mu}_i$ and $\mathbf{E}\{(\boldsymbol{X}_i - \boldsymbol{\mu}_i)(\boldsymbol{X}_j - \boldsymbol{\mu}_j)'\} = \Sigma_{ij}$, where $\boldsymbol{X}_i$ is of order $p_i \times 1$ and $s = \Sigma_{i=1}^q p_i$. In this paper, we consider approximations to the distributions of certain powers of the likelihood ratio statistics for testing the hypotheses H_1, H_2, H_3, H_4 and H_5 where

H_1: $\Sigma_{ij} = 0 \ (i \neq j = 1, \ldots, q)$,

H_2: $\Sigma = \sigma^2 \Sigma_0$ (σ^2 is unknown, Σ_0 is known),

H_3: $\Sigma = \Sigma_0$,

H_4: $\Sigma_{11} = \ldots = \Sigma_{qq}$ (under the assumption that H_1 is true and $p_1 = \ldots = p_q$),

$$\mathrm{H}_5: \begin{cases} \Sigma_{11} = \ldots = \Sigma_{q_1, q_1}, \\ \Sigma_{q_1+1, q_1+1} = \ldots = \Sigma_{q_2^*, q_2^*}, \\ \Sigma_{q_{k-1}^*+1, q_{k-1}^*+1} = \ldots = \Sigma_{q, q}, \end{cases}$$

where $q_0^* = 0$, $q_j^* = \Sigma_{i=1}^j q_i$ and $q_k^* = q$. These distributions are approximated with Pearson type distributions. For a description of the family of Pearson type distributions, the reader is referred to Kendall and Stuart [8].

Now let $(\boldsymbol{X}_{1j}', \ldots, \boldsymbol{X}_{qj}')$, $(j = 1, \ldots, N)$ be N independent observations on $\boldsymbol{X}' = (\boldsymbol{X}_1', \ldots, \boldsymbol{X}_q')$ and let $A = (A_{lm})$ where

$$A_{lm} = \sum_{j=1}^N (\boldsymbol{X}_{lj} - \bar{\boldsymbol{X}}_{l.})(\boldsymbol{X}_{mj} - \bar{\boldsymbol{X}}_{m.})', \qquad \bar{\boldsymbol{X}}_{l.} = \frac{1}{N} \sum_{j=1}^N \boldsymbol{X}_{lj},$$

and $l, m = 1, 2, \ldots, q$. Then, the likelihood ratio statistic for testing H_1 is known to be

$$V_1 = \frac{|A|}{\prod_{j=1}^{q} |A_{jj}|} \tag{2.1}$$

The moments of the statistic V_1 are given by

$$E(V_1^h) = \frac{\prod_{i=1}^{s} \Gamma[\frac{1}{2}(n+1-i)+h] \prod_{i=1}^{q} \left\{ \prod_{j=1}^{p_i} \Gamma[\frac{1}{2}(n+1-j)] \right\}}{\prod_{i=1}^{s} \Gamma[\frac{1}{2}(n+1-i)] \prod_{i=1}^{q} \left\{ \prod_{j=1}^{p_i} \Gamma[\frac{1}{2}(n+1-j)+h] \right\}}, \tag{2.2}$$

where $n = N - 1$ and $\Gamma(\cdot)$ is the complete gamma function. The statistic V_1 and its moments were derived by Wilks [22]. The distribution of V_1 is quite skew. Hence, we approximate the distribution of $V_1^{1/4}$ with Pearson's Type I distribution by using the first four moments.

Box [3] derived asymptotic expressions for a class of likelihood ratio test statistics in multivariate statistical analysis. This class inlcudes the statistics associated with testing the hypotheses H_1, H_2, H_4, and H_5. The number of terms given by Box is not sufficient to get the desired degree of accuracy in several practical situations. So, the authors [14] obtained terms up to $O(n^{-15})$ applying the method of Box; for details of these terms, the reader is referred to [14] by the authors.

Consul [4] and Mathai and Rathie [17] derived exact expressions for the distribution of V_1, and these expressions are very difficult to compute.

Table 1 gives a comparison of values obtained by the Pearson type approximation and the asymptotic expression of order n^{-13}. In Table 1, α_1

Table 1
Comparison of the Pearson Type Approximation
with the Asymptotic Expansion of Order n^{-13}
for the Distribution of V_1^* when $p_i = 1$

$n \backslash q$	3			5		
	c_1	α_1	α_2	c_1	α_1	α_2
10	1.913	0.05	0.0499	4.978	0.05	0.0488
15	1.187	0.05	0.0500	2.947	0.05	0.0497
20	0.860	0.05	0.0500	2.099	0.05	0.0499
30	0.555	0.05	0.0500	1.333	0.05	0.0500

is the value of α if we use the Pearson type approximation whereas α_2 is the value of α when we use the asymptotic expression of order n^{-13}, where α is given by $\mathbf{P}[V_1^* \leq c_1 \mid H_1] = (1-\alpha)$, and $V_1^* = -2\log V_1$.

Davis and Field [6] computed the percentage points of $-2\rho \log V_1$ for some values of the parameters by using the Cornish–Fisher type inversion (see Davis [5]) of Box's asymptotic series when

$$1-\rho = \left\{2\left(s^3 - \sum_{i=1}^{q} p_i^3\right) + 9\left(s^2 - \sum_{i=1}^{q} p_i^2\right)\right\} \Big/ 6N\left(s^2 - \sum_{i=1}^{q} p_i^2\right). \quad (2.3)$$

In Table 2, the entries under the columns L–C–K are the values of c_1 obtained by the authors with the Pearson type approximation whereas the entries under the columns D–F are the corresponding values obtained by Davis and Field [6].

Tables 1 and 2 indicate that the accuracy of the Pearson type approximation to the distribution of $V_1^{1/4}$ is sufficient for practical purposes. Hence, using the Pearson type approximation, we computed the values of c_1 for $p_i = p = 1, 2, 3$, $q = 3, 4, 5$, $\alpha = 0.05$, and $M = 1\,(1)\,20\,(2)\,30$, where $M = n - s - 3$. These values are given in Table 3. When $q = 2$, it is found that the Pearson type approximation is quite satisfactory, and the results are reported in Lee, Krishnaiah and Chang [13].

Let us now consider the likelihood ratio statistic for testing H$_2$: $\Sigma = \sigma^2\Sigma_0$, where σ^2 is unknown and Σ_0 is known. The statistic is given by

$$V_2 = \frac{|A\Sigma_0^{-1}|}{\{\operatorname{tr} A\Sigma_0^{-1}/s\}^s} \quad (2.4)$$

where $\operatorname{tr} B$ denotes the trace of B. The h-th moment of V_2 is given by

$$\mathbf{E}(V_2^h) = \frac{p^{hs}\,\Gamma(sn/2)}{\Gamma(sh + \frac{1}{2}sn)} \prod_{i=1}^{s} \frac{\Gamma[\frac{1}{2}(n+1-i)+h]}{\Gamma[\frac{1}{2}(n+1-i)]}. \quad (2.5)$$

Table 2
Comparison of the Pearson Type Approximation with Inversion of the Asymptotic Expression for the Percentage Points of V_1^* when $p_i = 1$, $\alpha = 0.05$

$n \backslash q$	4		5	
	L–C–K	D–F	L–C–K	D–F
10	3.238	3.238	4.978	4.977
15	1.967	1.967	2.949	2.948
20	1.414	1.414	2.099	2.099
24	1.154	1.154	1.706	1.707

Table 3*
Upper 5% Points of $-2 \log V_1$

(p, q)	(1, 3)	(2, 3)	(3, 3)	(1, 4)	(2, 4)	(3, 4)	(1, 5)	(2, 5)	(3, 5)
M									
1	3.023	6.371	10.037	4.376	9.417	14.932	5.784	12.600	20.006
2	2.534	5.509	8.857	3.721	8.285	13.408	4.978	11.228	18.192
3	2.180	4.857	7.937	3.238	7.406	12.190	4.371	10.146	16.716
4	1.913	4.345	7.198	2.867	6.701	11.189	3.900	9.261	15.477
5	1.704	3.933	6.587	2.572	6.122	10.347	3.520	8.525	14.425
6	1.537	3.593	6.075	2.333	5.637	9.627	3.208	7.900	13.515
7	1.400	3.307	5.639	2.134	5.225	9.004	2.949	7.364	12.717
8	1.284	3.064	5.262	1.967	4.870	8.460	2.727	6.897	12.011
9	1.187	2.854	4.933	1.825	4.560	7.980	2.537	6.487	11.386
10	1.103	2.672	4.643	1.701	4.289	7.552	2.372	6.125	10.821
11	1.030	2.511	4.386	1.593	4.048	7.169	2.227	5.801	10.313
12	0.967	2.369	4.157	1.498	3.833	6.823	2.099	5.510	9.852
13	0.910	2.242	3.950	1.414	3.639	6.510	1.985	5.247	9.432
14	0.860	2.128	3.763	1.338	3.465	6.225	1.882	5.009	9.047
15	0.815	2.025	3.593	1.271	3.306	5.964	1.790	4.791	8.691
16	0.775	1.932	3.438	1.210	3.162	5.725	1.706	4.592	8.364
17	0.739	1.847	3.296	1.154	3.030	5.503	1.631	4.409	8.061
18	0.705	1.769	3.165	1.104	2.908	5.299	1.561	4.240	7.779
19	0.675	1.697	3.044	1.057	2.796	5.110	1.497	4.084	7.518
20	0.647	1.631	2.932	1.015	2.692	4.933	1.438	3.938	7.272
22	0.597	1.513	2.732	0.939	2.506	4.615	1.333	3.677	6.828
24	0.555	1.412	2.557	0.874	2.344	4.336	1.243	3.448	6.435
26	0.518	1.323	2.403	0.817	2.202	4.088	1.163	3.247	6.086
28	0.486	1.244	2.267	0.767	2.076	3.868	1.094	3.067	5.772
30	0.458	1.175	2.145	0.723	1.964	3.671	1.032	2.906	5.491

* The entries in this table are the values of c_1 where $M = n - s - 3$ and $\mathbf{P}[-2 \log V_1 \leq c_1 \mid H_1] = (1 - \alpha)$.

The statistic V_2 and its moments were derived by Mauchly [15]. Using the first four moments, we approximated the distribution of $V_2^{1/4}$ with the Pearson Type I distribution. Expressions for the exact distribution of V_2 were given by Consul [4], Mathai and Rathie [16], and Nagarsenker and Pillai [18], but these expressions are very difficult to compute. In Table 4 the values under the column L–C–K are the values obtained by the authors using the Pearson type approximation whereas the corresponding exact values are taken from Nagarsenker and Pillai [18]. This table indicates that the accuracy of the Pearson type approximation is sufficient for practical

Table 4
Comparison of the Pearson Type Approximation with Exact Expression for the Distribution of V_2 ($\alpha = 0.05$)

$n \backslash s$	4		5		7	
	L–C–K	Exact	L–C–K	Exact	L–C–K	Exact
6	0.0169	0.0169	0.0013	0.0013	—	—
10	0.1297	0.1297	0.0492	0.0492	0.0029	0.0030
15	0.2812	0.2812	0.1608	0.1608	0.0368	0.0368
21	0.4173	0.4173	0.2877	0.2876	0.1111	0.1111
33	0.5833	0.5833	0.4663	0.4663	0.2665	0.2665
41	0.6507	0.6508	0.5453	0.5453	0.3515	0.3515

purposes. The values of α in this table are given by the relation $\mathbf{P}[V_2 \geq c_2 | \mathrm{H}_2] = (1 - \alpha)$.

We next consider the problem of testing the hypothesis H_3: $\Sigma = \Sigma_0$ where Σ_0 is specified. The likelihood ratio statistic for testing H_3 and the moments of this statistic were derived by Anderson [1]. The modified likelihood ratio test statistic (obtained by changing N to n in the likelihood ratio statistic) and its moments are as given below:

$$V_3 = (\mathrm{e}/n)^{sn/2} |A \Sigma_0^{-1}|^{n/2} \operatorname{etr}(-\tfrac{1}{2} A \Sigma_0^{-1}) \tag{2.6}$$

$$\mathbf{E}(V_3^h) = (2\mathrm{e}/n)^{shn/2} |\Sigma_0|^{nh/2} \cdot |I + h\Sigma_0|^{-n(1+h)/2} \frac{\prod_{i=1}^{s} \Gamma[\frac{1}{2}(n + nh + 1 - i)]}{\prod_{i=1}^{s} \Gamma[\frac{1}{2}(n + 1 - i)]}. \tag{2.7}$$

Using the first four moments, we approximated the distribution of $V_3^{1/34}$ with the Pearson Type 1 distribution. Korin [9] obtained an asymptotic expression of order n^{-15} for the distribution of $V_3^* = -2 \log V_3$. Using this expression, he computed percentage points of V_3^* for some values of the parameters. Nagarsenker and Pillai [19] obtained an expression for the distribution of V_3^*, but this expression is complicated from a computational point of view. Using this expression, they computed exact percentage points of V_3^*. In Table 5 we compare our values (given in the column L–C–K) obtained by using the Pearson type approximation with the exact values of Nagarsenker and Pillai, and the values obtained by Korin [9]. Table 5 indicates that the accuracy of the Pearson type approximation is sufficient for practical purposes.

Table 5
Comparison of the Pearson Type Approximation with Exact and Asymptotic Expressions for the Distribution of V_3^* ($\alpha = 0.05$)

s	4			6				10		
n	L–C–K	Exact	Korin	n	L–C–K	Exact	Korin	n	L–C–K	Exact
6	25.76	25.76	25.8	8	49.24	49.25	—	12	119.08	119.07
7	24.06	24.06	24.06	9	45.82	45.83	—	13	111.15	111.15
10	21.75	21.75	21.75	10	43.62	43.63	—	14	105.76	105.76
11	21.35	21.35	21.35	15	38.71	38.71	—	15	101.83	101.82
13	20.77	20.77	20.77	20	36.87	36.86	36.87	20	91.28	91.28
				25	35.89	35.88	35.89	25	86.51	86.52

Now let $\boldsymbol{X}_{ij}$ ($j = 1, \ldots, N_i$) be j-th independent observation on $\boldsymbol{X}_i$. Also, let $p_i = p (i = 1, \ldots, q)$ and

$$A_{ii} = \sum_{j=1}^{N_i} (\boldsymbol{X}_{ij} - \bar{\boldsymbol{X}}_{i.})(\boldsymbol{X}_{ij} - \bar{\boldsymbol{X}}_{i.})'.$$

Wilks [21] derived the likelihood ratio statistic for H_4: $\Sigma_{11} = \ldots = \Sigma_{qq}$ when $\Sigma_{ij} = 0$ ($i \neq j = 1, \ldots, q$) and derived its moments. Let $n_i = N_i - 1$, and $n = \Sigma_{i=1}^{q} n_i$. The modified likelihood ratio statistic V_4 (obtained by interchanging N_i with n_i in the likelihood ratio statistic) for H_4 and the moments of V_4 are given below:

$$V_4 = \frac{\prod_{g=1}^{q} |A_{gg}|^{n_g/2}}{\left|\sum_{g=1}^{q} A_{gg}\right|^{n/2}} \frac{n^{pn/2}}{\prod_{g=1}^{q} n_{n_g}^{pn_g/2}} \tag{2.8}$$

$$\begin{aligned} \mathbf{E}(V_4^h) = {} & \left(n^{phn/2} \Big/ \prod_{g=1}^{q} n_g^{phn_g/2}\right) \\ & \cdot \prod_{i=1}^{p} \left[\left\{\prod_{g=1}^{q} \Gamma[(n_g + hn_g + 1 - i)/2] / \Gamma[(n_g + 1 - i)/2]\right\}\right. \\ & \left. \cdot \{\Gamma[((n + 1 - i)/2] / \Gamma[(n + hn + 1 - i)/2]\}\right]. \end{aligned} \tag{2.9}$$

Korin [10] computed percentage points by using Box's asymptotic expression up to terms of order n^{-15}. Davis and Field [6] computed the percentage points by using the Cornish–Fisher type inversion of Box's asymptotic expression. In this paper, we approximated the distribution of $V_4^{1/b}$, with Pearson Type I distribution where b is a suitably chosen integer.

We chose b to be 30 when $p = 2, 3$, and 45 when $p = 4, 5$, and 70 when $p = 6$. In Table 6, we compared some of the percentage points obtained by us with the corresponding values obtained by Davis and Field [6] and Korin [10], when $n_i = n_0$. The percentage points c_4 obtained by us using the Pearson type approximation are given under the column L–C–K whereas the corresponding values obtained by Korin, and Davis–Field are given under the columns Korin and D–F, respectively. Table 6 indicates that the Pearson type approximation is satisfactory for practical purposes. In this table, α is defined by the equation $\mathbf{P}[V_4^* \leq c_4 \mid H_4] = (1-\alpha)$, where $V_4^* = -2 \log V_4$. When n_0 is large, Bishop [2] investigated the accuracy of approximating the distribution of the $(2/N)$-th power of the likelihood ratio criterion with the Beta distribution for a few cases, where $N = q(n_0 + 1)$.

Using the Pearson type approximation, we computed the values of c_4 when $n_0 = (p+1)(1)20(5)30$, $p = 2(1)5$, $q = 2(1)10$ and $\alpha = 0.05$. These percentage points are given in Table 7.

Finally, we discuss Pearson type approximations to the distribution of the likelihood ratio statistic for testing the hypothesis H_5 when $p_i = p$ $(i = 1, \ldots, q)$. This hypothesis is of interest in studying certain linear structures on the covariance matrices (see Krishnaiah and Lee [11]). The likelihood ratio statistic for testing H_5 is

Table 6*
Comparison of the Pearson Type Approximation
With the Asymptotic Expression for
the Distribution of V_4^* ($\alpha = 0.05$)

		$p = 2$			$p = 3$		
n_0	q	L–C–K	Korin	D–F	L–C–K	Korin	D–F
4	2	10.70	10.70	10.70	22.41	—	—
4	9	46.07	46.07	—	99.94	—	—
7	2	9.24	9.24	9.24	16.59	16.59	16.59
7	9	41.26	41.26	—	79.90	79.91	—
10	2	8.76	8.76	8.76	15.11	15.11	15.11
10	9	39.65	36.65*	—	74.58	74.57	—
15	2	8.42	—	8.42	14.15	—	14.15
15	9	38.50	—	—	71.05	—	—
20	2	8.26	—	8.26	13.72	—	13.72
20	9	37.95	—	—	69.45	—	—

* There is a typographical error in Korin's table. The correct value seems to be 39.65.

Table 7
Upper 5% Points of the Distribution of $-2 \log V_4$

$p = 2$

$n_0 \backslash q$	2	3	4	5	6	7	8	9	10
3	12.18	18.70	24.55	30.09	35.45	40.68	45.81	50.87	55.86
4	10.70	16.65	22.00	27.07	31.97	36.75	41.45	46.07	50.64
5	9.97	15.63	20.73	25.57	30.23	34.79	39.26	43.67	48.02
6	9.53	15.02	19.97	24.66	29.19	33.61	37.95	42.22	46.45
7	9.24	14.62	19.46	24.05	28.49	32.83	37.08	41.26	45.40
8	9.04	14.33	19.10	23.62	27.99	32.26	36.44	40.57	44.64
9	8.88	14.11	18.83	23.30	27.62	31.84	35.98	40.05	44.08
10	8.76	13.94	18.61	23.05	27.33	31.51	35.61	39.65	43.64
11	8.67	13.81	18.44	22.85	27.10	31.25	35.32	39.33	43.29
12	8.59	13.70	18.30	22.68	26.90	31.03	35.08	39.07	43.00
13	8.52	13.60	18.19	22.54	26.75	30.85	34.87	38.84	42.76
14	8.47	13.53	18.10	22.42	26.61	30.70	34.71	38.66	42.56
15	8.42	13.46	18.01	22.33	26.50	30.57	34.57	38.50	42.38
16	8.38	13.40	17.94	22.24	26.40	30.45	34.43	38.36	42.23
17	8.35	13.35	17.87	22.17	26.31	30.35	34.32	38.24	42.10
18	8.32	13.30	17.82	22.10	26.23	30.27	34.23	38.13	41.99
19	8.28	13.26	17.77	22.04	26.16	30.19	34.14	38.04	41.88
20	8.26	13.23	17.72	21.98	26.10	30.12	34.07	37.95	41.79
25	8.17	13.10	17.55	21.79	25.87	29.86	33.78	37.63	41.44
30	8.11	13.01	17.44	21.65	25.72	29.69	33.59	37.42	41.21

$p = 3$

$n_0 \backslash q$	2	3	4	5	6	7	8	9	10
4	22.41	35.00	46.58	57.68	68.50	79.11	89.60	99.94	110.21
5	19.19	30.52	40.95	50.95	60.60	70.26	79.69	89.03	98.27
6	17.57	28.24	38.06	47.49	56.67	65.69	74.58	83.39	92.09
7	16.59	26.84	36.29	45.37	54.20	62.89	71.44	79.90	88.30
8	15.93	25.90	35.10	43.93	52.54	60.99	69.32	77.57	85.73
9	15.46	25.22	34.24	42.90	51.33	59.62	67.78	75.86	83.87
10	15.11	24.71	33.59	42.11	50.42	58.57	66.62	74.58	82.46
11	14.83	24.31	33.08	41.50	49.71	57.76	65.71	73.57	81.36
12	14.61	23.99	32.67	41.00	49.13	57.11	64.97	72.75	80.45
13	14.43	23.73	32.33	40.60	48.65	56.56	64.36	72.09	79.72
14	14.28	23.50	32.05	40.26	48.26	56.11	63.86	71.53	79.11
15	14.15	23.32	31.81	39.97	47.92	55.73	63.43	71.05	78.60
16	14.04	23.16	31.60	39.72	47.63	55.40	63.06	70.64	78.14
17	13.94	23.02	31.43	39.50	47.38	55.11	62.73	70.27	77.76
18	13.86	22.89	31.26	39.31	47.16	54.86	62.45	69.97	77.41
19	13.79	22.78	31.13	39.15	46.96	54.64	62.21	69.69	77.11
20	13.72	22.69	31.01	39.00	46.79	54.44	61.98	69.45	76.84
25	13.48	22.33	30.55	38.44	46.15	53.70	61.16	68.54	75.84
30	13.32	22.10	30.25	38.09	45.73	53.22	60.62	67.94	75.18

Table 7 (continued)

$p = 4$

$n_0 \backslash q$	2	3	4	5	6	7	8	9	10
5	35.39	56.10	75.36	93.97	112.17	130.11	147.81	165.39	182.80
6	30.06	48.62	65.90	82.60	98.93	115.03	130.94	146.69	162.34
7	27.31	44.69	60.89	76.56	91.88	106.98	121.90	136.71	151.39
8	25.61	42.24	57.77	72.77	87.46	101.94	116.23	130.43	144.50
9	24.45	40.57	55.62	70.17	84.42	98.46	112.32	126.08	139.74
10	23.62	39.34	54.04	68.26	82.19	95.90	109.46	122.91	136.24
11	22.98	38.41	52.84	66.81	80.48	93.95	107.27	120.46	133.57
12	22.48	37.67	51.90	65.66	79.14	92.41	105.54	118.55	131.45
13	22.08	37.08	51.13	64.73	78.04	91.15	104.12	116.98	129.74
14	21.75	36.59	50.50	63.95	77.13	90.12	102.97	115.69	128.32
15	21.47	36.17	49.97	63.30	76.37	89.26	101.99	114.59	127.14
16	21.24	35.82	49.51	62.76	75.73	88.51	101.14	113.67	126.10
17	21.03	35.52	49.12	62.28	75.16	87.87	100.42	112.87	125.22
18	20.86	35.26	48.78	61.86	74.68	87.31	99.80	112.17	124.46
19	20.70	35.02	48.47	61.50	74.25	86.82	99.25	111.56	123.79
20	20.56	34.82	48.21	61.17	73.87	86.38	98.75	111.02	123.18
25	20.06	34.06	47.23	59.98	72.47	84.78	96.95	109.01	120.99
30	19.74	33.59	46.61	59.21	71.58	83.74	95.79	107.71	119.57

$p = 5$

$n_0 \backslash q$	2	3	4	5	6	7	8	9	10
6	51.11	81.99	110.92	138.98	166.54	193.71	220.66	247.37	273.88
7	43.40	71.06	97.03	122.22	146.95	171.34	195.49	219.47	243.30
8	39.29	65.15	89.45	113.03	136.18	159.04	181.65	204.14	226.48
9	36.71	61.39	84.62	107.17	129.30	151.17	172.80	194.27	215.64
10	34.93	58.78	81.25	103.06	124.48	145.64	166.56	187.37	208.02
11	33.62	56.85	78.75	100.02	120.92	141.54	161.98	182.24	202.37
12	32.62	55.37	76.83	97.68	118.15	138.38	158.38	178.23	198.03
13	31.83	54.19	75.30	95.82	115.96	135.86	155.54	175.10	194.51
14	31.19	53.23	74.05	94.29	114.16	133.80	153.21	172.49	191.68
15	30.66	52.44	73.01	93.02	112.66	132.07	151.29	170.36	189.38
16	30.22	51.76	72.14	91.94	111.41	130.61	149.66	166.53	187.32
17	29.83	51.19	71.39	91.03	110.34	129.38	148.25	166.99	185.61
18	29.51	50.69	70.74	90.23	109.39	128.29	147.03	165.65	184.10
19	29.22	50.26	70.17	89.54	108.57	127.36	145.97	164.45	182.81
20	28.97	49.88	69.67	88.93	107.85	126.52	145.02	163.38	181.65
25	28.05	48.48	67.86	86.70	105.21	123.51	141.62	159.60	177.49
30	27.48	47.61	66.71	85.29	103.56	121.60	139.47	157.22	174.87

$$V_5 = \frac{\prod_{i=1}^{q} |A_{ii}/n_i|^{n_i/2}}{\prod_{j=1}^{k} \left| \sum_{i=q_{j-1}^*+1}^{q_j^*} A_{ii}/n_j^* \right|^{n_j^*/2}} \tag{2.10}$$

where

$$n_j^* = \sum_{i=q_{j-1}^*+1}^{q_j^*} n_i .$$

The h-th moment of V_5 is given by

$$\mathbf{E}(V_5^h) = \frac{\prod_{\alpha=1}^{k} (n_\alpha^*)^{hn_\alpha^* p/2}}{\prod_{g=1}^{q} (n_\alpha)^{hn_g p/2}} \prod_{i=1}^{p} \prod_{\alpha=1}^{k} \left\{ \prod_{g=q_{\alpha-1}^*+1}^{q_\alpha^*} \frac{\Gamma[\frac{1}{2}(n_g + hn_g + 1 - i)]}{\Gamma[\frac{1}{2}(n_g + 1 - i)]} \cdot \frac{\Gamma[\frac{1}{2}(n_\alpha^* + 1 - i)]}{\Gamma[\frac{1}{2}(n_\alpha^* + hn_\alpha^* + 1 - i)]} \right\}. \tag{2.11}$$

As in the preceding section, the distribution of $V_5^{1/b}$ is approximated with the Pearson's Type I distribution where b is a properly chosen integer. We chose b to be equal to 6, 15, 20 and 30 accordingly as p is equal to 1, 2, 3 and 4 respectively. In Tables 8 and 9, we have $n_i = n_0$ and $q = kd$. In Table 8, α_1 is the value of α obtained if we use Pearson type approximation whereas α_2 is the value of α obtained by using Box's asymptotic series up to order n^{-13}, where $\mathbf{P}[V_5^* \leq c_5 | H_5] = (1 - \alpha)$, and $V_5^* = -2 \log V_5$. Table 8 indicates that the accuracy of the Pearson type approximation is sufficient

Table 8
Comparison of the Pearson Type Approximation with the Asymptotic Expression for the Distribution of V_5^* ($\alpha_1 = 0.05$)

			$p = 1$		$p = 2$		$p = 3$		$p = 4$	
n_0	q	k	c_5	α_2	c_5	α_2	c_5	α_2	c_5	α_2
10	6	2	8.20	0.0500	18.97	0.0500	34.63	0.0501	56.43	0.0503
10	6	3	9.90	0.0500	23.28	0.0500	42.79	0.0499	69.79	0.0498
20	6	2	8.01	0.0500	17.89	0.0500	31.45	0.0500	49.16	0.0500
20	6	3	9.70	0.0500	22.09	0.0500	39.30	0.0500	61.81	0.0499
30	6	2	7.94	0.0500	17.55	0.0500	30.54	0.0500	47.20	0.0500
30	6	3	9.63	0.0500	21.72	0.0500	38.28	0.0500	59.61	0.0500

Table 9

Upper 5% Points of V_5^*

$M \backslash p$	1				2				3				4			
	$k=2$		$k=3$		$k=2$		$k=3$		$k=2$		$k=3$		$k=2$		$k=3$	
	$q=4$	$q=6$	$q=6$	$q=9$	$q=4$	$q=6$	$q=6$	$q=9$	$q=4$	$q=6$	$q=6$	$q=9$	$q=4$	$q=6$	$q=6$	$q=9$
1	7.23	9.42	11.26	14.96	19.64	26.38	31.22	42.86	37.36	51.25	60.49	84.63	60.49	84.13	99.17	140.36
2	6.89	8.98	10.76	14.29	17.24	23.17	27.80	38.18	32.02	43.95	52.80	73.90	51.49	71.67	86.14	122.00
3	6.69	8.72	10.48	13.91	16.06	21.58	26.11	35.84	29.32	40.25	48.87	68.42	46.80	65.17	79.22	112.27
4	6.56	8.56	10.29	13.66	15.36	20.64	25.09	34.44	27.70	38.02	46.47	65.06	43.90	61.15	74.92	106.21
5	6.47	8.44	10.17	13.49	14.89	20.01	24.41	33.51	26.60	36.52	44.84	62.79	41.93	58.41	71.95	102.04
6	6.41	8.35	10.07	13.37	14.56	19.57	23.92	32.85	25.82	35.44	43.67	61.16	40.50	56.43	69.79	98.97
7	6.36	8.29	10.00	13.28	14.31	19.23	23.56	32.35	25.23	34.63	42.79	59.93	39.41	54.92	68.15	96.63
8	6.32	8.24	9.95	13.20	14.12	18.97	23.28	31.96	24.77	34.00	42.10	58.96	38.56	53.73	66.85	94.81
9	6.29	8.20	9.90	13.14	13.97	18.77	23.05	31.66	24.40	33.49	41.55	58.18	37.88	52.78	65.80	93.31
10	6.26	8.16	9.87	13.10	13.84	18.60	22.87	31.40	24.09	33.08	41.09	57.55	37.31	51.99	64.94	92.08
11	6.24	8.14	9.84	13.05	13.73	18.46	22.72	31.19	23.84	32.73	40.71	57.01	36.83	51.33	64.21	91.06
12	6.22	8.11	9.81	13.02	13.65	18.34	22.59	31.01	23.62	32.44	40.39	56.56	36.43	50.76	63.58	90.16
13	6.20	8.09	9.79	12.99	13.57	18.24	22.48	30.86	23.44	32.19	40.10	56.16	36.08	50.28	63.04	89.42
14	6.19	8.07	9.77	12.96	13.50	18.15	22.38	30.73	23.28	31.96	39.86	55.82	35.78	49.86	62.58	88.75
15	6.18	8.06	9.75	12.94	13.45	18.07	22.29	30.61	23.14	31.78	39.65	55.53	35.51	49.49	62.17	88.16
16	6.17	8.04	9.73	12.92	13.40	18.00	22.22	30.51	23.02	31.60	39.46	55.27	35.28	49.16	61.81	87.66
17	6.16	8.03	9.72	12.90	13.35	17.94	22.15	30.42	22.91	31.45	39.30	55.03	35.07	48.87	61.48	87.20
18	6.15	8.02	9.71	12.89	13.31	17.89	22.09	30.33	22.81	31.32	39.15	54.83	34.88	48.61	61.19	86.78
19	6.14	8.01	9.70	12.87	13.28	17.84	22.04	30.26	22.72	31.20	39.01	54.63	34.71	48.37	60.92	86.40
20	6.13	8.00	9.69	12.86	13.24	17.79	21.99	30.20	22.64	31.09	38.89	54.47	34.56	48.15	60.68	86.06
22	6.12	7.98	9.67	12.84	13.19	17.72	21.91	30.08	22.50	30.90	38.68	54.17	34.29	47.78	60.26	85.47
24	6.11	7.97	9.66	12.82	13.14	17.66	21.84	29.98	22.39	30.74	38.50	53.91	34.06	47.47	59.92	84.97
26	6.10	7.96	9.64	12.80	13.10	17.60	21.78	29.90	22.28	30.60	38.35	53.71	33.87	47.20	59.61	84.55
28	6.09	7.95	9.63	12.79	13.06	17.55	21.72	29.83	22.20	30.48	38.22	53.52	33.70	46.97	59.35	84.19
30	6.09	7.94	9.62	12.78	13.03	17.51	21.68	29.77	22.12	30.38	38.10	53.35	33.56	46.76	59.13	83.87

for practical purposes. Using the above approximation, we computed the values of c_5 where $\alpha = 0.05$, $k = 2, 3$, $M = 1(1)20(5)30$, $M = n_0 - p$, $p = 1, 2, 3, 4$. These values are given in Table 9. Tukey and Wilks [20] discussed the problem of approximating suitable powers of certain likelihood ratio test statistics with beta distribution. But, they did not discuss in detail as to how good this approximation is.

References

[1] Anderson, T.W. (1958). *An Introduction to Multivariate Statistical Analysis.* Wiley, New York.

[2] Bishop, D.J. (1939). On a comprehensive test for the homogeneity of variances and covariances in multivariate problems. *Biometrika*, **31**, 31–55.

[3] Box, G.E.P. (1949). A general distribution theory for a class of likelihood criteria. *Biometrika*, **36**, 317–346.

[4] Consul, P.C. (1969). The exact distributions of likelihood criteria for different hypotheses. In *Multivariate Analysis — II* (P.R. Krisnaiah, editor), Academic Press, 1969.

[5] Davis, A.W. (1971). Percentile approximations for a class of likelihood criteria. *Biometrika*, **58**, 349–356.

[6] Davis, A.W. and Field, J.B.F. (1971). Tables of some multivariate test criteria. Tech. Report No. 32, Division of Mathematical Statistics, CSIRO, Australia.

[7] Johnson, N.L., Nixon, E., Amos, D.E. and Pearson, E.S. (1963). Table of percentage points of Pearson curves, for given $\sqrt{\beta_1}$ and β_2, expressed in standard measure. *Biometrika*, **50**, 459–497.

[8] Kendall, M.G. and Stuart, A. (1947). *The Advanced Theory of Statistics*, (third edition). Hafner Publishing Company, New York.

[9] Korin, B.P. (1968). On the distribution of a statistic used for testing a covariance matrix. *Biometrika*, **55**, pp. 171–178.

[10] Korin, B.P. (1969). On testing of equality of k covariance matrices. *Biometrika*, **56**, 216–217.

[11] Krishnaiah, P.R. and Lee, J.C. (1976). On covariance structures. *Sankhyā* (to appear).

[12] Krishnaiah, P.R., Lee, J.C. and Chang, T.C. (1975). On the distributions of the likelihood ratio statistics for tests of certain covariance structures of complex multivariate normal populations. ARL TR 75–0169; also see *Biometrika* (1976).

[13] Lee, J.C., Krishnaiah, P.R. and Chang, T.C. (1975). Approximations to the distributions of the determinants of real and complex multivariate beta matrices. ARL 75–0168.

[14] Lee, J.C., Chang, T.C. and Krishnaiah, P.R. (1976). On the distribution of the likelihood ratio test statistic for compound symmetry. *S. African Statist. J.*

[15] Mauchly, J.W. (1940). Significant test for sphericity of a normal n-variate distribution. *Ann. Math. Statist.*, **1**, 204–209.

[16] Mathai, A.M. and Rathie, P.N. (1970). The exact distribution for the sphericity test, *J. Statis. Res.*, (Dacca), **4**, 140–159.

[17] Mathai, A.M. and Rathie, P.N. (1971). The problem of testing independence. *Statistica*, **31**, 673–688.

[18] Nagarsenker, B.M. and Pillai, K.C.S. (1973). The distribution of the sphericity test criterion. *J. Multivariate Anal.*, **3**, 226–235.

[19] Nagarsenker, B.N. and Pillai, K.C.S. (1973). Distribution of the likelihood ratio criterion for testing a hypothesis specifying a covariance matrix. *Biometrika*, **60**, 359–364.

[20] Tukey, J.W. and Wilks, S.S. (1946). Approximation of the distribution of the product of beta variables by a single beta variable. *Ann. Math. Statist.*, **17**, 318–324.

[21] Wilks, S.S. (1932). Certain generalizations in the analysis of variance. *Biometrika*, **24**, 471–494.

[22] Wilks, S.S. (1935). On the independence of k sets of normally distributed statistical variables. *Econometrica*, **3**, 309–325.

P.R. Krishnaiah, ed., *Multivariate Analysis–IV*

A CHARACTERIZATION OF A BIVARIATE GAMMA DISTRIBUTION

Eugene LUKACS*

Bowling Green State University, Bowling Green, Ohio, U.S.A.

The following theorem is proven: Let $\mathbf{X}_\alpha = (X_{1\alpha}, X_{2\alpha})$ $\alpha = 1, 2, \ldots, n$ be a sample of size n taken from a bivariate population whose distribution function $F(x_1, x_2)$ is nondegenerate and whose marginal distributions are also nondegenerate. Let $f(t_1, t_2)$ be the characteristic function of $F(x_1, x_2)$ and suppose that the components $X_{1\alpha}, X_{2\alpha}$, of $\mathbf{X}_\alpha$ are nonnegative random variables. Assume also that F has moments up to and including order 3. Three statistics S, T_1, T_2 are constructed which have the following property: The population distribution function F is a bivariate gamma distribution if and only if the statistic S as well as the statistics T_1 and T_2 have zero regression on $\mathbf{\Lambda} = (\Lambda_1, \Lambda_2)$ where $\Lambda_j = \Sigma_{\alpha=1}^{n} X_{j\alpha}$ $(j = 1, 2)$.

1. Introduction

Let $\mathbf{X} = (X_1, X_2, \ldots, X_p)$ be a p-dimensional, normally distributed random vector with density function

$$p(x_1, \ldots, x_p) = \frac{1}{(2\pi)^{p/2}\sqrt{|\mathbf{P}|}} \exp[-\tfrac{1}{2}\mathbf{x}\mathbf{P}^{-1}\mathbf{x}']$$

where P is the determinant of the correlation matrix $\mathbf{P} = \|\rho_{ij}\|$ of $\mathbf{X}$ with $\rho_{ii} = \mathbf{E}(X_i^2) = 1$ and $\rho_{ij} = \mathbf{E}(X_iX_j)$ $(i \neq j,\ i, j = 1, 2, \ldots, p)$. The characteristic function $g_1(\mathbf{t}) = g_1(t_1, \ldots, t_p)$ of the random vector $\mathbf{Z} = (X_1^2, \ldots, X_p^2)$ with non-negative components is then

$$g_1(\mathbf{t}) = \{\det\|\mathbf{I} - 2i\mathbf{P}\mathbf{T}\|\}^{-\frac{1}{2}}$$

Here $\mathbf{T}$ is the diagonal matrix whose diagonal elements are $t_1, t_2, \ldots, t_p$, while $\mathbf{I}$ is a diagonal matrix whose diagonal elements are all 1's. The distribution which corresponds to the characteristic function $g_1(\mathbf{t})$ is the multivariate extension of the chi-square distribution with one degree of freedom. The p-variate analogue to the chi-square distribution with n degrees of freedom has the characteristic function

$$g_n(\mathbf{t}) = \{\det\|\mathbf{I} - 2i\mathbf{P}\mathbf{T}\|\}^{-n/2}.$$

* Research supported by the National Science Foundation under grant MPS 72–04986.

The multivariate chi-square distribution with the above characteristic function was derived by Krishnamoorthy and Parthasarathy [1]. It is not known whether the characteristic function $g_n(\boldsymbol{t})$ is infinitely divisible (see [2]) that is whether the function

$$g_\alpha(\boldsymbol{t}) = \{\det \|\boldsymbol{I} - 2\mathrm{i}\boldsymbol{PT}\|\}^{-\alpha}$$

is a characteristic function for all $\alpha > 0$ and all positive integers p. This problem is at present solved only for the bivariate and trivariate case [4].

We put $p = 2$, $\rho_{12} = \rho_{21} = \rho$ so that

$$\boldsymbol{P} = \begin{Vmatrix} 1, \rho \\ \rho, 1 \end{Vmatrix}.$$

Then

$$g_\alpha(t_1, t_2) = \{1 - 2\mathrm{i}t_1 - 2\mathrm{i}t_2 - 4t_1t_2(1-\rho^2)\}^{-\alpha} \tag{1.1}$$

D. Vere-Jones [3] proved that the function $g_\alpha(t_1, t_2)$ is a characteristic function for every $\alpha > 0$. The functions $g_\alpha(t_1, t_2)$ are positive powers of the characteristic function $g_{n/2}(t_1, t_2)$ of a bivariate chi-square distribution with n degrees of freedom.

A bivariate gamma distribution is obtained from (1.1) by introducing scale factors. Its characteristic function is given by

$$g(t_1, t_2 \mid \theta_1, \theta_2, \lambda) = \left\{1 - \frac{\mathrm{i}t_1}{\theta_1} - \frac{\mathrm{i}t_2}{\theta_2} + \frac{1-\rho^2}{\theta_1\theta_2}(\mathrm{i}t_1)(\mathrm{i}t_2)\right\}^{-\lambda} \tag{1.2}$$

where $\lambda > 0$ and $\rho^2 \leq 1$.

In Section 2 we formulate a characterization theorem for the distribution which corresponds to the characteristic function (1.2). In Section 3 we derive differential equations which the logarithm of the characteristic function of the population must satisfy. We solve these equations and we prove the theorem in Section 4. In Section 5 we make some supplementary remarks concerning the p-variate ($p > 2$) case.

2. Formulation of the theorem

Let $\boldsymbol{Y} = (Y_1, Y_2, \cdots, Y_p)$ be a random vector and let Z be a random variable and suppose that $\boldsymbol{Y}$ and Z are defined on the same probability space and that the expectations $\mathbf{E}(\boldsymbol{Y}) = (\mathbf{E}(Y_1), \mathbf{E}(Y_2), \ldots, \mathbf{E}(Y_p))$ and $\mathbf{E}(Z)$ exist. The random variable Z is said to have constant regression on the vector $\boldsymbol{Y}$ if the relation

$$\mathbf{E}(Z \mid \boldsymbol{Y}) = \beta$$

holds almost everywhere. Here β is a real constant, if $\beta = 0$ then we say that Z has zero regression on $\boldsymbol{Y}$.

The most important tool for the proof of our theorem is the following lemma.

Lemma. *Let Z be a random variable and let* $\boldsymbol{Y}$ *be a p-dimensional random vector. Suppose that* $\mathbf{E}(Z)$ *and* $\mathbf{E}(\boldsymbol{Y})$ *exist. Z has constant regression on* $\boldsymbol{Y}$ *if and only if the relation*

$$\mathbf{E}(Z e^{it'\boldsymbol{Y}}) = \beta \mathbf{E}(e^{it'\boldsymbol{Y}})$$

holds for all real (nonrandom) vectors $\boldsymbol{t} = (t_1, t_2, \dots, t_p)$.

The lemma is an immediate generalization of a univariate result (see [1]) and is proved in exactly the same way.

In the following we shall use the greek letters α, β, and γ as subscripts for summation. These subscripts will always be taken from the first n positive integers with the following convention concerning the summations.

$$\sum_{\alpha,\beta,\gamma}^{(3)} h_{\alpha\beta\gamma} = \sum_{\substack{\alpha=1 \\ \alpha\neq\beta,}}^{n} \sum_{\substack{\beta=1 \\ \alpha\neq\gamma,}}^{n} \sum_{\substack{\gamma=1 \\ \beta\neq\gamma}}^{n} h_{\alpha\beta\gamma} \quad \text{and} \quad \sum_{\alpha,\beta}^{(2)} h_{\alpha\beta} = \sum_{\substack{\alpha=1 \\ \alpha\neq\beta}}^{n} \sum_{\beta=1}^{n} h_{\alpha\beta}$$

We also write $n^{(3)} = n(n-1)(n-2)$ and $n^{(2)} = n(n-1)$.

We consider a sample

$$\boldsymbol{X}_\alpha = (X_{1\alpha}, X_{2\alpha}) \quad (\alpha = 1, 2, \dots, n)$$

of size n from a bivariate population and we introduce the following four statistics

$$S = \frac{1}{n^{(3)}} \sum_{\alpha,\beta,\gamma}^{(3)} [X_{1\alpha}^2 X_{2\beta} X_{2\gamma} - X_{2\alpha}^2 X_{1\beta} X_{1\gamma}], \tag{2.1}$$

$$\begin{aligned} T_j = {} & \frac{1}{n^{(2)}} \sum_{\alpha,\beta}^{(2)} [X_{j\alpha}^3 X_{j\beta} - 2X_{j\alpha}^2 X_{j\beta}^2] \\ & + \frac{1}{n^{(3)}} \sum_{\alpha,\beta,\gamma}^{(3)} X_{j\alpha}^2 X_{j\beta} X_{j\gamma} \quad (j = 1, 2), \end{aligned} \tag{2.2}$$

$$\boldsymbol{\Lambda} = (\Lambda_1, \Lambda_2) \quad \text{with } \Lambda_j = \sum_{\alpha=1}^{n} X_{j\alpha} \quad (j = 1, 2). \tag{2.3}$$

The statistics S, T_1, and T_2 are one-dimensional random variables which depend on the components of $\boldsymbol{X}_\alpha$ while $\boldsymbol{\Lambda}$ is a two-dimensional random vector.

Theorem. *Let $\mathbf{X}_\alpha = (X_{1\alpha}, X_{2\alpha})$, $\alpha = 1, 2, \ldots, n$ be a sample of size n taken from a bivariate population whose distribution function $F(x_1, x_2)$ is non-degenerate and whose marginal distributions are also non-degenerate. Let $f(t_1, t_2)$ be the characteristic function of $F(x_1, x_2)$ and suppose that the components $X_{1\alpha}$, $X_{2\alpha}$ of $\mathbf{X}_\alpha$ are non-negative random variables. Assume also that the moments $F(x_1, x_2)$ exist up to and including order 3. The population distribution function is a bivariate gamma distribution (with characteristic function of the form* (1.2)) *if and only if the statistic S as well as the statistics T_1 and T_2 have zero regression on $\mathbf{\Lambda}$.*

3. Derivation of the differential equation

Let $\boldsymbol{t} = (t_1, t_2)$ be a non-random vector. It follows from the assumptions of the theorem and from the lemma that the relations

$$\begin{cases} \mathbf{E}(S e^{i\boldsymbol{t}'\mathbf{\Lambda}}) = 0, \\ \mathbf{E}(T_j e^{i\boldsymbol{t}'\mathbf{\Lambda}}) = 0 \quad (j = 1, 2) \end{cases} \tag{3.1}$$

must be satisfied for all real vectors $\boldsymbol{t}$. Written in greater detail this means that

$$\frac{1}{n^{(3)}} \mathbf{E}\left\{ \sum_{\alpha,\beta,\gamma}^{(3)} [X_{1\alpha}^2 X_{2\beta} X_{2\gamma} - X_{2\alpha}^2 X_{1\beta} X_{1\gamma}] e^{i\boldsymbol{t}'\mathbf{\Lambda}} \right\} = 0 \tag{3.2a}$$

$$\frac{1}{n^{(2)}} \mathbf{E}\left\{ \sum_{\alpha,\beta}^{(2)} [X_{j\alpha}^3 X_{j\beta} - 2X_{j\alpha}^2 X_{j\beta}^2] e^{i\boldsymbol{t}'\mathbf{\Lambda}} \right\}$$
$$+ \frac{1}{n^{(3)}} \mathbf{E}\left\{ \sum_{\alpha,\beta,\gamma}^{(3)} [X_{j\alpha}^2 X_{j\beta} X_{j\gamma}] e^{i\boldsymbol{t}'\mathbf{\Lambda}} \right\} = 0 \quad (j = 1, 2). \tag{3.2b}$$

We note that

$$\boldsymbol{t}'\mathbf{\Lambda} = t_1\Lambda_1 + t_2\Lambda_2 = t_1 \sum_{\alpha=1}^{n} X_{1\alpha} + t_2 \sum_{\alpha=1}^{n} X_{2\alpha}$$

so that

$$e^{i\boldsymbol{t}'\mathbf{\Lambda}} = \prod_{\alpha=1}^{n} \exp(it_1 X_{1\alpha} + it_2 X_{2\alpha}).$$

The characteristic function of the population distribution function is

$$f(t_1, t_2) = \mathbf{E}(e^{i\boldsymbol{t}'\mathbf{X}}) = \mathbf{E}[\exp(it_1 X_1 + it_2 X_2)] \tag{3.3}$$

and we write

$$f_j = \frac{\partial f}{\partial t_j} = i\mathbf{E}[X_j \exp(it_1X_1 + it_2X_2)] \quad (j = 1, 2), \tag{3.4a}$$

$$f_{jj} = \frac{\partial^2 f}{\partial t_j^2} = -\mathbf{E}[X_j^2 \exp(it_1X_1 + it_2X_2)] \quad (j = 1, 2), \tag{3.4b}$$

$$f_{jjj} = \frac{\partial^3 f}{\partial t_j^3} = -i\mathbf{E}[X_j^3 \exp(it_1X_1 + it_2X_2)] \quad (j = 1, 2). \tag{3.4c}$$

Since the vectors $\mathbf{X}_\alpha$ $(\alpha = 1, 2, \ldots, n)$ are independently and identically distributed we see easily that

$$\mathbf{E}(X_{1\alpha}^2 X_{2\beta} X_{2\gamma} e^{it'\Lambda}) = f_{11} f_2^2 (f)^{n-3} \tag{3.5a}$$

and similarily

$$\mathbf{E}(X_{2\alpha}^2 X_{1\beta} X_{1\gamma} e^{it'\Lambda}) = f_{22} f_1^2 (f)^{n-3}, \tag{3.5b}$$

$$\mathbf{E}(X_{j\alpha}^3 X_{j\beta} e^{it'\Lambda}) = f_{jjj} f_j (f)^{n-2} \quad (j = 1, 2), \tag{3.5c}$$

$$\mathbf{E}(X_{j\alpha}^2 X_{j\beta}^2 e^{it'\Lambda}) = f_{jj}^2 (f)^{n-2} \quad (j = 1, 2), \tag{3.5d}$$

$$\mathbf{E}(X_{j\alpha}^2 X_{j\beta} X_{j\gamma} e^{it'\Lambda}) = f_{jj} f_j^2 (f)^{n-3} \quad (j = 1, 2), \tag{3.5e}$$

provided that $\alpha \neq \beta$, $\alpha \neq \gamma$, $\beta \neq \gamma$. It follows from (3.2a), (3.5a) and (3.5b) that

$$f_{11} f_2^2 (f)^{n-3} - f_{22} f_1^2 (f)^{n-3} = 0. \tag{3.6}$$

In a similar way we use the two relations (3.2b) together with (3.5c), (3.5d) and (3.5e) and obtain

$$f_{jjj} f_j (f)^{n-2} + f_{jj} f_j^2 (f)^{n-3} - 2 f_{jj}^2 (f)^{n-2} = 0 \quad (j = 1, 2). \tag{3.7}$$

Since $f(t_1, t_2)$ is a characteristic function there exists a neighborhood $\mathcal{N}$ of the origin such that $f(t_1, t_2) \neq 0$ if $(t_1, t_2) \in \mathcal{N}$. We restrict (t_1, t_2) to $\mathcal{N}$ and can then rewrite equations (3.6) and (3.7) in the following form:

$$\left(\frac{f_{11}}{f}\right)\left(\frac{f_2}{f}\right)^2 - \left(\frac{f_{22}}{f}\right)\left(\frac{f_1}{f}\right)^2 = 0 \tag{3.8}$$

$$\frac{f_{jjj}}{f}\frac{f_j}{f} + \frac{f_{jj}}{f}\left(\frac{f_j}{f}\right)^2 - 2\left(\frac{f_{jj}}{f}\right)^2 = 0 \quad (j = 1, 2). \tag{3.9}$$

We introduce the logarithm of the characteristic function f by writing $\phi = \log f$, this function exists in $\mathcal{N}$. We also denote the partial derivatives of ϕ by lower subscripts. We express the partial derivatives of ϕ by the partial derivatives of f and obtain

$$\frac{f_j}{f}=\phi_j,\qquad \frac{f_{jj}}{f}=\phi_{jj}+(\phi_j)^2,\qquad \frac{f_{jjj}}{f}=\phi_{jjj}+3\phi_{jj}\phi_j+\phi_j^3\quad (j=1,2). \tag{3.10}$$

We substitute the expressions of (3.10) into (3.8) and (3.9) and get the differential equations

$$\phi_{11}\phi_2^2-\phi_{22}\phi_1^2=0, \tag{3.11}$$

$$\phi_{jjj}\phi_j-2\phi_{jj}^2=0\quad (j=1,2). \tag{3.12}$$

Our assumptions assure that $\phi_j\neq 0$ and $\phi_{jj}\neq 0$ so that we can rewrite equations (3.11) and (3.12) as

$$\phi_{11}/\phi_1^2=\phi_{22}/\phi_2^2, \tag{3.13}$$

$$\phi_{jjj}/\phi_{jj}=2\phi_{jj}/\phi_j\quad (j=1,2). \tag{3.14}$$

4. Proof of the theorem

We consider first the two equations (3.14). They can be written in the form

$$\frac{\partial}{\partial t_j}[\ln(\phi_{jj}/\phi_j^2)]=0\quad (j=1,2).$$

This means that ϕ_{jj}/ϕ_j^2 is independent of the variable t_j, or written in greater detail

$$\phi_{11}/\phi_1^2=A_2(t_2),\qquad \phi_{22}/\phi_2^2=A_1(t_1).$$

It follows then from (3.13) that

$$A_2(t_2)=A_1(t_1).$$

This is only possible if both functions A_1 and A_2 reduce to the same constant and we write

$$A_1(t_1)=A_2(t_2)=-a\quad \text{(say)}.$$

Therefore

$$\phi_{jj}/\phi_j^2=-a\quad (j=1,2)$$

or $\partial/\partial t_j\,(1/\phi_j)=a$ $(j=1,2)$. We integrate these equations and obtain

$$\phi_1 = [at_1 + B_2(t_2)]^{-1}, \tag{4.1a}$$

$$\phi_2 = [at_2 + B_1(t_1)]^{-1} \tag{4.1b}$$

Integrating equations (4.1a) and (4.1b) we see that

$$\phi(t_1, t_2) = \frac{1}{a} \ln \{[at_1 + B_2(t_2)]C_2(t_2)\} \tag{4.2a}$$

$$\phi(t_1, t_2) = \frac{1}{a} \ln \{[at_2 + B_1(t_1)]C_1(t_1)\} \tag{4.2b}$$

These are two different representations of the same function $\phi(t_1, t_2)$ so that the right-hand sides of equations (4.2a) and (4.2b) are necessarily equal and we conclude that

$$[at_1 + B_2(t_2)]C_2(t_2) = [at_2 + B_1(t_1)]C_1(t_1). \tag{4.3}$$

We differentiate (4.3) first with respect to t_1 and then with respect to t_2 and obtain the equation

$$C_2' = C_1'.$$

The function on the left-hand side of this equation depends only on t_2, the one on the right only on t_1; this equation is therefore only possible if both sides are equal to the same constant b, i.e. if

$$C_1' = b; \qquad C_2' = b. \tag{4.4}$$

Therefore

$$C_1(t_1) = bt_1 + c, \qquad C_2(t_2) = bt_2 + d \tag{4.4a}$$

where c and d are constants. We substitute this into (4.3) and get

$$[at_1 + B_2(t_2)(bt_2 + d) = [at_2 + B_1(t_1)]\,(bt_1 + c)$$

or

$$bB_2(t_2)t_2 + adt_1 + dB_2(t_2) = bB_1(t_1)t_1 + act_2 + cB_1(t_1).$$

Therefore

$$(bt_2 + d)B_2(t_2) - act_2 = (bt_1 + c)B_1(t_1) - adt_1. \tag{4.5}$$

Since the left-hand side is a function of t_2, while the right-hand side depends only on t_1, equation (4.5) can hold only if both sides are equal to the same constant, say k. Then

$$B_1(t_1) = \frac{k + adt_1}{bt_1 + c}, \qquad B_2(t_2) = \frac{k + act_2}{bt_2 + d}.$$

We see then from (4.2a) [or alternatively from (4.2b)] that

$$\phi(t_1, t_2) = \frac{1}{a} \ln\{abt_1t_2 + adt_1 + act_2 + k\}.$$

Since $\phi(0,0) = 0$ we have $k = 1$ and

$$\phi(t_1, t_2) = \frac{1}{a} \ln\{1 + adt_1 + act_2 + abt_1t_2\}. \tag{4.6}$$

We compute the first and second derivatives of $\phi(t_1, t_2)$ and set then $t_1 = t_2 = 0$. In this way we obtain

$$\left.\frac{\partial \phi}{\partial t_1}\right|_{00} = d, \qquad \left.\frac{\partial \phi}{\partial t_2}\right|_{00} = c$$

$$\left.\frac{\partial^2 \phi}{\partial t_1^2}\right|_{00} = -ad^2, \qquad \left.\frac{\partial^2 \phi}{\partial t_2^2}\right|_{00} = -ac^2, \qquad \left.\frac{\partial^2 \phi}{\partial t_1 \partial t_2}\right|_{00} = b - acd.$$

These quantities can be expressed in terms of the cumulants and one concludes easily that d and c are purely imaginary,

$$c = \mathrm{i}\gamma, \qquad d = \mathrm{i}\delta$$

and that $a < 0$. We write from now on

$$1/a = -\lambda \quad (\lambda > 0)$$

and rewrite (4.6) in the form

$$\phi(t_1, t_2) = -\lambda \ln\left\{1 - \frac{\delta}{\lambda}(\mathrm{i}t_1) - \frac{\gamma}{\lambda}(\mathrm{i}t_2) + \frac{b}{\lambda}(\mathrm{i}t_1)(\mathrm{i}t_2)\right\}. \tag{4.7}$$

Since $\phi(t_1, t_2)$ is the logarithm of the characteristic function of a random vector with non-negative components, we see that

$$\delta > 0 \quad \text{and} \quad \gamma > 0. \tag{4.7a}$$

The characteristic function which corresponds to (4.7) has the form[1]

$$f(t_1, t_2) = \left\{1 - \frac{\delta}{\lambda}(\mathrm{i}t_1) - \frac{\gamma}{\lambda}(\mathrm{i}t_2) + \frac{b}{\lambda}(\mathrm{i}t_1)(\mathrm{i}t_2)\right\}^{-\lambda}. \tag{4.8}$$

We transform the scale of the variables in (4.8) by putting

$$\frac{\delta}{2\lambda} t_1 = u_1, \qquad \frac{\gamma}{2\lambda} t_2 = u_2$$

[1] Formula (4.8) was derived under the restriction that $(t_1, t_2) \in \mathcal{N}$. However, it can be extended in the usual manner to all values of (t_1, t_2).

and obtain

$$g(u_1, u_2) = \left\{1 - 2iu_1 - 2iu_2 - 4\frac{\lambda b}{\delta\gamma} u_1 u_2\right\}^{-\lambda}.$$

The function $g(u_1, u_2)$ is a characteristic function and is infinitely divisible if

$$1 - \rho^2 = \frac{\lambda b}{\delta\gamma}$$

where ρ is a coefficient of correlation. That is if

$$0 \leq \frac{\lambda b}{\delta\gamma} \leq 1$$

or

$$0 \leq b \leq \frac{\delta\gamma}{\lambda} \quad (\gamma > 0,\ \delta > 0,\ \lambda > 0). \tag{4.9}$$

The fact that $g(u_1, u_2)$ is a characteristic function assures that $f(t_1, t_2)$ as given by (4.8) is the characteristic function of a bivariate gamma distribution. The conditions of the theorem are therefore sufficient, their necessity follows in the usual way from the lemma.

5. Remarks concerning the p-variate case

The statistics to be considered in the p-variate case are completely analogous to those of the bivariate case. Let

$$S_j = \frac{1}{n^{(3)}} \sum_{\alpha,\beta,\gamma}^{(3)} [X_{1\alpha}^2 X_{j\beta} X_{j\gamma} - X_{j\alpha}^2 X_{1\beta} X_{1\gamma}] \quad (j = 2, \ldots, p),$$

$$T_j = \frac{1}{n^{(2)}} \sum_{\alpha,\beta}^{(2)} [X_{j\alpha}^3 X_{j\beta} - 2X_{j\alpha}^2 X_{j\beta}^2]$$

$$+ \frac{1}{n^{(3)}} \sum_{\alpha,\beta,\gamma}^{(3)} X_{j\alpha}^2 X_{j\beta} X_{j\gamma} \quad (j = 1, 2, \ldots, p),$$

$$\Lambda = \left(\sum_{\alpha=1}^{n} X_{1\alpha}, \sum_{\alpha=1}^{n} X_{2\alpha}, \ldots, \sum_{\alpha=1}^{n} X_{p\alpha} \right).$$

One assumes again that the $p - 1$ statistics S_γ and the p statistics T_γ have zero regression on Λ. This yields again a system of differential equations for the logarithm of the characteristic function of the population distribu tion function. This system can be solved (by a somewhat tedious inductive procedure). The resulting solution of this system is the a-th power ($a < 0$)

of an expression which is linear in each of the variables $t_1, \ldots, t_p$ and it contains terms with single t_j as well as terms with products of $2, 3, \cdots, p$ of these variables. Conditions have to be imposed on the coefficients as well as on the exponent a which assure that the solution is a characteristic function. This part of the work can not be completed at present since for general p we do not know whether the p-variate extension of the chi-square distribution is infinitely divisible. Hence we can at present not say whether any restrictions must be imposed on the exponent a, beyond the assumption that it is negative.

References

[1] Krishnamoorthy, A.S. and Parthasarathy, M. (1951). A multivariate gamma distribution. *Ann. Math. Statist.* **22**, 549–557 [Erratum: *Ann. Math. Statist.* **33**, 229 (1960)].

[2] Lukacs, E. and Laha, R.G. (1964). *Applications of characteristic functions.* Charles Griffin & Company, London.

[3] Vere-Jones, D. (1967). The infinite divisibility of a bivariate gamma distribution. *Sankhyā A* **29**, 421–422.

[4] Griffith, R.C. (1970). Infinitely divisible multivariate gamma distributions. *Sankhyā A* **32**, 393–404.

P.R. Krishnaiah, ed., *Multivariate Analysis–IV*

USE OF HOTELLING'S GENERALIZED T_0^2 IN MULTIVARIATE TESTS

Claude McHENRY
and
A.M. KSHIRSAGAR*
Texas A&M University, Texas, U.S.A.

The problem of discrimination among several normal populations can be looked upon as the problem of the relationship between two sets of variables, one of which has a multinormal distribution and the other consists of dummy variables. The canonical variables corresponding to each set and the canonical correlations play an important role in this and the canonical variables corresponding to the dummy variables are sometimes called discriminant functions from the space of the dummy variables. Goodness of fit tests of one or more hypothetical discriminant functions were first considered by Williams and Bartlett and later by Kshirsagar. These tests are based on factorization of Wilks's Λ criterion. It is well known that Hotelling's generalized T_0^2 (or Hotelling-Lawley trace criterion) is a competitor to Wilks's Λ for testing the equality of means of several populations but it was not known whether this can be used for such goodness of fit tests also. This paper shows that tests based on T_0^2 can be derived for this purpose. Exact null distribution of T_0^2 can be derived. The discriminant functions tested are from the dummy variable space.

1. Introduction

Consider $k = q + 1$ p-variate normal populations π_α $(\alpha = 1, 2, \cdots, k)$ with means $\boldsymbol{\mu}_\alpha$ (column vector) and the same variance-covariance matrix Σ. If $\boldsymbol{x}$ denotes the column vector of the p variables, one can express $\mathbf{E}(\boldsymbol{x})$ as

$$\mathbf{E}(\boldsymbol{x}) = \boldsymbol{\mu}_k + \beta \boldsymbol{y}, \tag{1.1}$$

where

$$\beta = [\boldsymbol{\mu}_1 - \boldsymbol{\mu}_k \,|\, \boldsymbol{\mu}_2 - \boldsymbol{\mu}_k \,|\, \ldots \,|\, \boldsymbol{\mu}_q - \boldsymbol{\mu}_k], \tag{1.2}$$

$$\boldsymbol{y}' = [y_1, y_2, \ldots, y_q], \tag{1.3}$$

and y_α $(\alpha = 1, 2, \ldots, q)$ are dummy variables taking the values 1 or 0 depending on whether $\boldsymbol{x}$ belongs to π_α or not. By the singular decomposi-

* The authors are grateful to the Air Force Office of Scientific Research for support of the research, under contract no. AFOSR–74-2676, Project 9769–05.

tion theorem [2], the matrix β can be expressed as

$$\beta = A'\Delta C \tag{1.4}$$

where A and C are respectively $p \times p$ and $q \times q$ and are orthogonal i.e. $A'A = I_p$, $C'C = I_q$. Δ is $p \times q$ and has non-zero elements $\delta_1, \delta_2, \ldots, \delta_f$ on its leading diagonal and zeros elsewhere. δ_i^2 are the eigenvalues of $\beta\beta'$ or $\beta'\beta$ and f is the rank of β. We shall denote by $\boldsymbol{a}_i$, the i-th column of A' and by $\boldsymbol{c}_i$, the i-th column of C'. Further we partition C as

$$C = \underset{q}{\begin{bmatrix} C_1 \\ \hline C_2 \end{bmatrix}} \begin{matrix} f \\ q-f \end{matrix} \tag{1.5}$$

From (1.1) and (1.4), it is easy to see that the linear functions $\boldsymbol{a}_i'\boldsymbol{x}$ $(i = f+1, \ldots, p)$ have the same mean for all the populations but $\boldsymbol{a}_i'\boldsymbol{x}$ $(i = 1, \ldots, f)$ have different means, and are thus a set of discriminant functions.

Alternatively, transforming from $\boldsymbol{y}$ to $\boldsymbol{t}$ and $\boldsymbol{e}$ where

$$\boldsymbol{t} = [t_1, \ldots, t_f]' = C_1\boldsymbol{y} \tag{1.6}$$

and

$$\boldsymbol{e} = [e_1, \ldots, e_{q-f}]' = C_2\boldsymbol{y} \tag{1.7}$$

we can rewrite (1.1) as

$$\begin{aligned} \mathbf{E}(\boldsymbol{x}) &= \boldsymbol{\mu}_k + \beta C'C\boldsymbol{y} \\ &= \boldsymbol{\mu}_k + \Gamma_1\boldsymbol{t} + \Gamma_2\boldsymbol{e} \end{aligned} \tag{1.8}$$

where

$$\beta C' = [\underset{f}{\Gamma_1} | \underset{q-f}{\Gamma_2}]p$$

and $\Gamma_2 = 0$ by (1.4) and (1.5). Thus the original relationship between $\boldsymbol{x}$ and the vector $\boldsymbol{y}$ of q components is actually a relationship of $\boldsymbol{x}$ with only f linear combinations $\boldsymbol{t}$ and the remaining linear combinations $\boldsymbol{e}$ do not enter into it. On account of this $t_1, t_2, \ldots, t_f$ are called the "discriminant functions" from the dummy variables space, analogous to $\boldsymbol{a}_i'\boldsymbol{x}$ $(i = 1, \ldots, f)$, the discriminant functions from the $\boldsymbol{x}$-space. However, there is one point to be remembered here that since $\boldsymbol{y}$ or $\boldsymbol{t}$ has no physical meaning and since (1.8) can be written as

$$\mathbf{E}(\boldsymbol{x}) = \boldsymbol{\mu}_k + \Gamma_1 J'J\boldsymbol{t}$$

where J is any $f \times f$ orthogonal matrix, $\boldsymbol{t}$ is not unique and can be replaced by any such $J\boldsymbol{t}$. The columns of Γ_1 or Γ_2 are contrasts among the mean vectors $\boldsymbol{\mu}_\alpha$ and what we imply by saying $\boldsymbol{t}$ are discriminators is that the corresponding coefficient matrix Γ_1 consists of non-null contrasts among the μ_α's, while the remaining contrasts Γ_2 are all null. There is thus a correspondence between the non-null contrasts among the $\boldsymbol{\mu}_\alpha$'s and the discriminators from the dummy variables space.

If one is considering factorial experiments with measurements on several responses, the contrasts may be various factorial effects and interactions and (1.1) will be a response surface type relationship. (See for example [8], where Williams analyses data on lamb carcasses by using dummy variables. See also [3].)

In practice, however, neither f, the number of discriminators nor $\boldsymbol{t}$, the discriminators are known and inference has got to be drawn about these only from sample observations. If we have samples of sizes n_α ($\alpha = 1, \ldots, k$) from π_α, and we estimate the regression equation (1.2) from these, we shall get the following multivariate analysis of variance table:

Table 1

Source	Degrees of freedom (d.f.)	$p \times p$ matrix of sums of squares and products (s.s. and s.p.)
Regression of $\boldsymbol{x}$ on $\boldsymbol{y}$	q	$B = C_{xy}C_{yy}^{-1}C_{yx}$
Error	$n - q$	$W = C_{xx \cdot y}$
Total	n	$B + W = C_{xx}$

Here $n = n_1 + \ldots + n_k - 1$ and $C_{xx\cdot y} = C_{xx} - C_{xy}C_{yy}^{-1}C_{yx}$ while C_{xx}, C_{xy} etc. denote the matrix of the corrected s.s. and s.p. of all the observations on $\boldsymbol{x}$ and $\boldsymbol{y}$. If the experimenter is able to propose some s linearly independent functions $Q\boldsymbol{y}$ of $\boldsymbol{y}$ as possible discriminators from the dummy variables, we set up the following null hypothesis H:

H: The assigned s functions $Q\boldsymbol{y}$ are good enough as discriminant functions from the y-space.

Obviously H will be true if (i) $s = f$ *and* if (ii) $Q\boldsymbol{y}$ and $\boldsymbol{t}$ are the same or $Q\boldsymbol{y}$ are linear combinations of $t_1, \ldots, t_f$. An alternative way will be for the experimenter to propose some s contrasts among the means $\boldsymbol{\mu}_\alpha$ which he thinks are the only ones that are non-null and set up the hypothesis that the rest $q - s$ contrasts are all null. In either case, the hypothesis can be tested by considering the regression of $\boldsymbol{x}$ on $Q\boldsymbol{y}$. The matrix of regression s.s. and s.p. is

$$C_{xy}Q'(QC_{yy}Q')^{-1}QC_{yx} = C_{xt}C_{tt}^{-1}C_{tx} \quad (\text{df. } s) \tag{1.9}$$

where we have now denoted $Q\boldsymbol{y}$ by $\boldsymbol{t}$ because if the hypothesis H is true, $Q\boldsymbol{y}$ and $\boldsymbol{t}$ both account for the same contribution (1.9). The matrix representing the deviation from H is, therefore,

$$L = C_{xy}C_{yy}^{-1}C_{yx} - C_{xt}C_{tt}^{-1}C_{tx} \tag{1.10}$$

with $q - s$ d.f. Whether this is significant or not can be tested by comparing it with the error matrix W of Table 1. The usual multivariate criteria used for this purpose are

$$\text{Wilks's } \Lambda = |W|/|W + L|, \tag{1.11}$$

or

$$\text{Hotelling–Lawley's } T_0^2 = \text{tr } W^{-1}L. \tag{1.12}$$

The distributions of Λ, and T_0^2, under the null hypothesis H, depend only on p (the order of the matrices W, L), $q - s$ (the d.f. of L) and $n - s = (n - q) + (q - s)$ (the d.f. of $W + L$). We shall denote, therefore, these distributions as $\Lambda(n - s, p, q - s)$, and $T_0^2(n - s, p, q - s)$ distributions. The exact and approximate percentage points of these statistics are well known in the literature (see for example [7]). It is also well known that $T_0^2(a, b, c)$ and $T_0^2(a, c, b)$ are the same distributions [7].

As we have already seen H consists of 2 parts (i) $s = f$ and (ii) $Q\boldsymbol{y}$ and $\boldsymbol{t}$ are "basis" of the same vector sub-space in the $\boldsymbol{y}$-space. If H is rejected, we will naturally wish to know whether the rejection was due to inadequate number s of the proposed functions $Q\boldsymbol{y}$ or due to the inadequate specifications of their directions, or both. Wilks's Λ criterion has been factorized (see [3]) for this purpose into independent test criteria and these factors were then used to test the dimensionality aspect (whether $s = f$, the true dimensionality of the means $\boldsymbol{\mu}_\alpha$) and the direction aspect of H. However such a factorization or partitioning of the other criterion T_0^2 has not been demonstrated so far. In this paper, we propose to derive independent statistics of the T_0^2 type for these hypotheses, for the benefit of those who prefer T_0^2 to Λ.

2. Dimensionality and direction criteria based on T_0^2

If the hypothesis H is true, it can be shown (see [3]) that L and W are independent $p \times p$ Wishart matrices with $q - s$ and $n - q$ d.f. respectively (and the same parameter matrix Σ). If $q - s \geq p$, L will have the Wishart distribution but otherwise it has the "pseudo-Wishart" distribution, i.e. L is expressible as

$$L = ZZ' \tag{2.1}$$

where Z is $p \times (q - s)$ and the $q - s$ columns of Z have independent p-variate normal distributions with zero means and variance-covariance matrix Σ. We therefore consider the two cases $q - s \geq p$ and $q - s < p$ separately.

Case 1. $q - s \geq p$.

Let $(L + W)^{1/2}$ denote the lower triangular $p \times p$ matrix with positive diagonal elements, such that $(L + W)^{1/2}(L + W)'^{1/2} = L + W$. Define

$$M = (W + L)^{-1/2} W (W + L)'^{-1/2}. \tag{2.2}$$

The matrix M has then [6] the matrix-variate beta distribution

$$B_p(M \mid n - q \mid q - s)\mathrm{d}M = \text{const.} \mid M \mid^{(1/2)(n-q-p-1)} \mid I - M \mid^{(1/2)(q-s-p-1)} \mathrm{d}M$$
$$M > 0, \qquad I - M > 0. \tag{2.3}$$

From (1.11) to (1.13), we can see that Λ, and T_0^2, are expressible as

$$\Lambda = \mid M \mid, \qquad T_0^2 = \operatorname{tr} M^{-1} - p. \tag{2.4}$$

We have therefore the following lemma:

Lemma 1. *If a $p \times p$ symmetric matrix M has the $B_p(M \mid n - q \mid q - s)$ distribution, $\mid M \mid$, and* tr *$M^{-1} - p$, have respectively the $\Lambda(n - s, p, q - s)$, $T_0^2(n - s, p, q - s)$ distributions.*

Let H be the $p \times s$ matrix defined by

$$H = (W + L)^{-1/2} C_{xt} C_{tt}^{-1/2} \tag{2.6}$$

and let

$$U_1 = (H'H)^{-1/2} H'. \tag{2.7}$$

Observe that $U_1 U_1' = I_s$ and hence we can always construct a $p \times p$

orthogonal matrix U such that

$$U = \begin{bmatrix} U_1 \\ \hline U_2 \end{bmatrix} \begin{matrix} s \\ p-s \end{matrix}, \tag{2.8}$$

where U_2 is arbitrary. Let

$$G = UMU' = \begin{bmatrix} G_{11} & G_{12} \\ G_{21} & G_{22} \end{bmatrix} \begin{matrix} s \\ p-s \end{matrix}. \tag{2.9}$$
$$\quad s \quad\; p-s$$

Kshirsagar [3] has shown that since U is orthogonal, with elements independently distributed of M, G has the $B_p(G \mid n-q \mid q-s)$ distribution and hence using results in [6], we find that G_{11} and $G_{22\cdot 1} = G_{22} - G_{21}G_{11}^{-1}G_{12}$ are independent and have respectively the

$$B_s(G_{11} \mid n-q \mid q-s) \quad \text{and} \quad B_{p-s}(G_{22\cdot 1} \mid (n-q)-s \mid q-s) \tag{2.10}$$

distributions. Alternatively, G_{22} and $G_{11\cdot 2} = G_{11} - G_{12}G_{22}^{-1}G_{21}$ are also independent and have the

$$B_{p-s}(G_{22} \mid n-q \mid q-s) \quad \text{and} \quad B_s(G_{11\cdot 2} \mid n-q-(p-s) \mid q-s) \tag{2.11}$$

distributions. By applying Lemma 1 (or (2.5)), we can now easily see that

$$T_D^2 = \text{tr } G_{11}^{-1} - s \text{ has the } T_0^2(n-s, s, q-s) \text{ distribution} \tag{2.12}$$

$$T_{C|D}^2 = \text{tr } G_{22\cdot 1}^{-1} - (p-s) \text{ has the } T_0^2(n-2s, p-s, q-s) \text{ distribution} \tag{2.13}$$

independent of T_D^2. Alternatively,

$$T_C^2 = \text{tr } G_{22}^{-1} - (p-s) \text{ has the } T_0^2(n-s, p-s, q-s) \text{ distribution} \tag{2.14}$$

$$T_{D|C}^2 = \text{tr } G_{11\cdot 2}^{-1} - s \text{ has the } T_0^2(n-p, s, q-s) \text{ distribution} \tag{2.15}$$

independent of T_C^2.

One can readily see (from (2.7), (2.8), and (2.9)) that U_1 and hence G_{11} depends directly on $\boldsymbol{t}$, i.e. $Q\boldsymbol{y}$ the proposed functions while U_2 is an arbitrary completion of U_1 and hence U_2 or G_{22} only indirectly depend on $\boldsymbol{t}$ but its dimensions $p-s$ however depend on s, the number of the proposed functions in H. Hence T_D^2 will test the direction aspect of H, while $T_{C|D}^2$ is the "partial" dimensionality criterion for testing the adequacy of the number of the proposed functions only. Alternatively T_C^2 is the dimensionality statistic and $T_{D|C}^2$ is the "partial" direction factor. (For more details of such factorizations see [1].)

Case 2. $q - s < p$.

For this case, define the $p \times (q - s)$ matrix

$$D = (L + W)^{-1/2} Z, \tag{2.16}$$

where Z is given by (2.1). Let, further,

$$E = UD \tag{2.17}$$

where U is already defined in (2.8). From the distributions of $L + W$ and Z, it can be very easily shown (see [4], p. 269–271 or p. 292–297 for similar derivations) that D has the distribution

$$\text{const.}\ |I - DD'|^{(1/2)(n-q-p-1)} \mathrm{d}D.$$

Using (2.17), the distribution of E, which is a $p \times (q - s)$ matrix, comes out as

$$\text{const.}\ |I - EE'|^{(1/2)(n-q-p-1)} \mathrm{d}E. \tag{2.18}$$

From (2.8),

$$E_1 = U_1 D, \qquad E_2 = U_2 D, \tag{2.19}$$

where

$$E = \overset{q-s}{\begin{bmatrix} E_1 \\ E_2 \end{bmatrix}} \begin{matrix} s \\ p-s \end{matrix}. \tag{2.20}$$

Observe that

$$\begin{aligned} |I - EE'| &= |I - E'E| \\ &= |I - (E_1'E_1 + E_2'E_2)| \\ &= |I - E_1'E_1|\ |I - (I - E_1'E_1)^{-1} E_2'E_2| \\ &= |I - E_1'E_1|\ |I - F'F| \end{aligned} \tag{2.21}$$

where

$$(I - E_1'E_1)^{-1/2} E_2' = F'. \tag{2.22}$$

Transforming from E_1, E_2 in the distribution (2.18) of E, to E_1 and F given by (2.22), it can be readily seen that E_1 and F are independent and that the distribution of E_1 is

$$\text{const.}\ |I - E_1 E_1'|^{(1/2)(n-q-s-1)} \mathrm{d}E_1 \tag{2.23}$$

and that of F is

$$\text{const.}\ |I - F'F|^{(1/2)(n-q-p-1)} \mathrm{d}F. \tag{2.24}$$

If $(q-s)>s$, from Lemma 6, p. 68 of [4] it follows that

$$I-E_1E_1' \text{ has the } B_s(I-E_1E_1' \mid n-q \mid q-s) \text{ distribution} \tag{2.25}$$

and if $(q-s)<(p-s)$,

$$I-F'F \text{ has the } B_{q-s}(I-F'F \mid n-p-s \mid p-s) \text{ distribution,} \tag{2.26}$$

independent of $I-E_1E_1'$. On the other hand, if $s>(q-s)$

$$I-E_1'E_1 \text{ has the } B_{q-s}(I-E_1'E_1 \mid n-2s \mid s) \text{ distribution} \tag{2.27}$$

and if $(q-s)>(p-s)$,

$$I-FF' \text{ has the } B_{p-s}(I-FF' \mid n-q-s \mid q-s) \text{ distribution.} \tag{2.28}$$

Furthermore, $I-E_1'E_1$ and $I-FF'$ are independently distributed.

An application of Lemma 1 (or (2.5)) of this paper now shows that

$$\operatorname{tr}(I-E_1E_1')^{-1}-s \text{ has the } T_0^2(n-s, s, q-s) \text{ distribution,} \tag{2.29}$$

$$\operatorname{tr}(I-E_1'E_1)^{-1}-(q-s) \text{ has the } T_0^2(n-s, q-s, s) \text{ distribution,} \tag{2.30}$$

$$\operatorname{tr}(I-F'F)^{-1}-(q-s) \text{ has the } T_0^2(n-2s, q-s, p-s) \text{ distribution,} \tag{2.31}$$

and

$$\operatorname{tr}(I-FF')^{-1}-(p-s) \text{ has the } T_0^2(n-2s, p-s, q-s) \text{ distribution.} \tag{2.32}$$

Alternatively, (2.21) can also be expressed as

$$\begin{aligned} |I-E'E| &= |I-E_2'E_2|\,|I-(I-E_2'E_2)^{-1}E_1'E_1| \\ &= |I-E_2E_2|\,|I-RR'| \end{aligned} \tag{2.33}$$

where

$$R'=(I-E_2'E_2)^{-1/2}E_1' \tag{2.34}$$

and proceeding in exactly the same manner as before, one can easily establish that if $(p-s)>(q-s)$,

$$I-E_2'E_2 \text{ has the } B_{q-s}(I-E_2'E_2 \mid n-p \mid p-s) \text{ distribution} \tag{2.35}$$

or if $(q-s)>(p-s)$,

$$I-E_2E_2' \text{ has the } B_{p-s}(I-E_2E_2' \mid n-q \mid q-s) \text{ distribution.} \tag{2.36}$$

If $s<q-s$,

$$I-RR' \text{ has the } B_s(I-RR' \mid n-q-p+s \mid q-s) \text{ distribution} \tag{2.37}$$

or if $s>q-s$,

$$I-R'R \text{ has the } B_{q-s}(I-R'R \mid n-p-s \mid s) \text{ distribution} \tag{2.38}$$

and also by Lemma 1 of this paper,

$$\operatorname{tr}(I - E_2'E_2)^{-1} - (q - s) \text{ has the } T_0^2(n - s, q - s, p - s) \text{ distribution,} \tag{2.39}$$

$$\operatorname{tr}(I - E_2E_2')^{-1} - (p - s) \text{ has the } T_0^2(n - s, p - s, q - s) \text{ distribution,} \tag{2.40}$$

$$\operatorname{tr}(I - RR')^{-1} - s \text{ has the } T_0^2(n - p, s, q - s) \text{ distribution,}$$

and

$$\operatorname{tr}(I - R'R)^{-1} - (q - s) \text{ has the } T_0^2(n - p, q - s, s) \text{ distribution.}$$

From (2.2), (2.9), (2.16) and (2.17), we obtain

$$\begin{aligned} G = UMU' &= U(I - DD')U' \\ &= I - EE' \\ &= \left[\begin{array}{c|c} I - E_1E_1' & -E_1E_2' \\ \hline -E_2E_1' & I - E_2E_2' \end{array}\right] \end{aligned} \tag{2.41}$$

and so

$$G_{11} = I - E_1E_1', \tag{2.42}$$

and

$$\begin{aligned} G_{22\cdot1} &= (I - E_2E_2') - E_2E_1'(I - E_1E_1')^{-1}E_1E_2' \\ &= I - E_2(I - E_1'E_1)^{-1}E_2' \\ &= I - FF'. \end{aligned} \tag{2.43}$$

Hence

$$\begin{aligned} T_D^2 &= \operatorname{tr} G_{11}^{-1} - s = \operatorname{tr}(I - E_1E_1')^{-1} - s \\ &= \operatorname{tr}(I - E_1'E_1)^{-1} - (q - s) \end{aligned} \tag{2.44}$$

is distributed as $T_0^2(n - s, s, q - s)$;

$$\begin{aligned} T_{C|D}^2 &= \operatorname{tr} G_{22\cdot1}^{-1} - (p - s) = \operatorname{tr}(I - F'F)^{-1} - (q - s) \\ &= \operatorname{tr}(I - FF')^{-1} - (p - s) \end{aligned} \tag{2.45}$$

is distributed as $T_0^2(n - 2s, p - s, q - s)$.

Also

$$\begin{aligned} T_C^2 &= \operatorname{tr} G_{22}^{-1} - (p - s) = \operatorname{tr}(I - E_2'E_2)^{-1} - (q - s) \\ &= \operatorname{tr}(I - E_2E_2')^{-1} - (p - s) \end{aligned} \tag{2.46}$$

is distributed as $T_0^2(n - s, q - s, p - s)$;

and

$$\begin{aligned} T_{D|C}^2 &= \operatorname{tr} G_{11\cdot2}^{-1} - s = \operatorname{tr}(I - RR')^{-1} - s \\ &= \operatorname{tr}(I - R'R)^{-1} - (q - s) \end{aligned} \tag{2.47}$$

is distributed as $T_0^2(n - p, s, q - s)$.

The method of derivations may be different, but the final distributions of T_D^2, $T_{C|D}^2$, T_C^2, $T_{D|C}^2$ under the null hypothesis are therefore the same whether $q-s>p$ or $q-s<p$ and in practice, this distinction need not be made.

However we must now express all the statistics in terms of the original quantities $C_{xt}L$, and W, for use in practical application. This is accomplished in the next section.

3. T_D^2, $T_{C|D}^2$, T_C^2, $T_{D|C}^2$ in terms of C_{xt}, W, and L

Observe, from (2.8) and (2.9) that

$$G_{11}=U_1MU_1',\qquad G_{12}=U_1MU_2',\qquad G_{22}=U_2MU_2' \tag{3.1}$$

and that

$$G^{-1}=UM^{-1}U'=\left[\begin{array}{c|c} G_{11\cdot 2}^{-1} & -G_{11}^{-1}G_{12}G_{22\cdot 1}^{-1} \\ \hline G_{22}^{-1}G_{21}G_{11\cdot 2}^{-1} & G_{22\cdot 1}^{-1}\end{array}\right]. \tag{3.2}$$

It therefore follows that

$$U_1M^{-1}U_1'=G_{11\cdot 2}^{-1} \tag{3.3}$$

and

$$U_2M^{-1}U_2'=G_{22\cdot 1}^{-1}. \tag{3.4}$$

Hence, using a little algebra (which is omitted here), we can show that

$$\begin{aligned} T_D^2 &= \operatorname{tr} G_{11}^{-1}-s \\ &= \operatorname{tr}\{C_{tx}(L+W)^{-1}C_{xt}[C_{tx}(L+W)^{-1}W(L+W)^{-1}C_{xt}]\}^{-1}-s \end{aligned} \tag{3.5}$$

$$\begin{aligned} T_{C|D}^2 &= -(p-s)+\operatorname{tr} G_{22\cdot 1}^{-1} \\ &= -(p-s)+\operatorname{tr} U_2M^{-1}U_2' \\ &= s+\operatorname{tr} W^{-1}L-\operatorname{tr}[C_{tx}(L+W)^{-1}C_{xt}]^{-1}C_{tx}W^{-1}C_{xt}. \end{aligned} \tag{3.6}$$

Similarly,

$$\begin{aligned} T_{D|C}^2 &= \operatorname{tr} G_{11\cdot 2}^{-1}-s \\ &= -s+\operatorname{tr} U_1M^{-1}U_1' \\ &= -s+\operatorname{tr}[C_{tx}(L+W)^{-1}C_{xt}]^{-1}C_{tx}W^{-1}C_{xt}. \end{aligned} \tag{3.7}$$

For obtaining T_C^2, we observe that (from (3.4))

$$G_{22\cdot 1}=(U_2M^{-1}U_2')^{-1} \tag{3.8}$$

so,

$$\begin{aligned} G_{22} &= (U_2M^{-1}U_2')^{-1} + G_{21}G_{11}^{-1}G_{12} \\ &= [I + G_{21}G_{11}^{-1}G_{12}U_2M^{-1}U_2'](U_2M^{-1}U_2')^{-1}. \end{aligned} \tag{3.9}$$

Hence,

$$\begin{aligned} G_{22}^{-1} = (U_2M^{-1}U_2')[I - G_{21}G_{11}^{-1}\{I + G_{12}U_2M^{-1}U_2'G_{21}G_{11}^{-1}\}^{-1} \\ G_{12}U_2M^{-1}U_2'], \end{aligned} \tag{3.10}$$

and so tr G_{22}^{-1} simplifies, after some algebra and use of $U_1'U_1 + U_2'U_2 = I$, to

$$\mathrm{tr}\ M^{-1} - \mathrm{tr}\ U_1M^{-2}U_1'(U_1M^{-1}U_1')^{-1}. \tag{3.11}$$

Therefore using (2.5) and (2.6)

$$\begin{aligned} T_C^2 &= \mathrm{tr}\ G_{22}^{-1} - (p - s) \\ &= \mathrm{tr}\ W^{-1}L - \mathrm{tr}[C_{tx}W^{-1}LW^{-1}C_{xt}(C_{tx}W^{-1}C_{xt})^{-1}]. \end{aligned} \tag{3.12}$$

References

[1] Bartlett, M.S. (1951). The goodness of fit of a single hypothetical discriminant function in the case of several groups. *Ann. Eugen.* **16,** 199.

[2] Good, I.J. (1969). Some applications of the singular decomposition of a matrix. *Technometrics* **11,** 823.

[3] Kshirsagar, A.M. (1971). Goodness of fit of a discriminant function from the vector space of dummy variables. *J. Roy. Stat. Soc. B* **33,** 111.

[4] Kshirsagar, A.M. (1972). *Multivariate analysis.* Marcel Dekker and Company, New York.

[5] McHenry, Claude (1974). Direction and collinearity tests based on Hotelling's generalized T_0^2 and Pillai's *V.* Unpublished Ph.D. Dissertation, Texas A&M University, College Station, Texas.

[6] Mitra, S.K. (1970). A density-free approach to the matrix variate beta distribution. *Sankhyā A* **32,** 81.

[7] Pearson, E.S., and Hartley, H.O. (1972). *Biometrika Tables for Statisticians, Volume* 2. Cambridge University Press, Cambridge, England.

[8] Williams, E.J. (1967). The analysis of association among many variates, *J. Roy. Stat. Soc., B.* **20,** p. 199.

PART II

ESTIMATION, DECISION PROCEDURES AND DESIGN OF EXPERIMENTS

P.R. Krishnaiah, ed., *Multivariate Analysis–IV*

CONDITIONAL CONFIDENCE AND ESTIMATED CONFIDENCE IN MULTIDECISION PROBLEMS (WITH APPLICATIONS TO SELECTION AND RANKING)*

J. KIEFER

Cornell University, Ithaca, N.Y., U.S.A.

The framework introduced earlier by the author is applied to certain multidecision models. Basic is the principle of exhibiting a measure of conclusiveness Γ_ω of the decision (when ω is true). Moreover, this Γ_ω should be highly "data-dependent", taking advantage of "lucky" observed values and thus perhaps giving a frequentist response to criticisms of Neyman–Pearson methodology raised by Bayesians. As is the case for unconditional confidence coefficients, it is also convenient to have Γ_ω almost independent of ω. The construction of admissible procedures that meet these requirements, characterized earlier, is implemented in various ranking and selection problems, including nonparametric ones, of both the Gupta and Bechhofer forms.

1. Introduction

This is one of a series of papers in a study of conditional confidence procedures and estimated confidence coefficients. There is a considerable literature of conditioning in statistical inference, but no previous methodical presentation of a frequentist non-Bayesian framework that considers possible criteria of goodness of such procedures, and methods for constructing them, in general statistical settings.

The present development was motivated by the desire to find an answer in Neyman–Pearson–Wald (NPW) frequentist terms, to a criticism that has been raised frequently against the NPW approach by authors supporting other foundational approaches. These critics are often disturbed at the prospect of making a decision that is not accompanied by some data-dependent measure of "conclusiveness" of the experimental outcome. For example, in the ranking problem for deciding which of two normal rv's X_i with unit variance has mean $+1$ and which has mean -1, where the standard symmetric NP test makes a decision accompanied only by the

* Research under NSF Grant MPS 72–04998 A02.

assessment of error probabilities $\Phi(-2^{1/2})$, some statisticians may be disturbed by an intuitive feeling that they are much surer of the conclusion when $X_1 - X_2 = 10$ than when $X_1 - X_2 = 0.5$. Authors adhering to Bayesian, likelihood, fiducial, or evidential foundations have criticized other aspects of the NP approach as well; but this notion of wanting a measure of conclusiveness that depends on the experimental outcome has also appealed to many practitioners, as is evident in the old and continued practice of stating the level at which a significance test "just rejects" a null hypothesis. Intuitively, even for testing between two simple hypotheses one tries to make use of "lucky outcomes" to make stronger assertions.

The principle adopted herein, in developing a methodology that may satisfy the objection mentioned in the previous paragraph for some workers, is that our data-dependent measure of conclusiveness should have a frequentist (law of large numbers) meaning similar to that emphasized by Neyman for classical NP tests and confidence intervals. As is discussed in [3] and [6], this implies that, except in rare cases of symmetry, our approach and form of conclusion cannot agree with those of the critics mentioned above. The conditional confidence procedures are suggested as possibilities that may sometimes have appeal to practitioners. No attempt is made to give a prescription for deciding whether to use any of them, or which one to use. In particular, the fineness of the conditioning partition is to be chosen by the practitioner; one possibility is the degenerate unconditional conditioning, which is the classical NPW framework; if nondegenerate conditioning is used, it should yield a highly variable conditional confidence, or one might as well use an unconditional procedure, instead.

We assume the underlying measure space of possible outcomes of the experiment, $(\mathscr{X}, \mathscr{B}, \nu)$, to be of countable type with compactly generated σ-finite ν with respect to which the possible states of nature $\Omega = \{\omega\}$ have densities f_ω. This implies existence of regular conditional probabilities in the sequel. (See, e.g., [8].) We write X for the rv representing the experiment with outcome in $\mathscr{X}$.

In the decision space D, we assume that for each ω there is specified a nonempty subset D_ω of decisions that are "correct" when ω is true. We write $\Omega_d = \{\omega : d \in D_\omega\}$. The confidence "flavor" is best exhibited in terms of the simple loss function implied by consideration in these terms; other possible treatments will be described in Section 7.

A (nonrandomized) decision rule $\delta : \mathscr{X} \to D$ is required to have $\delta^{-1}(D_\omega) \in \mathscr{B}$ for each ω. The apparent neglect of randomization is intended to aid in clarity of exposition. It is not a genuine neglect, since $\mathscr{X}$

can be regarded as the product of a more primitive sample space and a randomization space.

Any subfield $\mathscr{B}_0$ of $\mathscr{B}$ may be called a *conditioning subfield*, but it is convenient to consider only those subfields generated by statistics on $(\mathscr{X}, \mathscr{B})$. If Z is such a statistic, $\mathscr{B}_0$ is the largest subfield inducing the same partition of $\mathscr{X}$ that Z induces [1]. Writing $Z(\mathscr{X}) = B$, we denote that partition by $\{C^b,\ b \in B\}$. We also write, for a given δ and $\mathscr{B}_0$,

$$C_d^b = C^b \cap \delta^{-1}(d), \qquad C_\omega = \delta^{-1}(D_\omega). \tag{1.1}$$

The C_ω are not necessarily disjoint, but the C_d^b $(b \in B,\ d \in D)$ are, and constitute the *partition* C of $\mathscr{X}$. Subscripts always index D or Ω; superscripts index B; symbols such as overbars distinguish different partitions and their operating characteristics.

A *conditional confidence procedure* is a pair $(\delta, \mathscr{B}_0)$ or (δ, Z) or, equivalently, the corresponding partition C. Its associated *conditional confidence function* is the set $\Gamma = \{\Gamma_\omega,\ \omega \in \Omega\}$ of conditional probabilities

$$\Gamma_\omega = P_\omega\{C_\omega \mid \mathscr{B}_0\}, \qquad \omega \in \Omega. \tag{1.2}$$

For each ω, this $\mathscr{B}_0$-measurable function on $\mathscr{X}$ can be regarded as a function of Z, and we write Γ_ω^b for its value on the set $Z = b$. A *conditional confidence statement* associated with $(\delta, \mathscr{B}_0)$ is a pair (δ, Γ). It is used as follows: if $X = x_0$, and thus $Z(x_0) = z_0$, we state that "for each ω, we have conditional confidence $\Gamma_\omega(x_0) = \Gamma_\omega^{z_0}$ of being correct if ω is true". We make decision $\delta(x_0)$.

Such statements have a conditional frequentist interpretation analogous to that of the NPW setting $\mathscr{B}_0 = \{\mathscr{X}, \phi\}$, where Γ_ω is simply the probability of a correct decision when ω is true. In a sequence of n independent experiments, if ω_i is true and X_i is observed in the ith experiment, then the proportion of correct decisions made will be close to $n^{-1}\Sigma_1^n \Gamma_{\omega_i}(X_i)$ with probability near one when n is large. This is true even if different $(\delta, \mathscr{B}_0)$'s are used in the experiments, but the frequentist meaning is of most intuitive value in cases where there is no dependence of Γ_ω on ω, just as in the familiar NP treatment of confidence intervals or composite hypotheses with specified minimum power. Thus, if there is a function ψ on $\mathscr{X}$ such that $\Gamma_\omega(x) \geq$ (resp. $=$) $\psi(x)$ for all ω and all x, we may make the more succinct statement that "we have conditional confidence at least (resp., exactly) $\psi(x_0)$ that $\delta(x_0)$ is a correct decision;" if $X = x_0$. The frequentist interpretation in terms of the law of large numbers is then simpler, and its practical meaning is made clearer if we look only at those experiments in which $\Gamma(X_i) \geq$ (resp. $=$) 0.95 (for example): a correct decision is very likely made in almost 95% or more (resp. almost 95%) of such experiments, if their number is large.

Whenever $\Gamma_\omega^Z(x)$ can be chosen not to depend on ω, we write Γ^b in place of Γ_ω^b. The examples that arise often enjoy a symmetry that permits construction of invariant C's with Γ_ω independent of ω throughout $\mathscr{X}$. We shall give examples in which the conditioning partition $\{C^b\}$ is "as fine as possible" while permitting this construction, and also in which it consists of the coarsest non-trivial partitioning, of two elements. Intermediate examples will also be given.

The operating characteristic of C can be thought of as the collection consisting of each sub-df of Γ_ω on C_ω, with corresponding sub-df's that refer to incorrect confidence statements. This is treated in detail in [6], and its use is indicated in (2), just below.

For finite (or denumerably infinite) B, we interpret (1.2) through the simple conditional probability formula: if $P_\omega\{C^b\} > 0$,

$$\Gamma_\omega^b = P_\omega\{C^b \cap C_\omega\}/P_\omega\{C^b\}. \tag{1.3}$$

The normal law with mean μ and variance σ^2 is denoted $\mathcal{N}(\mu, \sigma^2)$. The standard $\mathcal{N}(0, 1)$ df and density are denoted Φ and ϕ.

To summarize briefly the main ideas behind our conditioning approach and the selection of a conditional confidence procedure, we want:

(1) a data-dependent measure of conclusiveness with frequentist interpretability;

(2) admissibility of the procedure as derived from some notion that a "good" procedure has a high (resp., low) probability of yielding a high value of the measure Γ_ω when ω is true (resp., false);

(3) high variability of that measure (of Γ_ω^b, as a function of b), to achieve a wide range of possible strengths of conclusiveness in the inferential statement that is lacking in an unconditional procedure (and without which we might as well use the latter);

(4) small variability of Γ_ω^b in ω, for the practical reason of making simpler confidence statements, and to avoid certain theoretical anomalies (alluded to later).

Discussion of the relationship with other approaches, shortcomings of various goodness criteria, general theoretical developments, etc., are contained in [3] and [6]. In particular, it is shown there that, for various admissibility criteria of the type (2) above, any unconditionally admissible δ, when conditioned aribtrarily, yields an admissible conditional procedure (but that, except when Ω has only 2 elements, there are other admissible conditional procedures, as well). In the present paper we omit theoretical details and give illustrations of the approach in fixed sample size ranking and selection settings. In addition to considering conditional procedures, we discuss the technique of *estimating* a conditional or unconditional Γ_ω that depends strongly on ω, and without which one may be forced to use an unsatisfactory lower bound ψ. This is also considered from a frequentist justification.

2. Bechhofer's indifference zone approach

For definiteness, we consider the problem of selecting exactly one of three normal populations, it being desired to select the one with the largest mean, the three variances and sample sizes being assumed equal. The simpler case of two populations is studied extensively in [3], and more than three populations offer only added computational and notational complexity, but no new conceptual difficulties. (A three-population example with perhaps less realistic D_ω's is treated in [3].)

After transformation, the model reduces to $X = (X_1, X_2, X_3)$ with $X_i \sim \mathcal{N}(\theta_i, 1)$; the X_i are independent, and d_i is the decision that "θ_i is largest." We write $\theta_{\max} = \max_i \theta_i$. There are a number of different possibilities for specifying parameter regions where only d_1 (for example) is correct, and we choose one of the simplest: $\theta_1 > \theta_{\max} - \bar{\Delta}$, where $\bar{\Delta} > 0$ is specified. Thus, we still view the selection as correct if the chosen population has mean "not too far" ($< \bar{\Delta}$) from the largest mean. This means that, if $\theta_1 > \theta_2 > \theta_3$,

$$D_\omega = \begin{cases} d_1 & \text{if } \theta_1 - \bar{\Delta} > \theta_2, \theta_3, \\ \{d_1, d_2\} & \text{if } \theta_1 > \theta_2 > \theta_1 - \bar{\Delta} \geq \theta_3, \\ \{d_1, d_2, d_3\} & \text{if } \theta_1 > \theta_2 > \theta_3 \geq \theta_1 - \bar{\Delta}, \end{cases} \tag{2.1}$$

and similarly for other possible orderings of the θ_i. Bechhofer's formulation is geared to guaranteeing a specified probability P^* of being correct when one d_i is correct (first line of (2.1)). Alternatively, one can think of this formulation for the above "correctness" structure when Ω is restricted to the set where one θ_i exceeds the others by at least $\bar{\Delta}$.

We hereafter denote by $\Omega(j)$ the subset of Ω where j of the d_i's are correct.

The Bechhofer unconditional procedure [2] selects the d_i for which X_i is largest. We shall use that δ in what follows. Write $X_{[1]} \le X_{[2]} \le X_{[3]}$ for the ordered X_i's, and similarly for the θ_i's. For $i = 1$ or 2, write $b_i = x_{[3]} - x_{[i]}$ and $\Delta_i = \theta_{[3]} - \theta_{[i]}$.

Example 2.1. *Two-element B.* We first consider the simplest possible conditioning, that given by a partition $\{C^1, C^2\}$, where C^2 consists of the more conclusive, "lucky", observations. A simple such possibility is $C^2 = \{b_2 \ge c\}$, where $c > 0$. (Questions of admissibility and other goodness properties of such partitions are discussed in [3] and [6].) This yields, in $\Omega(1)$, in an obvious notation,

$$\Gamma^2_{(\Delta_1,\Delta_2)} \ge \Gamma^2_{(\bar\Delta,\bar\Delta)}$$

$$= \frac{\mathbf{P}\{Y_1 > -\bar\Delta + \max(Y_2, Y_3)\}}{\mathbf{P}\{Y_1 > c - \bar\Delta + \max(Y_2, Y_3)\} + 2\mathbf{P}\{Y_1 > c + \max(Y_2, Y_3 + \bar\Delta)\}}, \tag{2.2}$$

where the Y_i are independent and $\mathcal{N}(0,1)$. To reduce repetitiveness, we omit discussion of the inequality of (2.2) and of the expression for Γ in $\Omega(2)$, since similar calculations arise in other examples we shall treat. $\Gamma^1_{(\Delta_1,\Delta_2)}$ can be computed similarly, and tables of the $\Gamma^b_{(\bar\Delta,\bar\Delta)}$ as a function of c and $\bar\Delta$ can be obtained from bivariate normal tables or from tables that appear in the work of Bechhofer and his school. (In more complex settings one cannot expect that the lower bound on Γ^b_ω will be obtained at the same configuration for each b, as it is in the present example.) Since Bechhofer's lower bound P^* on the probability of a correct selection for his unconditional procedure on the set $\Delta_2 \ge \bar\Delta$ satisfies

$$P^* = \Gamma^1_{(\bar\Delta,\bar\Delta)} P_{(\bar\Delta,\bar\Delta)}\{C^1\} + \Gamma^2_{(\bar\Delta,\bar\Delta)} P_{(\bar\Delta,\bar\Delta)}\{C^2\}, \tag{2.3}$$

we see the way in which the Γ^b's must balance under the "least favorable (LF) configuration", $(\Delta_1, \Delta_2) = (\bar\Delta, \bar\Delta)$. For example, if $P^* = 0.9$ and we choose to use a conditional procedure to attain the higher value $\Gamma^2_{(\bar\Delta,\bar\Delta)} = 0.99$ for the more conclusive set C^2 of (b_1, b_2)-values, and if C^2 occurs with probability 0.3 under the LF configuration (and hence $P_{(\bar\Delta,\bar\Delta)}$ {correct selection, $\Gamma^2_{(\bar\Delta,\bar\Delta)} = 0.99\} = 0.3(0.99)$), then $\Gamma^1_{(\bar\Delta,\bar\Delta)} = [0.9 - 0.3(0.99)]/0.7 = 0.86$. Practitioners may encounter settings in which they are willing to have their confidence in a correct selection reduced from 0.9 to 0.86 with probability 0.7 (in the least favorable case) in order to achieve the more highly conclusive value 0.99 in the case of a "lucky sample," which occurs with probability 0.3.

Those who feel uncomfortable with conditional inference may prefer to use the 6-decision unconditional procedure with decision sets C_i^b, where in C_i^1 the decision is "weakly conclusive choice of population i" and C_i^2 stands for a strongly conclusive choice, and where the properties of the procedure are assessed entirely in terms of unconditional probabilities $P_\omega\{C_i^b\}$. (The author would certainly use that procedure in many circumstances.) The theoretical relationship of this procedure to the corresponding conditional one, in terms of admissibility, is treated in [6]. From a foundational viewpoint, such a procedure is in the NPW frequentist mold; but its frequency interpretation is not in terms of a data-dependent proportion of correct decisions with the interpretation Γ_ω^b had, since $P_\omega\{C_i^b\}$ for a single b does not have the "confidence" flavor and (summing over b) $P_\omega\{C_\omega\}$ is the confidence of an unconditional 3-decision procedure.

Example 2.2. *Continuum B.* We now turn to a possible "continuum" conditioning. The "finest" invariant B is $\{(b_1, b_2)\colon 0 \le b_1 \le b_2\}$. Each point in B corresponds to six possible orderings of the X_i's. Writing out the densities of these six orderings (e.g., when $\theta_1 = 0$, $\theta_2 = -\Delta_1$, $\theta_3 = -\Delta_2$, since the results are translation and permutation invariant), we obtain

$$\Gamma_{(\Delta_1,\Delta_2)}^{(b_1,b_2)} = \begin{cases} [1 + (e^{b_1\Delta_1} + e^{b_1\Delta_2} + e^{b_2\Delta_1} + e^{b_2\Delta_2})/(e^{b_1\Delta_1+b_2\Delta_2} + e^{b_1\Delta_2+b_2\Delta_1})]^{-1} & \text{in } \Omega(1), \\ [1 + (e^{b_1\Delta_1} + e^{b_2\Delta_2})/(e^{b_1\Delta_2} + e^{b_2\Delta_1} + e^{b_1\Delta_1+b_2\Delta_2} + e^{b_1\Delta_2+b_2\Delta_1})]^{-1} & \text{in } \Omega(2), \\ 1 & \text{in } \Omega(3). \end{cases} \tag{2.4}$$

It is not hard to see that each of these expressions is nondecreasing in $\Delta_1 \ge \Delta_2$ (fixed); putting $\Delta_1 = \Delta_2$, the resulting expressions are nondecreasing in Δ_2. The infimum of (2.4) on $\Omega(1)$ is thus attained at $\Delta_1 = \Delta_2 = \bar{\Delta}$, and this is the limit of a sequence of points (Δ_1, Δ_2) at which the infimum on $\Omega(2)$ is approached. Thus,

$$\Gamma_{(\Delta_1,\Delta_2)}^{(b_1,b_2)} \begin{cases} \ge \Gamma_{(\bar{\Delta},\bar{\Delta})}^{(b_1,b_2)} = [1 + (e^{b_1\bar{\Delta}} + e^{b_2\bar{\Delta}})/e^{b_1\bar{\Delta}+b_2\bar{\Delta}}]^{-1} & \text{on } \Omega(1), \\ \ge \frac{1}{2} + \frac{1}{2}(1 + e^{-b_1\bar{\Delta}} + e^{-b_2\bar{\Delta}})^{-1} & \text{on } \Omega(2), \\ = 1 & \text{on } \Omega(3). \end{cases} \tag{2.5}$$

The first line of the right-hand side of (2.5) is the smallest of the three, and thus can be used as the function ψ of Section 1.

In some respects, this finest symmetric partition is the most natural one, and its use obviates the choice of c in Example 2.1, or the choice among B's of various sizes.

Example 2.3. *A mixed procedure.* The procedure of Example 2.2 may appeal to some practitioners who want their conditional confidence statement to reflect the fact that, for b_1 and b_2 near 0, they do not feel very conclusive about their selection. Others may want to achieve the benefits of the procedure of Example 2.2 for large b_1 and b_2 without completely sacrificing the value of $\Gamma^1_{(\bar{\Delta},\bar{\Delta})}$ of the procedure of Example 2.1 when b_1 and b_2 are small. One possible procedure, with the same δ, is given by

$$Z(x_1, x_2, x_3) = \begin{cases} 0 & \text{if } b_2 < c, \\ \\ (b_1, b_2) & \text{if } b_2 \geq c, \end{cases} \tag{2.6}$$

for which Γ^0_ω is the Γ^1_ω of Example 2.1 (corresponding to (2.2)) and $\Gamma^{(b_1,b_2)}_\omega$ is given by (2.5) for $b_2 \geq c$. Other possibilities with similar properties will occur to the reader.

Other B's. It may be desirable to have a finite number of C^b's but more than 2. On the other hand, in place of the procedure of Example 2.2 one may find it desirable to use a continuum B procedure with a simpler B (especially if the number of populations is large). An example of the latter is $B = \{b_2 \colon 0 \leq b_2\}$. The result, derived in [3], is, in $\Omega(1)$,

$$\Gamma^{b_2}_{(\bar{\Delta},\bar{\Delta})} = \left\{1 + \frac{\Phi(-6^{-1/2}b_2)\phi(2^{-1/2}(b_2+\bar{\Delta})) + \Phi(-6^{-1/2}(b_2+\bar{\Delta}))\phi(2^{-1/2}b_2)}{\Phi(6^{-1/2}(\bar{\Delta}-b_2))\phi(2^{-1/2}(\bar{\Delta}-b_2))}\right\}^{-1}. \tag{2.7}$$

Ranking problems. We shall not take the space to give the straightforward details of the parallels of the previous procedures for other "goals". Briefly, suppose, for example, that we want a complete ranking of the three θ_i's, and condition as in Example 2.2 for Bechhofer's procedure of ranking the θ_i's in the same order as the X_i's, and compute Γ^b_ω on the set where $\theta_{[3]} - \theta_{[2]} \geq \Delta^*_2$, $\theta_{[2]} - \theta_{[1]} \geq \Delta^*_1$, where the Δ^*_i are specified; on this set, only the fully correct ranking is a "correct decision". The expression for Γ is then obtained by moving the last exponential in the first line of (2.4) into the numerator; it is more convenient to represent the result with the substitution $\Delta_1 = \Delta_2 + \Delta'_1$ where $\Delta'_1 = \theta_{[2]} - \theta_{[1]}$. The LF configuration is then seen to be $\Delta_2 = \Delta^*_2$, $\Delta'_1 = \Delta^*_1$, yielding an appropriate lower bound as in (2.5). This must be considered in combination with the evaluation of Γ on other parameter sets analogous to those of (2.1), for whatever schedule of "correct decisions" is adopted.

3. Gupta's subset selection approach

In this approach of Gupta (1956), the unconditional format allowed for seven possible decisions, the nonempty subsets of $\{1, 2, 3\}$; the subscript of the largest θ_i is asserted to be in that subset. We consider the normal probability model stated at the outset of Section 2. One of the rules δ studied by Gupta, and which we consider here, is

$$\delta(x_1, x_2, x_3) = \{i\colon x_{\max} - x_i \leq c\}, \tag{3.1}$$

where $c > 0$.

We discuss only the analogue of Example 2.2, again with $B = \{(b_1, b_2)\colon 0 \leq b_2 \leq b_1 < \infty\}$. The formulas for Γ over the several regions of interest are given by (2.4), whether "correctness" corresponds to (2.1) (with, for example, $\{d_1, d_2\}$ there being replaced by $\{1, 2\}$ here) with $\bar{\Delta} > 0$, or to the original Gupta formulation with $\bar{\Delta} = 0$ in (2.1). But there is an important difference in the form of assertion possible here, from that in Example 2.2.

In Example 2.2, we would never *know* that $\omega \in \Omega(3)$, and are led to use the $\Gamma^{(b_1,b_2)}_{(\bar{\Delta},\bar{\Delta})}$ of $\Omega(1)$ as lower bound ψ, as described there. In the present subset selection formulation, though, if $\delta(x)$ contains two elements we know $\omega \in \Omega(2) \cup \Omega(3)$ and can use the smaller $\Gamma^{(b_1,b_2)}_{(\bar{\Delta},\bar{\Delta})}$ of $\Omega(2)$ as ψ; and if $\delta(x) = \{1, 2, 3\}$, we know $\omega \in \Omega(3)$ and can take $\psi(x) = 1$.

Some practitioners may feel uncomfortable about asserting "conditional confidence 1" in the correctness of $\delta(x)$ whenever $b_1 \leq c$. The author does not feel that the value $\Gamma = 1$ here is defective in the way it is in certain examples discussed in [3] and [6], which in the present context of Section 2 would amount to taking $B = \{1, 2, 3\}$ and $C^b = \{x_b = x_{\max}\}$ a.e. For the latter ill-advised conditioning, one can make a "$\Gamma = 1$" statement about a *decision which might be wrong*; this anomaly is reflected in the great variation of Γ^b_ω as a function of ω, for that conditioning. In the present example of Gupta's procedure δ, when $\delta(x) = \{1, 2, 3\}$ so that $\Gamma(x) = 1$, *the decision must be correct.* Hence, many practitioners may feel more comfortable about saying $\Gamma = 1$ in this event, than in using the unconditional procedure that yields a constant value $\psi < 1$ that refers also to this case of a decision that *must be correct*; in fact, the latter type of behavior is one of the subjects of criticism of classical unconditional confidence statements by authors such as Pratt (1961). Thus, it does not seem unreasonable to the author to make the assertion $\Gamma = 1$ when $\delta(x) = \{1, 2, 3\}$ in the present example. (In the above "ill-advised conditioning", $\psi = 0$.)

Although the other types of conditioning partitions mentioned in Section

2 can be used here, one motivation for the procedure of Example 2.3 is now absent: the ψ of the conditional subset selection procedure discussed here does not approach $\frac{1}{3}$ as $b_1 \to 0$, as does that given by the first line of (2.5); the form of (3.1) guarantees a more satisfactory confidence statement.

4. Conditioning in other selection and ranking problems

The conditioning partitions $\{C^b\}$ in the previous examples were adopted without the justification, discussed in [3], [6], in terms of operating characteristics (the law of Γ_ω on C_ω when ω is true, with analogues for incorrect decisions). We mention here only some intuitive ideas on the choice of $\{C^b\}$.

Aş mentioned in Section 1, we try to choose C to make Γ_ω^b highly variable in b, since the desire for a data-dependent measure of conclusiveness motivated the development. Some optimum properties, in terms of such variability, are discussed in [3] in simpler examples with conditionings analogous to those chosen here. On the other hand, lack of strong dependence of Γ_ω^b on ω, for each fixed b, or at least existence of a useful lower bound ψ (since such independence of ω is impossible for nontrivial procedures in the examples we have treated), is very convenient; it also eliminates anomalies related to that mentioned in the next to last paragraph of Section 3, in the possibility of incorrectly asserting a $\Gamma_{\omega'}$ value close to 1 with appreciable probability under an $\omega'' \neq \omega'$.

In symmetric examples in which there is a group leaving the problem invariant that is transitive on Ω, the maximal invariant (or a function of it) will be an ancillary statistic that can often be used for a convenient conditioning. In the setting of Section 2, (b_1, b_2) is such a statistic if the original parameter space is replaced by $\{(\theta_1, \theta_2, \theta_3)\colon \Delta_1 = \bar{\Delta}_1, \Delta_2 = \bar{\Delta}_2\}$ for fixed $\bar{\Delta}_i$. For the actual Ω of Section 2, the use of conditioning based on functions of (b_1, b_2) achieves constant Γ_ω^b on orbits given by fixing the Δ_i, and thus yields the convenient ψ from the least favorable configuration.

In other settings no such invariant structure may be present, and the guessing of a good $\{C^b\}$, and calculation of useful ψ, may be difficult. As an example, suppose the X_i of Sections 2 and 3 are now independent binomial (n, θ_i) rv's. The only symmetry is that associated with population labels (symmetric group) and the transformation $X_i \to n - X_i$ (reflection); even if only two populations are present, a parameter set such as $|\theta_1 - \theta_2| = \bar{\Delta}$ admits no transitive action as it did in the normal case.

A conditioning used for decades in a variety of problems involving binomial populations is that based on $b = \Sigma X_i$; generally this was used in a spirit opposite to ours, in conditional calculations that made Γ_ω^b (or an analogue) as independent of b as possible, in achieving some unconditional aim. Gupta, Huang and Huang (1974) have used this conditioning in subset selection (and other) problems, where, however, a specified minimum unconditional probability of correct selection is again the goal, and the decision rule is based in part on ΣX_i so as to yield the desired unconditional results through convenient conditional calculations.

By way of illustration of conditional confidence, we now describe, for two binomial populations, what can be achieved by such a conditioning. We treat the Bechhofer goal of selecting the population with largest θ_i. The conditional law of $Y = X_2 - X_1$, given $X_1 + X_2 = b$, is well known, and it follows from calculations in [5] that the Bechhofer rule (with equal randomization probabilities if $Y = 0$) achieves

$$\Gamma_\omega^b = [1 + \rho_n(\lambda, b)]^{-1} \tag{4.1}$$

where $\lambda = \theta_{[2]}(1 - \theta_{[1]})/\theta_{[1]}(1 - \theta_{[2]})$ and

$$\rho_n(\lambda, b) = \frac{\sum_{r<\frac{1}{2}t} \binom{n}{t-r} \binom{n}{r} \lambda^r + h_n(\lambda, b)}{\sum_{r>\frac{1}{2}t} \binom{n}{t-r} \binom{n}{r} \lambda^r + h_n(\lambda, b)};$$

$$h_n(\lambda, b) = \begin{cases} 0 & \text{if } b \text{ is odd,} \\ \frac{1}{2}\binom{n}{b/2}^2 \lambda^{b/2} & \text{if } b \text{ is even.} \end{cases} \tag{4.2}$$

This Γ_ω^b, for fixed λ, is *roughly* decreasing in $|b - \frac{1}{2}n|$, as one might expect; there is a fine-structure oscillation, even values of b generally producing smaller values of Γ than both neighboring odd values because of the randomization term h_n. (Thus, one might conveniently lump together *at least* the four values $b = 2j$, $2j + 1$, $n - 2j$, $n - 2j - 1$, into a coarser conditioning partition.)

As a numerical example, when $n = 9$ one calculates that, if $\lambda = 2$, Γ_ω^b ranges from $\frac{2}{3}$ when $b = 1$ (or $\frac{1}{2}$ for the trivial $b = 0$) to 0.78 when $b = 9$. This is not much variability, and it is not hard to see that a conditioning based on values of $|X_1 - X_2|$ can achieve much more in this respect. (An example is treated in [3].) We have gone through this example to illustrate that a conditioning that is attractive for calculations, and which has proved useful in calculating procedures for various unconditional aims, may also yield simple calculations for our framework, but may often produce procedures whose variability of Γ is less satisfactory than for other conditionings.

For large sample sizes, approximate computations of conditional confidence in examples like this last one can often be considerably simplified by the use of the central limit theorem (local version for fine B) or large deviation theory. An example is treated in [3].

5. Nonparametric models

It is possible to apply the conditional confidence approach to a variety of nonparametric settings. As in parametric models, a good conditional procedure can generally be constructed by imposing an appropriate conditioning partition $\{C^b\}$ on a satisfactory unconditional procedure δ. In many problems it is possible thereby to obtain a useful ψ for fairly large Ω with the use of traditional tools such as sample quantiles and simple rank statistics. An example is the decision as to whether iid rv's X_i with continuous df have median >0 or ≤ 0, based on $\Sigma_i \operatorname{sgn} X_i$. Obviously, coarse conditionings will have better "robustness" properties than fine ones. We now illustrate the more detailed analysis that can accompany a stronger structural assumption on Ω.

Suppose X_{ij} are independent rv's, $1 \leq j \leq M_i$, $1 \leq i \leq k$, where the X_{ij} have law F_i. Write $N = \Sigma_i M_i$. The stronger structure we assume is that the F_i belong to the same family of continuous Lehmann alternatives [7]; that is, the elements of Ω can be labeled $\omega = (F, \theta_1, \theta_2, \ldots, \theta_k)$ where F is a continuous df on the reals, $\theta_i > 0$ for all i, and $F_i = F^{\theta_i}$. (It is trivial but unnecessary to specify F further so as to make this representation unique.) We consider the formulation of Section 2, of selecting the population for which θ_i is largest. When $k = 3$, for specified $\bar{\Delta} > 1$ we adopt a correct decision structure corresponding to (2.1):

$$D_\omega = \begin{cases} d_1 & \text{if } \min(\theta_1/\theta_2, \theta_1/\theta_3) > \bar{\Delta}, \\ \{d_1, d_2\} & \text{if } \theta_1/\theta_3 \geq \bar{\Delta} > \theta_1/\theta_2 > 1, \\ \{d_1, d_2, d_3\} & \text{if } \bar{\Delta} > \theta_1/\theta_3 \geq \theta_1/\theta_2 > 1, \end{cases} \tag{5.1}$$

and similarly for other orderings. Order the X_{ij}, N in number, and let $Z_{it} = 0$ if the tth largest observation is an X_{ij}, and $Z_{it} = 1$ otherwise, for $1 \leq t \leq N$. Also write $V_{it} = \Sigma_{s=1}^{t} Z_{is}$. Finally, write $T_i = \Sigma_{t=1}^{N} V_{it}/t$. (These definitions are employed in order to be able to use tables of [11] easily, below.)

The use of rank order statistics for inference in models where Ω consists of Lehmann alternatives has been discussed extensively. In particular, Savage (1956) gives tables when $k=2$ and N is small, of the probabilities of various rank orders, which depend only on θ_1/θ_2; and he discusses the use of T_1 for testing $\theta_1=\theta_2$.

For $k>2$, analogous formulas are known, but tables like Savage's can be used for the present ranking problem, as we now illustrate. Suppose $k=3$ and all $M_i=2$, so that $N=6$. A possible unconditional δ selects the population for which T_i is smallest. The possible values of T_i are listed in Table 1 of [11], p. 607, case "$N=6$, $m=2$, $n=4$" there. We consider a B of two elements, as in Example 2.2, and let C^2 be the event that $\min_i T_i$ is its smallest possible value, 2.1, this being the event that the selected population yielded the two largest of the 6 observed values. It is not hard to verify that, for this C, the LF configuration for both Γ^b's is again that in which one θ_i exceeds the other two by exactly $\bar{\Delta}$. Moreover, because of the small value of N in the present example, δ reduces to choosing the population that yielded the largest observation. Using the formula for the probability of a rank order when $k=2$, one can show that, in the present case $k=3$,

$$\begin{aligned} \Gamma^2_{\mathrm{LF}} &= [1+2(2+3\bar{\Delta})/\bar{\Delta}^4(2\bar{\Delta}+3)]^{-1}, \\ P_{\mathrm{LF}}\{C^2, \text{correct decision}\} &= 6\bar{\Delta}^4/(2+\bar{\Delta})(1+\bar{\Delta})(2+3\bar{\Delta})(1+2\bar{\Delta}). \end{aligned} \tag{5.2}$$

Thus, for example, if $\bar{\Delta}=3.62$ (chosen for use of a column of Savage's tables) the unconditional δ has confidence ≥ 0.805 (LF probability of correct selection). The conditional procedure just described has $\Gamma^2_{\mathrm{LF}}=0.986$, $P_{\mathrm{LF}}\{C^2, \text{correct}\}=0.374$, and similarly $\Gamma^1_{\mathrm{LF}}=0.69$, $P_{\mathrm{LF}}\{C^1, \text{correct}\}=0.431$. The practitioner can compare this operating characteristic with that of the unconditional procedure based on the same δ, to choose whether it is worth lowering the conditional confidence from 0.805 to 0.69 on a set where a correct selection is made with LF probability 0.431, in order to increase it to 0.986 on a set of corresponding probability 0.374.

For larger k and n_i the computations are lengthier; for sufficiently large M_i asymptotic approximations can be made, as indicated earlier.

6. Estimating Γ

In some selection problems of the type considered in Section 2 it may be appropriate to consider the format (2.1) with $\bar{\Delta}=0$. In that case the lower bound ψ is $\frac{1}{3}$, a rather useless fact. Intuitively, one may feel that large b_2

makes one feel very confident in the decision, and it should be possible to quantify this. Thus, for either a conditional or unconditional procedure we may want to estimate the Γ_ω which is unknown and for which we have no useful lower bound.

This is the spirit of estimating the power or risk function, a subject that has received attention (e.g., in [10]) for unconditional procedures. For arithmetical simplicity, we summarize some of the features in the simpler case $k=2$. In that case the problem, in terms of $Y=2^{-1/2}(X_1-X_2)$, becomes that of deciding whether a $\mathcal{N}(\theta,1)$ rv has $\theta>0$ or $\theta<0$, using the symmetric δ based on sgn Y. For the unconditional procedure, $\Gamma_\theta = \Phi(|\theta|)$. Intuitively, one may want to estimate this by $\Phi(|Y|)$. This is not an unbiased estimator, and a simple analyticity argument shows that none exists. The estimator $\Phi(|Y|)$ overestimates $\Phi(|\theta|)$ with probability $\Phi(-2|\theta|)+\frac{1}{2}$ and is thus not median unbiased, either. However, other simple estimators of Γ_θ can be constructed which are conservative from either the bias or median bias viewpoint; e.g., for which the expectation is $\leq \Gamma_\theta$. (For example, $\Phi(|Y|-c)$, for c sufficiently large, is of this nature.) What may seem more satisfactory to some practitioners is the estimation of Γ_θ in terms of what may be termed a *secondary confidence statement*, for example in terms of a one-sided lower confidence limit with associated confidence function (possibly also depending on θ!) As an illustration, let τ be nondecreasing on the non-negative reals, with $0<\tau<1$. When θ is the true parameter value, Y^2 has df $G_{|\theta|}$ (say), the non-central chi-square df with noncentrality parameter $E_\theta Y^2-1=\theta^2$ and one degree of freedom. Let $h(|\theta|)=G_{|\theta|}^{-1}(\tau(|\theta|))$ for all $|\theta|\geq 0$. Also, denote by h^{-1} the inverse of h on $[h(0),\infty)$ with $h^{-1}=0$ otherwise. Then the one-sided lower confidence limit $\Phi(h^{-1}(|Y|))$ on $\Gamma_\theta=\Phi(|\theta|)$ is calculated, in standard fashion, to have confidence function (correctness probability)

$$\begin{aligned}\mathbf{P}_\theta\{\Phi(h^{-1}(|Y|))\leq\Phi(|\theta|)\} &= \mathbf{P}_\theta\{|Y|\leq h(|\theta|)\}\\ &= G_{|\theta|}(h(|\theta|))=\tau(|\theta|).\end{aligned} \tag{6.1}$$

Thus, if τ is chosen to be a constant γ_0, we obtain a lower confidence bound on Γ_θ with confidence coefficient γ_0 (median unbiased estimator if $\gamma_0=\frac{1}{2}$); since it may seem unsatisfactory for $1-\Phi(h^{-1}(|Y|))$ often to be much smaller than $1-\gamma_0$, one may want to let τ vary accordingly, e.g., $\tau=\Phi^{1/2}$.

Although the exposition and illustration of the previous paragraph were given in terms of unconditional procedures, exactly the same considerations occur for estimating Γ_ω^b for conditional procedures. We remark only that one can examine such properties as expectations or covering probabilities as computed either conditionally or unconditionally for estimators of Γ_ω^b.

The frequentist interpretation, conditional or unconditional, of such estimators is clear. For example, if $E_\omega\{\hat{\Gamma}(X) | Z(X) = b\} \leq \Gamma_\omega^b$ for all ω, b, then the average of the values $\hat{\Gamma}(X_i)$, over a large number of experiments in which $Z(X_i) = b_0$, is with high probability exceeded or almost attained by the proportion of correct decisions in those experiments.

There are certainly many multidecision problems for which an unconditional procedure with estimated confidence will seem to some workers to answer the desire for a data-dependent highly variable measure of conclusiveness (with frequentist interpretation) as satisfactorily as does any conditional procedure. The two-decision normal problem just considered is such a setting, and it is worthwhile to point out the differences in what the two ideas — conditional confidence and estimated confidence — achieve. Although a useful conservatively biased estimator $\hat{\Gamma}(|y|)$ of the confidence $\Phi(|\theta|)$ of the unconditional procedure will approach 1 as $|y| \to \infty$, such an estimator of $\Gamma_\theta^{|y|}$ for our *conditional* procedure will do so more rapidly. Still, some of those who are uneasy with the conditioning framework will prefer the unconditional procedure with estimated Γ in that case. However, in settings where unconditional Γ_θ is bounded away from 1 but where likelihood ratios are unbounded (as in the example of the second paragraph of Section 1, or in the example just above but with Ω restricted to configurations where $|\theta_1 - \theta_2| = \bar{\Delta}$), the conditional procedure can take advantage of lucky observations to state an estimated confidence close to 1, which the unconditional procedure cannot (the probability of a correct selection being *known* to be $\Phi(\bar{\Delta})$, in the example). Th greater conceptual simplicity of estimating Γ for an unconditional procedure, over both conditioning and also estimating Γ, is of course an attraction of the former. A weighing of these features will indicate, I believe, different settings where it is attractive to use one of the two techniques, or both, or neither.

7. Other variations

We have discussed conditional inference using the simple zero-one loss function that is probably of greatest intuitive value in developing concepts of conditional as well as unconditional inference. Other loss structures are considered in [3], [6]. In particular, admissibility results are obtained for problems in which, as in the formulation of Section 3, the geometric size of the confidence set is variable, and in which a penalty is assessed that depends on that size. Roughly, it is again the case that a procedure which was unconditionally admissible for such a model is still admissible after conditioning in any way, and there are also other admissible conditional procedures. The sequential framework, discussed briefly in [3], leads to more difficult probabilistic calculations, because the sets C^b are now obtained from a stopping rule. This will be treated in detail in another paper.

References

[1] Bahadur, R.R. and Lehmann, E.L. (1955). Two comments on "Sufficiency and statistical decision functions". *Ann. Math. Statist.* **26**, 139–142.
[2] Bechhofer, R.E. (1954). A single-sample multiple decision procedure for ranking means of normal populations with known variances. *Ann. Math. Statist.* **25**, 16–39.
[3] Brownie, C. and Kiefer, J. (1975). Conditional confidence statements. [To appear.]
[4] Gupta, S.S. (1956). On a decision rule for a problem in ranking means. *Inst. Stat. Mimeo Ser.* No. 150, Univ. of N.C., Chapel Hill.
[5] Gupta, S.S., Huang, D.-Y., and Huang, W.-T. (1974). On ranking and selection procedures and tests of homogeneity for binomial populations. Purdue Mimeo Series # 375.
[6] Kiefer, J. (1975). Admissibility of conditional confidence procedures. *Ann. Statist.*, to appear.
[7] Lehmann, E.L. (1953). The power of rank tests. *Ann. Math. Statist.* **24**, 23–43.
[8] Neveu, J. (1965). *Mathematical Foundations of the Calculus of Probabilities.* Holden-Day (San Francisco).
[9] Pratt, J.W. (1961). *Review of Lehmann's Testing Statistical Hypotheses. J.A.S.A.* **56**, 163–166.
[10] Sandved, E. (1968). Ancillary statistics and estimation of the loss in estimation problems. *Ann. Math. Statist.* **39**, 1755–1758.
[11] Savage, I.R. (1956). Contributions to the theory of rank order statistics — the two-sample case. *Ann. Math. Statist.* **27**, 590–615.

P.R. Krishnaiah, ed., *Multivariate Analysis–IV*
© North-Holland Publishing Company (1977) 159–174

EMPIRICAL SAMPLING STUDY OF A GOODNESS OF FIT STATISTIC FOR DENSITY FUNCTION ESTIMATION

P.A.W. LEWIS*, L.H. LIU,
D.W. ROBINSON and M. ROSENBLATT**

Naval Postgraduate School and
University of California, San Diego, Calif., U.S.A.

The distribution of a measure of the distance between a probability density function and its estimate is examined through empirical sampling methods. The estimate of the density function is that proposed by Rosenblatt using sums of weight functions centered at the observed values of the random variables. The weight function in all cases was triangular, but both uniform and Cauchy densities were tried for different sample sizes and bandwidths. The simulated distributions look as if they could be approximated by Gamma distributions, in many cases. Some assessment can also be made of the rate of convergence of the moments and the distribution of the measure to the limiting moments and distribution, respectively.

1. Introduction

There are several recently proposed classes of empirical probability density function [1, 4, 5, 7] all generally considered to be superior to the classical histogram estimates. The class considered in this paper is based on independent observations, i.e. $X_1, X_2, \ldots, X_n$ are independent and identically distributed random variables with continuous unknown density function $f(x)$. The method used to estimate $f(x)$ is that proposed by Rosenblatt; denoting the estimate by $f_n(x)$, we define

$$f_n(x) = \frac{1}{nb(n)} \sum_{j=1}^{n} W\left[\frac{x - X_j}{b(n)}\right],$$

where $W(u)$ is a bounded non-negative integrable weight function with

$$\int_{-\infty}^{\infty} W(u)\mathrm{d}u = 1,$$

* Research supported by National Science Foundation AG–476.
** Research supported by Office of Naval Research.

and $b(n)$ is a positive bandwidth function which tends to zero as $n \to \infty$, but is such that $o[b(n)] = 1/n$. Thus we might have $b(n) \sim n^{-1/2}$, for example.

We note that all estimates of this form are themselves density functions for a given set of observations; that is,

$$f_n(x) \geq 0, \qquad \int_{-\infty}^{\infty} f_n(x)\mathrm{d}x = 1.$$

Since the X_j's are random variables, $f_n(x)$ is a continuous parameter stochastic process, but it is clearly non-stationary.

The estimate $f_n(x)$ can be shown to be locally biased for any value of x under relatively mild conditions [4]. Our object in this paper is to investigate a global measure of how good $f_n(x)$ is as an estimate of $f(x)$. The measure was originally proposed by Bickel and Rosenblatt [2] and is given by

$$\beta(n) = \int \frac{[f_n(x) - f(x)]^2}{f(x)} \mathrm{d}x.$$

Since the value of $\beta(n)$ will vary with each realization of $X_1, \ldots, X_n$, it is a statistic or function of the n random variables. A possible application for such a statistic would be in goodness-of-fit type tests, in an analogous manner to the more familiar Kolmogorov–Smirnov test.

Bickel and Rosenblatt [2] have established that if $b(n) = o[n^{-2/9}]$ as $n \to \infty$ and if $a(x)$ is a bounded, piecewise smooth integrable function with f sufficiently smooth and bounded, then

$$b(n)^{-1/2}\left[nb(n)\int [f_n(x) - f(x)]^2 a(x)\mathrm{d}x - \int f(x)a(x)\mathrm{d}x \int W(z)^2\mathrm{d}z\right]$$

is asymptotically normally distributed with zero mean and variance

$$2W^{(4)}(0)\int a(x)^2 f(x)^2 \mathrm{d}x,$$

as $n \to \infty$, where $W^{(4)}(0)$ is the fourth convolution of W with itself. Thus, $\beta(n)$ has an asymptotically normal distribution, regardless of the underlying density $f(x)$.

A problem in this situation is that, unlike the Kolmogorov–Smirnov test statistic, the statistic $\beta(n)$ is not distribution-free. Further, its exact distribution for any finite value of n does not seem to be mathematically tractable. We thus examined some representative cases through simulation, hoping that $\beta(n)$ would be fairly robust with rapid convergence to the asymptotic distribution. It was also hoped that the simulations would cast light on these conjectures and perhaps suggest some unexpected results.

2. Simulation

The primary object of the simulation was to investigate the distribution of the statistic $\beta(n)$:

$$\beta(n) = \int \frac{[f_n(x) - f(x)]^2}{f(x)} \, dx,$$

over a suitable range of integration. We performed simulations with synthetic sampling from both uniform and Cauchy distributions; the triangular weight function

$$W(u) = \begin{cases} 1 - |u| & \text{if } |u| \leq 1, \\ 0 & \text{otherwise} \end{cases}$$

was used to evaluate $f_n(x)$ in both cases. We found little difference as far as $\beta(n)$ was concerned between the triangular and other "smoother" (e.g., quadratic) weight functions for our samples of from 100 to 1500 deviates.

2.1. *Uniform random variables*

In the case of uniform (0, 1) random variables, we have

$$f(x) = \begin{cases} 1 & \text{if } 0 \leq x \leq 1, \\ 0 & \text{otherwise.} \end{cases}$$

Thus, $\beta(n)$ becomes,

$$\beta(n) = \int_{b(n)}^{1-b(n)} [f_n(x) - 1]^2 dx. \tag{2.1}$$

The limits of integration are from $b(n)$ to $1 - b(n)$ instead of from 0 to 1 to avoid the marked bias of $f_n(x)$ near 0 and 1. As long as $b(n) \leq x \leq 1 - b(n)$, though, $f_n(x)$ is unbiased:

$$\begin{aligned} \mathbf{E}[f_n(x)] &= \frac{1}{b(n)} \int_0^1 W\left[\frac{x-y}{b(n)}\right] dy \\ &= \frac{1}{b(n)} \int_{x-b(n)}^{x+b(n)} \left[1 - \frac{|x-y|}{b(n)}\right] dy \end{aligned}$$

$$= \frac{1}{b(n)} \left[\int_{x-b(n)}^{x+b(n)} dy - \int_{x-b(n)}^{x} \frac{x-y}{b(n)} dy - \int_{x}^{x+b(n)} \frac{y-x}{b(n)} dy \right]$$

$$= \frac{1}{b(n)} \left[2b(n) - \frac{1}{b(n)} \frac{[b(n)]^2}{2} - \frac{1}{b(n)} \frac{[b(n)]^2}{2} \right]$$

$$= 1.$$

Also, for the same range of x,

$$\mathrm{Var}[f_n(x)] = \mathrm{Var}\left[\frac{1}{nb(n)} \sum_{j=1}^{n} W\left[\frac{x - X_j}{b(n)} \right] \right]$$

$$= \frac{1}{n^2 b(n)^2} \sum_{j=1}^{n} \mathrm{Var}\, W\left[\frac{x - X_j}{b(n)} \right]$$

$$= \frac{1}{nb(n)^2} \mathrm{Var}\, W\left[\frac{x - X_j}{b(n)} \right]$$

$$= \frac{1}{nb(n)^2} \left[\left[\int_0^1 W^2 \left[\frac{x-y}{b(n)} \right] dy - \left[\int_0^1 W\left[\frac{x-y}{b(n)} \right] dy \right]^2 \right] \right].$$

Since $f_n(x)$ is a piecewise linear function when a triangular weight function is used, the integral in (2.1) can be evaluated in principle but the work becomes prohibitive for even moderate sample sizes. We thus approximated the integral using Simpson's rule with 100 equal subintervals. The results were found to be satisfactory in the sense that the value did not change appreciably when a finer grid (up to 500 subintervals) was used. In general, we found that a larger sample size required a finer grid; apparently the value of $f_n(x)$ changes more rapidly over a small interval when n is large.

We used three different bandwidths in the unifrom case: $3/n^{1/2}$, $1/n^{1/2}$ and $1/n$. For each bandwidth sample sizes of 100, 200, 500, 1000 and 1500 were investigated so that a total of 15 experiments were carried out. Each experiment consisted of 2000 independent replications each of which resulted in the calculation of a single value of $\beta(n)$ using (2.1). The replications for a given experiment were divided into five sections of 400 observations each so that variability of the simulation results could be assessed between sections.

Besides the 400 observed values of $\beta(n)$, the computer output for each of the 75 sections included a histogram, an empirical log-survivor function plot, an empirical CDF plot and a normal probability plot. A histogram and an empirical log-survivor plot were also computed for the pooled sample of 2000 for each experiment. These plots are all reproduced in reference [3]; some of the more interesting cases are included in Section 4.

It was found that a better picture of the distribution of the data resulted when the empirical density function of the $\beta(n)$'s was plotted over the histogram plot. A fairly wide bandwidth was needed to suppress large fluctuations in $f_n(x)$; it was found that $b(n) = R/n^{1/2}$ was a fairly robust choice. (R denotes the sample range [maximum value–minimum value] of the $\beta(n)$ sample.) The solid lines in the Figures in Section 4 are empirical density estimates using this bandwidth and the triangular weight function.

2.2. Cauchy random variables

The Cauchy density function is

$$f(x) = \frac{1}{\pi(1+x^2)}\,.$$

We used the same density estimator as in the uniform case:

$$f_n(x) = \frac{1}{nb(n)} \sum_{j=1}^{n} W\left[\frac{x - X_j}{b(n)}\right],$$

and again the triangular weight function. We chose a range of integration $(-3, +3)$:

$$\beta(n) = \int_{-3}^{+3} \frac{[f_n(x) - f(x)]^2}{f(x)}\,\mathrm{d}x.$$

This range comprises 80% of the probability mass for this distribution. Again, Simpson's rule was used to approximate the integral; in this case a grid of 600 subintervals was selected after examining 100, 300, 600 and 900 subinterval grids.

The Cauchy distribution was chosen because for finite n $f_n(x)$ has a bias component; this component usually decreases with bandwidth for a fixed value of n, although the pointwise variance of $f_n(x)$ increases with decreasing bandwidth. It seems likely that the variance of $\beta(n)$ would also decrease under these conditions, as indeed it was observed to do.

Three bandwidths were also employed in the Cauchy case: $1/n^{1/2}$, $3/n^{1/2}$ and $20/n^{1/2}$, the last one representing a case in which bias in the estimator $f_n(x)$ plays a major role in the distribution of $\beta(n)$. The same five sample sizes were used here for each bandwidth as were used for the uniform simulations; output from the fifteen Cauchy experiments was obtained just as in the uniform case.

3. Tabular results and Gamma fits

Using the asymptotic result obtained by Bickel and Rosenblatt [5], for a uniform random variable the quantity

$$b(n)^{-1/2}\left\{nb(n)\int_{b(n)}^{1-b(n)}|f_n(x)-1|^2\mathrm{d}x-[1-2b(n)]\int W(u)^2\mathrm{d}u\right\}$$

is asymptotically normally distributed with mean 0 and variance

$$2W^{(4)}(0)[1-2b(n)]$$

as $n\to\infty$ if $nb(n)\to\infty$ and $b(n)=o(n^{-2/9})$. For the triangular weight function,

$$\int W(u)^2\mathrm{d}u=\tfrac{2}{3}$$

and $W^{(4)}(0)$, the fourth convolution of W with itself at zero, is 302/630.

From the above expressions, we get

$$\mathbf{E}[\beta(n)]=\mathbf{E}\left[\int_{b(n)}^{1-b(n)}|f_n(x)-1|^2\mathrm{d}x\right]\sim\tfrac{2}{3}\,\frac{1-2b(n)}{nb(n)},$$

$$\mathrm{Var}[\beta(n)]=\mathrm{Var}\left[\int_{b(n)}^{1-b(n)}|f_n(x)-1|^2\mathrm{d}x\right]\sim\frac{2W^{(4)}(0)[1-2b(n)]}{n^2b(n)}.$$

Comparisons of the simulated values for the uniform experiments with the conjectured ones are tabulated in Table 1 (means) and Table 2 (variances). Especially for small bandwidth the agreement between the asymptotic and simulated variances is very good even for small n (n = 100). The same is true for expected value, although convergence is slower than for the variance and again slower for large bandwidth.

In contrast to the moments, the distribution of $\beta(n)$ converges very slowly. The complete results (reference [3]) reveal that the histograms and empirical density functions of the $\beta(n)$'s are all skewed to the right; see Figures 1 to 9 in Section 4 for examples.

The form of the histograms as well as the log-survivor plots suggested that the $\beta(n)$ statistic is approximately Gamma (θ, k) distributed, where

Table 1
Comparison of estimated mean values and asymptotic mean values of $\beta(n)$ for different bandwidths and sample sizes

n	$b(n)=3/\sqrt{n}$	$\mathbf{E}(\beta(n))$	$\mathbf{E}(\beta(n))/(1-2b(n))$	
			Conjectured	Computer output
100	0.3000	0.0089	0.0222	0.0127
200	0.2121	0.0090	0.0157	0.0109
500	0.1342	0.0073	0.0099	0.0075
1000	0.0949	0.0057	0.0070	0.0058
1500	0.0775	0.0048	0.0057	0.0051
	$b(n)=1/\sqrt{n}$			
100	0.1000	0.0533	0.0667	0.0583
200	0.0707	0.0405	0.0471	0.0415
500	0.0447	0.0271	0.0298	0.0269
1000	0.0316	0.0197	0.0211	0.0197
1500	0.0258	0.0163	0.0172	0.0168

Table 2
Comparison of estimated standard deviation values and asymptotic standard deviation values of $\beta(n)$ for different bandwidths and sample sizes

n	$b(n)=3/\sqrt{n}$	$\sigma(\beta(n))$	$\sigma(\beta(n))/(1-2b(n))$	
			Conjectured	Computer output
100	0.3000	0.0113	0.0283	0.0115
200	0.2121	0.0081	0.0141	0.0088
500	0.1342	0.0046	0.0063	0.0047
1000	0.0949	0.0029	0.0036	0.0030
1500	0.0775	0.0022	0.0026	0.0023
	$b(n)=1/\sqrt{n}$			
100	0.1000	0.0277	0.0346	0.0315
200	0.0707	0.0171	0.0199	0.0189
500	0.0447	0.0088	0.0097	0.0092
1000	0.0316	0.0053	0.0057	0.0056
1500	0.0258	0.0040	0.0042	0.0043

the Gamma density is given by

$$f(x; k, \theta) = \frac{(x/\theta)^{k-1} e^{-x/\theta}}{\theta(k-1)!},$$

and the mean and variance are

$$\mathbf{E}[X] = k\theta; \qquad \mathrm{Var}[X] = k\theta^2.$$

Table 3
Estimated parameters for fitted Gamma distribution for $\beta(n)$

Distribution	$b(n)$	n	$\bar{k}$	$\bar{\theta}$
		100	3.969 ± 0.206	0.01390 ± .00095
		200	5.780 ± 0.659	0.00715 ± .00095
	$1/\sqrt{n}$	500	8.881 ± 0.889	0.00311 ± .00029
		1000	13.011 ± 0.796	0.00153 ± .00008
		1500	17.316 ± 1.467	0.00095 ± .00008
		100	1.153 ± 0.048	0.00967 ± .00058
		200	1.718 ± 0.174	0.00588 ± .00078
Uniform	$3/\sqrt{n}$	500	2.707 ± 0.241	0.00281 ± .00026
		1000	4.028 ± 0.241	0.00145 ± .00007
		1500	5.248 ± 0.423	0.00096 ± .00008
		100	40.337 ± 2.555	0.01616 ± .00117
		200	39.511 ± 2.347	0.01675 ± .00106
	$1/\sqrt{n}$	500	33.649 ± 1.820	0.01958 ± .00111
		1000	32.033 ± 3.305	0.02059 ± .00244
		1500	31.712 ± 1.999	0.02088 ± .00124
		100	22.362 ± 1.488	0.01745 ± .00114
		200	32.305 ± 2.022	0.00864 ± .00054
	$1/\sqrt{n}$	500	60.147 ± 4.009	0.00293 ± .00022
		1000	79.897 ± 6.608	0.00157 ± .00014
		1500	101.100 ± 7.783	0.00102 ± .00007
		100	9.272 ± 0.406	0.01331 ± .00062
		200	12.744 ± 0.645	0.00709 ± .00037
Cauchy	$3/\sqrt{n}$	500	20.701 ± 1.673	0.00277 ± .00022
		1000	29.303 ± 2.541	0.00140 ± .00012
		1500	34.265 ± 3.963	0.00099 ± .00010
		100	7.103 ± 0.217	0.00776 ± .00035
		200	4.144 ± 0.069	0.00619 ± .00009
	$20/\sqrt{n}$	500	3.445 ± 0.161	0.00312 ± .00016
		1000	4.211 ± 0.357	0.00152 ± .00009
		1500	5.385 ± 0.335	0.00095 ± .00005

Accordingly, estimates $\bar{k}$ and $\bar{\theta}$ of k and θ for each experiment were obtained from the sample of 2000 $\beta(n)$'s. Shenton and Bowman's almost unbiased estimators for the Gamma distribution [6] were used; these give reasonable results when $k \geq 0.5$, as in this case. The estimate values are tabulated in Table 3, also tabulated are estimates of the standard deviation of $\bar{k}$ and $\bar{\theta}$ which were obtained from the five sections in each experiment. A parametric density estimate is thus obtained for the $\beta(n)$ sample; it may be compared with the non-parametric estimate $f_n(x)$ by examining the graphs in Section 4, where the Gamma density function is plotted with a dashed line.

4. Graphical results and general discussion

The graphs for the following experiments have been reproduced from [3] because they give the greatest insight into the distribution of $\beta(n)$; these graphical results are more informative than the Tabulated means, variance and Gamma fits of the previous section.

Figure	Random variable	n	$b(n)$	$\bar{k}$
1	Uniform	200	$3/n^{1/2}$	1.718
2	Uniform	500	$1/n^{1/2}$	8.881
3	Uniform	1500	$1/n^{1/2}$	17.316
4	Uniform	200	$1/n$	39.511
5	Cauchy	100	$1/n^{1/2}$	22.362
6	Cauchy	100	$3/n^{1/2}$	9.272
7	Cauchy	1500	$20/n^{1/2}$	5.385
8	Uniform	1500	$3/n^{1/2}$	5.248
9	Uniform	100	$1/n^{1/2}$	3.969

In interpreting the graphs we can be guided by crude heuristics. In the case of a density estimate $f_n(x)$ with bandwidth $b(n)$ there is dependence within a range of order $b(n)$ and an approach to independence for points separated by a distance of order larger than $b(n)$. Thus in the case of uniform random variables the integral $\beta(n)$ could be thought of as having the equivalent of the order of $[1-2b(n)]/b(n)$ independent summands. In the first case (Figure 1, $n=200$, $b(n)=3/\sqrt{n}$, $\bar{k}=1.718$) we obtain

$$(1-3\sqrt{2}/10)/[3/(10\sqrt{2})] = 2.71.$$

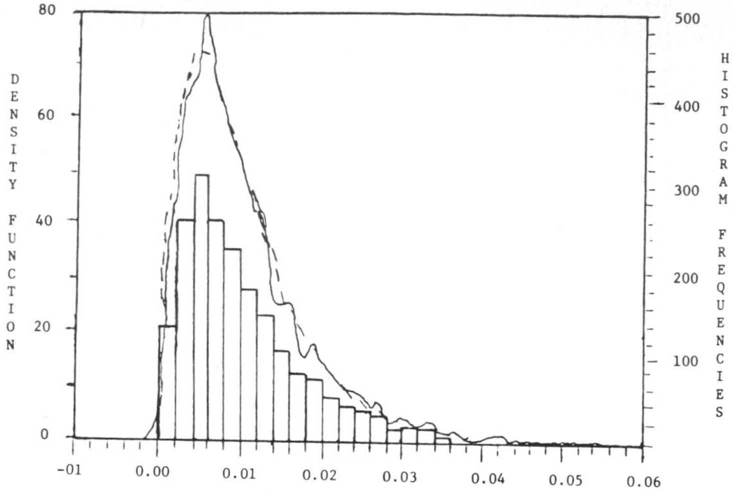

Fig. 1. Distribution of the statistic $\beta(n)$ for a uniform random variable with $n = 200$ and bandwidth $3/\sqrt{n}$. The solid line shows the Rosenblatt empirical density function of the $\beta(n)$'s while the dashed line is a fitted Gamma density function with $\bar{k} = 1.718$ and $\bar{\theta} = 0.00588$.

This is rather small so that one does not expect a good Gaussian fit. We give $\bar{k}$ from the previous section since $2\bar{k}$ may be interpreted as an equivalent number of degrees of freedom; the larger the fitted $\bar{k}$, the closer we are to normality. In a loose sense it is clear that a gamma fit is likely to be more appropriate and this is confirmed by looking at the graphs.

In the second case (Figure 2; $n = 500$, $b(n) = 1/\sqrt{n}$, $\bar{k} = 8.881$) we have

$$(1 - \sqrt{2}/10)\,10\sqrt{2} = 12.14,$$

which is a bit larger. It is interesting to note that the estimated (smoothed) density function of $\beta(n)$ gives us greater insight apparently in all cases. Here we see the beginning of an approach to asymptotic normality though it is still suggested that a Gamma fit might be appropriate. The next case (Figure 3; $n = 1500$, $b(n) = 1/\sqrt{n}$, $\bar{k} = 17.316$) with

$$[1 - 2/(5\sqrt{15})]\,10\sqrt{15} = 36.73$$

shows a closer approach to normality. It may be seen that the major departure between the parametric and non-parametric density estimates

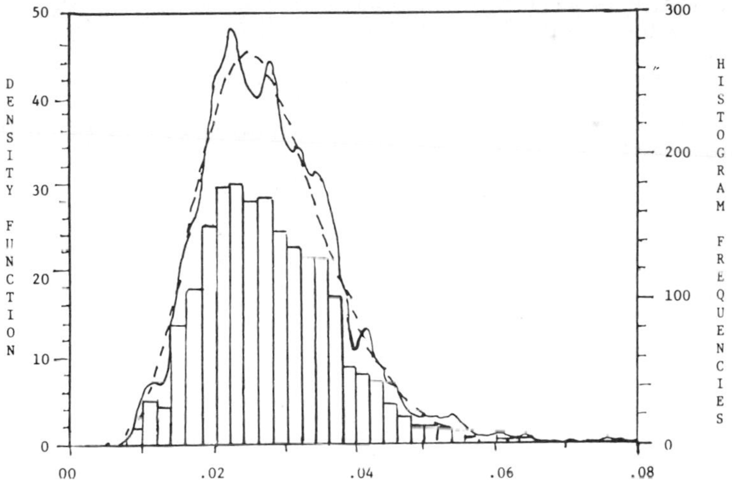

Fig. 2. Distribution of the statistic $\beta(n)$ for a uniform random variable with $n = 500$ and bandwidth $1/\sqrt{n}$. The solid line shows the Rosenblatt empirical density function of the $\beta(n)$'s while the dashed line is a fitted Gamma density function with $\bar{k} = 8.881$ and $\bar{\theta} = 0.00311$.

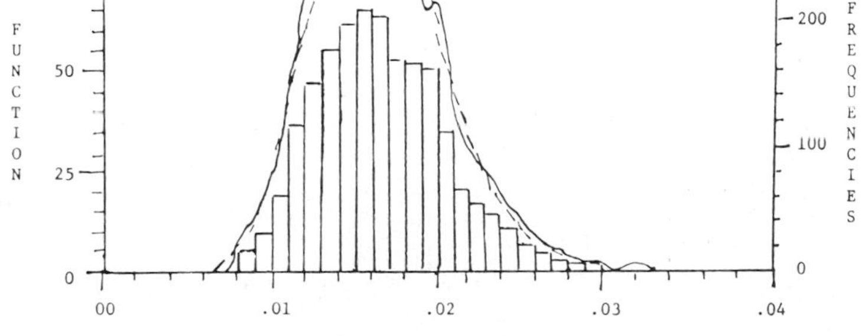

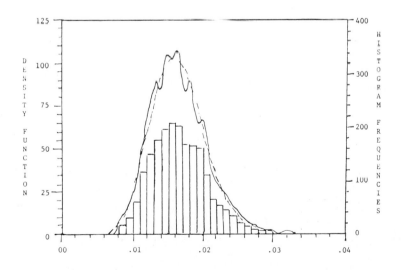

Fig. 3. Distribution of the statistic $\beta(n)$ for a uniform random variable with $n = 1500$ and bandwidth $1/\sqrt{n}$. The solid line shows the Rosenblatt empirical density function of the $\beta(n)$'s while the dashed line is a fitted Gamma density function with $\bar{k} = 17.316$ and $\bar{\theta} = 0.00095$.

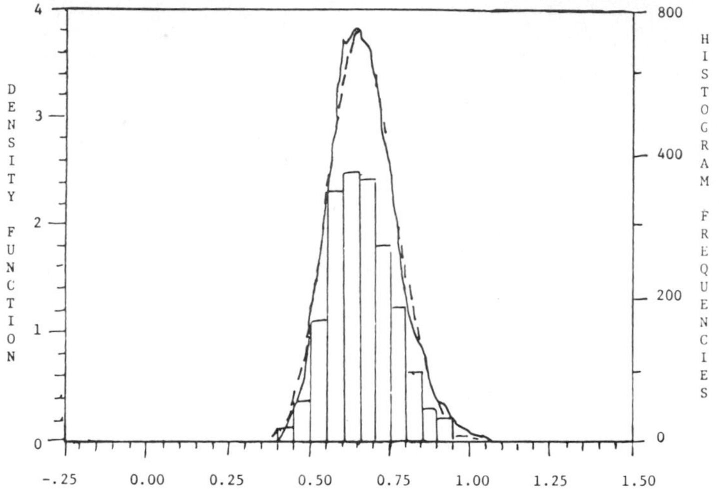

Fig. 4. Distribution of the statistic $\beta(n)$ for a uniform random variable with $n = 200$ and bandwidth $1/n$. The solid line shows the Rosenblatt empirical density function of the $\beta(n)$'s while the dashed line is a fitted Gamma density function with $\bar{k} = 39.511$ and $\bar{\theta} = 0.01675$.

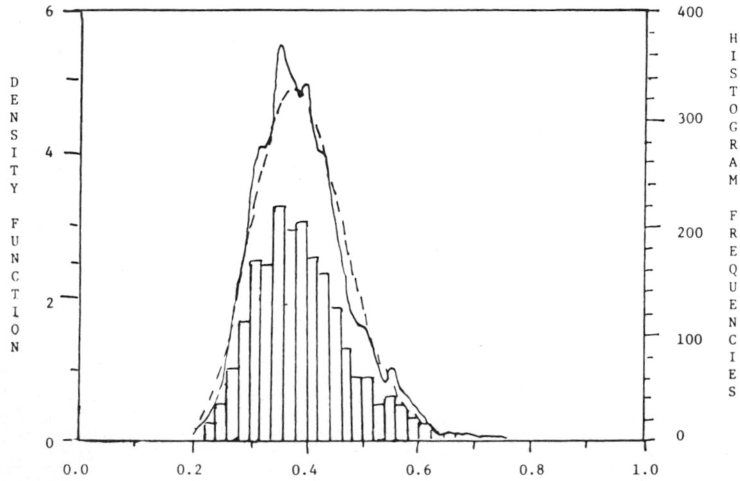

Fig. 5. Distribution of the statistic $\beta(n)$ for a Cauchy random variable with $n = 100$ and bandwidth $1/\sqrt{n}$. The solid line shows the Rosenblatt empirical density function of the $\beta(n)$'s while the dashed line is a fitted Gamma density function with $\bar{k} = 22.362$ and $\bar{\theta} = 0.01745$.

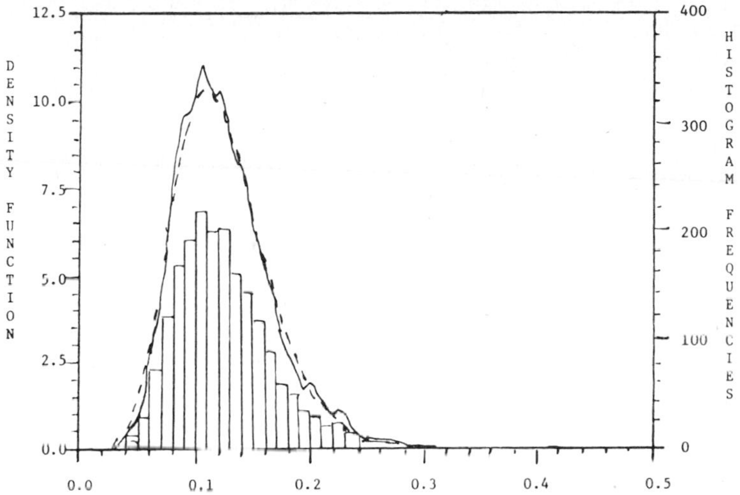

Fig. 6. Distribution of the statistic $\beta(n)$ for a Cauchy random variable with $n = 100$ and bandwidth $3/\sqrt{n}$. The solid line shows the Rosenblatt empirical density function of the $\beta(n)$'s while the dashed line is a fitted Gamma density function with $\bar{k} = 9.272$ and $\bar{\theta} = 0.01331$.

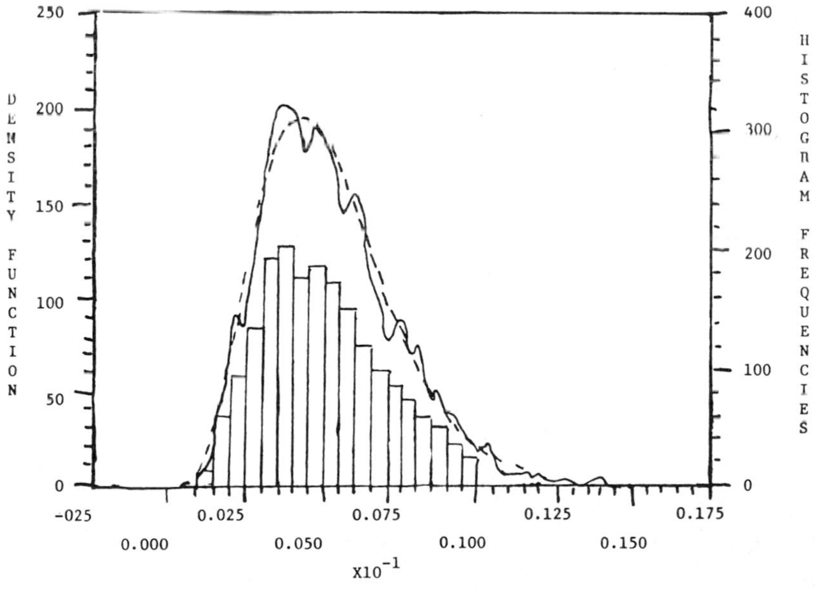

Fig. 7. Distribution of the statistic $\beta(n)$ for a Cauchy random variable with $n = 1500$ and bandwidth $20/\sqrt{n}$. The solid line shows the Rosenblatt empirical density function of the $\beta(n)$'s while the dashed line is a fitted Gamma density function with $\bar{k} = 5.385$ and $\bar{\theta} = 0.00095$.

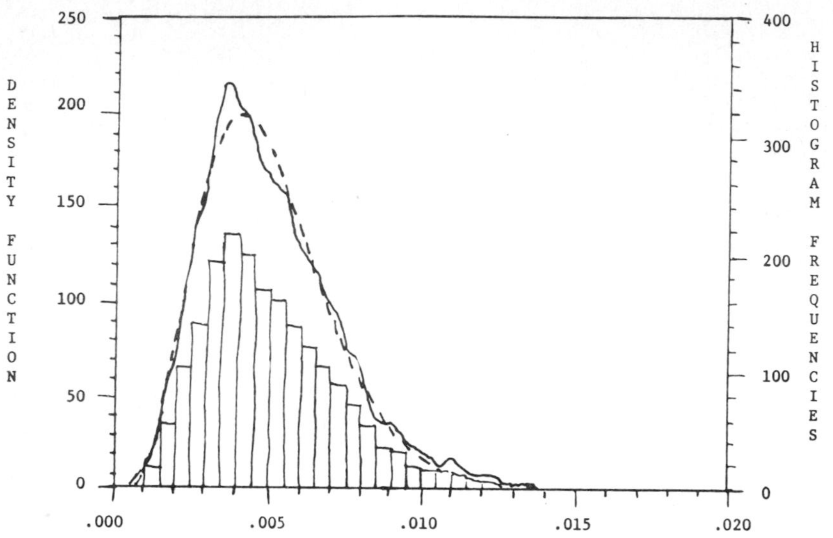

Fig. 8. Distribution of the statistic $\beta(n)$ for a uniform random variable with $n = 1500$ and bandwidth $3/\sqrt{n}$. The solid line shows the Rosenblatt empirical density function of the $\beta(n)$'s while the dashed line is a fitted Gamma density function with $\bar{k} = 5.248$ and $\bar{\theta} = 0.00096$.

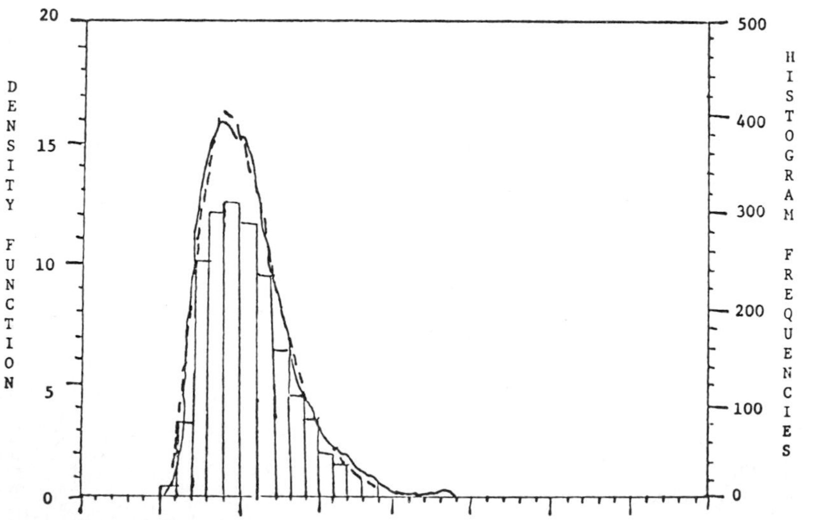

Fig. 9. Distribution of the statistic $\beta(n)$ for a uniform random variable with $n = 100$ and bandwidth $1/\sqrt{n}$. The solid line shows the Rosenblatt empirical density function of the $\beta(n)$'s while the dashed line is a fitted Gamma density function with $\bar{k} = 3.969$ and $\bar{\theta} = 0.01390$.

occurs in the vicinity of the mode where $f_n(x)$ tends to fluctuate about the true value. The fit in the tails appears excellent in all cases.

The next uniform case (Figure 4; $n = 200$, $b(n) = 1/n$, $\bar{k} = 39.511$) is strictly speaking outside the range of results suggested by the paper of Bickel and Rosenblatt [2]. Here $f_n(x)$ is asymptotically compound Poisson rather than asymptotically normal. Nonetheless we notice that it looks as if a Gaussian fit would be very good and this is consistent with the magnitude of our crude index

$$(1 - 0.01)200 = 198.$$

It would be interesting for someone to prove the suggested asymptotic normality.

In the simulation of sampling from a uniform distribution, the density estimator has no bias. To investigate the effect of bias, we repeated the uniform experiments for Cauchy-distributed random variables, integrated over the range $-3 + b(n)$ to $3 - b(n)$. The first case (Figure 5; $n = 100$, $b(n) = 1/\sqrt{n}$, $\bar{k} = 22.362$) has index

$$(6 - 0.2)10 = 58$$

and one notices that a Gaussian fit looks very good. The next case (Figure 6; $n = 100$, $b(n) = 3/\sqrt{n}$, $\bar{k} = 9.272$) has index

$$(6 - 0.6)10/3 = 17.66,$$

and a Gaussian fit looks fair but not good. In the last Cauchy case one expects substantial bias (Figure 7; $n = 1500$, $b(n) = 20/\sqrt{n}$, $\bar{k} = 5.385$) and the crude index is

$$(6 - 4/\sqrt{15})\sqrt{15}/2 = 9.62.$$

A Gamma fit is suggested. Altogether the effects of bias do not seem to be that extreme when sampling from the Cauchy distribution but this may be due to the fact that the Cauchy density is a very smooth function.

The last two cases involve sampling from the uniform distribution again but with different sample sizes and bandwidths. Figure 8 is for $n = 1500$, $b(n) = 3/\sqrt{n}$ and $\bar{k} = 5.248$, while Figure 9 is for $n = 100$ and $b(n) = 1/\sqrt{n}$ for which $\bar{k} = 3.969$.

The problem in using $\beta(n)$ as a measure of goodness of fit in the non-limiting Gamma case is to determine k and θ. If one wishes to fit the Gamma distribution using the method of moments, one can use the fact that the mean and variance of $\beta(n)$ should be approximately (on asymptotic grounds) $W^{(2)}(0)$ and $b(n)W^{(4)}(0)$, respectively. One might then use

$$k^* = \frac{[W^{(2)}(0)]^2}{b(n)\,W^{(4)}(0)},$$

$$\theta^* = \frac{k^*}{W^{(2)}(0)}$$

as estimates of k and θ. The results in Section 3 suggest that this procedure should produce adequate results except when there is appreciable bias in the density function estimate.

References

[1] Bartlett, M.S. (1963). Statistical estimation of density functions. *Sankhyā*, Ser. **A25**, p. 245–254.

[2] Bickel, P.J. and Rosenblatt, M. (1973). On some global measures of the deviations of density function estimates. *The Annals of Mathematical Statistics*, **1**, pp. 1071–1095.

[3] Liu, L.H. (1974). Empirical sampling investigation of a global measure of fit of probability density functions. M.S. Thesis, Naval Postgraduate School, Monterey.

[4] Rosenblatt, M. (1956). Remarks on some non-parametric estimates of a density function. *The Annals of Mathematical Statistics*, **27**.

[5] Rosenblatt, M. (1971). Curve estimates. *The Annals of Mathematical Statistics*, **42**.

[6] Shenton, L.R., and Bowman, K.O. (1973). Comments on the Gamma distribution and uses in rainfall data. *Third Conference on Probability and Statistics in Atmospheric Science*, AMS.

[7] Wegman, E.J. (1972). Non-parametric probability density estimation: I. A summary of available methods. *Technometrics*, **14**.

P.R. Krishnaiah, ed., *Multivariate Analysis–IV*

CORRELATIONAL ANALYSIS WHEN SOME VARIANCES AND COVARIANCES ARE KNOWN

I. OLKIN

Stanford University, Stanford, Calif., U.S.A.

and

M. SYLVAN

University of California, Santa Cruz, Calif., U.S.A.

This work examines maximum likelihood estimators (M.L.E.'s) and likelihood ratio tests for two special multivariate cases. In both cases a random matrix S having a Wishart distribution with parameter Σ, $p+q$ degrees of freedom and sample size n, is observed. Occasionally (for both cases) it is assumed that, in addition to S, the random vector $\bar{x} = (\dot{\bar{x}}, \ddot{\bar{x}})$ having a multivariate normal distribution with mean $\mu = (\dot{\mu}, \ddot{\mu})$ and covariance matrix Σ is observed.

The first of the two cases is characterized by the assumption that a principal submatrix, Σ_{22}, of Σ is known. For the second case it is assumed that Σ_{22} is unknown, but that in addition to S the random matrix W having a Wishart distribution with parameter Σ_{22}, q degrees of freedom and sample size m, is observed.

For both cases the M.L.E. of the correlation coefficient is examined in detail. Distributions, exact and asymptotic, null and non-null are computed. Functions of the estimators whose asymptotic variance is not a function of the parameter are derived. Density functions and confidence curves are numerically evaluated and presented. Finally the M.L.E. for the multiple correlation is examined in the same detail.

1. Introduction

If we have a bivariate normal distribution and wish to estimate the correlation coefficient, ρ, then we would normally use the sample correlation coefficient, $r = s_{12}/s_1 s_2$, or some function of it as an estimator. Now suppose that the value of σ_2 is known. A first attempt might be to use $r^* = s_{12}/(s_1\sigma_2)$, but then r^* need not be in the interval $[-1, 1]$. It is somewhat surprising that an estimate for ρ that entails this information is not readily available.

In the present paper we consider this problem in the more general context when there are p variates, and some variances and covariances are known. It is also noted that this model may be embedded into a more general framework that permits extra or missing observations.

2. Preliminaries

We may take as our starting point the $(p+q)\times(p+q)$ sample cross-product matrix, S, having a Wishart distribution $\mathscr{W}(\Sigma; p+q, n)$, with density function

$$p(S;\Sigma) = c(p+q,n)(\det S)^{\frac{1}{2}(n-(p+q)-1)}(\det \Sigma)^{-\frac{1}{2}n}\, e^{-\frac{1}{2}\operatorname{tr}\Sigma^{-1}S}, \tag{2.1}$$

for $S>0$, $\Sigma>0$, where

$$c(k,n) = \left[\pi^{k(k-1)/2}2^{nk/2}\prod_{1}^{k}\Gamma\left(\frac{n-i+1}{2}\right)\right]^{-1}.$$

The problem is to find the maximum likelihood estimator (MLE) of Σ_{11} and Σ_{12} when $\Sigma_{22}=I$. (Note that the assumption that $\Sigma_{22}=\Sigma_{22}^{0}$ is known may be translated to the assumption $\Sigma_{22}=I$ without loss in generality.)

Because Σ^{-1} arises in the density, a reparameterization will help simplify many of the expressions.

We partition the matrices Σ and S:

$$\Sigma = \begin{pmatrix}\Sigma_{11} & \Sigma_{12}\\ \Sigma_{21} & \Sigma_{22}\end{pmatrix}, \qquad S = \begin{pmatrix}S_{11} & S_{12}\\ S_{21} & S_{22}\end{pmatrix},$$

where Σ_{11}, S_{11}, are of order $p\times p$, and Σ_{22}, S_{22}, are of order $q\times q$. We then define

$$\Gamma = \begin{pmatrix}\Gamma_{11} & \Gamma_{12}\\ \Gamma_{12}' & \Gamma_{22}\end{pmatrix}$$

by

$$\begin{aligned}\Gamma_{11} &= \Sigma_{11}-\Sigma_{12}\Sigma_{22}^{-1}\Sigma_{21} = (\Sigma^{11})^{-1},\\ \Gamma_{12} &= \Sigma_{12}\Sigma_{22}^{-1/2},\\ \Gamma_{22} &= \Sigma_{22},\end{aligned} \tag{2.2}$$

where $\Sigma_{22}^{1/2}$ is the unique positive definite square root. (Throughout this paper the square root means the positive definite square root.) It then follows that

$$\begin{aligned}\Sigma_{11} &= \Gamma_{11}+\Gamma_{12}\Gamma_{21},\\ \Sigma_{12} &= \Gamma_{12}\Gamma_{22}^{1/2},\\ \Sigma_{22} &= \Gamma_{22}.\end{aligned} \tag{2.3}$$

If we make a transformation from S to

$$G = \begin{pmatrix} G_{11} & G_{12} \\ G_{21} & G_{22} \end{pmatrix},$$

$$G_{11} = S_{11} - S_{12} S_{22}^{-1} S_{21}, \qquad G_{12} = S_{12} S_{22}^{-1/2}, \qquad G_{22} = S_{22}, \tag{2.4}$$

then the parameter Γ enters naturally, as may be seen from the following lemma.

Lemma 2.1. *If $\mathcal{L}(S) = \mathcal{W}(\Sigma; p+q, n)$ and Γ and G are defined by* (2.2) *and* (2.4), *respectively, then for G_{11} fixed, G_{11} and G_{22} are (conditionally) independently distributed. The conditional (and unconditional) distribution of G_{11} is $\mathcal{L}(G_{11}) = \mathcal{W}(\Gamma_{11}; p, n-q)$, the columns of $G_{12}G_{22}^{-1}$ are mutually independent, each having a multivariate normal distribution with covariance matrix Γ_{11} and $E(G_{12}G_{22}) = \Gamma_{12}G_{22}^{1/2}$. Further, G_{11} and G_{22} are independently distributed with $\mathcal{L}(G_{22}) = \mathcal{W}(\Sigma_{22}; q, n)$.*

3. Maximum likelihood estimates of Σ_{11} and Σ_{12} when $\Sigma_{22} = I$

The transformation from (S_{11}, S_{12}, S_{22}) to (G_{11}, G_{12}, G_{22}) is one-to-one, so that we may use the distribution of G as our starting point. Also, by using (2.3), the MLE of Γ will yield the MLE of Σ. Because, the reparameterization is such that $\Sigma_{22} = \Gamma_{22}$, the condition $\Sigma_{22} = I$ is equivalent to $\Gamma_{22} = I$. The condition that Σ be positive definite when $\Sigma_{22} = I$ is the condition that Γ_{11} be positive definite.

Theorem 3.1. *If $\mathcal{L}(S) = \mathcal{W}(\Sigma; p+q, n)$ and $\Sigma_{22} = I$, the* MLE *of Σ_{11} and Σ_{12} are given by*

$$\hat{\Sigma}_{11} = \frac{1}{n}(S_{11} - S_{12}S_{22}^{-1}S_{21}) + S_{12}S_{22}^{-2}S_{21},$$

$$\hat{\Sigma}_{12} = S_{12}S_{22}^{-1}.$$

Proof. Using Lemma 2.1 with $\Gamma_{22} = I$, the MLE of Γ_{11} is immediate, namely,

$$\hat{\Gamma}_{11} = \frac{1}{n} G_{11} = \frac{1}{n}(S_{11} - S_{12}S_{22}^{-1}S_{21}).$$

The MLE of Γ_{12} is obtained by minimizing (with respect to Γ_{12})

$$\operatorname{tr}(G_{12}-\Gamma_{12}G_{22}^{1/2})'\Gamma_{11}^{-1}(G_{12}-\Gamma_{12}G_{22}^{1/2}),$$

to yield

$$\hat{\Gamma}_{12}=G_{12}G_{22}^{-1/2}=(S_{12}S_{22}^{-1/2})(S_{22}^{-1/2})=S_{12}S_{22}^{-1}.$$

The proof is completed by using (2.3) to transform from Γ_{11}, Γ_{12} to Σ_{11}, Σ_{12}.

When $p=1$ and q is arbitrary, $\Sigma_{12}=\sigma_1$ is a vector, $\Sigma_{11}=\sigma_{11}$ a scalar, the multiple correlation coefficient between X_1 and $X_2,\dots,X_q$ is defined by

$$\rho_{1\cdot(2\dots q)}^2=\frac{\sigma_1\Sigma_{22}^{-1}\sigma_1'}{\sigma_{11}}.$$

Thus the MLE of ρ^2 when $\Sigma_{22}=I$ is (with $S_{12}\equiv s_1$, $S_{11}=s_{11}$):

$$\hat{\rho}_{1\cdot(2\dots q)}^2=\frac{s_1S_{22}^{-2}s_1'}{\dfrac{1}{n}(s_{11}-s_1S_{22}^{-1}s_1')+s_1S_{22}^{-2}s_1'}. \tag{3.1}$$

If we let $V=S/n$ be the sample covariance matrix, then

$$\hat{\rho}_{1\cdot(2\dots q)}^2=\frac{v_1V_{22}^{-2}v_1'}{v_{11}-v_1V_{22}^{-1}v_1'+v_1V_{22}^{-2}v_1'} \tag{3.2}$$

is the MLE of $\rho_{1\cdot(2\dots q)}^2$.

Note that because $v_{11}-v_1V_{22}^{-1}v_1'\geq 0$, $\hat{\rho}_{1\cdot(2\dots q)}^2\leq 1$. Also, as $n\to\infty$, plim $V=\Sigma$, and since $\Sigma_{22}=I$,

$$\operatorname{plim}\hat{\rho}_{1\cdot(2\dots q)}^2=\frac{\sigma_1\Sigma_{22}^{-2}\sigma_1'}{\sigma_{11}-\sigma_1\Sigma_{22}^{-1}\sigma_1'+\sigma_1\Sigma_{22}^{-2}\sigma_1'}=\frac{\sigma_1\sigma_1'}{\sigma_{11}}=\rho_{1\cdot(2\dots q)}^2.$$

4. The correlation coefficient for the bivariate case

Because the Pearson product moment correlation plays such an important role, we discuss the bivariate case separately. This is the case when $p=q=1$ and σ_{22} is known (say $\sigma_{22}=1$). The exact and asymptotic distributions, both central and noncentral, of the MLE of the correlation coefficient are presented. Because the asymptotic variance depends on the true correlation, we also provide a variance stabilizing transformation. Confidence intervals and some efficiency comparisons are also given.

4.1. *Distribution of the MLE of the correlation coefficient*

From (3.2), the MLE of ρ is

$$\hat\rho = \frac{v_{12}}{[v_{22}(v_{11}v_{22} - v_{12}^2) + v_{12}^2]^{1/2}}. \tag{4.1}$$

To find the distribution of $\hat\rho$, we start with the density of V, namely,

$$p(v_{11}, v_{22}, v_{12}) = c(\det V)^{\frac{1}{2}(n-3)} e^{-\frac{1}{2}n \operatorname{tr} \Lambda V}, \tag{4.2}$$

where $\Lambda = \Sigma^{-1}$, and c is a normalizing constant. We first transform from v_{12} to $\hat\rho$. Using (4.1) we obtain

$$\det V = \frac{v_{11}v_{22}(1-\hat\rho^2)}{1-\hat\rho^2 + \hat\rho^2 v_{22}},$$

and

$$\frac{\partial v_{12}}{\partial \hat\rho} = \frac{v_{11}^{1/2} v_{22}}{(1-\hat\rho^2+\hat\rho^2 v_{22})^{3/2}},$$

so that (4.2) becomes

$$p(v_{11}, v_{22}, \hat\rho) = c\,\frac{v_{11}^{(n-2)/2} v_{22}^{(n-1)/2}(1-\hat\rho^2)^{(n-3)/2}}{(1-\hat\rho^2+\hat\rho^2 v_{22})^{n/2}}$$

$$\times \exp\left\{-\frac{n}{2(1-\rho^2)}\left[\frac{v_{11}}{\sigma_{11}} - \frac{2v_{11}^{1/2} v_{22}\rho\hat\rho}{\sigma_{11}^{1/2}(1-\hat\rho^2+v_{22}\hat\rho^2)^{1/2}} + v_{22}\right]\right\}, \tag{4.3}$$

for $v_{11} > 0$, $v_{22} > 0$, $\hat\rho^2 < 1$, where $\rho = \sigma_{12}/\sigma_{11}^{1/2}$ and

$$c^{-1} = \left(\frac{2}{n}\right)^n \pi^{1/2}\Gamma\left(\frac{n}{2}\right)\Gamma\left(\frac{n-1}{2}\right)\sigma_{11}^{n/2}(1-\rho^2)^{n/2}.$$

In the null case when $\rho = 0$, (4.3) reduces to

$$p(v_{11}, v_{22}, \hat\rho) = \frac{c\,v_{11}^{(n-2)/2} v_{22}^{(n-1)/2}(1-\hat\rho^2)^{(n-3)/2}}{(1-\hat\rho^2+\hat\rho^2 v_{22})^{n/2}} e^{-(n/2)[v_{11}/\sigma_{11}+v_{22}]},$$

in which case we may integrate out over v_{11} to obtain

$$p(v_{22}, \hat\rho) = c_1 \frac{v_{22}^{(n-1)/2}(1-\hat\rho^2)^{(n-3)/2}}{(1-\hat\rho^2+v_{22}\hat\rho^2)^{n/2}} e^{-(n/2)v_{22}}, \tag{4.4}$$

where $c_1 = (\frac{1}{2}n)^{\frac{1}{2}n}/[\pi^{1/2}\Gamma((n-1)/2)]$. The density of $\hat\rho$ is now obtained by integrating (4.4) over v_{22}. This involves several facts concerning the confluent hypergeometric function $\psi(g, h; x)$, defined by

$\psi(g, h; x) =$

$$\frac{1}{x^g} \sum_{j=0}^{\infty} \frac{\Gamma(g+j)}{\Gamma(g)} \frac{\Gamma(g-h+1+j)}{\Gamma(h)} \frac{(-1/x)^d}{d!} \frac{1}{x^g} {}_2F_0(g, g-h+1; -1/x).$$

(For a reference, see Bateman (1954), p. 255–257, 285.) This function has an integral expression which we use

$$\int_0^{\infty} \frac{u^a}{(1+Au)^b} e^{-u} \, du = \frac{\Gamma(a+1)}{A^{a+1}} \psi(a+1, a-b+2; 1/A) \tag{4.5}$$

Consequently, a direct application of (4.5) to (4.4) yields

$$p(\hat{\rho}) = \frac{(n-1)}{(2n\pi)^{\frac{1}{2}}} (1-\hat{\rho}^2)^{-3/2} \sum_{j=0}^{\infty} \frac{\Gamma(\frac{1}{2}(n+1)+j)}{\Gamma(\frac{1}{2}(n+1))} \frac{\Gamma(\frac{1}{2}n+j)}{\Gamma(\frac{1}{2}n)} \left[-\frac{2\hat{\rho}^2}{n(1-\hat{\rho}^2)}\right]^j \tag{4.6}$$

for $-1 < \hat{\rho} < 1$.

To obtain the distribution of $\hat{\rho}$ in the nonnull case (that is, when $\rho \neq 0$) we start with (4.3). As a first step we expand the exponential term involving $v_{11}^{1/2}$ in a power series, and then integrate term by term with respect to v_{11}. This yields the joint distribution of v_{22} and $\hat{\rho}^2$, from which we then integrate out over v_{22}:

$$p(\hat{\rho}) = (1-\rho^2)^{(n-3)/2} \sum_{j=0}^{\infty} \frac{d_j}{j!} \frac{(\rho\hat{\rho})^j}{(1-\rho^2)^{j/2}} \int_0^{\infty} \frac{v_{22}^{(n+2j-1)/2}}{(1-\hat{\rho}^2+\hat{\rho}^2 v_{22})^{(n+j)/2}} e^{-(n/2)(v_{22}/(1-\rho^2))},$$

where

$$d_j = n^{\frac{1}{2}(n-j)} 2^{\frac{1}{2}(j-n)} \Gamma\left(\frac{n+j}{2}\right) \Big/ \left[\sqrt{\pi}\,\Gamma\left(\frac{n}{2}\right) \Gamma\left(\frac{n-1}{2}\right)\right].$$

Again using the representation (4.5) in terms of the ψ function, we obtain

$$p(\hat{\rho}) = \sum \frac{d_j \Gamma\left(\dfrac{n+2j+1}{2}\right)}{j!} \rho^j (1-\rho^2)^{\frac{1}{2}j} \frac{(1-\hat{\rho}^2)^{\frac{1}{2}(n+j-2)}}{\rho}$$
$$\times \hat{\rho}^{-(n+j+1)} \psi\left(\frac{n+2j+1}{2}, \frac{j+5}{2}, \frac{1-\hat{\rho}^2}{\hat{\rho}^2}\right), \tag{4.7}$$

for $-1 < \hat{\rho} < 1$.

The density function of $\hat{\rho}$ can be evaluated by numerical methods; graphs for $n = 10$, 50 and $\rho = 0.00$, 0.05, 0.08 are given in Figures 1 and 2. Figure 3 provides 95% confidence limits for ρ for $n = 4$, 5, 6, 7, 8, 10, 12, 15, and 25.

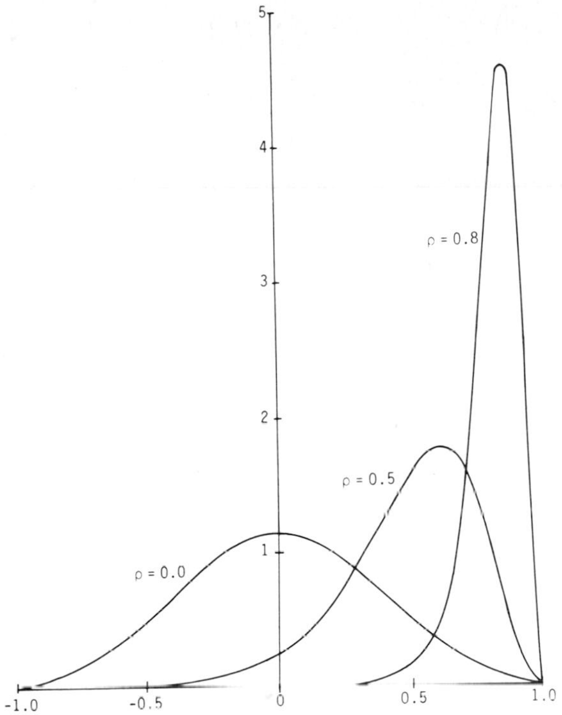

Figure 1. Frequency curves for $\hat{\rho}$ when $n = 10$.

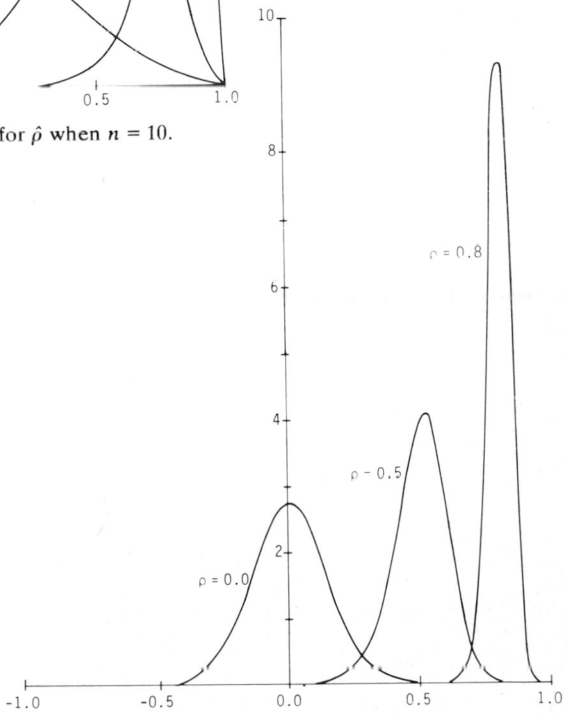

Figure 2. Frequency curves for $\hat{\rho}$ when $n = 50$.

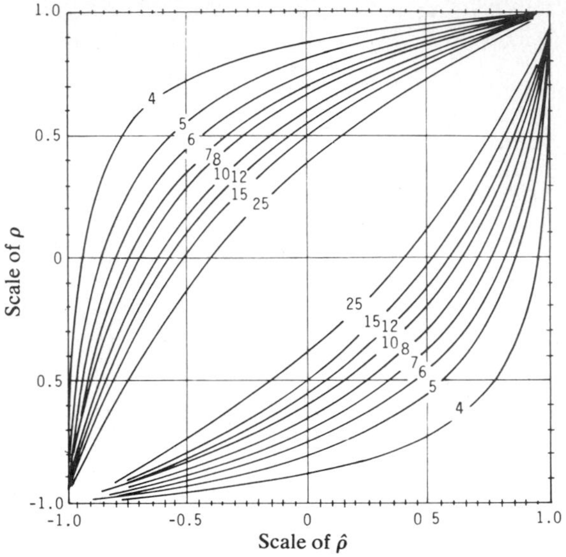

Figure 3. 95% confidence curves for the correlation coefficient (numbers on curves indicate sample size).

4.2. *Asymptotic distribution*

Using the standard delta method, the asymptotic covariance matrix of

$$\sqrt{n}\left(\frac{v_{11}}{\sigma_{11}},\ v_{22},\ \frac{v_{12}}{\sqrt{\sigma_{11}}}\right)$$

is

$$V_\infty = \begin{pmatrix} 2 & 2\rho^2 & 2\rho \\ 2\rho^2 & 2 & 2\rho \\ 2\rho & 2\rho & 1+\rho^2 \end{pmatrix},$$

(see Anderson (1958), p. 76) so that the asymptotic variance of $\hat{\rho}$ is

$$v_\infty(\hat{\rho}) = \left(\frac{\partial\hat{\rho}}{\partial v_{11}}, \frac{\partial\hat{\rho}}{\partial v_{22}}, \frac{\partial\hat{\rho}}{\partial v_{12}}\right) V_\infty \left(\frac{\partial\hat{\rho}}{\partial v_{11}}, \frac{\partial\hat{\rho}}{\partial v_{22}}, \frac{\partial\hat{\rho}}{\partial v_{12}}\right)', \tag{4.8}$$

evaluated at $\sigma_{11}, \sigma_{22} = 1$, $\sigma_{12} = \rho\sqrt{\sigma_{11}}$. Since

$$\frac{\partial\hat{\rho}}{\partial v_{11}} = -\tfrac{1}{2}\rho, \qquad \frac{\partial\hat{\rho}}{\partial v_{22}} = -\tfrac{1}{2}\rho(2-\rho^2), \qquad \frac{\partial\hat{\rho}}{\partial v_{12}} = 1,$$

a straightforward evaluation of (4.8) yields

$$v_\infty(\hat\rho) = \frac{1}{n}(1-\rho^2)^2(1-\tfrac{1}{2}\rho^2).$$

Consequently, we have the asymptotic result that

$$\sqrt{n}(\hat\rho-\rho) \sim \mathcal{N}(0, (1-\rho^2)^2(1-\tfrac{1}{2}\rho^2)).$$

Recall that the usual estimator of ρ, $r = v_{12}/(v_{11}v_{22})^{1/2}$ has the asymptotic distribution:

$$\sqrt{n}(r-\rho) \sim \mathcal{N}(0, (1-\rho^2)^2),$$

so that the gain by using $\hat\rho$ is in the factor $1-\frac{1}{2}\rho^2$, which ranges from $\frac{1}{2}$ to 1.

4.3. Variance stabilizing transformation

Because the variance of the asymptotic distribution of $\hat\rho$ depends on ρ, we seek a function $g(\hat\rho)$ such that $v_\infty(g(\hat\rho))$ is not a function of ρ. Since

$$v_\infty(g(\hat\rho)) = \left(\frac{\partial g(\hat\rho)}{\partial\hat\rho}\bigg|_{\hat\rho=\rho}\right)^2 v_\infty(\hat\rho),$$

we need to solve the differential equation $dg/d\rho = (1-\rho^2)(1-\frac{1}{2}\rho^2)^{1/2}$. The solution is given in the following.

Theorem 4.1.

$$g(x) = \frac{1}{\sqrt{2}}\log\left\{\frac{\sqrt{2-x^2}+x}{\sqrt{2-x^2}-x}\right\}.$$

The asymptotic distribution of $\sqrt{n}(g(\hat\rho)-g(\rho))$ is $\mathcal{N}(0,1)$.

5. The multiple correlation coefficient

In this section we obtain the distribution of the MLE of $\rho^2_{1\cdot(2\ldots q)}$ and its asymptotic distribution. However, instead of $\hat\rho$, it will be simpler to study the statistic

$$h = \frac{\hat\rho^2}{1-\hat\rho^2} = \frac{v_1 V_{22}^{-2} v_1'}{v_{11} - v_1 V_{22}^{-1} v_1'},$$

where the sample covariance matrix V is partitioned as

$$V = \begin{pmatrix} v_{11} & v_1 \\ v_1' & V_{22} \end{pmatrix},$$

with v_{11} a scalar and V_{22}: $q \times q$.

By invariance, we may assume without loss in generality that

$$\Sigma = \begin{pmatrix} 1 & \beta \\ \beta' & I_q \end{pmatrix}, \qquad \beta = (\rho, 0, \ldots, 0),$$

where $\rho = \rho_{1\cdot(2\ldots q)}$ is the population multiple correlation coefficient.

In the null case when $\rho = 0$, $\Sigma = I$ and we start with the Wishart distribution

$$p(V) = c(q+1, n)(\det V)^{\frac{1}{2}(n-q-2)} \operatorname{etr}(-\tfrac{1}{2}n V).$$

$$= c(q+1, n)(\det V_{22})^{\frac{1}{2}(n-q-2)}(v_{11} - v_1 V_{22}^{-1} v_1')^{\frac{1}{2}(n-q-2)} \exp -\tfrac{1}{2}n(v_{11} + \operatorname{tr} V_{22}).$$

Now let $u = v_{11} - v_1 V_{22}^{-1} v_1'$ and $w = v_1 V_{22}^{-1}$ be a transformation from (v_{11}, v_1) to (u, w). The Jacobian is $(\det V_{22})$ and we obtain

$$p(u, w, V_{22}) = c(q+1, n)(\det V_{22})^{\frac{1}{2}(n-q)} u^{\frac{1}{2}(n-q-2)}$$

$$\times \exp -\tfrac{1}{2}n(u + wV_{22}w' + \operatorname{tr} V_{22}). \quad (5.1)$$

In this form we may integrate (5.1) over $V_{22} > 0$ to obtain

$$p(u, w) = c_1 \frac{u^{\frac{1}{2}(n-q-2)} e^{-\frac{1}{2}nu}}{(1 + ww')^{\frac{1}{2}(n+1)}}, \quad (5.2)$$

where $c_1 = c(q+1, n)/[c(q, n+1)n^{\frac{1}{2}q(n+1)}]$.

Recall that we wish to find the distribution of $h = ww'/u$. We first use a result of Hsu (1939) (see Anderson (1958), p. 319) to find the distribution of $t = ww'$, namely

$$p(u, t) = c_2 u^{\frac{1}{2}(n-q-2)} e^{-\frac{1}{2}nu} \frac{t^{\frac{1}{2}(q-2)}}{(1+t)^{\frac{1}{2}(n+1)}},$$

so that the distribution of $h = t/u$ is

$$p(h) = c_2 h^{\frac{1}{2}q-1} \int_0^\infty \frac{u^{\frac{1}{2}n-1}}{(1+hu)^{\frac{1}{2}(n+1)}} e^{-\frac{1}{2}nu} du,$$

which may again be expressed in the form of a hypergeometric function:

$$p(h) = Lh^{\frac{1}{2}q-1} {}_2F_0(n/2, n+1/2; -2h/n),$$

for $0 < h < \infty$, where

$$L = \left(\frac{2}{n}\right)^{\frac{1}{2}q} \cdot \frac{\Gamma\left(\frac{n}{2}\right)\Gamma\left(\frac{n+1}{2}\right)}{\Gamma\left(\frac{q}{2}\right)\Gamma\left(\frac{n-q+1}{2}\right)\Gamma\left(\frac{n-q}{2}\right)}.$$

5.1. Nonnull distribution

The derivation of the distribution of h in the nonnull case, i.e., when $\rho \neq 0$, is somewhat more difficult. The density function of (v_{11}, v_1, V_{22}) when

$$\Sigma = \begin{pmatrix} 1 & \beta \\ \beta & I \end{pmatrix}$$

is

$$p(v_{11}, v_1, V_{22}) = \frac{c(q+1, n)(\det V_{22})^{\frac{1}{2}(n-q-2)}}{(1-\rho^2)^{\frac{1}{2}n}} (v_{11} - v_1 V_{22}^{-1} v_1')^{\frac{1}{2}(n-q-2)} \tag{5.3}$$

$$\times \operatorname{etr}\left(-\frac{n}{2} V\Lambda\right),$$

where

$$\Sigma^{-1} \equiv \Lambda = \begin{pmatrix} \lambda_{11} & \lambda_1 \\ \lambda_1' & \Lambda_{22} \end{pmatrix}, \qquad \lambda_{11} = (1-\rho^2)^{-1}, \qquad \lambda_1 = -\beta/(1-\rho^2),$$

$$\Lambda_{22} = (I - \beta'\beta)^{-1} = [(1-\rho^2)I + \beta'\beta]/(1-\rho^2).$$

We again let $u = v_{11} - v_1 V_{22}^{-1} v_1'$, $w = v_1 V_{22}^{-1}$ to obtain

$$p(u, w, V_{22}) = \frac{c(q+1, n)}{(1-\rho^2)^{\frac{1}{2}n}} u^{\frac{1}{2}(n-q-2)}(\det V_{22})^{\frac{1}{2}(n-q)}$$

$$\times \exp \frac{-n}{2(1-\rho^2)} [u + \operatorname{tr} V_{22} Q],$$

where

$$Q = (1-\rho^2) I_q + (w-\beta)'(w-\beta).$$

Integration over $V_{22} > 0$ then yields

$$p(u, w) = c_1 u^{\frac{1}{2}(n-q-2)} e^{-\frac{1}{2}(nu/(1-\rho^2))} (\det Q)^{-\frac{1}{2}(n+1)},$$

where

$$c_1 = \frac{c(q+1, n)(1-\rho^2)^{\frac{1}{2}[n(q-1)+1]}}{n^{\frac{1}{2}q(n+1)} c(q, n+1)}.$$

Note that

$$\det Q = (1-\rho^2)^q \left[1 + \frac{(w_1-\rho)^2 + \sum_2^q w_i^2}{1-\rho^2}\right].$$

We may invoke Hsu's result [Anderson (1958) p. 319] to obtain the joint distribution of u, w_{11} and $t = \Sigma_2^q w_i^2$:

$$p(u, w_1, t) = \frac{c_2 u^{\frac{1}{2}(n-q-2)} e^{-\frac{1}{2}nu/(1-\rho^2)} t^{\frac{1}{2}(q-3)}}{\left[1 + \frac{(w_1-\rho)^2 + t}{1-\rho^2}\right]^{\frac{1}{2}(n+1)}},$$

where

$$c_2 = \frac{c_1}{(1-\rho^2)^{\frac{1}{2}(n+1)}} \frac{\pi^{(q-1)/2}}{\Gamma\left(\frac{q-1}{2}\right)}.$$

Now let $z = (w_1^2 + t)/u$ be a transformation from t to z. Since we wish to obtain the distribution of z, we eventually integrate out over u and w_1. The joint distribution of u, w_1, and z is

$$p(u, w_1, z) = c_3 \frac{u^{\frac{1}{2}(n-q)} e^{-\frac{1}{2}nu/(1-\rho^2)} (zu - w_1^2)^{\frac{1}{2}(q-3)}}{(1 + zu - 2\rho w_1)^{\frac{1}{2}(n+1)}}, \tag{5.4}$$

where $c_3 = (1-\rho^2)^{\frac{1}{2}(n+1)} c_2$. Note that $z > 0$, $u > 0$ and $zu > w_1^2$ (which implies that $1 + zu > 2\rho w_1$). In (5.4) we integrate over w_1 after expanding the denominator in a power series:

$$\int_{zu > w_1^2} \frac{(zu - w_1^2)^{\frac{1}{2}(q-3)}}{(1 + zu - 2\rho w_1)^{\frac{1}{2}(n+1)}} dw_1$$

$$= \frac{1}{(1+zu)^{\frac{1}{2}(n+1)}} \sum_{j=0}^{\infty} \frac{\Gamma\left(\frac{n+1}{2} + j\right)}{\Gamma\left(\frac{n+1}{2}\right)} \left(\frac{2\rho}{1+zu}\right)^j \frac{1}{j!} \int_{zu > w_1^2} (zu - w_1^2)^{\frac{1}{2}(q-3)} w_1^j \, dw_1.$$

But

$$\int_{zu > w_1^2} (zu - w_1^2)^{\frac{1}{2}(q-3)} w_1^j \, dw_1 = (zu)^{\frac{1}{2}(q-2+j)} \int_{-1}^{1} (1 - x^2)^{\frac{1}{2}(q-3)} x^j \, dx,$$

which vanishes for j odd, and which is equal to $B(\frac{1}{2}(2k+1), \frac{1}{2}(q-1))$ for $j = 2k$.

Consequently, we obtain

$$p(u,z)=\frac{c_3 u^{\frac{1}{2}(n-q)}e^{-\frac{1}{2}nu/(1-\rho^2)}}{(1+zu)^{\frac{1}{2}(n+1)}}\sum_{k=0}^{\infty}\frac{\Gamma\left(\frac{n+1}{2}+2k\right)}{\Gamma\left(\frac{n+1}{2}\right)}$$

$$\times\left(\frac{2\rho}{1+zu}\right)^{2k}\frac{1}{(2k)!}(zu)^{\frac{1}{2}(q-2+2k)}B\left(\frac{2k+1}{2},\frac{q-1}{2}\right),$$

and hence

$$p(z)=c_3\sum_{k=0}^{\infty}\frac{\Gamma\left(\frac{n+1}{2}+2k\right)}{\Gamma\left(\frac{n+1}{2}\right)}\frac{(2\rho)^{2k}}{(2k)!}B\left(\frac{2k+1}{2},\frac{q-1}{2}\right)z^{\frac{1}{2}(q-2+2k)}$$

$$\times\int_0^{\infty}\frac{u^{\frac{1}{2}(n-2+2k)}e^{-\frac{1}{2}nu/(1-\rho^2)}}{(1+zu)^{\frac{1}{2}(n+1)+2k}}du. \tag{5.5}$$

The integral in (5.5) may be expressed as a hypergeometric function to yield

$$p(z)=\sum_{k=0}^{\infty}d_k\frac{\rho^{2k}}{(1-\rho^2)^{\frac{1}{2}(n+2k)}}z^{\frac{1}{2}(q+2k-2)}\psi\left(\frac{n+2k}{2},\frac{3-2k}{2},\frac{n}{(2z(1-\rho^2))}\right), \tag{5.6}$$

where

$$d_k=c_3\frac{\Gamma\left(\frac{n+1}{2}+2k\right)}{\Gamma\left(\frac{n+1}{2}\right)}\frac{2^{2k}}{(2k)!}B\left(\frac{2k+1}{2},\frac{q-1}{2}\right)\left(\frac{2}{n}\right)^{\frac{1}{2}(n+2k)}\Gamma\left(\frac{n+2k}{2}\right).$$

5.2. Asymptotic distribution of the multiple correlation coefficient

The asymptotic distribution of

$$h=\frac{R^2}{1-R^2}=\frac{v_1V_{22}^{-2}v_1'}{v_{11}-v_1V_{22}^{-1}v_1'}$$

is obtained with the aid of the following lemma (Madansky and Olkin (1969) p. 269):

Lemma 5.1. *If the random matrix V has a Wishart distribution $\mathscr{L}(V)=W(\Sigma/n;p,n)$, and $f(V)$ is a real-valued function with first order partial* *derivatives* Σ, *then* $\sqrt{n}(f(V)-f(\Sigma))$ *has an asymptotic normal distribution* $\mathcal{N}(0, 2\,\mathrm{tr}\, F\Sigma F\Sigma)$, *where* $F=[F_{ij}]$ *with*

$$F_{ii} = \left.\frac{\partial f(V)}{\partial v_{ii}}\right|_{\Sigma}, \qquad F_{ij} = \tfrac{1}{2}\left.\frac{\partial f(V)}{\partial v_{ij}}\right|_{\Sigma}.$$

By a direct computation of the derivatives evaluated at

$$\Sigma = \begin{pmatrix} 1 & \beta \\ \beta' & I \end{pmatrix},$$

we obtain

$$F = \frac{1}{(1-\rho^2)}\begin{pmatrix} -\rho^2 & \beta \\ \beta' & -(2-\rho^2)\beta'\beta \end{pmatrix},$$

from which

$$\mathrm{tr}\, F\Sigma F\Sigma = 2\rho^2(1-\tfrac{1}{2}\rho^2)/(1-\rho^2)^2.$$

Consequently, the asymptotic distribution of $\sqrt{n}(R-\rho_{1\cdot(2\ldots q)})$ is $\mathcal{N}(0, v_\infty(R))$, where

$$v_\infty(R) = \left(\frac{\partial R}{\partial h}\right)^2 \frac{4\rho^2(1-\frac{1}{2}\rho^2)}{(1-\rho^2)^2} = (1-\rho^2)^2(1-\tfrac{1}{2}\rho^2).$$

This result is independent of q, and is the same result as obtained in the bivariate case. Thus, we may use the same variance stabilizing transformation as in the bivariate case.

6. Estimating the covariance matrix and correlation when additional observations on some variables are available

Consider the situation where we observe a sample cross-product matrix S, with $\mathcal{L}(S) = W(\Sigma; p+q, n)$, and in addition, we observe a $q \times q$ sample cross-product matrix W with $\mathcal{L}(W) = W(\Sigma_{22}; q, m)$. The key point is that as $m \to \infty$, the additional observations act as if Σ_{22} were known.

6.1. MLE of Σ

If we transform from S to G as in Section 2, and reparameterize from Σ to Γ, we find that

$$\hat{\Gamma}_{11} = S_{11}/n,$$

$$\hat{\Gamma}_{12} = S_{12}S_{22}^{-1}(S_{22}+W)^{1/2}/(m+n)^{1/2},$$

$$\hat{\Gamma}_{22} = (S_{22}+W)/(m+n),$$

from which, by (2.3), we obtain

$$\hat{\Sigma}_{11} = \frac{(S_{11} - S_{12}S_{22}^{-1}S_{21})}{n} + \frac{S_{12}S_{22}^{-1}(S_{22} + W)S_{22}^{-1}S_{21}}{m+n},$$

$$\hat{\Sigma}_{12} = \frac{S_{12} + S_{12}S_{22}^{-1}W}{m+n}, \tag{6.1}$$

$$\hat{\Sigma}_{22} = \frac{S_{22} + W}{m+n}.$$

6.2. MLE of the correlation coefficient

The MLE of the correlation coefficient when extra observations on one variable are available is obtained from (6.1) with $p = q = 1$:

$$\tilde{\rho} = \frac{\dfrac{s_{12}}{s_{22}}\left(\dfrac{s_{22} + w}{m+n}\right)^{1/2}}{\left(\dfrac{1}{n}(s_{11} - s_{12}^2/s_{22}) + \dfrac{s_{12}^2(s_{22} + w)}{(m+n)s_{22}^2}\right)^{1/2}}, \tag{6.2}$$

where

$$S = \begin{pmatrix} s_{11} & s_{12} \\ s_{12} & s_{22} \end{pmatrix}.$$

If we let $V = S/n$ and $z = w/m$, then (6.2) becomes

$$\tilde{\rho} = \frac{v_{12}(\bar{k}v_{22} + kz)^{1/2}}{[v_{22}(v_{11}v_{22} - v_{12}^2) + v_{12}^2(\bar{k}v_{22} + kz)]^{1/2}}, \tag{6.3}$$

where $k = m/(m+n)$, $\bar{k} = 1 - k = n/(m+n)$

Remark. Note that when $m = 0$, so that $k = 0$, $\tilde{\rho}$ reduces to the standard product moment correlation. When $m = \infty$, so that $k = 1$, then $\operatorname{plim} z = \sigma_{22} = 1$ and $\tilde{\rho}$ reduces to (4.1). In carrying out the asymptotics we assume $\lim_{n\to\infty} m(m+n)^{-1} = k$.

The actual distribution of $\tilde{\rho}$ is complicated, and of questionable import. Instead, we obtain the asymptotic distribution of $\tilde{\rho}$. First note that $\operatorname{plim}\tilde{\rho} = \rho$ and that $\hat{\rho}$ evaluated at Σ is equal to ρ.

Using the standard delta method, $\sqrt{n}(\tilde{\rho} - \rho)$ has an asymptotic normal distribution $\mathcal{N}(0, v_\infty(\tilde{\rho}))$, where

$$v_\infty(\tilde{\rho}) = \left(\frac{\partial\tilde{\rho}}{\partial v_{11}}, \frac{\partial\tilde{\rho}}{\partial v_{22}}, \frac{\partial\tilde{\rho}}{\partial v_{12}}, \frac{\partial\tilde{\rho}}{\partial z}\right) V_\infty \left(\frac{\partial\tilde{\rho}}{\partial v_{11}}, \frac{\partial\tilde{\rho}}{\partial v_{22}}, \frac{\partial\tilde{\rho}}{\partial v_{12}}, \frac{\partial\tilde{\rho}}{\partial z}\right)',$$

evaluated at Σ, where

$$V_\infty = \begin{pmatrix} 2 & 2\rho^2 & 2\rho & 0 \\ 2\rho^2 & 2 & 2\rho & 0 \\ 2\rho & 2\rho & 1+\rho^2 & 0 \\ 0 & 0 & 0 & 2 \end{pmatrix}.$$

Remark. In carrying out this computation we may use invariance, and assume that $\sigma_{11} = \sigma_{22} = 1$.

The partial derivatives, evaluated at Σ, are

$$\frac{\partial \tilde{\rho}}{\partial v_{11}} = -\tfrac{1}{2}\rho, \qquad \frac{\partial \tilde{\rho}}{\partial v_{22}} = -\tfrac{1}{2}\rho[1 + k(1-\rho^2)],$$

$$\frac{\partial \tilde{\rho}}{\partial v_{12}} = 1, \qquad \frac{\partial \tilde{\rho}}{\partial z} = \tfrac{1}{2}\rho k(1-\rho^2),$$

from which we obtain $v_\infty(\tilde{\rho}) = (1-\rho^2)^2(1 - k\bar{k}\rho^2)$.

Consequently, we have the result that $\sqrt{n}(\tilde{\rho} - \rho)$ is asymptotically normal $\mathcal{N}(0, v_\infty(\tilde{\rho}))$. To obtain a variance stabilizing transformation, we solve the differential equation

$$\frac{\mathrm{d}g(x)}{\mathrm{d}x} = \frac{1}{(1-x^2)(1-k\bar{k}x^2)^{1/2}},$$

to obtain

$$g(x) = \frac{1}{2(1-k\bar{k})^{\frac{1}{2}}} \log \left[\frac{(1-k\bar{k}x^2)^{\frac{1}{2}} + x(1-k\bar{k})^{\frac{1}{2}}}{(1-k\bar{k}x^2)^{\frac{1}{2}} - x(1-k\bar{k})^{\frac{1}{2}}} \right].$$

Consequently, $\sqrt{n}(g(\tilde{\rho}) - g(\rho))$ is asymptotically $\mathcal{N}(0, 1)$.

7. A testing for the mean vector

In the classical Hotelling's T^2 model, we have N independent observations on the p-dimensional random vector $x \sim \mathcal{N}(\mu, \Sigma)$ and wish to test $H: \mu = 0$ versus $A: \mu \neq 0$. The likelihood ratio statistic (LR) is given by

$$\lambda^{2/N} = \frac{1}{1 + N\bar{x}V^{-1}\bar{x}'}.$$

Now suppose that we know $\Sigma_{22} = I$. Of course, we still could use the

T^2-statistic above. However, we may also calculate the LR test; to do this we reparameterize the problem as in Section 2, in which case the MLE are readily obtained. The result is

$$\lambda^{2/N} = \frac{1 + N\ddot{\bar{x}} V_{22}^{-1} \ddot{\bar{x}}'}{1 + N\bar{x} V^{-1} \bar{x}'} e^{-N\bar{x}\bar{x}'},$$

where $\bar{x} = (\dot{\bar{x}}, \ddot{\bar{x}})$ and $\ddot{\bar{x}}$ is a q-dimensional vector.

The distribution of this statistic is complicated. But asymptotically $-2\log\lambda \sim \chi_p^2$. We mention that a variety of other tests on means and/or covariances have been obtained by Sylvan (1969).

Acknowledgment

The authors are grateful to the referees for their many suggestions and comments.

References

[1] Anderson, T.W. (1958). *An Introduction to Multivariate Analysis*, John Wiley and Sons, New York.

[2] Bateman Manuscript Project (1954). *Higher Transcendental Functions*, Volume I, McGraw-Hill Book Company, Inc., New York.

[3] Madansky, A. and Olkin, I. (1969). Approximate Confidence Regions for Constraint Parameters, p. 261–286, *Multivariate Analysis — II*, ed. by P.R. Krishnaiah, Academic Press, New York.

[4] Sylvan, M. (1969). Estimation and hypothesis testing for Wishart matrices when part of the covariance matrix is known. Technical Report No. 38, Stanford University, Stanford, California.

P.R. Krishnaiah, ed., *Multivariate Analysis–IV*
© North-Holland Publishing Company (1977) 193–208

PREDICTION OF FUTURE OBSERVATIONS WITH SPECIAL REFERENCE TO LINEAR MODELS

C. Radhakrishna RAO

Indian Statistical Institute, New Delhi, India

Formulae have been obtained for predicting future observations in a linear model. The performances of different formulae are compared by applying them on empirical data relating to biological growth. The method of principal components is used to estimate the coefficients of a linear model when they are not specified. Some comments have been made on the general problem of predicting future observations in a stochastic process when the parameters characterizing the process are unknown.

1. General theory

Let $(y_1, \ldots, y_p, y_{p+1})$ be a $(p+1)$-vector random variable whose distribution depends on a parameter θ which may be a vector. We consider the problem of predicting y_{p+1} given $y_1, \ldots, y_p$, when the value of θ is unknown.

Example. $y_1, \ldots, y_p$ are observed weights of a growing individual at p time points. What is the predicted value of y_{p+1} the weight of the individual at the $(p+1)$-th future time point?

We consider different situations:
(i) Prediction for a specific individual.
(ii) Prediction for an individual drawn at random from a specified population.
(iii) Simultaneous prediction for a number of individuals.
We shall first discuss some general principles for solving these problems and develop the formulae for the special case of linear models.

1.1. *Prediction for a specific individual*

Let the probability density of $y_1, \ldots, y_{p+1}$ given a specific individual characterized by an unknown parameter θ be denoted by

$$P(y_1, \ldots, y_{p+1} \mid \theta) \tag{1.1.1}$$

which can be written as

$$P_2(y_{p+1} \mid y_1, \ldots, y_p, \theta)P_1(y_1, \ldots, y_p \mid \theta) \tag{1.1.2}$$

of which the first factor is the conditional density of y_{p+1} given $y_1, \ldots, y_p, \theta$, which may be called the predictive density function (PDF) of y_{p+1}. If θ is known then the PDF of y_{p+1} provides all the information for drawing inference on y_{p+1}. If not there are various possibilities.

(a) *Pivotal ancillary statistic*

The method advocated by Fisher consists in finding a function of $y_1, \ldots, y_{p+1}$ whose distribution is independent of the unknown parameter θ. Such a statistic, $T = f(y_1, \ldots, y_{p+1})$, is called a pivotal ancillary. Since the distribution of T is independent of θ, it should be possible to obtain exact confidence limits to y_{p+1} given $y_1, \ldots, y_p$. A simple example is as follows.

Example. Let $y_1, \ldots, y_p$ be independent observations from $N(\mu, \sigma^2)$. Then

$$T = \frac{y_{p+1} - \bar{y}}{s\sqrt{(p+1)/p}} \tag{1.1.3}$$

where $\bar{y}$ and s^2 are the usual unbiased estimators of μ and σ^2, has a t distribution on $(p-1)$ degrees of freedom. Confidence interval for y_{p+1} can be obtained by equating T to upper and lower percentile points of t.

(b) *Empirical PDF*

The density function (1.1.2) consists of two parts of which the first is the PDF which involves the unknown parameter θ. The second part is the density of $y_1, \ldots, y_p$ given θ. Using the second part we can obtain an estimate of θ as a function of the observations $y_1, \ldots, y_p$ by the maximum likelihood or some appropriate method. If $\hat{\theta}$ is an estimate of θ, then the estimated or empirical probability density function (EPDF) is

$$P_2(y_{p+1} \mid y_1, \ldots, y_p, \hat{\theta}). \tag{1.1.4}$$

We may choose the mode or the mean of the distribution (1.1.4) as a point estimate of y_{p+1}.

It may be seen that computing $\hat{\theta}$ first and choosing the mode of (1.1.4) may not be equivalent to obtaining the maximum likelihood estimates of the unknowns y_{p+1} and θ directly from the density function (1.1.1), which seems to be an alternative way of predicting y_{p+1}. The relative performances of the two methods may be examined in particular cases.

1.2. *Prediction for an individual drawn at random from a specified population*

Suppose it is known that an individual for whom the value of y_{p+1} has to be predicted belongs to a specified population. How can this information be used in addition to $y_1, \ldots, y_p$ observed on the individual to improve prediction?

Let θ in (1.1.1), the parameter characterising an individual, have a specified probability density

$$q(\theta \mid \phi) \tag{1.2.1}$$

where ϕ may be an unknown parameter. Note that $q(\theta \mid \phi)$ is not in the nature of an arbitrarily or conveniently chosen prior distribution of a parameter θ. It has reference to the actual distribution of θ in the population of individuals under consideration. Then the probability density of $y_1, \ldots, y_{p+1}$ and θ is

$$P(y_1, \ldots, y_{p+1} \mid \theta)\, q(\theta \mid \phi). \tag{1.2.2}$$

Integrating over θ, the probability density of $y_1, \ldots, y_{p+1}$ is

$$P_3(y_1, \ldots, y_{p+1} \mid \phi) = P_5(y_{p+1} \mid y_1, \ldots, y_p, \phi)\, P_4(y_1, \ldots, y_p \mid \phi) \tag{1.2.3}$$

where the first factor

$$P_5(y_{p+1} \mid y_1, \ldots, y_p, \phi) \tag{1.2.4}$$

is the PDF of y_{p+1}. Let us suppose that we have past data on k individuals from the specified population

$$\begin{array}{l} y_{11}, \ldots, y_{p+1,1} \\ y_{12}, \ldots, y_{p+1,2} \\ \ldots \quad \ldots \quad \ldots \\ y_{1k}, \ldots, y_{p+1,k} \end{array} \tag{1.2.5}$$

The probability density at the observed values (1.2.5) on k individuals and $y_1, \ldots, y_p$ on the current individual is

$$P_3(y_{11}, \ldots, y_{p+1,1} \mid \phi) \ldots P_3(y_{1k}, \ldots, y_{p+1,k} \mid \phi)\, P_4(y_1, \ldots, y_p \mid \phi) \tag{1.2.6}$$

from which ϕ can be estimated by the method of maximum likelihood. If $\hat{\phi}$ denotes the estimate, then the EPDF is

$$P_5(y_{p+1} \mid y_1, \ldots, y_p, \hat{\phi}). \tag{1.2.7}$$

A point estimate of y_{p+1} may be obtained by computing the mode or the mean of the distribution of (1.2.7).

The data (1.2.5) on individuals observed in the past need not be of the same type as that observed on the current individual. Any data giving information on the parameter ϕ and providing a good estimate of it will do.

Note 1. As observed in Section 1, we may consider the density

$$P_3(y_{11},\ldots,y_{p+1,1}\,|\,\phi)\ldots P_3(y_{1k},\ldots,y_{p+1,k}\,|\,\phi)\,P_3(y_1,\ldots,y_{p+1}\,|\,\phi) \quad (1.2.8)$$

and obtain the unknowns y_{p+1} and ϕ by maximizing (1.2.8), which is an alternative way of predicting y_{p+1}. This may differ from the mode of (1.2.7). The relative performances of these estimates may be examined in particular cases.

Note 2. We may write (1.2.2) in the form

$$P_6(y_{p+1},\theta\,|\,y_1,\ldots,y_p,\phi)\,P_4(y_1,\ldots,y_p\,|\,\phi)$$

from which we can obtain the joint EPDF of y_{p+1} and θ, by substituting the estimate $\hat{\phi}$ from (1.2.6). Such an EPDF may be useful in drawing simultaneous inference on y_{p+1} and θ.

1.3. *Simultaneous prediction for a number of individuals*

Let $y_{1i},\ldots,y_{pi}$ be p measurements on the i-th individual and $y_{p+1,i}$, the $(p+1)$-th measurement to be predicted, $i=1,\ldots,k$. The population from which the k individuals are drawn is not specified. If θ_i is the characteristic of the i-th individual, then the PDF of $y_{p+1,i}$ is

$$P_5(y_{p+1,i}\,|\,y_{1i},\ldots,y_{pi},\theta_i),\qquad i=1,\ldots,k. \quad (1.3.1)$$

We need estimates of $\theta_1,\ldots,\theta_k$ based on the observations

$$y_{1i},\ldots,y_{pi},\qquad i=1,\ldots,k \quad (1.3.2)$$

to obtain EPDF's of $y_{p+1,i}$, $i=1,\ldots,k$. The problem is thus reduced to that of simultaneous estimation of the parameters θ_i, $i=1,\ldots,k$.

We shall determine the criterion for simultaneous estimation in the simple case where $y_{1i},\ldots,y_{p+1,i}$ are independent for given θ_i and $\mathbf{E}(y_{ji})=\theta_i$ for all j. Let $\hat{\theta}_i$ be an estimate of θ_i, based on the observations (1.3.2), in which case $\hat{\theta}_i$ is a predictor of $y_{p+1,i}$. Under the mean square error criterion we find

$$\mathbf{E}\sum_1^k (y_{p+1,i}-\hat{\theta}_i)^2=\mathbf{E}\sum_1^k (y_{p+1,i}-\theta_i)^2+\mathbf{E}\sum_1^k (\hat{\theta}_i-\theta_i)^2. \quad (1.3.3)$$

To minimize (1.3.3) for given θ_i, one has to find estimates $\hat{\theta}_i$ to minimize $\mathbf{E}\Sigma(\hat{\theta}_i - \theta_i)^2$. Thus we are led to a compound decision problem involving the simultaneous estimation of the parameters $\theta_1, \ldots, \theta_k$, for which good solutions exist James and Stein (1961), Rao (1974a, 1975), Efron (1974).

Note 1. If we assume that θ_i are random variables drawn from the distribution (1.2.1), then the EPDF of $y_{p+1,i}$ is

$$P_5(y_{p+1,i} \mid y_{1i}, \ldots, y_{pi}, \hat{\phi}) \tag{1.3.4}$$

where ϕ is estimated by maximising the likelihood

$$\prod_{i=1}^{k} P_4(y_{1i}, \ldots, y_{pi}, \phi). \tag{1.3.5}$$

The predicted value of $y_{p+1,i}$ may be obtained as the mode or the mean of (1.3.4). Alternatively one may consider the joint density function

$$\prod_{i=1}^{k} P_3(y_{1i}, \ldots, y_{p+1,i} \mid \phi) \tag{1.3.6}$$

and estimate all the unknowns $y_{p+1,i}$, $i = 1, \ldots, k$ and ϕ by maximising (1.3.6).

1.4. *Individual versus decision maker's loss*

In Sections 1.1–1.3 we have discussed the problem of prediction in three different situations. A decision maker who is required to predict in a routine way for individuals referred to him from a specified population will try to minimize his overall loss and may, therefore, use the methods developed in Sections 1.2 and 1.3. For any predictor $\hat{y}_{p+1}$ of y_{p+1}, the expected loss to the decision maker is

$$L = \underset{\theta}{\mathbf{E}} \; \underset{y_{p+1}}{\mathbf{E}} \left[(y_{p+1} - \hat{y}_{p+1})^2 \mid \theta\right] \tag{1.4.1}$$

where the second expectation over y_{p+1} for given θ

$$L(\theta) = \underset{y_{p+1}}{\mathbf{E}} \left[(y_{p+1} - \hat{y}_{p+1})^2 \mid \theta\right] \tag{1.4.2}$$

is the loss to an individual with characteristic θ. In some situations minimization of overall loss may be achieved by the choice of a predictor which imposes an unduly heavy loss on certain individuals. For instance, consider a predictor $\tilde{y}_{p+1}$ as an alternative to $\hat{y}_{p+1}$ in (1.4.1), with the overall loss $\tilde{L}$ and individual loss $\tilde{L}(\theta)$ such that

$$\begin{aligned} L &< \tilde{L}, \\ L(\theta) &\leq \tilde{L}(\theta) \quad \text{for } \theta \leq \delta, \\ L(\theta) &> \tilde{L}(\theta) \quad \text{for } \theta > \delta, \end{aligned} \tag{1.4.3}$$

i.e., the predictor $\tilde{y}_{p+1}$ is better than y_{p+1} for $\theta > \delta$, but worse for values of $\theta \leq \delta$. In such a case, the choice of a predictor depends on to what extent we give importance to the loss for individuals with $\theta > \delta$. Do there exist predictors which provide a proper balance between overall loss and individual's loss? The problem is a difficult one and needs a careful study. A simple method is suggested for this purpose in Rao (1975), but its performance characteristic has not been studied.

We shall apply the general principles mentioned in Sections 1.1–1.4 to the special case of prediction of future observations in linear models.

2. Prediction in linear models

Let us consider the Gauss–Markoff model

$$Y = X\beta + E, \tag{2.1}$$

where Y and E are $p \times 1$, X is $p \times m$ and β is $m \times 1$ matrices. Further let y be a vector of k random variables (we shall take $k = 1$ without loss of generality as shown later) with a Gauss–Markoff structure

$$y = x\beta + e, \tag{2.2}$$

where β is the same parameter as in (2.1). The problem we consider is that of predicting y by a linear function of Y depending on the nature of information available on X, β, E and e.

We note that the problem of finding an optimum predictor $\hat{y}$ under the loss function

$$(y - \hat{y})'G(y - \hat{y}), \tag{2.3}$$

where G is a positive definite matrix is equivalent to that of optimum prediction of each component of y under a quadratic loss function. Thus the solution under the loss function (2.3) is independent of G. We shall, therefore, consider y to be a single future observation to be predicted.

Let the dispersion matrix of (E, e) given β be written in the partitioned form, apart from a multiplier σ^2,

$$\begin{pmatrix} V_{11} & V_{12} \\ V_{21} & V_{22} \end{pmatrix}, \tag{2.4}$$

where V_{ij} and σ^2 do not depend on β. Further let β be a random variable with mean μ and dispersion matrix $\sigma^2 F$. We shall assume that (V_{ij}), F and X are all of full rank to avoid some complications. Necessary modifications could be made using generalized inverse of matrices when one or more of these matrices do not have full rank.

The problem of predicting future observations in linear models has been studied extensively. Valuable contributions have been made by Geisser (1970, 1971, 1974), Lee (1972, 1975), Lee and Geisser (1972, 1975), Lindley and Smith (1972), Toutenburg (1970) and others. The present study is based on a recent work by the author on simultaneous estimation of parameters (Rao (1974a, 1975)) and the earlier study on growth curves (Rao (1958)).

2.1. *Best linear predictor (BLP)*

When all parameters are known, the best linear predictor (BLP) of y under quadratic loss function is the regression of y on Y

$$x\mu + (xFX' + V_{21})(XFX' + V_{11})^{-1}(Y - X\mu) \tag{2.1.1}$$

and the associated prediction mean square error is

$$\sigma^2[V_{22} + xFx' - (xFX' + V_{21})(XFX' + V_{11})^{-1}(XFx' + V_{12})]. \tag{2.1.2}$$

Let us denote by β^l, the least squares estimator of β from (2.1)

$$\begin{aligned} \beta^l &= (X'V_{11}^{-1}X)^{-1}X'V_{11}^{-1}Y \\ &= UX'V_{11}^{-1}Y, \quad \text{where } U = (X'V_{11}^{-1}X)^{-1}. \end{aligned} \tag{2.1.3}$$

Then (2.1.1) can be written as the sum of three expressions

$$\begin{aligned} &x\beta^l \\ &-(x - V_{21}V_{11}^{-1}X)U(F+U)^{-1}(\beta^l - \mu), \\ &+ V_{21}V_{11}^{-1}(Y - X\beta^l). \end{aligned} \tag{2.1.4}$$

The prediction mean square error (2.1.2) can also be written, apart from the multiplier σ^2, as

$$V_{22} - xUX'V_{11}^{-1}V_{12} - V_{21}V_{11}^{-1}XUx' + xUx'$$

$$- (x - V_{21}V_{11}^{-1}X)U(F+U)^{-1}U(x' - X'V_{11}^{-1}V_{12}) \tag{2.1.5}$$

$$- V_{21}(V_{11}^{-1} - V_{11}^{-1}XUX'V_{11}^{-1})V_{12}$$

corresponding to the three terms in (2.1.4). The expression (2.1.1) can be also written as

$$x\mu + (xF + V_{21}V_{11}^{-1}XU)(F+U)^{-1}(\beta' - \mu) \tag{2.1.6}$$

$$+ V_{21}V_{11}^{-1}(Y - X\beta') \tag{2.1.7}$$

where (2.1.6) is the regression of y on β' and (2.1.7) is the regression of y on the residual $(Y - X\beta')$.

The BLP (2.1.1, 2.1.4, 2.1.6, 2.1.7) depends on all the parameters (V_{ij}), μ, σ^2 and F whose values may not be known in any particular situation. If past data on the linear model (2.1) are available, it may be possible to estimate the unknowns, substitute the estimates for parameters in the formula for the BLP and thus obtain an empirical best linear predictor (EBLP). The method of constructing EBLP is explained in Section 2.2.

2.2. *Simultaneous prediction*

Suppose that we have k linear models

$$Y_i = X\beta_i + E_i, \qquad i = 1, \ldots, k \tag{2.2.1}$$

and in each case one future observation with the structure

$$y_i = x_i\beta_i + e_i, \qquad i = 1, \ldots, k \tag{2.2.2}$$

has to be predicted. Let us further suppose that (V_{ij}) is known but μ, σ^2 and F are unknown, and no additional information other than observations on Y_i, $i = 1, \ldots, k$, is available. In such a case we can use the EBLP by estimating μ, σ^2 and F from (2.2.1). The problem is similar to that of simultaneous estimation of parameters considered in earlier papers by the author Rao (1974a, 1975). By analogy with the formula for estimation of parameters (see Rao (1975)) we may write the EBLP for the i-th model as

$$y_i = x\beta_i^l - cW(x - V_{21}V_{11}^{-1}X)UB^{-1}(\beta_i^l - \hat{\mu})$$

$$+ V_{21}V_{11}^{-1}(Y_i - X\beta_i^l), \tag{2.2.3}$$

$$i = 1, \ldots, k$$

where

$$c = (k - m - 2)/(kp - km + 2)$$

$$k\hat{\mu} = \sum_1^k \beta_i^l$$

$$W = \sum_1^k (Y_i' V_{11}^{-1} Y_i - Y_i' V_{11}^{-1} X \beta_i^l)$$

$$B = \sum_1^k (\beta_i^l - \hat{\mu})(\beta_i^l - \hat{\mu})'$$

$$U = (X' V_{11}^{-1} X)^{-1}.$$

When $V_{21} = 0$, the formula (2.2.3) reduces to

$$y_i = x[\beta_i^l - cWUB^{-1}(\beta_i^l - \mu)] \tag{2.2.4}$$

which is the same as for the simultaneous estimation of the parametric functions $x\beta_i$, $i = 1, \dots, k$. It is easily shown (see Rao (1975)) that the total mean square error for (2.2.4) is less than that for the predictors $x\beta_i^l$, $i = 1, \dots, k$.

2.3. *Best linear unbiased predictor (BLUP)*

The BLP is a non homogeneous linear function $L_0 + L'Y$ such that $\mathbf{E}(y - L_0 - L'Y)^2$ is a minimum. If in addition we require that

$$\mathbf{E}(y - L_0 - L'Y \mid \beta) = 0 \tag{2.3.1}$$

for any β, or that $\mathbf{E}(y - L_0 - L'Y) = 0$ independently of μ, then the linear predictor reduces to

$$x\beta^l + V_{21} V_{11}^{-1}(Y - X\beta^l) \tag{2.3.2}$$

which is the same as the BLP (2.1.4), without the second term. The expression (2.3.1) is also independent of F, and thus does not involve any unknown parameters. Thus the prediction formula does not depend on past data. We call (2.3.2) the best linear unbiased predictor BLUP. The prediction mean square error is, apart from the multiplier σ^2,

$$V_{22} - xUX'V_{11}^{-1}V_{12} - V_{21}V_{11}^{-1}XUx + xUx' - V_{21}(V_{11}^{-1} - V_{11}^{-1}XUX'V_{11}^{-1})V_{12}. \tag{2.3.3}$$

The problem of BLUP was considered earlier by Toutenburg (1970). Related concepts in the estimation of parameters have been considered by the author in Rao (1965, 1973, Section 4a.11).

2.4. *Best homogeneous linear predictor*

The BLUP is a homogeneous linear function of Y while the BLP has a constant term in addition. We shall now find the best predictor of y in the class of homogeneous linear functions of Y, without demanding the unbiasedness condition (2.3.1). Now

$$\mathbf{E}[(y - L'Y)^2 \mid \beta] = (x - L'X)\beta\beta'(x' - X'L) + \sigma^2(V_{22} + L'V_{11}L - 2L'V_{12}) \tag{2.4.1}$$

$$= \sigma^2[L'(X\gamma\gamma'X' + V_{11})L - 2L'X\gamma\gamma'x' - 2L'V_{12}] + \sigma^2(x\gamma\gamma'x' + V_{22}) \tag{2.4.2}$$

where $\gamma = \sigma^{-1}\beta$. The optimum choice of L which minimizes (2.4.2) for given γ is

$$L = (X\gamma\gamma'X' + V_{11})^{-1}(X\gamma\gamma'x' + V_{12}). \tag{2.4.3}$$

Then the best homogeneous linear predictor of y is

$$(V_{21} + x\gamma\gamma'X')(X\gamma\gamma'X' + V_{11})^{-1}Y. \tag{2.4.4}$$

If γ is not known, we can estimate it by

$$\hat{\gamma} = \hat{\sigma}^{-1}(X'V_{11}^{-1}X)^{-1}X'V_{11}^{-1}Y = \hat{\sigma}^{-1}\beta' \tag{2.4.5}$$

where

$$(n - m)\hat{\sigma}^2 = Y'V_{11}^{-1}Y - Y'V_{11}^{-1}X\beta'. \tag{2.4.6}$$

Then the empirical BHLP (EBHLP) is

$$(V_{21} + x\hat{\gamma}\hat{\gamma}'X')(X\hat{\gamma}\hat{\gamma}'X' + V_{11})^{-1}Y \tag{2.4.7}$$

which is no longer linear in Y. The reader is referred to earlier work by Rao (1971) and Bibby (1972), and the recent contributions by Shinozaki (1975) in the context of estimation of parameters.

3. Prediction of growth

3.1. *A polynomial trend model for growth*

It is usual to assume, in studying growth over a short period of time, a model of the form

$$y_t = \beta_0 + \beta_1 t + \dots + \beta_{m-1}t^{m-1} + E_t \qquad t = 1, \dots, p \tag{3.1.1}$$

where y_t is the measurement at time t, and $\beta_0, \ldots, \beta_{m-1}$ are coefficients in the polynomial trend specific to an individual and E_t is in the nature of an error. In such a case we have a linear model for the observations and to predict a future observation such as y_{p+1}, the methods developed in Section 2 could be used.

3.2. *An alternative model for growth*

The computations by Lee (1972) showed that much of the information for predicting future growth of an individual is contained in the data on his past growth itself and no appreciable gain results by utilizing the information on growth of other individuals observed in the past. This may be due to the fact that the information on past data was not properly utilized. For instance, no attempt was made to determine an appropriate growth model by examining past data. In this section we shall indicate a method of obtaining a suitable model which may be an improvement over the polynomial trend model in some situations.

Let us assume a growth model of the form

$$y_t = a_0 + a_1 g_{1t} + \ldots + a_{m-1,1} g_{m-1,t} + E_t \tag{3.2.1}$$

$$t = 1, \ldots, p$$

where $g_{1t}, \ldots, g_{m-1,t}$ are unknown functions of time and E_t is in the nature of an error, the coefficients $a_0, \ldots, a_{m-1}$ being specific to an individual. Let $z_t = y_{t+1} - y_t$ and $h_{it} = g_{it+1} - g_{it}$ and $\eta_t = E_{t+1} - E_t$, in which case

$$z_t = a_1 h_{1t} + \ldots + a_{m-1} h_{m-1,t} + \eta_t, \tag{3.2.2}$$

$$t = 1, \ldots, p-1.$$

We assume that $\eta_1, \ldots, \eta_{p-1}$ are uncorrelated and have the same variance. Let past data be available on k individual growth curves giving the values

$$z_{1i}, \ldots, z_{p-1,i} \tag{3.2.3}$$

of z_t for the i-th individual, $i = 1, \ldots, k$. Further let S be the matrix whose (i, j)-th element is

$$S_{ij} = \sum_{r=1}^{k} z_{ir} z_{jr}. \tag{3.2.4}$$

Then $(h_{11}, \ldots, h_{1p-1}), \ldots, (h_{m-1,1}, \ldots, h_{m-1,p-1})$ may be estimated by $(m-1)$ principal components (eigen vectors) corresponding to the first $(m-1)$ dominant eigen values of S. With h_{ij} so estimated the methods developed in Section 2 could be used, using the linear model (3.2.2).

The prediction formulae developed in Section 2 have been applied to the data in Table 1 on ramus heights of 20 boys at 8, $8\frac{1}{2}$, 9 and $9\frac{1}{2}$ years of age given by Grizzle and Allen (1969), using the polynomial model (3.1.1) and the estimated model (3.2.2). The predicted value for $9\frac{1}{2}$ years of age on the basis of the earlier three measurements is computed and compared with the observed value in each individual case (see Table 2). The mean square error is also computed for each prediction formula.

The following prediction formulae are considered:

(i) BLUP (1): For each individual case, a straight line is fitted (by the least squares method) to the heights at ages 8, $8\frac{1}{2}$ and 9 and the extrapolated value on the linear trend at $9\frac{1}{2}$ is taken as the predicted value (see formula 2.3.2).

(ii) EBLP (1): The formula (2.2.3) for simultaneous prediction based on the first three measurements only and assuming a linear trend is used. The mean square error for EBLP (1) is slightly smaller than that for BLUP (1) as expected.

(iii) BLUP (2): A transformation of the time axis is obtained using the totals of measurements (correct to nearest integer) at each age over the 20 boys as indicated in Rao (1958) (see Table 1).

Actual time	8	$8\frac{1}{2}$	9	$9\frac{1}{2}$
Transformed time (Total)	973	992	1012	1029

A straight line is fitted to the first three measurements using the transformed time points (instead of the actual time points as in BLUP (1) and EBLP (1)) and the extrapolated value at the fourth transformed time point is taken as the predicted value. In practice the transformation on time is obtained on past data using observations at all time points and used in the prediction of an unobserved value at a time point for the current individual.

(iv) EBLP (2): The formula (2.2.3) for simultaneous prediction is applied using the first three measurements and the transformed time points as in BLUP(2). Again the mean square error for EBLP(2) is smaller than that for BLUP(2) as expected. The mean square error for EBLP(2) is smaller than that for EBLP(1) indicating that a straight line trend with respect to transformed time provides a better model than what is considered in (i) and (ii).

(v) BLUP (3): Transformation of the time axis is obtained as explained in Section 3.2. First the values of $z_1 = y_2 - y_1$, $z_2 = y_3 - y_2$, $z_3 = y_4 - y_3$

(where y_1, y_2, y_3 and y_4 are the observed measurements at the four time points) are computed and the S matrix as defined in (3.2.4) is obtained:

$$S = \begin{pmatrix} 26.45 & 20.74 & 16.85 \\ 20.74 & 38.38 & 21.53 \\ 16.85 & 21.53 & 24.45 \end{pmatrix}.$$

The first eigenvector of S is proportional to (4.36, 5.73, 4.29), giving the estimated model

$$z_1 = 4.36b + E_1, \qquad z_2 = 5.73b + E_2, \qquad z_3 = 4.29b + E_3. \quad (3.2.5)$$

The formula (2.3.2) is applied by estimating b from z_1, z_2 and predicting z_3 by $4.29b$. The predicted value of y_4 is taken as $z_3 + y_3$.

(vi) EBLP (3): The formula (2.2.3) for simultaneous prediction is applied on the model (3.2.5). EBLP(3) has the smallest mean square error indicating an improvement in the specification of the growth model by using eigen vectors of the matrix S to transform the time axis.

Table 1
Ramus heights of 20 boys at different ages

Ind.	Age			
	8	$8\frac{1}{2}$	9	$9\frac{1}{2}$
1	47.8	48.8	49.0	49.7
2	46.4	47.3	47.7	48.4
3	46.3	46.8	47.8	48.5
4	45.1	45.3	46.1	47.2
5	47.6	48.5	48.9	49.3
6	52.5	53.2	53.3	53.7
7	51.2	53.0	54.3	54.5
8	49.8	50.0	50.3	52.7
9	48.1	50.8	52.3	54.4
10	45.0	47.0	47.3	48.3
11	51.2	51.4	51.8	51.9
12	48.5	49.2	53.0	55.5
13	52.1	52.8	53.7	55.0
14	48.2	48.9	49.3	49.8
15	49.6	50.4	51.2	51.8
16	50.7	51.7	52.7	53.3
17	47.2	47.7	48.4	49.5
18	53.3	54.6	55.1	55.3
19	46.2	47.5	48.1	48.4
20	46.3	47.6	51.3	51.8
Total	973.1	992.5	1011.6	1029.0

It may be noted that in (i) and (ii) the prediction was made solely on the basis of the first three measurements. In (iii)–(vi), the last measurement was used to construct an appropriate transformation of the time axis. There will be some reduction in the mean square on this account alone. However, there seems to be some reduction due to an improvement in the specification of the growth model. This could have been examined by constructing such a transformation using part of the data and judging the performance of the estimated model on the rest of the data.

(vii) EBHLP: The formula (2.4.7) is applied using only the first three measurements.

(viii) In addition to the seven formulae described above, predicted values are also obtained by fitting a quadratic function of time to the first three measurements and extrapolating for the last time point. The values

Table 2
Observed and predicted values of ramus height at $9\frac{1}{2}$ years

Ind.	Obs. height	Predicted value						
		BLUP(1)	EBLP(1)	BLUP(2)	EBLP(2)	BLUP(3)	EBLP(3)	EBHLP
1	49.7	49.733	49.803	49.670	49.736	49.456	49.628	49.732
2	48.4	48.433	48.495	48.365	48.424	48.214	48.357	48.432
3	48.5	48.467	48.510	48.388	48.429	48.455	48.528	48.465
4	47.2	46.500	46.592	46.447	46.535	46.551	46.726	46.499
5	49.3	49.633	49.693	49.565	49.622	49.414	49.557	49.632
6	53.7	53.800	53.901	53.758	53.853	53.600	53.849	53.799
7	54.5	55.933	55.819	55.770	55.661	55.566	55.336	55.932
8	52.7	50.533	50.666	50.507	50.633	50.515	50.806	50.532
9	54.4	54.600	54.385	54.379	54.176	53.985	53.547	54.599
10	48.3	48.733	48.702	48.612	48.583	48.164	48.134	48.732
11	51.9	52.067	52.189	52.035	52.150	51.767	52.330	52.065
12	55.5	54.733	54.491	54.496	54.267	55.055	54.433	54.732
13	55.0	54.467	54.493	54.382	54.407	54.379	54.444	54.465
14	49.8	49.900	49.978	49.842	49.916	49.742	49.921	49.899
15	51.8	52.000	51.789	51.916	51.944	51.868	51.935	51.999
16	53.3	53.700	53.690	53.595	53.585	53.535	53.519	53.699
17	49.5	48.967	49.037	48.903	48.970	48.912	49.056	48.965
18	55.3	56.133	56.139	56.039	56.043	55.806	55.854	56.132
19	48.4	49.167	49.172	49.067	49.072	48.853	48.878	49.165
20	51.8	53.400	53.113	53.137	52.865	53.524	52.819	53.399
Mean square error		0.641	0.570	0.591	0.540	0.554	0.496	0.644

so obtained are found to deviate from the observed measurements by large amounts indicating that a straight line trend is more appropriate for the growth data under study. The predicted values under a quadratic trend are not shown in Table 2.

It is seen that the performances of the various formulae are nearly the same, although there is slight reduction in the mean square error when formulae for simultaneous prediction are used with a suitable transformation of the time axis. In simultaneous prediction, predicted values are under valued for individuals with a large growth rate and over valued for individuals with a small growth rate (see Rao (1974b) for similar comments on simultaneous estimation). It may be noted that a large component of error in prediction is due to inadequacy of the growth models used. In such a case minor refinements in prediction formulae based on an assumed model are not of much help. It will pay to increase our efforts in understanding the mechanism of growth and building up appropriate growth models for prediction on the basis of observed data on growth.

References

[1] Bibby, J. (1972). Minimum mean square error estimation, ridge regression, and some unanswered questions. *Proc. European meeting of statisticians*, Budapest, pp. 107–120.

[2] Efron, B. (1974). Biased versus unbiased estimation. Tech. Report 1. Stat. Dept., Stanford University.

[3] Geisser, S. (1970). Bayesian analysis of growth curves. *Sankhya A*, **32**, 53–64

[4] Geisser, S. (1971). The inferential use of predictive distributions. *Foundations of Statistical Inference*. Holt, Reinhart and Winston, pp. 456–469.

[5] Geisser, S. (1975). The predictive sample reuse method with applications. *J. Am. Statist. Assoc.*, **70**, 320–328.

[6] Grizzle, J.E. and Allen, D.M. (1969). Analysis of growth and dose response curves. *Biometrics*, **25**, 357–82.

[7] James, W. and Stein, C. (1961). Estimation with quadratic loss. *Proc. Fourth Berkeley Symposium Math. Stat. and Prob.* University of California, 361–379.

[8] Lindley, D.V. and Smith, A.F.M. (1972). Bayesian estimates for the linear model (with discussion). *J. Roy. Statist. Soc.* B, **34**, 1–41.

[9] Lee, J.C. (1972). On the generalized growth model. Dissertation submitted to the University of New York at Buffalo for the Ph.D. degree.

[10] Lee, J.C. (1975). On growth curve with a special covariance structure. Tech. Report.

[11] Lee, J.C. and Geisser, S. (1972). Growth curve prediction. *Sankhya A*, 393–412.

[12] Lee, J.C. and Geisser, S. (1975). Applications of growth curve prediction. *Sankhya* (in press).

[13] Rao, C.R. (1958). Some statistical methods for comparison of growth curves. *Biometrics*, **14**, 1–17.

[14] Rao, C.R. (1965, 1973). *Linear Statistical Inference and its Applications*. John Wiley and Sons, New York.

[15] Rao, C.R. (1971). Unified theory of linear estimation. *Sankhya* (A) **33**, 370–396.
[16] Rao, C.R. (1974a). Characterization of prior distributions and solution to a compound decision problem. Discussion Paper No. 101, Indian Statistical Institute, New Delhi.
[17] Rao, C.R. (1974b). Some thoughts on regression and prediction — Part I, *Gujarat Statistical Review*, **1**, 7–32. [Also *Sankhya* (C) **37**, 102–120.]
[18] Rao, C.R. (1975). Simultaneous estimation of parameters in different linear models and applications to biometric problems. *Biometrics*, **31**, 545–554.
[19] Shinozaki, M. (1975). A study of generalized inverse of a matrix and estimation with quadratic loss function. Ph.D. dissertation (Keio University, Japan).
[20] Toutenburg, H. (1970). Vorhersage im allegemeinen linearen regression modell mit stochastishen regressoren. *Op. Forschung und Math. Stat.*, **2**, 105–116.

P.R. Krishnaiah, ed., *Multivariate Analysis–IV*
© North-Holland Publishing Company (1977) 209–223

PROBLEMS AND APPROACHES IN DESIGN OF EXPERIMENTS FOR ESTIMATION AND TESTING IN NON-LINEAR MODELS*

S. ZACKS

Case Western Reserve University, Cleveland, Ohio, U.S.A.

Consider a parametric family $\mathscr{F} = \{F(y; \theta, x)\}$ of a distribution functions of a random variable Y, which depends on a known and controllable real (or vector) value x (the experimental level) and on a vector of unknown parameters θ. The experimental level x belongs to a specified experimental region, $\mathscr{X}$, and $\theta \in \Theta$. The objective of the experiment is to estimate a certain nonlinear parametric function $\omega(\theta)$. An experiment consisting of N trials, at the levels $\boldsymbol{x} = (x_1, \ldots, x_N)$ is considered to be optimal if it maximizes the total amount of "information" on $\omega(\theta)$. We consider the notion of "information" in the context of Lindley, Fisher and Kullback–Leibler. The basic theory relating the information function to the estimation or testing problems concerning $\omega(\theta)$ is discussed. It is shown that generally, the optimal designs depend on the values of the unknown parameters θ. We review and compare certain approaches in the literature to overcome the dependence on θ. More specifically, we consider fixed sample, two-stage and sequential designs based on maximum likelihood and a Bayes estimation of θ.

1. Introduction

In the present paper we consider design problems of the following nature. A family $\mathscr{F}$ of density functions $f(y; \theta, x)$ is under consideration. θ designates an unknown parameter, which belongs to a parameter space Θ: ; and x denotes an experimental level, which belongs to an experimental domain $\mathscr{X}$ and is under the control of the statistician. Our primary concern is with non-linear statistical models, such as the ones encountered in reliability theory, in bioassay methods, in economic growth models etc. Moreover, we will assume that the family $\mathscr{F}$ satisfies, for each x in $\mathscr{X}$, the usual regularity conditions of the Cramér–Rao type for the existence of the Fisher information functions. Under those regularity conditions, the Fisher information function (in the real parameter case) is the inverse of the asymptotic variance of the best asymptotically normal estimators (see Zacks [11, p. 247]). Thus, designs which yield maximal total Fisher

* Partially supported by Project NR 042–276, of the Office of Naval Research at Case Western Reserve University.

information are associated with minimal asymptotic variance. We notice that most of the common experimental design methods in normal linear cases are oriented essentially to the maximization of the Fisher information functions or related functions. Fisher [9] considered designs which are constructed with respect to this information function in his famous Dilution Series experiments (see also Cochran [4]). Chernoff [3] suggested also to consider in non-linear models the criterion of maximizing the Fisher information. The problem is generally, in non-linear non-normal cases, that the Fisher information function (matrix) depends also on the unknown parameter θ, and one cannot determine the experimental level, x^0, which maximizes the Fisher information without knowing the actual value of θ. Chernoff [3] suggested to consider locally optimal designs, to maximize the Fisher information in a neighborhood of a specified value θ^0. In Cochran's paper [4] a review is given of various approaches to the problem. In the present paper we consider the design problem in both Bayesian and non-Bayesian framework. If one is ready to consider the unknown parameter as random variables having a specified prior distribution, $H(\theta)$, then the corresponding Bayes designs are determined so that the anticipated (expectation with respect to $H(\theta)$) Fisher information function is maximized. If N experiments ($N \geq 2$) are to be designed the total anticipated Fisher information is maximized by Bayes sequential designs. In non-normal cases it is generally very difficult to determine the optimal Bayes sequential designs. However, two-stage Bayes designs can already increase the efficiency of the designs considerably. In Section 3 we illustrate single stage and the two-stage Bayes designs for a dichotomous exponential model of the type encountered in discrete time inspections of reliability systems. We also provide there (Section 3.2) a non-Bayesian multistage design for bioassay models. As shown in Section 3 the determination of the optimal two-stage Bayes design is not a simple matter and it involves a considerable amount of computer work. A function of relative efficiency is defined for these Bayes designs from a non-Bayesian point of view. This function is based on the ratio of the actual amount of information at a given value of θ, to the maximal possible information at that θ. We illustrate numerically that the relative efficiency of two-stage Bayes designs may be considerably larger than that of single stage designs. Another example considered in Section 3 is that of designing quantal bioassays. We discuss there an adaptive sequential design, for the determination of tolerance doses. This is another method of overcoming the problems of unknown parameters.

2. Exponential families with parameters depending on the experimental levels

Consider a family of density functions (or probability functions) of the exponential type, i.e.,

$$f(y;\alpha,\beta,x)=h(y;x)\exp\{y\omega(\alpha+\beta x)+\psi(\alpha+\beta x)\},\tag{2.1}$$

where x is the experimental level, α and β are unknown parameters, $\omega(u)$ and $\psi(u)$ are analytic functions. For example, suppose that Y is a binomial random variable, with probability of success, θ, which is a function of $\alpha+\beta x$, i.e., $\theta\equiv\theta(\alpha+\beta x)$. Then,

$$\omega(\alpha+\beta x)=\log(\theta(\alpha+\beta x)/(1-\theta(\alpha+\beta x)),\ \psi(\alpha+\beta x)=\log(1-\theta(\alpha+\beta x)).$$

The function $\theta(u)$ is known in bioassays as the tolerance distribution. It is specified by the statistical model. In reliability life testing the lifetime of the experimental unit, Y, may have an exponential distribution, with mean life between failures $\theta(\alpha+\beta x)=(\alpha+\beta x)^{-1}$. In this case $\omega(\alpha+\beta x)=-(\alpha+\beta x)$ and $\psi(\alpha+\beta x)=\log(\alpha+\beta x)$. It is well known that the expected value of Y and its variance can be expressed in this exponential model as

$$\mathbf{E}\{Y\mid\alpha+\beta x\}=-\frac{\psi'(\alpha+\beta x)}{\omega'(\alpha+\beta x)}\tag{2.2}$$

and

$$V\{Y\mid\alpha+\beta x\}=\frac{1}{\omega'^{3}(\alpha+\beta x)}\{\psi'(\alpha+\beta x)\cdot\omega''(\alpha+\beta x)-\psi''(\alpha+\beta x)\omega'(\alpha+\beta x)\},\tag{2.3}$$

where $\psi'(u)$, $\omega'(u)$, $\psi''(u)$ and $\omega''(u)$ are the first and second order derivatives of $\psi(u)$ and $\omega(u)$, respectively. Given k experiments at levels $x_1,\ldots,x_k$, yielding the observations $y_1,\ldots,y_k$, the likelihood function of α and β is

$$L(\alpha,\beta\mid\mathbf{x},\mathbf{y})=\exp\left\{\sum_{i=1}^{k}y_i\omega(\alpha+\beta x_i)+\sum_{i=1}^{k}\psi(\alpha+\beta x_i)\right\}.\tag{2.4}$$

By differentiating the logarithm of the likelihood function partially with respect to α and β and determining the variances and covariance of these derivatives we obtain the Fisher information matrix

$$II(\alpha, \beta) = \begin{pmatrix} I_{11} & I_{12} \\ I_{21} & I_{22} \end{pmatrix}, \tag{2.5}$$

where

$$I_{11} = \mathbf{E}\left\{\left[\frac{\partial}{\partial \alpha} \log L(\alpha, \beta \mid \boldsymbol{x}, \boldsymbol{Y})\right]^2\right\} = \sum_{i=1}^{k} W(\alpha + \beta x_i), \tag{2.6}$$

$$I_{12} = I_{21} = \mathbf{E}\left\{\frac{\partial}{\partial \alpha} \log L(\alpha, \beta \mid \boldsymbol{x}, \boldsymbol{Y}) \cdot \frac{\partial}{\partial \beta} \log L(\alpha, \beta \mid \boldsymbol{x}, \boldsymbol{Y})\right\} = \sum_{i=1}^{k} x_i W(\alpha + \beta x_j) \tag{2.7}$$

and

$$I_{22} = \mathbf{E}\left\{\left[\frac{\partial}{\partial \beta} \log L(\alpha, \beta \mid \boldsymbol{x}, \boldsymbol{Y})\right]^2\right\} = \sum_{j=1}^{k} x_j^2 W(\alpha + \beta x_j), \tag{2.8}$$

where

$$W(\alpha + \beta x_j) = (\omega'(\alpha + \beta x_j))^2 V\{Y \mid \alpha + \beta x_j\}, \tag{2.9}$$

$j = 1, \dots, k$ are the weights associated with the experimental levels. The inverse of the Fisher information function is the covariance matrix of the best asymptotically normal estimators of α and β (i.e., the maximum likelihood estimators (see Zacks [11; p. 244–252]). For this reason one could consider also designs which minimize certain functions of the inverse of the Fisher information function, which is

$$II^{-1}(\alpha, \beta) = \begin{bmatrix} \frac{1}{T_W} + \frac{x^2}{\mathrm{SSD}} & -\frac{\bar{x}}{\mathrm{SSD}} \\ -\frac{\bar{x}}{\mathrm{SSD}} & \frac{1}{\mathrm{SSD}} \end{bmatrix}, \tag{2.10}$$

where $T_W = \sum_{i=1}^{k} W(\alpha + \beta x_i)$ is the sum of the weights, $\bar{x}$ is the weighted average of the experimental level, $\bar{x} = \sum_{i=1}^{k} x_i W(\alpha + \beta x_i)/T_W$ and SSD is the weighted sum of squares of deviations

$$\mathrm{SSD} = \sum_{i=1}^{k} W(\alpha + \beta x_i)(x_i - \bar{x})^2. \tag{2.11}$$

Consider for example the criterion of minimizing the determinant of (2.10) which is

$$|II^{-1}(\alpha, \beta)| = \frac{1}{T_W \cdot \mathrm{SSD}} = \frac{1}{Q(\alpha, \beta; \boldsymbol{x})} \tag{2.12}$$

where

$$Q(\alpha, \beta; x) = \sum_{i=1}^{k} W(\alpha + \beta x_i) \cdot \sum_{i=1}^{k} x_i^2 W(\alpha + \beta x_i) - \Big(\sum_{k=1}^{k} x_i W(\alpha + \beta x_i)\Big)^2. \tag{2.13}$$

A D-optimal design $\boldsymbol{x}^0 = (x_1, \ldots, x_k)$ is a vector of x-levels which minimizes $|II^{-1}(\alpha, \beta)|$ or maximizes $Q(\alpha, \beta; \boldsymbol{x})$ (see Fedrov [7]). One could consider also other types of design criteria. The common problem with all such optimal designs is that they depend on α and β. We can consider, however, a Bayes approach in which we optimize the expected value of $Q(\alpha, \beta; \boldsymbol{x})$, with respect to some prior distribution of (α, β). In two-stage or multi-stage Bayes designs we can determine the x vector for a new stage of experimentation by maximizing the expected value of $Q(\alpha, \beta; \boldsymbol{x})$ with respect to the posterior distribution of (α, β) given the results from the previous stages. In a non-Bayesian multi-stage design we can estimate α and β after each stage by their maximum likelihood estimators $\hat{\alpha}$ and $\hat{\beta}$ (using the results from all the previous stages) and design the next stage by maximizing $Q(\hat{\alpha}, \hat{\beta}; \boldsymbol{x})$. Such a procedure is expected to yield consistent estimators of α and β and thus the multi-stage designs should converge to the optimal design. In the following section we provide some examples.

3. Examples of non-linear designs for estimation

3.1. Single and two-stage Bayes designs for reliability experiments with exponential life time distributions

In the present section we consider the following reliability experiment. The life time of a component follows an exponential distribution with mean time between failures $\beta = 1/\tau$. This parameter is unknown and an experiment is designed in order to estimate τ. The experiment under consideration is of the attribute type, as the one discussed by Ehrenfeld [6]. The structure of this experiment is the following. N components are subjected to the same environmental conditions for a period of x units of time. After x units of time the number S_N of components which have survived the test is counted. There is no information about the failure times of the other $N - S_N$ components. The question is how large should x be? We notice that S_N has a binomial distribution with success (survival) probability $e^{-\beta x}$, $0 < \beta < \infty$. In terms of the model developed in the previous section we have

N experiments, all performed at the level x. The parameters of the exponential model are

$$\omega(\beta x) = \log(1 - e^{-\beta x}) + \beta x \quad \text{and} \quad \psi(\beta x) = -\beta x.$$

Here $\alpha = 0$ and we have only one unknown parameter, β, which is the intensity of the exponential failure times. In the sequel we let $\theta = \beta$. The Fisher information function of θ is

$$I_N(\theta, x) = N x^2 e^{-\theta x}(1 - e^{-\theta x})^{-1}. \tag{3.1}$$

The value of x which maximizes (3.1) depends on θ, and is given approximately by $x^0(\theta) \approx 1.6/\theta$. In a Bayes single stage design we choose an x which maximizes the anticipated (prior expectation) value of $I_N(\theta, x)$, namely

$$\mathbf{E}_H\{I_N(\theta, x)\} = N x^2 \int_0^\infty e^{-\theta x}(1 - e^{-\theta x})^{-1} h(\theta)\, d\theta. \tag{3.2}$$

If θ has a prior gamma distribution, $\mathcal{G}(1/\tau, \upsilon)$, $0 < \tau$, $\upsilon < \infty$; where τ is the scale parameter and υ is the shape parameter, the anticipated information function assumes the form:

$$\mathbf{E}_H\{I_N(\theta, x)\} = N x^2 \sum_{j=1}^{\infty} (1 + \tau x j)^{-\upsilon}. \tag{3.3}$$

We have to require that $\upsilon > 1$, otherwise (3.3) is not finite. Differentiating (3.3) with respect to x we obtain

$$\frac{\partial}{\partial x} \mathbf{E}_H\{I_1(\theta, x)\} = 2x \sum_{j=1}^{\infty} [1 + \tau x j]^{-\upsilon} - \tau \upsilon x^2 \sum_{j=1}^{\infty} j[1 + \tau x j]^{-(\upsilon+1)}. \tag{3.4}$$

As before, this derivative exists at every x when $\upsilon \geq 1 + \delta$, $\delta > 0$. Furthermore, we can write

$$\frac{\partial}{\partial x} \mathbf{E}_H\{I_1(\theta, x)\} = x \sum_{j=1}^{\infty} \frac{2 - (\upsilon - 2)\tau x j}{(1 + \tau x j)^{\upsilon+1}}. \tag{3.5}$$

Hence, if $\upsilon \leq 2$ the anticipated information, $\mathbf{E}_H\{I_1(\theta, x)\}$ increases monotonically with x and the optimal design is associated with the largest possible x. If $\mathcal{X} = (0, \infty)$ there exists no optimal design. On the other hand, if $\upsilon > 2$, x^0 is the root of the equation

$$\sum_{j=1}^{\infty} \frac{2 - (\upsilon - 1)\tau x j}{(1 + \tau x j)^{\upsilon+1}} = 0. \tag{3.6}$$

Without loss of generality we can assume that $\tau = 1$. The root x^0 for $\tau \neq 1$ is given by dividing x^0 by τ. The anticipated information for a general τ is obtained by dividing that for $\tau = 1$ by τ^2.

In Table 1 we present the solution x^0 for several cases of prior gamma distributions, with $\tau = 1$ and $\upsilon = 3(1)7$. The corresponding values of the maximal anticipated information are also given, for the case of $N = 1$. Generally, one has to multiply the values in the table by N.

Table 1
The Bayes single-stage designs for
prior gamma distributions with $\upsilon = 3(1)7$ and $\tau = 1$

υ	3	4	5	6	7
x^0	1.518	0.775	0.519	0.390	0.312
$E_H\{I_1(\theta, x^0)\}$	0.215	0.084	0.045	0.028	0.019

The variance of the prior gamma distribution is increasing with υ i.e., $\text{Var}_H\{\theta\} = \upsilon^2\tau$. Thus, as seen in Table 1, the increase in the prior variance (the value of υ) induces a decrease in the corresponding maximal anticipated information. We also observe in Table 1 that the x^0 value corresponding to a given υ is approximately equal to $1.6/(\upsilon - 2)$. Thus, if we let $\theta^*(\tau, \upsilon) = \tau(\upsilon - 2)$, $\upsilon > 2$, we obtain that $x^0 \approx 1.6/\theta^*(\tau, \upsilon)$. This is an interesting result since for $\upsilon \leq 2$ there is no finite solution, and if θ is known the optimal solution is $1.6/\theta$. We should therefore estimate θ by $\theta^* = \tau(\upsilon - 2)$ rather than by the common Bayes estimator $\theta = \tau\upsilon$. This conclusion is reinforced by the relative efficiency function which is discussed later.

Two-stage designs are constructed in the following manner. At the first stage of the design we determine the size, n, of the first sample, and the design level x. After observing that n units at x, we count the number S_n of survivors in the first stage. The prior distribution of θ, $H(\theta)$, is then converted to the corresponding posterior distribution given (n, x_1, S_n), i.e., $H(\theta \mid n, x_1, S_n)$. The design level x_2 for the second stage is determined so that the anticipated information per observation, under the posterior distribution its maximized. The whole second sample of size $(N - n)$ is observed at x_2. The general objective is to determine n, x_1 and x_2 so that the total expected information

$$J_H(n, x_1) = \mathbf{E}_H\{nx_1^2 e^{-\theta x_1}(1 - e^{-\theta x_1})^{-1} + \\ + \mathbf{E}_H\{(N-n)x_2^2 e^{-\theta x_2}(1 - e^{-\theta x_2})^{-1} \mid n, x_1, S_n\}\} \tag{3.7}$$

is maximized.

The method of solution follows the following three steps:

(i) Determine x_2^0 to maximize $\mathbf{E}_{\theta \mid n, x_1, S_n}\{I_{N-n}(\theta; x_2)\}$, where $\mathbf{E}_{\theta \mid S_n, x_1}\{\cdot\}$ designates the anticipated information at x_2 according to the posterior distribution $H(\theta \mid S_n, x_1)$. The level x_2 is a function of the sufficient statistic (n, x_1, S_n).

(ii) Determine $J_H(n, x_1) = \mathbf{E}_H\{I_n(\theta; x_1) + \mathbf{E}_{\theta \mid S_n, x_1}\{I_{N-n}(\theta; x_2^0)\}\}$.

(iii) Determine x_1^0 and n^0 to maximize $J_H(n, x_1)$.

We carry out the solution of the design problem according to these three main steps. We mention first that if the prior distribution of θ has a density $h(\theta)$ then its posterior density, given (n, x_1, S_n) is

$$h(\theta \mid n, x_1, S_n) = \frac{h(\theta) e^{-\theta x_1 S_n}(1 - e^{-\theta x_1})^{(n-S_n)}}{\int_0^\infty h(\theta) e^{-\theta x_1 S_n}(1 - e^{-\theta x_1})^{n-S_n} d\theta}. \tag{3.8}$$

Thus, in the case of a gamma prior distribution, $\mathcal{G}(1/\tau, \nu)$, the posterior density is

$$h(\theta \mid n, x_1, S_n, \tau, \nu) =$$

$$= \frac{1}{\tau^\nu \Gamma(\nu)} \cdot \frac{\sum_{j=0}^{n-S_n} \binom{n-S_n}{j} (-1)^j \theta^{\nu-1} \exp\left\{-\theta\left[x_1(S_n + j) + \frac{1}{\tau}\right]\right\}}{\sum_{j=0}^{n-S_n} \binom{n-S_n}{j} (-1)^j [1 + \tau x_1(S+j)]^{-\nu}}. \tag{3.9}$$

After the first stage sample has been observed, the anticipated information function (per observation) given (n, x_1, S_n) is

$$\mathbf{E}_{\theta \mid n, x_1, S_n}\{I_1(\theta, x_2)\} =$$

$$= x_2^2 \frac{\sum_{j=0}^{n-S_n} (-1)^j \binom{n-S_n}{j} \sum_{k=1}^{\infty} [1 + \tau x_1(S_n + j) + \tau x_2 k]^{-\nu}}{\sum_{j=0}^{n-S_n} (-1)^j \binom{n-S_n}{j} [1 + \tau x_1(S_n + j)]^{-\nu}}. \tag{3.10}$$

It is generally quite difficult to determine the value of x_2 which maximizes (3.10) analytically. One could use available search techniques and work directly on the anticipated information (3.10) to obtain (even approximately) the optimal second stage design x_2^0, given (n, x_1, S_n).

For the evaluation of $J_H(n, x_1)$ we notice that the priorly marginal probability function of S_n, given (n, x_1) is

$$\begin{aligned} f_{S_n \mid n, x_1}(j) &= \frac{\binom{n}{j}}{\tau^{\upsilon}\Gamma(\upsilon)} \int_0^{\infty} \theta^{\upsilon-1} e^{-\theta x_1 j}(1-e^{-\theta x_1})^{n-j} e^{-\theta/\tau}\, d\theta \\ &= \binom{n}{j} \sum_{i=0}^{n-j} (-1)^i \binom{n-j}{i} [1+\tau x_1(j+i)]^{-\upsilon}. \end{aligned} \tag{3.11}$$

Multiplying (3.11) by (3.10) and summing over j we obtain

$$\begin{aligned} &\mathbf{E}_H\{\mathbf{E}_{\theta \mid S_n, x_1, n}\{I(\theta, x_2^0(n, j))\}\} = \\ &= \sum_{j=0}^{n} \binom{n}{j} (x_2^0(n, j))^2 \sum_{i=0}^{n-j} (-1)^i \binom{n-j}{i} \sum_{k=1}^{\infty} [1+\tau x_1(j+i)+\tau x_2^0(n, j)k]^{-\upsilon}. \end{aligned} \tag{3.12}$$

Finally,

$$\begin{aligned} J_H(n, x_1) &= n x_1^2 \sum_{k=1}^{\infty} [1+\tau x_1 k]^{-\upsilon} \\ &+ (N-n) \sum_{j=0}^{n} \binom{n}{j} [x_2^0(n, j)]^2 \sum_{i=0}^{n-j} (-1)^i \binom{n-j}{i} \sum_{k=1}^{\infty} [1+\tau x_1(j+i) \\ &\qquad + \tau x_2^0(n, j)k]^{-\upsilon}. \end{aligned} \tag{3.13}$$

The problem is to determine a value of n, $n = 0, \ldots, N$, and x_1 which minimize (3.13). It is relatively simple to compute (3.13) on a computer, therefore, the optimal values n^0 and x_1^0 can be determined numerically by computing the function $J_H(n, x_1)$ for each n, $n = 1, \ldots, N-1$, over a grid of x points and thus locating a neighborhood of x points at which $J_H(n, x_0)$ assumes its maximum over x. We then choose the n value for which the absolute maximum is located. We notice that for $n = 0$ and $n = N$ the two-stage design reduces to a single-stage design for which the solution is simpler.

In Table 2 we present the values of $J_H(n, x)$ for the case of $\upsilon = 3$, $\tau = 1$ and $N = 50$. We computed this function only for $n = 10, 20$ and 25 and for $x = 1.00\,(0.25)\,2.00$. We have restricted the computations to values of n up to 25 due to the excessive computing time required for larger values of n, and because we do not expect the optimal first sample size to be larger than half of the total sample size.

Table 2
The anticipated total information in a two-stage procedure, for $\upsilon = 3$, $\tau = 1$, $N = 50$

n \ x	1.00	1.25	1.50	1.75	2.00
10	12.821	12.894	12.899	12.883	12.823
20	12.360	12.527	12.559	12.501	12.407
25	9.070	8.918	N.C.*	N.C.	N.C.

* N.C. means "not computed".

As seen in Table 2, the maximal value of $J_H(n, x)$ over the range of our computations is at $n = 10$ and $x = 1.5$. The total anticipated information for such a design is 12.899. The corresponding maximal anticipated information for a one-stage design with $N = 50$ is (see Table 1) 10.5. Thus, the two-stage design with $n = 10$ and $x_1 = 1.5$ increases the total anticipated information over that of a one-stage design by more than 23%.

Table 3
The rel. eff. function for $\tau = 1$ and $\upsilon = 3(1)7$

θ \ υ	3	4	5	6	7
0.25	0.4818	0.2711	0.1876	0.1433	0.1162
0.50	0.7830	0.4899	0.3510	0.2727	0.2232
0.75	0.9432	0.6618	0.4917	0.3888	0.3216
1.00	0.9986	0.7923	0.6114	0.4924	0.4116
1.25	0.9807	0.8865	0.7117	0.5841	0.4937
1.50	0.9152	0.9494	0.7943	0.6647	0.5681
1.75	0.8226	0.9857	0.8607	0.7348	0.6352
2.00	0.7180	0.9996	0.9124	0.7952	0.6954
2.25	0.6120	0.9950	0.9507	0.8463	0.7491
2.50	0.5115	0.9756	0.9772	0.8890	0.7966
2.75	0.4204	0.9446	0.9931	0.9238	0.8382
3.00	0.3406	0.9047	0.9997	0.9513	0.8742
3.25	0.2726	0.8583	0.9980	0.9721	0.9050
3.50	0.2158	0.8077	0.9893	0.9868	0.9309
3.75	0.1693	0.7544	0.9745	0.9958	0.9522
4.00	0.1316	0.7000	0.9545	0.9997	0.9692
4.25	0.1016	0.6457	0.9301	0.9991	0.9822
4.50	0.0779	0.5924	0.9023	0.9943	0.9915
4.75	0.0594	0.5408	0.8716	0.9858	0.9972
5.00	0.0450	0.4914	0.8388	0.9740	0.9998

The relative efficiency of the single-stage Bayes designs can be defined as the ratio of the information value of the design, under θ, divided by the maximal possible information for that θ.

$$\text{rel. eff.}(\theta \mid \tau, \upsilon) = I(\theta; x^0(\tau, \upsilon)/I(\theta, x^*(\theta)), \tag{3.14}$$

where $x^0(\tau, \upsilon)$ designates the Bayes solution and $x^*(\theta) = 1.6/\theta$.

As seen in Table 3, the maximal efficiency is attained at $\theta = \tau(\upsilon - 2)$.

In analogy to the relative efficiency function of the Bayes single stage procedure we define here the relative function of the Bayes two-stage procedure as

$$\begin{aligned} \text{rel. eff.}_{\cdot 2}(\theta) = {} & \Big\{\frac{n^0}{N}(x_1^0)^2 e^{-\theta x_1^0}(1 - e^{-\theta x_1^0})^{-1} + \\ & + \left(1 - \frac{n^0}{N}\right) \sum_{j=0}^{n} \binom{n^0}{j} e^{-\theta x_1^0 j}(1 - e^{-\theta x_1^0})^{n^0 - j} \cdot (x_2^0(j, x_1^0))^2 e^{-\theta x_2^0(j, x_1^0)} \cdot \\ & \cdot (1 - e^{-\theta x_2^0(j, x_1^0)})^{-1}\Big\} \Big/ \{(x^*(\theta))^2 e^{-\theta x^*(\theta)}(1 - e^{-\theta x^*(\theta)})^{-1}\}. \end{aligned} \tag{3.15}$$

The numerator of (3.15) is the expected total information (per observation) of the two-stage procedure, where the expectations is with respect to the binomial distribution of S_n under θ. The sample size and x_1 are the Bayes optimal ones, (n^0, x_1^0). The denominator of (3.15) is the maximal information (per observation) of a single-stage procedure, which is available only if θ is known. In Table 4 we present these relative efficiency functions for $\tau = 1$, $\upsilon = 3$, and $N = 50$. According to Table 2, we take $n^0 = 10$, $x_1^0 = 1.518$ and the $x_2^0(j; x_1^0)$ values $(j = 0, \ldots, 10)$ are those which yield maximal anticipated information for the second stage.

Table 4
The relative efficiency of a two-stage design ($\upsilon = 3$, $\tau = 1$, $N = 50$, $n = 10$, $x_1^0 = 1.518$)

θ	0.25	0.50	0.75	1.00	1.50	2.00	2.50	3.00	4.00	5.00
rel. eff.	0.683	0.916	0.945	0.939	0.935	0.915	0.864	0.788	0.604	0.426

3.2. *Quantal bioassay*

Optimal design of bioassays in quantal response situations has been the subject of several studies (Finney [8; pp. 496–7], Brown [2], Healy [10],

Dixon [5] and others). We can describe the problem in the following terms. A certain preparation is tested on biological systems (units) at different dosages, x. At each dose we test the preparation independently on n units and record the number of responses J. The probability of response at dosage x is a function $F(\alpha + \beta x)$, called the tolerance distribution. It is required that $F(z)$ will have the properties of an absolutely continuous distribution function. Models commonly applied in the literature are:

the logistic: $F(z) = 1/(1 + e^{-z})$,
the normal: $F(z) = \Phi(z)$, which is the standard normal integral,
extreme values: $F(z) = \exp\{-e^{-z}\}$,

and others. The two unknown parameters α and β are related to the mean, μ, and standard deviation, σ, of the tolerance distribution according to $\alpha = -\mu/\sigma$, $\beta = 1/\sigma$.

Given the outcomes of the trials at $x_1, \ldots, x_k$, with the samples of size $n_1, \ldots, n_k$ and number of responses $J_1, \ldots, J_k$, the likelihood function of α and β is

$$L(\alpha, \beta \mid \boldsymbol{j}, \boldsymbol{x}, \boldsymbol{n}) = \prod_{i=1}^{k} \left[\frac{F(\alpha + \beta x_i)}{1 - F(\alpha + \beta x_i)}\right]^{J_i} \cdot \prod_{i=1}^{k} [1 - F(\alpha + \beta x_i)]^{n_i}. \tag{3.16}$$

or

$$\log L(\alpha, \beta \mid \boldsymbol{J}, \boldsymbol{x}, \boldsymbol{n}) = \sum_{i=1}^{k} J_i \log \frac{F(\alpha + \beta x_i)}{1 - F(\alpha + \beta x_i)} + \sum_{i=1}^{k} n_i \log(1 - F(\alpha + \beta x_i)). \tag{3.17}$$

The maximum likelihood (M.L.) estimation of α and β require the simultaneous solution of the non-linear equations

$$\begin{cases} \displaystyle\sum_{j=1}^{k} J_j \frac{f(\alpha + \beta x_j)}{F(\alpha + \beta x_j)\bar{F}(\alpha + \beta x_j)} = \sum_{j=1}^{k} n_j \frac{f(\alpha + \beta x_j)}{\bar{F}(\alpha + \beta x_j)} \\ \displaystyle\sum_{j=1}^{k} J_j x_j \frac{f(\alpha + \beta x_j)}{F(\alpha + \beta x_j)\bar{F}(\alpha + \beta x_j)} = \sum_{j=1}^{k} n_j x_j \frac{f(\alpha + \beta x_j)}{\bar{F}(\alpha + \beta x_j)} \end{cases} \tag{3.18}$$

where $f(z) = F'(z)$ is the probability density function of the standardized distribution $F(z)$, and $\bar{F}(z) = 1 - F(z)$. The solution of the M.L. equations (3.18) can be done iteratively, by the Newton–Raphson method. According to this method we start with an initial solution $(\hat{\alpha}, \hat{\beta})$ then solve the linear equations

$$\begin{bmatrix} \sum_{j=1}^{k} n_j G'(\hat{\alpha}+\hat{\beta}x_j) & \sum_{j=1}^{k} n_j x_j G'(\hat{\alpha}+\hat{\beta}x_j) \\ & \sum_{j=1}^{k} n_j x_j^2 G'(\hat{\alpha}+\hat{\beta}x_j) \end{bmatrix} \begin{bmatrix} \delta_\alpha \\ \delta_\beta \end{bmatrix} = -\begin{bmatrix} \sum_j n_j G(\hat{\alpha}+\hat{\beta}x_j) \\ \sum_j n_j x_j G(\hat{\alpha}+\hat{\beta}x_j) \end{bmatrix}, \tag{3.19}$$

where

$$\hat{p}_j = J_j/n_j, \qquad G(\hat{z}_j) = \frac{f(\hat{z}_j)(\hat{p}_j - F(\hat{z}_j))}{F(\hat{z}_j)\bar{F}(\hat{z}_j)},$$

$\hat{z}_j = \hat{\alpha} + \hat{\beta}x_j$ and $G'(z) = (\mathrm{d}/\mathrm{d}z)G(z)$. We then set $\hat{\alpha} \leftarrow \hat{\alpha} + \delta\alpha$ and $\hat{\beta} \leftarrow \hat{\beta} + \delta\beta$ and resolve the system of linear equations, until $\hat{\alpha}$ and $\hat{\beta}$ do not change significantly.

Let σ_1^2, σ_{12} and σ_2^2 denote the large sample variance of the MLE $\hat{\alpha}$, covariance of $\hat{\alpha}$ and $\hat{\beta}$ and variance of $\hat{\beta}$. According to the theory of MLE's (see Zacks [11; pp. 247]) these parameters are given by the inverse of the Fisher information matrix (2.10). The weights in this problem are given by

$$W(\alpha + \beta x_i) = n_i \frac{f^2(\alpha + \beta x_i)}{F(\alpha + \beta x_i) \cdot \bar{F}(\alpha + \beta x_i)}, \quad i = 1, \ldots, k. \tag{3.20}$$

One of the interesting problems in quantal response analysis is to estimate the fractiles of the responses distribution $F(\alpha + \beta x)$. In other words, for a given proportion of tolerance, γ, we wish to estimate the value ξ_γ so that $F(\alpha + \beta\xi_\gamma) = 1 - \gamma$. If we denote by $z_{1-\gamma}$ the $(1-\gamma)$-th fractile of the standard distribution $F(z)$ then $\xi_\gamma = (z_{1-\gamma} - \alpha)/\beta$, assuming obviously that $0 < \beta < \infty$. The MLE of ξ_γ is $\hat{\xi}_\gamma = (z_{1-\gamma} - \hat{\alpha})/\hat{\beta}$, where $\hat{\alpha}$ and $\hat{\beta}$ are MLE's of α and β, respectively. Consider the problem of designing multi-stage experiments which minimize the asymptotic variance of the MLE $\hat{\xi}_\gamma$. The asymptotic variance of $\hat{\xi}_\gamma$, for positive β values, as obtained from the inverse of the Fisher information matrix (see Zacks [13; pp. 226]), is

$$V = \frac{1}{\beta^2}\sigma_1^2 + \frac{(z_{1-\gamma} - \alpha)^2}{\beta^4}\sigma_2^2 + 2\,\frac{z_{1-\gamma} - \alpha}{\beta^3}\sigma_{12} \tag{3.21}$$

where σ_1^2, σ_2^2 and σ_{12} are the asymptotic variances of α and β and their covariance. Accordingly, for every α and β,

$$\beta^2 V = \frac{1}{T_W} + \frac{\bar{x}^2}{\mathrm{SSD}} + \frac{\xi_\gamma^2}{\mathrm{SSD}} - 2\xi_\gamma \frac{\bar{x}}{\mathrm{SSD}} = \frac{1}{T_W} + \frac{(\bar{x} - \xi_\gamma)^2}{\mathrm{SSD}}, \tag{3.22}$$

where T_W, $\bar{x}$ and SSD are defined in Section 2. Consider a multistage experiment of the following nature. In the first stage we perform the experiment at some pre-assigned dosages to obtain initial MLE of α, β and ξ_γ. We then design for the next stage dosages (one or more) so that the current estimate of $\beta^2 V$, is

$$\frac{1}{\hat{T}_W} + \frac{(\bar{x} - \xi_\gamma)^2}{\hat{\mathrm{S}}\mathrm{SD}}$$

is minimized. Here, $\hat{T}_W$, $\bar{x}$ and $\hat{\mathrm{S}}\mathrm{SD}$ are computed according to the previous fomulae by substituting $\hat{\alpha}$ and $\hat{\beta}$ for the unknown α and β. Moreover, we can try in progressive stages only two dosages x_1 and x_2 at each stage, such that

$$(x_1 W_1 + x_2 W_2)/(W_1 + W_2) = \hat{\xi}_\gamma, \tag{3.23}$$

where

$$W_i = W(\hat{\alpha} + \hat{\beta}x_i), \qquad i = 1, 2. \tag{3.24}$$

Obviously, $x_1 < \hat{\xi}_\gamma < x_2$. In such a case, $\hat{\mathrm{S}}\mathrm{SD} > 0$, $(\bar{x} - \hat{\xi}_\gamma)^2/\hat{\mathrm{S}}\mathrm{SD} = 0$, and the asymptotic variance is proportional to $1/\hat{T}_W = (W_1 + W_2)^{-1}$. We can then require that x_1 and x_2 will be determined so that $\hat{T}_W$ is maximized. It turns out that in order to maximize $\hat{T}_W$ we have to set $x_1 = x_2 = \hat{\xi}_\gamma$. This, however, is not consistent with the requirement that $x_1 < \hat{\xi}_\gamma < x_2$. We therefore modify the procedure and perform the experiment in the initial stage at a few different dosages then, at each stage we perform the experiment at one dose only, namely $x = \hat{\xi}_\gamma$. After each stage we recompute $\hat{\alpha}$ and $\hat{\beta}$, to adjust for the new information. In the following example (cf. Table 5) we simulate 20 stages for the Insulin bioassay, reported by Finney [8; pp. 477], under the Normit model, in which the first stage consists of the 9 dosages of Finney's example. In each stage we perform $n = 50$ trials at the same dose. The initial stage yields the MLE's $\hat{\alpha} = -3.3160$ and $\hat{\beta} = 3.1567$.

Table 5
MLE of α and β in a multistage design for estimating the 0.5-tolerance dose

$\hat{\alpha}$	$\hat{\beta}$	σ^2	σ_{12}	σ_2^2
−3.3160	3.1567	0.1337	−.1228	0.1183
−3.3026	3.1605	0.1327	−.1226	0.1182
−3.2890	3.1626	0.1318	−.1223	0.1181
−3.3176	3.1625	0.1325	−.1231	0.1184
−3.3176	3.1625	0.1323	−.1231	0.1184
−3.3129	3.1638	0.1320	−.1231	0.1184
−3.3176	3.1629	0.1320	−.1233	0.1184
−3.2938	3.1706	0.1311	−.1229	0.1182
−3.2779	3.1696	0.1304	−.1225	0.1180
−3.2896	3.1718	0.1305	−.1227	0.1182
−3.2697	3.1698	0.1297	−.1223	0.1179
−3.3042	3.1787	0.1306	−.1231	0.1185
−3.3042	3.1787	0.1305	−.1231	0.1185
−3.2965	3.1786	0.1301	−.1230	0.1184
−3.2965	3.1786	0.1300	−.1230	0.1184
−3.3004	3.1790	0.1300	−.1230	0.1185
−3.2932	3.1785	0.1297	−.1229	0.1184
−3.2932	3.1785	0.1296	−.1229	0.1184
−3.2859	3.1774	0.1293	−.1227	0.1183
−3.2631	3.1726	0.1286	−.1222	0.1178
−3.2308	3.1608	0.1275	−.1213	0.1171

The (log) dose at each stage is $\hat{\xi}_\gamma = -\hat{\alpha}/\hat{\beta}$.

References

[1] Box, G.E.P. and Hill, W.J. (1967). "Discrimination Among Mechanistic Models", *Technometrics*, **9**, 57–71.

[2] Brown, B.W. Jr. (1966). "Planning A Quantal Assay of Potency", *Biometrics*, **22**, 322–329.

[3] Chernoff, H. (1953). "Locally Optimal Designs For Estimating Parameters", *Annals of Mathematical Statistics*, **24**, 586–602.

[4] Cochran, W.G. (1972). "Experiments For Non-Linear Functions", Technical Report No. 39, Project NR 042–097, Department of Statistics, Harvard University.

[5] Dixon, W.J. (1970). "Quantal Response Variable Experimentation: The Up and Down Method". In *Statistics in Endocrinology*, ed. J.W. McArthur and T. Coltan, M.I.T. Press, pp. 251–268.

[6] Ehrenfeld, S. (1962). "Some Experimental Design Problems In Attribute Life Testing", *Jour. American Statistical Association* **57**, 668–79.

[7] Fedorov, V.V. (1972). *Theory of Optimal Experiments*, New York: Academic Press.

[8] Finney, D.J. (1964). *Statistical Methods In Biological Assay*, 2nd. Ed. Griffin and Co., London.

[9] Fisher, R.A. (1922). "On the Mathematical Foundations of Theoretical Statistics", *Phil. Trans. Royal Society*, London, A, **222**, 309–68.

[10] Healy, M.J.R. (1950). "The Planning of Probit Assays", *Biometrics*, **6**, 424–434.

[11] Zacks, S. (1971). *The Theory of Statistical Inference*, New York: John Wiley and Sons.

PART III

TIME SERIES AND STOCHASTIC PROCESSES

P.R. Krishnaiah, ed., *Multivariate Analysis–IV*

ASYMPTOTIC THEORY OF TOTAL TIME ON TEST PROCESSES WITH APPLICATIONS TO LIFE TESTING

Richard E. BARLOW*
University of California, Berkeley, Calif., U.S.A.
and
Frank PROSCHAN**
Florida State University, Tallahassee, Fla., U.S.A.

In estimating life distributions and failure rates, a basic function is the total time on test process, i.e., the total number of units of time observed on all operating units as a function of the time elapsed since the start of the test. In this paper asymptotic results are obtained for the total time on test process and for the interval failure rate estimators; these failure rates turn out to be asymptotically uncorrelated. Applications are obtained for piecewise exponential life distributions. Finally, the total time on test concept is extended to the multivariate case.

0. Introduction and summary

A unifying concept in the statistical theory of reliability and life testing is that of *total time on test*. Let n units (organisms or devices) be placed on life test at age 0 and let $N(u)$ be the number of units alive (on test) at age u (time u). Then $T_n(x) = \int_0^x N(u)\mathrm{d}u$ is the *total time on test* at time x. (Retrospective life data can be analyzed by referring all failure times back to birth dates of the units and plotting data with respect to the age axis rather than the real time axis. However, for convenience, we assume all units are placed on life test at time $t = 0$.)

We call $\{T_n(x); x \geq 0\}$ the total time on test process. Since $N(u)/n = \bar{F}_n(u)$ where $\bar{F}_n = 1 - F_n$ is the empirical survival function, we see that $(1/n)T_n(x) \to \int_0^x \bar{F}(u)\mathrm{d}u$ almost surely as $n \to \infty$. The normed total time on test process

$$\sqrt{n}\left(\frac{1}{n}T_n(x) - \int_0^x \bar{F}(u)\mathrm{d}u\right) = -\int_0^x \sqrt{n}[F_n(u) - F(u)]\mathrm{d}u$$

*This research has been partially supported by the Office of Naval Research under Contract N00014–75–C–0781 and Contract with the Nuclear Regulatory Agency.

** This research was supported by the Air Force Office of Scientific Research under Grant AFOSR 74–2581B.

converges in probability to $-\int_0^x U[F(u)]\mathrm{d}u$, where $\{U(t); 0 \le t \le 1\}$ is the Brownian Bridge process. This is a standard result concerning empirical processes (see [4], pp. 141–142).

Total time on test statistics arise in certain maximum likelihood estimation problems. For example, suppose

$$\bar{F}(x) = \exp - \left[\sum_{j=1}^{i-1} \lambda_j (x_j - x_{j-1}) + \lambda_i (x - x_{i-1}) \right]$$

for $x_{i-1} \le x \le x_i$, where the nodes $0 \equiv x_0 < x_1 < \ldots < x_k \le \infty$ are assumed known. Given n independent observations from F and n_i observed failures in $[x_{i-1}, x_i)$, it is easy to verify that

$$\hat{\lambda}_{n_i} = n_i \bigg/ \int_{x_{i-1}}^{x_i} N(u)\mathrm{d}u$$

is the maximum likelihood estimate of λ_i. This model was first treated by Harris, Meier and Tukey [6] for incomplete, discrete, retrospective data. (Cf. also [9].)

Another application occurs in estimating the failure rate function, $r(x) = f(x)/\bar{F}(x)$ assumed monotone (say increasing) where f is the density of F. The maximum likelihood estimate of F in this case is piecewise exponential [7]. The maximum likelihood estimate of the failure rate function is found by inverting the slopes of the least concave majorant to the total time on test process. [Cf. [1], pp. 231–242.]

In [2], scaled total time on test data plots were used to identify probability distribution models and to test for exponential versus increasing (decreasing) failure rate.

The integral notation for total time on test suggests an immediate extension of the concept to cases of incomplete data. Tests for exponentiality versus monotone failure rate when data is incomplete are discussed in [3].

In Section 1 we introduce the concept of interval failure rate and obtain the asymptotic joint distribution of k disjoint interval failure rate estimators. These results also apply to estimating failure rates for piecewise exponential distributions.

In Section 2 we briefly indicate extensions of total time on test processes to multivariate failure data. In Section 3 we discuss the asymptotic distribution of various scaled total time on test processes of interest.

1. Asymptotic properties of interval failure rate estimators

Let $0 \equiv x_0 < x_1 < \ldots < x_k \leq \infty$ be a partition of $[0, \infty)$. As before, let $N(u)$ be the number of units on test at time u. Then

$$\hat{\lambda}_{n_1} = n_i \Big/ \int_{x_{i-1}}^{x_i} N(u)\mathrm{d}u = [F_n(x_i) - F_n(x_{i-1})] \Big/ \int_{x_{i-1}}^{x_i} \bar{F}_n(u)\mathrm{d}u$$

is the ratio of the number of failures in $[x_{i-1}, x_i)$ to the total time on test in the interval. Letting $n \to \infty$, we have

$$\lim_{n\to\infty} \hat{\lambda}_{n_i} \underset{\text{a.s.}}{=} [F(x_i) - F(x_{i-1})] \Big/ \int_{x_{i-1}}^{x_i} \bar{F}(u)\mathrm{d}u.$$

$$\overset{\text{def}}{=} \lambda_i.$$

We call λ_i the interval failure rate for interval $[x_{i-1}, x_i)$. Note that if F has density f, $x_{i-1} \leq x \leq x_i$, and we let the interval shrink to x, then

$$\lim_{x_i - x_{i-1} \downarrow 0} \frac{[F(x_i) - F(x_{i-1})]/(x_i - x_{i-1})}{\int_{x_{i-1}}^{x_i} \bar{F}(u)\mathrm{d}u/(x_i - x_{i-1})} = \frac{f(x)}{\bar{F}(x)} = r(x),$$

the failure rate at time x. If F has constant failure rate, say c, on $[x_{i-1}, x_i)$, then $\lambda_i = c$. If F has increasing (decreasing) failure rate, then $\lambda_i \leq (\geq) \lambda_j$ for $i \leq j$. To see this, observe that if r is increasing, then, for $0 \leq u \leq x_i - x_{i-1}$ and $0 \leq v \leq x_j - x_{j-1}$ we have

$$\begin{vmatrix} f(x_{i-1} + u) & f(x_{i-1} + (x_{j-1} - x_{i-1}) + v) \\ \bar{F}(x_{i-1} + u) & \bar{F}(x_{i-1} + (x_{j-1} - x_{i-1}) + v) \end{vmatrix} \leq 0.$$

Now integrate on u between 0 and $x_i - x_{i-1}$ and then integrate on v between 0 and $x_j - x_{j-1}$ to obtain $\lambda_i \leq \lambda_j$.

We next obtain the asymptotic joint distribution of the normed estimators of the interval failure rates,

$$\sqrt{n}(\hat{\lambda}_{n_1} - \lambda_1), \sqrt{n}(\hat{\lambda}_{n_2} - \lambda_2), \ldots, \sqrt{n}(\hat{\lambda}_{n_k} - \lambda_k).$$

As one would expect, they are asymptotically jointly normally distributed. Surprisingly, however, they are also asymptotically uncorrelated for arbitrary F satisfying $F(0^-) = 0$. This fact, together with the formula for the asymptotic estimator variance allows the computation of probability limits on the failure rate function for the piecewise exponential model. This should be useful for moderate to large sample size data.

Preliminaries

A stochastic process $\{W(t), t \geq 0\}$ is called a Wiener process with drift coefficient equal to 1 if $W(0) = 0$, $\{W(t), t \geq 0\}$ has stationary, independent increments and $W(t)$ is normally distributed with mean 0 and variance t for all $t \geq 0$. A process $\{U(t), t \geq 0\}$ is called a Brownian Bridge process on $[0, 1]$ where $U(t) = W(t) - tW(1)$, $0 \leq t \leq 1$. Note that such a process is normal, has all sample paths continuous, $E\,U(t) = 0$ for $0 \leq t \leq 1$ and that its covariance function is $s\Lambda t - st$.

We will use the fact that the empirical process $\sqrt{n}[F_n(x) - F(x)]$ converges weakly to $\{U[F(x)], x \geq 0\}$ where $\{U(t), 0 \leq t \leq 1\}$ is the Brownian Bridge (cf. [4], p. 141).

Lemma 1.1. *Let* $0 < x_1 < \ldots < x_k \leq \infty$ *be a partitioning of* $[0, \infty)$ *and* $\lambda_i = p_i / \int_{x_{i-1}}^{x_i} \bar{F}(u)\mathrm{d}u$ *for arbitrary* F $(F(0^-) = 0)$, *where* $p_i = F(x_i) - F(x_{i-1})$. *Then*

$$\sqrt{n}[\hat{\lambda}_{n_i} - \lambda_i] \xrightarrow[n \to \infty]{P}$$

$$\frac{\lambda_i^2}{p_i^2}\left[p_i \int_{x_{i-1}}^{x_i} U[F(u)]\mathrm{d}u + \{U[F(x_i)] - U[F(x_{i-1})]\} \int_{x_{i-1}}^{x_i} \bar{F}(u)\mathrm{d}u\right].$$

Proof.

$$\sqrt{n}(\hat{\lambda}_{n_i} - \lambda_i) = \sqrt{n}\left[\frac{n_i/n}{\int_{x_{i-1}}^{x_i} \bar{F}_n(u)\mathrm{d}u} - \frac{p_i}{\int_{x_{i-1}}^{x_i} \bar{F}(u)\mathrm{d}u}\right]$$

$$= \frac{1}{\int_{x_{i-1}}^{x_i} \bar{F}_n(u)\mathrm{d}u \int_{x_{i-1}}^{x_i} \bar{F}(u)\mathrm{d}u}\left[p_i \int_{x_{i-1}}^{x_i} \sqrt{n}[F_n(u) - F(u)]\mathrm{d}u \right.$$

$$\left. + \sqrt{n}\left(\frac{n_i}{n} - p_i\right) \int_{x_{i-1}}^{x_i} \bar{F}(u)\mathrm{d}u\right]$$

$$\xrightarrow[n \to \infty]{P} \frac{\lambda_i^2}{p_i^2}\left[p_i \int_{x_{i-1}}^{x_i} U[F(u)]\mathrm{d}u + \{U[F(x_i)] - U[F(x_{i-1})]\} \int_{x_{i-1}}^{x_i} \bar{F}(u)\mathrm{d}u\right]$$

since

$$\sqrt{n}\left(\frac{n_i}{n} - p_i\right) = \sqrt{n}(F_n(x_i) - F(x_i)] - \sqrt{n}[F_n(x_{i-1}) - F(x_{i-1})]$$

and

$$\sqrt{n}[F_n(u) - F(u)] \xrightarrow[n \to \infty]{P} U[F(u)].$$

Theorem 1.2. *For arbitrary F,*

$$\sqrt{n}(\hat{\lambda}_{n_1}-\lambda_1),\ \sqrt{n}(\hat{\lambda}_{n_2}-\lambda_2),\ldots,\sqrt{n}(\hat{\lambda}_{n_k}-\lambda_k)$$

are, asymptotically, independent $N(0,\sigma_i^2)$ $(i=1,2,\ldots,k)$ *random variables, where*

$$\begin{aligned}\sigma_i^2=\frac{\lambda_i^4}{p_i^4}\Big[2p_i^2&\int\int_{x_{i-1}\le u<v<x_i}F(u)\bar{F}(v)\mathrm{d}u\,\mathrm{d}v\\&+F(x_i)\bar{F}(x_i)\Big[\int_{x_{i-1}}^{x_i}\bar{F}(u)\mathrm{d}u\Big]^2\\&+F(x_{i-1})\bar{F}(x_{i-1})\Big[\int_{x_{i-1}}^{x_i}\bar{F}(u)\mathrm{d}u\Big]^2\\&+2p_i\bar{F}(x_i)\int_{x_{i-1}}^{x_i}\bar{F}(u)\mathrm{d}u\int_{x_{i-1}}^{x_i}F(u)\mathrm{d}u\\&-2p_iF(x_{i-1})\Big[\int_{x_{i-1}}^{x_i}\bar{F}(u)\mathrm{d}u\Big]^2\\&-2F(x_{i-1})\bar{F}(x_i)\Big[\int_{x_{i-1}}^{x_i}\bar{F}(u)\mathrm{d}u\Big]^2\Big].\end{aligned}\tag{1.1}$$

Note that when F is piecewise exponential, σ_i depends only on $x_1<x_2<\ldots<x_i$ and $\lambda_1,\lambda_2,\ldots,\lambda_i$.

Proof. We use Lemma 1.1 and the invariance principle (cf. [4], p. 72) to compute the asymptotic covariance of $\sqrt{n}(\hat{\lambda}_{n_i}-\lambda_i)$ and $\sqrt{n}(\lambda_{n_j}-\lambda_j)$ for $i\neq j$. From Lemma 1.1,

$$\begin{aligned}&\lim_{n\to\infty}\mathrm{Cov}\,(\sqrt{n}(\hat{\lambda}_{n_i}-\lambda_i),\sqrt{n}(\hat{\lambda}_{n_j}-\lambda_j))\\&=\frac{\lambda_i^2\lambda_j^2}{p_i^2p_j^2}\mathrm{Cov}\Big[\{U[F(x_i)]-U[F(x_{i-1})]\}\int_{x_{i-1}}^{x_i}\bar{F}(u)\mathrm{d}u+p_i\int_{x_{i-1}}^{x_i}U[F(u)]\mathrm{d}u,\\&\qquad\{U[F(x_j)]-U[F(x_{j-1})]\}\int_{x_{j-1}}^{x_j}\bar{F}(u)\mathrm{d}u+p_j\int_{x_{j-1}}^{x_j}U[F(u)]\mathrm{d}u\Big].\end{aligned}$$

For convenience, assume $i<j$ and use the fact that $\mathrm{Cov}\,[U[F(x_i)],U[F(x_j)]]=F(x_i)\bar{F}(x_j)$. A straighforward calculation shows that the limiting covariance is

$$\frac{\lambda_i^2\lambda_j^2}{p_i^2p_j^2}\Big[F(x_i)\bar{F}(x_j)\int_{x_{i-1}}^{x_i}\bar{F}(u)\mathrm{d}u\int_{x_{j-1}}^{x_j}\bar{F}(u)\mathrm{d}u$$

$$-F(x_i)\bar{F}(x_{j-1})\int_{x_{i-1}}^{x_i}\bar{F}(u)\mathrm{d}u\int_{x_{j-1}}^{x_j}\bar{F}(v)\mathrm{d}v$$

$$+p_j\int_{x_{i-1}}^{x_i}\bar{F}(u)\mathrm{d}u\int_{x_{j-1}}^{x_j}F(x_i)\bar{F}(u)\mathrm{d}u$$

$$-F(x_{i-1})\bar{F}(x_j)\int_{x_{i-1}}^{x_i}\bar{F}(u)\mathrm{d}u\int_{x_{j-1}}^{x_j}\bar{F}(u)\mathrm{d}u$$

$$+F(x_{i-1})\bar{F}(x_{j-1})\int_{x_{i-1}}^{x_i}\bar{F}(u)\mathrm{d}u\int_{x_{j-1}}^{x_j}\bar{F}(v)\mathrm{d}v$$

$$-p_jF(x_{i-1})\int_{x_{j-1}}^{x_j}\bar{F}(u)\mathrm{d}u\int_{x_{i-1}}^{x_i}\bar{F}(v)\mathrm{d}v$$

$$+p_i\bar{F}(x_j)\int_{x_{i-1}}^{x_i}F(u)\mathrm{d}u\int_{x_{j-1}}^{x_j}\bar{F}(v)\mathrm{d}v$$

$$-p_i\bar{F}(x_{j-1})\int_{x_{i-1}}^{x_i}F(u)\mathrm{d}u\int_{x_{j-1}}^{x_j}\bar{F}(v)\mathrm{d}v$$

$$+p_ip_j\int_{x_{i-1}}^{x_i}F(u)\mathrm{d}u\int_{x_{j-1}}^{x_j}\bar{F}(v)\mathrm{d}v\Big].$$

Substituting $F(x_j)-F(x_{j-1})$ for p_j we see that the first three terms in the square brackets cancel each other. The same is true for the next three terms and also for the last three terms. Hence $\sqrt{n}(\hat{\lambda}_{n_i}-\lambda_i)$ and $\sqrt{n}(\hat{\lambda}_{n_j}-\lambda_j)$ are asymptotically uncorrelated for $i\neq j$.

A similar calculation provides (1.1). (Cf. Crow and Shimi [11].)

Theorem 1.2 generalizes results of Miller [9] for the case where $k=2$ and F is a two piece exponential distribution with known truncation time.

Total time on test data plots as described in [2] may visually suggest nodal values $0<x_1<x_2<\ldots<x_k$. Assuming these are nodal values for a k piece exponential model, we can use Theorem 1.2 to compute an approximate 95% confidence interval

$$\left[\hat{\lambda}_{n_i}-\frac{2\sigma_i}{\sqrt{n}},\ \hat{\lambda}_{n_i}+\frac{2\sigma_i}{\sqrt{n}}\right]$$

for λ_i. We can compute σ_i by replacing $\lambda_s, 1\le s\le i$ by its estimate $\hat{\lambda}_{n_s}, 1\le s\le i$ in the expression

$$\bar{F}(x_j) = \exp\left[-\sum_{s=1}^{j} \lambda_s (x_s - x_{s-1})\right] \quad \text{for } 1 \le j \le i.$$

In any case, $\hat{\lambda}_{n_i}$ estimates the failure rate $[F(x_i) - F(x_{i-1})]/\int_{x_{i-1}}^{x_i} \bar{F}(u)\mathrm{d}u$ with variance σ_i^2 for arbitrary F.

Chiang [5] in his Lemma 3, page 628 shows that for

$$\hat{q}_i = [F_n(x_i) - F_n(x_{i-1})]/\bar{F}_n(x_i),$$

$\operatorname{Cov}[\hat{q}_i, \hat{q}_j] = 0$ for $i \neq j$. However, $\hat{q}_i$ and $\hat{q}_j$ are not independent for finite n. Chiang also calculates the asymptotic distribution of the observed expectation of life at age a, called $\hat{e}_a$. This is our

$$(\hat{\lambda}_{n_k})^{-1} = \int_{x_{k-1}}^{\infty} \bar{F}_n(u)\mathrm{d}u / \bar{F}_n(x_{k-1}),$$

where $x_k = \infty$.

2. Multivariate total time on test processes

The generalization of the concept of total time on test to $k \ge 2$ dimensions leads us to a new generalization of the univariate failure rate. Let (X, Y) have joint life distribution $F(x, y)$ and let $(X_1, Y_1), \ldots, (X_n, Y_n)$ be n independent observations on (X, Y). Let $N(u, v)$ be the number of pairs (X_i, Y_i) such that $X_i \ge u$ and $Y_i \ge v$. Then $N(u, v)$ is the number of pairs "on test" at joint times u and v. We call $T_n(x, y) = \int_0^x \int_0^y N(u, v)\mathrm{d}u\,\mathrm{d}v$ the *joint total time on test*. Let $\bar{F}_n(x, y) = N(x, y)/n$ be the joint empirical survival probability. Then

$$\frac{1}{n} T_n(x, y) = \int_0^x \int_0^y \bar{F}_n(u, v)\mathrm{d}u\,\mathrm{d}v$$

and

$$\lim_{n \to \infty} \frac{1}{n} T_n(x, y) = \int_0^x \int_0^y \bar{F}(u, v)\mathrm{d}u\,\mathrm{d}v$$

as in the univariate case, where $\bar{F}(u, v) = \mathbf{P}(X > u, Y > v)$.

Lemma 2.1.

$$\int_0^\infty \int_0^\infty \bar{F}_n(u, v)\mathrm{d}u\,\mathrm{d}v = \frac{1}{n} \sum_{i=1}^{n} X_i Y_i.$$

Proof. Integrating by parts yields

$$\int_0^\infty \int_0^\infty \bar{F}_n(u,v)\mathrm{d}u\,\mathrm{d}v = -uv\int_{v-0}^{\infty}\left\{u\bar{F}_n(u,v)\Big|_0^\infty - \int_0^\infty u\frac{\mathrm{d}}{\mathrm{d}u}\bar{F}_n(u,v)\right\}\mathrm{d}v$$

$$= -uv\int_{u=0}^{\infty}\frac{\mathrm{d}}{\mathrm{d}u}\bar{F}_n(u,v)\Big|_{v=0}^{\infty}$$

$$+\int_0^\infty\int_0^\infty uv\frac{\partial}{\partial u\partial v}\bar{F}_n(u,v)\mathrm{d}u\,\mathrm{d}v$$

$$= \int_0^\infty uv\bar{F}_n(\mathrm{d}u,\mathrm{d}v).$$

Now $\bar{F}_n(x,y) = 1 - F_{1n}(x) - F_{2n}(y) + F_n(x,y)$ so that $\bar{F}_n(\mathrm{d}u,\mathrm{d}v) = F_n(\mathrm{d}u,\mathrm{d}v)$, and thus

$$\int_0^\infty\int_0^\infty \bar{F}_n(u,v)\mathrm{d}u\,\mathrm{d}v = \int_0^\infty\int_0^\infty uvF_n(\mathrm{d}u,\mathrm{d}v) = \frac{1}{n}\sum_{i=1}^{n} X_iY_i.$$

It follows from Lemma 2.1 and the strong law of large numbers that $\int_0^\infty\int_0^\infty \bar{F}(u,v)\mathrm{d}u\,\mathrm{d}v = EXY$.

To obtain the analogue in two dimensions of the univariate interval failure rate consider the following figure.

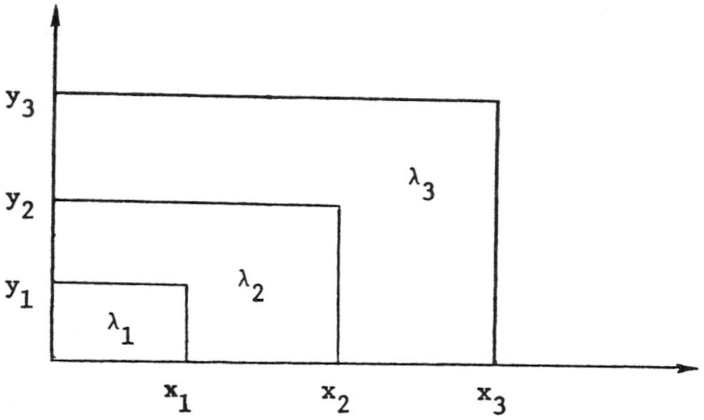

Define the sample L shaped section failure rate, $\hat{\lambda}_{n_i}$, to be the number of failures in the L shaped section divided by the total time on test in that section. Then

$\hat{\lambda}_{n_i} =$

$$\frac{\int_{x_{i-1}}^{x_i}\int_0^{y_i}\mathrm{d}F_n(u,v) + \int_0^{x_i}\int_{y_{i-1}}^{y_i}\mathrm{d}F_n(u,v) - \int_{x_{i-1}}^{x_i}\int_{y_{i-1}}^{y_i}\mathrm{d}F_n(u,v)}{\int_{x_{i-1}}^{x_i}\int_0^{y_i}\bar{F}_n(u,v)\mathrm{d}u\,\mathrm{d}v + \int_0^{x_i}\int_{y_{i-1}}^{y_i}\bar{F}_n(u,v)\mathrm{d}u\,\mathrm{d}v - \int_{x_{i-1}}^{x_i}\int_{y_{i-1}}^{y_i}\bar{F}_n(u,v)\mathrm{d}u\,\mathrm{d}v}$$

Letting $n \to \infty$, we obtain

$$\lambda_i = \frac{\int_{x_{i-1}}^{x_i}\int_0^{y_i} dF(u,v) + \int_0^{x_i}\int_{y_{i-1}}^{y_i} dF(u,v) - \int_{x_{i-1}}^{x_i}\int_{y_{i-1}}^{y_i} dF(u,v)}{\int_{x_{i-1}}^{x_i}\int_0^{y_i} \bar{F}(y,v)\,du\,dv + \int_0^{x_i}\int_{y_{i-1}}^{y_i} \bar{F}(u,v)\,du\,dv - \int_{x_{i-1}}^{x_i}\int_{y_{i-1}}^{y_i} \bar{F}(u,v)\,du\,dv}. \tag{2.1}$$

Now let $x = t\cos\alpha$, $y = t\sin\alpha$, and fix α, $0 < \alpha < \pi/2$. Let $x_i - x_{i-1} = \Delta t\cos\alpha$ and $y_i - y_{i-1} - \Delta t\sin\alpha$. Then dividing numerator and denominator by Δt and letting $\Delta t \downarrow 0$ we obtain

$$\lambda(t;\alpha) = \frac{\dfrac{\partial}{\partial x}F(x,y)\cos\alpha + \dfrac{\partial}{\partial y}F(x,y)\sin\alpha}{\int_0^y \bar{F}(x,v)\,dv\cos\alpha + \int_0^x \bar{F}(u,y)\,du\sin\alpha}$$

assuming of course, the existence of first partial derivatives. We call $\lambda(t;\alpha)$ the *directional failure rate of F at t in direction α*. The extension to $k > 2$ is easy to see.

If $\bar{F}(x,y) = e^{-\lambda_1 x}e^{-\lambda_2 y}$, independent exponentials, then it is easy to check that $\lambda(t;\alpha) \equiv \lambda_1\lambda_2$. If $F_1 = F_2$, $\bar{F}(x,y) = \bar{F}_1(x)\bar{F}_2(x)$, and F_1, F_2 have increasing failure rate (are IFR), then $\lambda(t;\pi/4)$ is increasing in $t \geq 0$ since if F_1 is IFR, then $\int_0^x \bar{F}_1(u)du/\bar{F}_1(x)$ is decreasing and so

$$\lambda(x,\pi/4) = F_1(x)\frac{f_1(x)}{\bar{F}_1(x)}\Big/\int_0^x \bar{F}_1(u)\,du$$

is increasing in $x \geq 0$.

If $F_1 \neq F_2$, and F_1, F_2 are IFR, then it does not follow that $\lambda(t;\pi/4)$ is increasing in $t \geq 0$. To see this, let $F_1(x) = x$ for $0 \leq x \leq x_1$ and $F_1(x) = 1$ for $x \geq x_1$. Let $F_2(y) = y$ for $0 \leq y \leq y_1$ and $F_2(y) = 1$ for $y \geq y_1$. It is easy to verify that F_1 and F_2 are IFR. Also, F_1 and F_2 can be approximated by absolutely continuous IFR distributions. If $x_1 \neq y_1$, then

$$\lambda(t;\pi/4) = \frac{f_1(x)F_2(y) + f_2(y)F_1(x)}{\bar{F}_1(x)\int_0^y \bar{F}_2(v)\,dv + \bar{F}_2(y)\int_0^x \bar{F}_1(u)\,du}$$

where $x = y = t/\sqrt{2}$, will be bimodal in $t \geq 0$ and hence not monotone.

Marshall (1975) discusses properties of the *hazard gradient*; i.e. if

$$R(t_1, t_2, \ldots, t_k) = -\log \bar{F}(t_1, t_2, \ldots, t_k),$$

then $(\partial R(t)/\partial t_1, \partial R(t)/\partial t_2, \ldots, \partial R(t)/\partial t_k)$ is the hazard gradient where $t = (t_1, t_2, \ldots, t_k)$. Unlike $\lambda(t; \alpha)$, the hazard gradient is nondirectional.

3. Asymptotic total time on test processes

In [2], scaled total time on test processes were used to identify probability models. We now give a rigorous derivation of the asymptotic results. Let $X_{n1} \leq X_{n2} \leq \ldots \leq X_{nn}$ be the order statistics from absolutely continuous distribution F. Let $\nu_n(u)$ be a discrete measure putting mass $1/n$ at $u = i/n$, $i = 1, 2, \ldots, n$. Then

$$\frac{\frac{1}{n} T_n(X_{ni})}{\frac{1}{n} T_n(X_{nn})} = \frac{1}{n}\left[\sum_{j=1}^{i} X_{nj} + (n-i)X_{ni}\right] \Big/ \bar{X} \tag{3.1}$$

or

$$\frac{\frac{1}{n} T_n(X_{ni})}{\frac{1}{n} T_n(X_{nn})} = \int_0^{i/n} \frac{X_{n[nu]}}{\bar{X}} \mathrm{d}\nu_n(u) + (1 - i/n)\frac{X_{ni}}{\bar{X}} \tag{3.2}$$

where $[nu]$ is the greatest integer in nu.

Now let $\mu = \int_0^\infty x \, \mathrm{d}F(x) < \infty$ and note that $X_{ni} \to F^{-1}(t)$ almost surely as $n \to \infty$ and $i/n \to t$.

Integrating by parts, we have

$$\int_0^{F^{-1}(t)} \frac{\bar{F}(x)}{\mu} \mathrm{d}x = \int_0^t \frac{F^{-1}(u)}{\mu} \mathrm{d}u + (1-t)\frac{F^{-1}(t)}{\mu}.$$

Hence, we see that

$$\begin{aligned}\sqrt{n}\left[\frac{T_n(X_{ni})}{T_n(X_{nn})} - \frac{\int_0^{F^{-1}(i/n)} \bar{F}(u)\mathrm{d}u}{\mu}\right] &= \int_0^{i/n} \sqrt{n}\left\{\frac{X_{n[nu]}}{\bar{X}} - \frac{F^{-1}(u)}{\mu}\right\} \mathrm{d}\nu_n(u) \\ &\quad + \int_0^{i/n} \sqrt{n}\frac{F^{-1}(u)}{\mu} \mathrm{d}[\nu_n(u) - u] \\ &\quad + (1 - i/n)\sqrt{n}\left\{\frac{X_{ni}}{\bar{X}} - \frac{F^{-1}(i/n)}{\mu}\right\}.\end{aligned} \tag{3.3}$$

Assume $g = F^{-1}$ has a nonzero continuous derivative, g', on $(0, 1)$, then Shorack [10] shows that

$$\sqrt{n}\left\{\frac{X_{n[nu]}}{\bar{X}}-\frac{F^{-1}(t)}{\mu}\right\}\xrightarrow[\substack{n\to\infty \\ [nu]\to t}]{P}-\frac{g'(t)}{\mu}U(t)-\frac{g(t)}{\mu^2}Z \tag{3.4}$$

where U is the Brownian Bridge process on $[0,1]$ and $Z=\int_0^\infty U[F(x)]\mathrm{d}x$ is $N(0,\mathrm{Var}\,(F))$. Since the second term in (3.3) converges deterministically to zero it follows from Shorack's results and [4], Theorem 5.1, p. 30 that

$$\sqrt{n}\left[\frac{T_n(X_{ni})}{T_n(X_{nn})}-\frac{\int_0^{F^{-1}(i/n)}\bar{F}(u)\mathrm{d}u}{\mu}\right]\xrightarrow[n\to\infty]{P}\int_0^t S(u)\mathrm{d}u+(1-t)S(t)=V(t),\text{ say,} \tag{3.5}$$

where

$$S(t)=-\frac{g'(t)}{\mu}U(t)-\frac{g(t)}{\mu^2}Z.$$

By a direct, but tedious calculation, it can be shown that, under exponentiality,

$$\mathrm{Cov}\,[V(s),V(t)]=s\wedge t-st.$$

Hence, in this case, the normed total time on test process, scaled by the sample mean, is the Brownian Bridge process.

References

[1] Barlow, R.E., Bartholomew, D.J., Bremner, J.M. and Brunk, H.D. (1972). *Statistical inference under order restrictions*, John Wiley and Sons.

[2] Barlow, R.E. and Campo, R. (1975). "Total Time on Test Processes and Applications to Failure Data Analysis," in *Reliability and Fault Tree Analysis*, edited by R.E. Barlow, J. Fussell and N. Singpurwalla, Conference volume published by SIAM, Philadelphia.

[3] Barlow, R.E. and Proschan, F. (1969). "A Note on Tests for Monotone Failure Rate Based on Incomplete Data," *Annals of Mathematical Statistics*, Vol. **40**, pp. 595–600.

[4] Billingsley, P. (1968). *Convergence of probability measures*, John Wiley and Sons.

[5] Chiang, C.L. (1960). "A Stochastic Study of the Life Table and Its Applications: I. Probability Distributions of the Biometric Functions," *Biometrics*, Vol. **16**, pp. 618–635.

[6] Harris, T.E., Meier, P. and Tukey, J.W. (1950). "Timing of the Distribution of Events Between Observations," *Human Biology*, Vol. **22**, pp. 249–270.

[7] Marshall, A.W. and Proschan, F. (1965). "Maximum Likelihood Estimation for Distributions with Monotone Failure Rate," *Annals of Mathematical Statistics*, Vol. **27**, pp. 887–906.

[8] Marshall, A.W. (1975). "Some Comments on the Hazard Gradient," *Stochastic Processes and Their Applications*, Vol. **3**, pp. 293–300.

[9] Miller, R.G. (1960). "Early Failures in Life Testing," *Journal of the American Statistical Association*, Vol. **55**, pp. 491–502.

[10] Shorack, G. (1972). "Convergence of Quantile and Spacings Processes with Applications," *Annals of Mathematical Statistics*, Vol. **43**, pp. 1400–1411.

[11] Crow, L.H. and Shimi, I.N. (1971). "Maximum Likelihood Estimation of a Monotone Step-Function Failure Rate From Time-Truncated Testing," Florida State University Tech. Report M197.

P.R. Krishnaiah, ed., *Multivariate Analysis–IV*
© North-Holland Publishing Company (1977) 239—245

TOPICS ON NONLINEAR FILTERING THEORY

Takeyuki HIDA
Nagoya University, Nagoya, Japan

Being inspired by nonlinear filtering or prediction theory, we discuss nonlinear functionals of Brownian motion, in particular quadratic functionals. The first topic is a characterization of quadratic functionals by the use of a characteristic function. Another topic is to analyze such nonlinear functionals by using integral representation and by applying the theory of functional analysis due to P. Lévy.

0. Introduction

We are interested in the filtering problems for stochastic processes each random variable of which is a Brownian functional, that is, a functional of a Brownian motion. These problems naturally require the analysis of nonlinear Brownian functionals. We shall, in this paper, present some topics that would be helpful to our filtering problems. It is hoped that such a discussion could be a first step of our approach.

We shall begin with an expression of Brownian functionals in terms of white noise, which is, formally speaking, the derivative of Brownian motion. Given a characteristic functional

$$C(\xi) = \exp[-\tfrac{1}{2}\|\xi\|^2] \tag{1}$$

on the Schwartz space $\mathcal{S}$, we can find a probability measure μ on $\mathcal{S}^*$ the dual space of $\mathcal{S}$ in such a way that

$$C(\xi) = \int_{\mathcal{S}^*} e^{i\langle x, \xi\rangle} d\mu(x), \tag{2}$$

where $\langle x, \xi\rangle$, $x \in \mathcal{S}^*$, $\xi \in \mathcal{S}$, is the canonical bilinear form connecting $\mathcal{S}$ and $\mathcal{S}^*$. The probability measure space $(\mathcal{S}^*, \mu)$ is called *white noise*, where each member x in $\mathcal{S}^*$ may be viewed as a sample function of the derivative of Brownian motion $\dot{B}(t) = dB(t)/dt$.

The complex Hilbert space $(L^2) \equiv L^2(\mathcal{S}^*, \mu)$ is the class of complex valued Brownian functionals with finite variance. We shall restrict our attention mainly to a subspace of (L^2) which is a class of quadratic functionals of white noise or of Brownian motion. For one thing, such a class seems to be the simplest one of nonlinear functionals. We shall first, in Section 2, discuss probability distributions of random variables that are quadratic functionals of Brownian motion. They can be characterized in terms of characteristic functions. We then proceed, in Section 3, to analyse such quadratic functionals where the integral representation of (L^2)-functionals is a powerful tool. We shall also make use of the theory of functional analysis developed by P. Lévy [5].

It is noted that our approach has been inspired by the work [4] and other results in the field of applications.

1. Integral representation of (L^2)

Some known results on the integral representation will be summarized in this section. We introduce a transformation $\mathcal{T}$ of (L^2) to a class of functionals on $\mathcal{S}$:

$$(\mathcal{T}\varphi)(\xi) = \int_{\mathcal{S}^*} e^{i\langle x, \xi\rangle}\varphi(x)d\mu(x), \qquad \varphi \in (L^2). \tag{3}$$

The collection

$$\mathcal{F} = \{(\mathcal{T}\varphi)(\xi); \varphi \in (L^2)\}$$

is a vector space having $C(\xi - \xi_0)$, ξ_0 fixed, as its member. We can introduce an inner product $((\,,\,))$ to $\mathcal{F}$ satisfying

$$((f(\cdot), C(\cdot - \xi))) = f(\xi) \quad \text{for every } f \in \mathcal{F}, \tag{4}$$

so that $\mathcal{F}$ is a reproducing kernel Hilbert space with the reproducing kernel $C(\xi - \eta)$, $(\xi, \eta) \in \mathcal{S} \times \mathcal{S}$. The transformation $\mathcal{T}$ is now a linear isomorphism of Hilbert spaces (L^2) and $\mathcal{F}$.

Let $\mathcal{F}_n$ $(n \geqq 0)$ be the closed linear subspace of $\mathcal{F}$ with the reproducing kernel $C_n(\xi, \eta)$:

$$C_n(\xi, \eta) = \frac{(\xi, \eta)^n}{n!} C(\xi)C(\eta), \qquad (\xi, \eta) \in \mathcal{S} \times \mathcal{S}. \tag{5}$$

The subspace $\mathcal{F}_n$ may be given alternatively in the form

$$\mathcal{F}_n = \{((f(\cdot), C_n(\cdot, \xi))); f \in \mathcal{F}\}.$$

Set $\mathcal{H}_n = \mathcal{T}^{-1}(\mathcal{F}_n)$, $n \geqslant 0$. The subspace $\mathcal{H}_n$ of (L^2) is called the *multiple*

Wiener integral of degree n. It can be proved that $\mathcal{F}$ and (L^2) admit the direct sum decompositions:

$$\mathcal{F} = \sum_{n=1}^{\infty} \oplus \mathcal{F}_n, \tag{6}$$

$$(L^2) = \sum_{n=0}^{\infty} \oplus \mathcal{H}_n. \tag{7}$$

The *integral representation* of a Brownian functional in $\mathcal{H}_n$ is illustrated as follows: For $\varphi \in \mathcal{H}_n$, we have an expression

$$(\mathcal{T}\varphi)(\xi) = i^n C(\xi) \int_{\mathbf{R}^n} \cdots \int F_\varphi(u_1, \ldots, u_n)\xi(u_1)\ldots\xi(u_n)du_1\ldots du_n, \tag{8}$$

under F_φ is taken to be a function in $\widehat{L^2(\mathbf{R}^n)} = \{F; F \in L^2(\mathbf{R}^n),\ F$ is symmetric$\}$. With this choice we have

$$n! \| F_\varphi \|^2_{L^2(\mathbf{R}^n)} = \| \varphi \|^2_{(L^2)}. \tag{9}$$

The $\widehat{L^2(\mathbf{R}^n)}$-function F is called the *kernel* of the integral representation of φ. For details we refer to [1].

The right-hand side of the expression (8) is a *regular* functional on $\mathcal{S}$ (after P. Lévy [5]). Being inspired by Lévy's approach we are able to introduce the concept of generalized Brownian functional by the help of integral representation theory (see [3]). Let $H^m(\mathbf{R}^n)$ be the Sobolev space of order m on $\mathbf{R}^n$ with the norm $\| \|_m$ and the inner product $\langle , \rangle_m$. Set

$$\mathcal{H}_n^{(n)} = \{\varphi ; F_\varphi \in H^{(n+1)/2}(\mathbf{R}^n) \cap \widehat{L^2(\mathbf{R}^n)}\}.$$

If we introduce a norm $\| \|_n^{n/2}$ in $\mathcal{H}_n^{(n)}$ by

$$\| \varphi \|_n = (n!)^{1/2} \| F \|_{n/2},$$

then $\mathcal{H}_n^{(n)}$ is a Hilbert space and the injection

$$\mathcal{H}_n^{(n)} \rightarrow \mathcal{H}_n$$

is continuous. The dual space of $\mathcal{H}_n^{(n)}$ will be denoted by $\mathcal{H}_n^{(-n)}$, which is viewed as a space of generalized Brownian functionals. The canonical bilinear form connecting $\mathcal{H}_n^{(n)}$ and $\mathcal{H}_n^{(-n)}$ is denoted by $(,)_n$. The transformation $\mathcal{T}$ can be extended to $\mathcal{H}_n^{(-n)}$ in a natural manner. Indeed, associated with G in $H^{-(n+1)/2}(\mathbf{R}^n)$ is a generalized Brownian functional ψ in $\mathcal{H}_n^{(-n)}$ satisfying

$$n! \langle G, F \rangle_{(n+1)/2} = (\psi, \varphi)_n, \tag{10}$$

where φ is in $\mathcal{H}_n^{(n)}$ and F, being in $H^{(n+1)/2}(\mathbf{R}^n)$, is the kernel of the integral representation of φ.

An example of $\mathcal{H}_2^{(-2)}$-functional is given in terms of the integral representation. A functional

$$i^2 C(\xi) \int f(u)\xi(u)^2 du, \qquad f \in L^2(\mathbf{R}^1) \tag{11}$$

is in $\mathcal{F}_2^{(-2)}$ and determines a Brownian functional whose formal expression is

$$\int f(u)\{\dot{B}(u)^2 - \mathbf{E}(\dot{B}(u)^2)\} du.$$

2. Quadratic functionals

This section is devoted to a characterization of random variables, assuming them to be in (L^2), by using analytic properties of the characteristic function.

Let $\varphi(x)$ be a real valued Brownian functional in $\mathcal{H}_2$ and let $F(u, v)$ be the kernel of the integral representation of φ. The $L^2(\mathbf{R}^2)$-function $F(u, v)$ may be viewed as the kernel of a symmetric integral operator of Hilbert–Schmidt type, so that we are given the eigensystem $\{\lambda_n, \eta_n;\, n \geq 1\}$:

$$\lambda_n \int F(u, v)\eta_n(v) dv = \eta_n(u), \qquad n \geq 1, \tag{12}$$

with the property that

$$\sum_{n=1}^{\infty} \lambda_n^{-2} < \infty.$$

The eigenfunction expansion

$$F(u, v) = \sum_n \lambda_n^{-1} \eta_n(u)\eta_n(v)$$

implies the expansion of $\varphi(x)$:

$$\varphi(x) = \sum_{n=1}^{\infty} \lambda_n^{-1}(\langle x, \eta_n\rangle^2 - 1),$$

which gives us an explicit expression for the characteristic function of the random variable $\varphi(x)$ on $(\mathcal{S}^*, \mu)$. Namely, the characteristic function $\chi(z)$ of $\varphi(x)$ is expressed in the form

$$\begin{aligned} \chi(z) &= \prod_n \{(1 - 2iz\lambda_n^{-1})^{-\frac{1}{2}} e^{-iz\lambda_n^{-1}}\} \\ &= \delta(2iz;\, F)^{-\frac{1}{2}}, \qquad z \text{ real}, \end{aligned} \tag{13}$$

where $\delta(\lambda; F)$ is the modified Fredholm determinant of F (see for example [2]).

With this expression in mind we are now able to characterize Brownian functionals in $\mathcal{H}_2$ in terms of a characteristic function.

Theorem 1. *Let $\varphi(x)$ be a real valued Brownian functional in* (L^2), *and let $\chi(z)$, $z \in \mathbf{R}$, be the characteristic function of $\varphi(x)$. Suppose that $\chi(z)^{-2}$ is extended uniquely to the complex plane* $\mathbf{C}$ *to be an entire function and satisfies*

(i) *The zero points of $\chi(z)^{-2}$ are non-zero, purely imaginary and their distribution has genus* 1.

(ii) *The order* of $\chi(z)^{-2}$ *is* 2.

Then, there exist two independent random variables $\varphi_1(x)$ and $\varphi_2(x)$ such that $\varphi_1(x)$ is Gaussian, $\varphi_2(x)$ is in $\mathcal{H}_2$ and the given $\varphi(x)$ has the same probability distribution as the sum $\varphi_1(x) + \varphi_2(x)$.

Proof. Let a_n's be zero points of $\chi(z)^{-2}$. By the assumption (i), $a_n = \pm i|a_n| \neq 0$ and

$$\sum |a_n|^{-2} < \infty.$$

Thus, in the Hadamard's factorization

$$\chi(z)^{-2} = z^m e^{h(z)} \prod_n \left(1 - \frac{z}{a_n}\right) \exp\left[\frac{z}{a_n} + \frac{1}{2}\left(\frac{z}{a_n}\right)^2 + \ldots + \frac{1}{m_n}\left(\frac{z}{a_n}\right)^{m_n}\right] \quad (14)$$

we may take m_n to be 0 or 1. Set $a_n = \lambda_n / 2i$, then the above expression is the same as (13) up to the factor $z^m e^{h(z)}$. Again by assumption, m has to be 0.

While the assumption (ii) assures that $h(z)$ is a polynomial of degree at most 2, since the canonical product of (14) is an entire function of order at most 2 (Hadamard's theorem). We are now able to obtain

$$\chi(z)^{-2} = \exp[az^2 + bz + c] \cdot \prod_n \left(1 - \frac{z}{a_n}\right) \exp\left[\frac{z}{a_n}\right].$$

(Note that m_n may be taken to be 1 for every n.) Since $\chi(z)$ is a characteristic function, we have

$$c = 0; \qquad a \text{ real, positive;} \qquad b = -2im, \qquad m \text{ real.}$$

Thus we have proved the theorem.

3. Analysis on the space $\mathscr{F}_2^{(-2)}$

We are now ready to discuss some analysis on the Hilbert space $\mathscr{F}_n^{(-n)}$, in particular on $\mathscr{F}_2^{(-2)}$. The results are proposed to be a tool for nonlinear filtering theory.

In the integral representation of $\mathscr{H}_n$-functionals there always appears a factor $\mathrm{i}^n C(\xi)$, therefore we may ignore it. In the case of $\mathscr{H}_2$, we are thus given a collection of regular functionals

$$\tilde{\mathscr{F}}_2 = \left\{ U(\xi) = \int\int F(u, v)\xi(u)\xi(v)\mathrm{d}u\,\mathrm{d}v\,;\ F \in \widehat{L^2(\mathbf{R}^2)} \right\}. \tag{15}$$

Let $U(\xi)$ be in $\tilde{\mathscr{F}}_2$ with corresponding F in $H^{\frac{3}{2}}(\mathbf{R}^2)$. Use the Gâteaux derivative to obtain the functional derivative

$$U'_\xi(t; \xi) = \int F(t, u)\xi(u)\mathrm{d}u. \tag{16}$$

This makes sense for each t since F is in Sobolev space $H^{\frac{3}{2}}(\mathbf{R}^2)$. The linear functional $U'_\xi(t; \xi)$ with parameter $t \in \mathbf{R}$ stands for a Gaussian process through the integral representation, but neglecting the factor $\mathrm{i}C(\xi)$. The Gaussian process is expressible as a stochastic integral of the form

$$X(t) = \int F(t, u)\mathrm{d}B(u). \tag{17}$$

Consider a converse operation. Given a Gaussian process expressed in the form (17), where F is a symmetric, trace class $H^{\frac{3}{2}}(\mathbf{R}^2)$-function and where we take $\langle x, \chi_{[t\wedge 0,\, t\vee 0]}\rangle$ as a version of Brownian motion $B(t)$. Multiply by $\dot{B}(t)$ to obtain $\dot{B}(t)\cdot X(t)$. Such a product has meaning as a generalized Brownian functional and the integral representation gives us a functional

$$V(t; \xi) = \xi(t)\int F(t, u)\xi(u)\mathrm{d}u + F(t, t) \tag{18}$$

(see [3] for details).

If we integrate $V(t; \xi)$ with respect to $\mathrm{d}t$, then we are given a functional in $\tilde{\mathscr{F}}_2$ plus constant the trace of F. Thus we have established the converse of Gâteaux derivative. In terms of (L^2), we can say that a Gaussian process is obtained from an $\mathscr{H}_2^{(2)}$-functional and the original functional can be formed multiplying by $\dot{B}(t)$ and then being integrated. It should be noted that a nonlinear operation is involved in our calculus and that integration is carried on in $\mathscr{H}_2^{(-2)}$ space without any difficulty although the integrand is not a martingale.

Summing up, we have proved

Theorem 2. *We have a Gaussian process from an $\mathscr{H}_2^{(2)}$-functional with a trace class kernel of the integral representation by the use of Gâteaux functional derivative in the space $\tilde{\mathscr{F}}_2$. The converse operation consists of multiplication by $\dot{B}(t)$ and integration with respect to* d*t up to constant.*

Finally we have to speak of analytic properties of generalized Brownian functionals in terms of functional derivatives. The space $\tilde{\mathscr{F}}_2^{(-2)}$ is defined by neglecting the factor $i^2C(\xi)$ of $\mathscr{F}_2^{(-2)}$-functional. Any functional $U(\xi)$ in $\tilde{\mathscr{F}}_2$ is, as we have seen, always regular quadratic functional and is *harmonic* in Lévy's sense (see [5], I-ère Partie), while those in $\tilde{\mathscr{F}}_2^{(-2)}$ are not always harmonic. For instance, a functional $U(\xi)$ satisfying (11) is not harmonic: Indeed, applying the Lévy's Laplacian Δ we obtain

$$\Delta U(\xi) = 2\int f(t)\mathrm{d}t.$$

References

[1] Hida, T. (1970). *Stationary stochastic processes*, Princeton Univ. Press.
[2] Hida, T. (1971). Quadratic functional of Brownian motion, *J. of Multivariate Analysis*, **1**, 58–69.
[3] Hida, T. (1975). Analysis of Brownian functionals, Preprint. International Symp. on Stochastic Systems. Lexington, Kentucky. *Math. Prog. Study* **5**, (1976) 53–59.
[4] Hida, T. and Kallianpur, G. (1975). Square of Gaussian Markov process and nonlinear prediction. *J. Multivariate Analysis*, **5**, 451–461.
[5] Lévy, P. (1951). *Problème concrets d'analyse fonctionnelle*. Gauthier-Villars.

P.R. Krishnaiah, ed., *Multivariate Analysis–IV*
© North-Holland Publishing Company (1977) 247–265

ON A CAUSAL AND CAUSALLY INVERTIBLE REPRESENTATION OF EQUIVALENT GAUSSIAN PROCESSES

Masuyuki HITSUDA

Nagoya Institute of Technology, Nagoya, Japan

and

Hisao WATANABE*

Kyushu University, Fukuoka, Japan

In this paper, the author obtained a causal and causally invertible representation of a Gaussian process $\{Y(t), 0 \leq t \leq 1\}$ which is equivalent to a given separable and non-deterministic Gaussian process $\{X(t), 0 \leq t \leq 1\}$ on a probability measure space $(\Omega, \mathcal{F}, P)$ and a representation of the density which connects these two Gaussian processes.

1. Introduction

The purpose of this paper is to obtain a causal and causally invertible representation of a Gaussian process $\{Y(t), 0 \leq t \leq 1\}$ which is equivalent to a given Gaussian process $\{X(t), 0 \leq t \leq 1\}$ on a probability measure space $(\Omega, \mathcal{F}, P)$, and to give a representation of the density which connects these two Gaussian processes. For the Wiener process, such a problem has been solved by Hitsuda [5], [6], when a given process is expressed by $X(t) = \int_0^t F(t, u)\mathrm{d}B(u)$, for each $t \in [0, 1]$. Here, we shall generalize these results to a general class of Gaussian processes. Kallianpur and Oodiara [8] have studied the same problem by a different approach. However, their approach is from the Hilbert space point of view, namely, in the use of the Gohberg and Krein theorem [3].

We solve this problem by a martingale technique which has been used by Hitsuda [5]. By this technique, at the same time we can obtain a representation of the Radon–Nikodym derivatives. Also we mention a recent paper by Kallianpur [7] in which similar results have been obtained.

*Visiting Professor, University of California, Los Angeles. Research supported in part under AFOSR Grant No. 4–442570–22514, USAF.

Our method is as follows: we suppose that Gaussian process $\{x(t), 0 \leq t \leq 1\}$ on a probability measure space $(\Omega, \mathcal{F}, P)$ is separable and non-deterministic and admits the following canonical representation (see [4])

$$X(t) = \sum_{i=1}^{N} \int_0^t F_i(t, u) \mathrm{d}B^{(i)}(u) + \sum_{\{j;\, t_j \leq t\}} \sum_{l=1}^{L_j} b_j^l(t) B_{t_j}^l \tag{1}$$

$$(N \leq \infty, L_i \leq \infty),$$

where stochastic processes $\{B^{(i)}(u), i = 1, 2, \ldots, N., 0 \leq t \leq 1\}$ and random variable $\{B_{t_j}^l, t_j \in \Lambda, l = 1, 2, \ldots L_j\}$ satisfy the following conditions:

(I.1) $\{B^{(i)}(u), i = 1, 2, \ldots N\}$ are processes with mean zero and independent increments such that $\mathbf{E}((\mathrm{d}B^{(i)}(u)^2) = m^i(\mathrm{d}u)$ are continuous measures with the property $m^i(\mathrm{d}u) \gg m^{i+1}(\mathrm{d}u)$, $(i = 1, 2, \ldots N)$.[1]

(I.2) $\{B_{t_j}^l, l = 1, 2, \ldots L_j, t_j \in \Lambda\}$ are standard Gaussian random variables and Λ is a countable set.

(I.3) All the $\{B^{(i)}(u), i = 1, 2, \ldots N\}$ and $\{B_{t_j}^l, l = 1, 2, \ldots L_j, t_j \in \Lambda\}$ are mutually independent.

(I.4)[2] $\mathcal{X}(t) \equiv \sigma(x(u), u \leq t) = \mathcal{B}(t) \equiv \sigma(B^{(i)}(u), u \leq t, \quad i = 1, 2, \ldots, N, B_{t_j}^l, t_j < t, t_j \in \Lambda, l = 1, 2, \cdots L_j B_{t_j}^1, t_j = t, b_j^1(t) \neq 0)$.

Furthermore, we suppose that $X = \{x(t), 0 \leq t \leq 1\}$ and $Y = \{Y(t), 0 \leq t \leq 1\}$ are equivalent on a probability measure space $(\Omega, \mathcal{F}, P)$, namely, there is a measure $\tilde{P}$ such that P and $\tilde{P}$ are mutually absolutely continuous and the probability law of Y with respect to the measure $\tilde{P}$ is the same as that of X with respect to the measure P, then $Y(t)$ takes the following form:

$$Y(t) = \sum_{i=1}^{N} \int_0^t F_i(t, u) \mathrm{d}\tilde{B}^{(i)}(u) + \sum_{\{j;\, t_j \leq t\}} \sum_{l=1}^{L_j} b_j^l(t) \tilde{B}_{t_j}^l \tag{2}$$

where the system

$$\{\tilde{B}^{(i)}(t), i = 1, 2, \ldots, N; \tilde{B}_{t_j}^l, t_j \in \Lambda, l = 1, 2, \ldots, L_j\}$$

has the same properties with respect to the measure $\tilde{P}$ as the system

$$\{B^{(i)}(t), i = 1, 2, \ldots, N, B_{t_j}^l, t_j \in \Lambda l = 1, 2, \ldots, L_j\}$$

do with respect to the measure P.

[1] $m^i \gg m^{i+1}$ implies that the measure m^{i+1} is absolutely continuous with respect to the measure m^i.

[2] By $\sigma(\cdot)$ we mean the σ-algebra generated by the system of random variables indicated in $(\cdot)$.

Thus, the study of relations between two equivalent Gaussian stochastic processes X and Y can be reduced to that of relations between two stochastic processes

$$\mathfrak{B}(t)=\{B^{(1)}(t), B^{(2)}(t), \ldots, ; \chi_{[t_1,1)}(t)B^1_{t_1}, \chi_{(t_1,t]}(t)B^2_{t_1}, \ldots; \chi_{[t_2,1)}(t)B^1_{t_2,\ldots}\},$$

$$\tilde{\mathfrak{B}}(t)=\{\tilde{B}^{(1)}(t), \tilde{B}^{(2)}(t), \ldots, \chi_{[t_1,1)}(t)\tilde{B}^1_{t_1}, \chi_{(t_1,1]}(t)\tilde{B}^2_{t_1}, \ldots; \chi_{[t_2,1)}\tilde{B}^1_{t_2}, \ldots\}.$$

Our main result is the following theorem.

Theorem. *Let $Y=(Y(t), 0\leqslant t\leqslant 1)$ be an equivalent process to a Gaussian process $X=(X(t), 0\leqslant t\leqslant 1)$ on a probability measure space $(\Omega, \mathcal{F}, P)$. We suppose that X admits the canonical representation (1). Then, we can construct newly a Gaussian process on the probability measure space $(\Omega, \mathcal{F}, P)$ which has the same properties as X. If we write such a process by X again, Y is represented canonically as follows:*

$$Y(t)=\sum_{i=1}^{N}\int_0^t F_i(t,u)(\mathrm{d}B^{(i)}(u)-\psi^i(u))m^i(\mathrm{d}u)$$
$$+\sum_{\{j;\,t_j\leqslant t\}}\sum_{l=1}^{L_j} b^l_j(t)(1+\gamma^l_j)(B^l_{t_j}-\Psi^l_j), \tag{3}$$

where random functions $\psi^i(u)$ and random variables Ψ^l_j in (3) are given in the following form,

$$\psi^i(u)=\sum_{k=1}^{N}\int_0^u l^i_k(u,v)\mathrm{d}B^{(k)}(v)+\sum_{t_j\leqslant u}\sum_{l=1}^{L_j} c^{i,j}_l(u)B^l_{t_j},$$

$$\Psi^l_j=\sum_{k=1}^{N}\int_0^{t_j} l^l_{j,k}(v)\mathrm{d}B^{(k)}(v)+\sum_{t_m\leqslant t_j}\sum_{k=1}^{L_j} c^{l,m}_{j,k}B^k_{t_m},$$

$$c^{l,j}_{j,k}=0 \quad \text{for} \quad k\geqslant l,$$

$l^i_k(u,v)$ is a Volterra kernel ($l^i_k(u,v)=0$ for $u<v$),

$$\sum_j\sum_l\Big\{\sum_m\sum_k (c^{l,m}_{j,k})^2+\sum_k\int_0^1 (l^l_{j,k}(v))^2\mathrm{d}m^k(v)\Big\}$$
$$+\int_0^1\sum_i\sum_k\Big\{\int_0^u (l^i_k(u,v))^2 m^k(\mathrm{d}v)+\sum_{t_j\leqslant u}\sum_{l=1}^{L_j}(c^{i,j}_l(u))^2\Big\}m^i(\mathrm{d}u)<+\infty,$$

$$\sum_j\sum_l(\gamma^l_j)^2<+\infty \quad \text{and} \quad \gamma^l_j>-1 \quad (j=1,2,\ldots,;\ l=1\ldots L_j).$$

Furthermore, we have

$$\mathscr{B}(t) = \mathscr{X}(t) = \mathscr{Y}(t) = \sigma(Y(s), s \leq t) = \tilde{\mathscr{B}}(t),$$
$$\mathscr{B}(t\pm) = \mathscr{X}(t\pm) = \mathscr{Y}(t\pm) = \tilde{\mathscr{B}}(t\pm),^3$$

$$\frac{d\tilde{P}}{dP}(\omega) = \prod_{i=1}^{\infty} \exp\left\{\int_0^1 \psi^i(u) dB^{(i)}(u) - \tfrac{1}{2}\int_0^1 (\psi^i(u))^2 m^i(du)\right\}$$
$$\times \prod_j \prod_{l=1}^{L_j} (1+\gamma_j^l) \exp(-\tfrac{1}{2}\{(1+\gamma_j^l)^2 (B_{t_j}^l - \Psi_j^l)^2 - (B_{t_j}^l)^2\}).$$

2. Proof

Since Y has the same law with respect to the measure $\tilde{P}$ as X with respect to the measure P, we have the following lemma.

Lemma 1. *It holds*

$$\tilde{\mathscr{B}}(t) \equiv \sigma(\mathfrak{B}(s), s \leq t) = \mathscr{Y}(t) \equiv \sigma(Y(s), s \leq t),$$

and

$$\tilde{\mathscr{B}}(t-) = \mathscr{Y}(t-), \tilde{\mathscr{B}}(t+) = \mathscr{Y}(t+).$$

We note that each component of $\mathfrak{B}(t)$ is a Gaussian random variable with respect to the measure P. In fact, from the construction of the canonical representation (2), we can see that each component of $\mathfrak{B}(t)$ is linearly spanned by $(Y(s), s \leq t)$. But, each $Y(u)$ is a Gaussian random variable with respect to the measure P.

Lemma 2. *For $t \notin \Lambda$, we have*

$$\tilde{\mathscr{B}}(t) = \tilde{\mathscr{B}}(t+) = \tilde{\mathscr{B}}(t-),$$

with $\tilde{P}$-measure 1. (*P-measure* 1.)

Proof. The proof is analogous to that of Theorem 4.3 in [10], p. 97. Let

$$\tilde{\mathbf{E}}\left(\frac{dP}{d\tilde{p}}(\omega) | \tilde{\mathscr{B}}(t)\right) = M_t,$$

$$\tilde{\mathbf{E}}\left(\frac{dP}{d\tilde{p}}(\omega) | \tilde{\mathscr{B}}(t+)\right) = M_{t+}$$

[3] For example; $\mathscr{X}(t+) = \bigcap_{u>t} \mathscr{X}(u)$, $\mathscr{X}(t-) = \sigma(\bigcup_{u<t} \mathscr{X}(u))$, etc.

and

$$\tilde{\mathbf{E}}\left(\frac{dP}{d\tilde{P}}(\omega)\,|\,\tilde{\mathcal{B}}(t-)\right)=M_{t-}.$$

By Lemma 2, for $t\notin\Lambda$, $M_t=M_{t-}=M_{t+}$. Gaps may occur only at $t\in\Lambda$.

Let t_j be an arbitrary element of Λ. Put $M_{t_j}/M_{t_j-}=N(t_j-)$.

By the definition,

$$M_{t_j}=\lim_{n\to\infty}\frac{p(X(s_1),X(s_2),\ldots,X(s_n))}{\tilde{p}(X(s_1),X(s_2),\ldots,X(s_n))},$$

where p and $\tilde{p}$ are Gaussian probability densities (with repect to the Lebesgue measure) of random variables $X(s_1),X(s_2),\ldots,X(s_n)$ with respect to the measures P and $\tilde{P}$ respectively and $\{s_1,s_2,\ldots,s_n,\ldots\}$ is a dense subset of set $\{s\leqslant t\}$. Therefore, we can see that $N(t_j-)$ is the ratio of conditional Gaussian density function of $\tilde{B}^1_{t_j}$ with respect to the conditional measure $P(\cdot\,|\,\tilde{\mathcal{B}}(t_j-))$ and $\tilde{P}(\cdot\,|\,\tilde{\mathcal{B}}(t_j-))$

Now, if we put

$\mathbf{E}(\tilde{B}^1_{t_j}\,|\,\tilde{\mathcal{B}}(t_j-))=\phi^1_{t_j}\in H(t_j-)\equiv$ linear manifold spanned by $\{Y(s)s<t_j\}$

and

$$\mathbf{E}(\tilde{B}^1_{t_j}-\phi^1_{t_j})^2-(1+\gamma^1_j)^2,$$

where we can choose γ^1_j such that $\gamma^1_j>-1$, then we have[4]

$$B^1_{t_j}=(1+\gamma^1_j)^{-1}(\tilde{B}^1_{t_j}-\phi^1_{t_j})$$

and $N(t_j-)$ is expressed in the following manner.

$$\begin{aligned}N(t_j-)&=(1+\gamma^1_j)^{-1}\exp[-\tfrac{1}{2}\{(1+\gamma^1_j)^{-2}(B^1_{t_j}-\phi^1_j)^2-(B^1_{t_j})\}]\\&=(1+\gamma^1_j)^{-1}\exp[-\tfrac{1}{2}\{(B^1_{t_j})^2-(1+\gamma^1_j)^2(B^1_{t_j}-\Phi^1_j)\}],\end{aligned}$$

where $\Phi^1_j=(-\phi^1_j)(1+\gamma^1_j)^{-1}$.

By the same method as being done in the representation of $N(t_j-)$, we shall express

$$\frac{M_{t_j+}}{M_{t_j}}=N(t_j+).$$

We define σ-fields as follows:

$$\mathcal{Y}(t_j,1)\equiv\mathcal{Y}(t_j),\ \mathcal{Y}(t_j,2)\equiv\sigma(\mathcal{Y}(t_j)\cup\sigma(\tilde{B}^2_{t_j})),$$

$$\mathcal{Y}(t_j,l)\equiv\sigma(\mathcal{Y}(t_j)\cup\sigma(\tilde{B}^2_{t_j},\ldots,\tilde{B}^l_{t_j})).$$

[4]We may use the same notation as in (1), without confusion.

Furthermore, we define

$$\mathbf{E}(M_{t_j+} \mid \mathscr{Y}(t_j, l)) \equiv M_{t_j, l}$$

and

$$\frac{M_{t_j, l+1}}{M_{t_j, l}} = N(t_j, l+1) \quad (l = 1, 2, \dots).$$

Obviously, we have

$$N(t_j +) = \lim_{l \to L_j} \frac{M_{t_j, l}}{M_{t_j}} = \prod_{l=2}^{L_j} N(t_j, l).$$

If we put

$$\mathbf{E}(\tilde{B}^l_{t_j} \mid y(t_j, l-1)) = \phi^l_j \qquad \mathbf{E}((\tilde{B}^l_{t_j} - \phi^l_j)^2) = (1 + \gamma^l_j)^2,$$

$\tilde{B}^l_{t_j}$ can be written as $\tilde{B}^l_{t_j} = (1 + \gamma^l_j) B^l_{t_j} + \phi^l_j$. Therefore, $N(t_j, l)$ can be written as follows

$$\begin{aligned} N(t_j, l) &= (1 + \gamma^l_j)^{-1}[-\tfrac{1}{2}\{(1 + \gamma^l_j)^{-2}(\tilde{B}^l_{t_j} - \phi^l_j)^2 - (\tilde{B}^l_{t_j})^2\}] \\ &= (1 + \gamma^l_j)^{-1} \exp[-\tfrac{1}{2}\{(B^l_{t_j})^2 - (1 + \gamma^l_j)^2 (B^l_{t_j} + \Phi^l_j)^2\}], \end{aligned}$$

where $\Phi^l_j = \phi^l_j (1 + \gamma^l_j)^{-1}$.

From the construction of the family of random variables $\{B^l_{t_j}\}$, we can easily see that they are a family of independent standard Gaussian random variables with respect to the measure P. So that, two families of Gaussian random variables $\{\tilde{B}^l_{t_j}\}$ and $\{B^l_{t_j}\}$ are equivalent.

Since we can write ϕ^l_j as $\phi^l_j = \Sigma_{l', k'} a^{l,k'}_{j,l'} B^{l'}_{t_{k'}} + \tilde{\phi}^l_j$ by some constant $a^{l,k'}_{j,l'}$ and random variables $\tilde{\phi}^l_j$ which are independent of $\{B^l_{t_j}\}$, we have

$$\begin{aligned} \tilde{B}^l_{t_j} &= (1 + \gamma^l_j) B^l_{t_j} + \phi^l_j \\ &= (1 + \gamma^l_j) B^l_{t_j} + \sum_{k'} \sum_{l'} a^{l,k'}_{j,l'} B^{l'}_{k'} + \tilde{\phi}^l_j \\ &= (I + L) B^l_{t_j} + \tilde{\phi}^l_j, \end{aligned}$$

where I (identity) and L are operators from $L^2(\Omega, \mathscr{F}, P)$ into itself. From the equivalence of $\{\tilde{B}^l_{t_j}\}$ and $\{B^l_{t_j}\}$, it follows (for example, see [11]) that

$$\sum_{j,l} (\gamma^l_j)^2 < \infty, \qquad \sum_{l,j} \sum_{l',k'} (a^{l,k'}_{j,l'})^2 < \infty, \qquad \sum_{j,l} (\phi^l_j)^2 < \infty$$

with $\tilde{P}$ (also P)-measure 1. Consequently, we have

$$\sum_{j,l} (\gamma^l_j)^2 < \infty, \qquad \sum_{j,l} (\phi^l_j)^2 < \infty \tag{4}$$

with $\tilde{P}$-measure 1.

Here, if we apply to (4) a theorem in [1] (Theorem 8 (p. 317)), we have

$$\sum_{j,k} \tilde{\tilde{\mathbf{E}}}((\phi_j^l)^2) < \infty \tag{5}$$

for Gaussian measures which are absolutely continuous with respect to the measure P, where $\tilde{\tilde{\mathbf{E}}}$ is the expectation taken with respect to the measure $\tilde{\tilde{P}}$.

Therefore, the series

$$\sum_{t_j \in \Lambda} \sum_{j=1}^{L_j} \{-\log(1+\gamma_j^l)^2 + (1+\gamma_j^l)^2 - 1 + \tilde{\tilde{\mathbf{E}}}((\phi_{t_j}^l)^2)\} \tag{6}$$

is convergent.

Using the above discussions, we can see that the following quantities are well-defined,

$$M_t^d \equiv \prod_{t_j \leq t} N(t_j -) \prod_{t_j < t} N(t_j +),$$

$$M_{t-}^d \equiv \prod_{t_j < t} N(t_j -) \prod_{t_j < t} N(t_j +),$$

$$M_{t+}^d = \prod_{t_j \leq t} N(t_j \quad) \prod_{t_j \leq t} N(t_j +).$$

Lemma 3. *M_t^d is a martingale with respect to the system $(\mathcal{Y}(+), \tilde{P})$ and $(\mathcal{Y}(t-), P)$.*

Proof. At first, we remark that

$$\mathbf{P}(M_t^d > 0) = 1.$$

Furthermore, we note that

$$\tilde{\mathbf{E}}(N(t_j, l) \mid \mathcal{Y}(t_j, l-1))$$

$$= (2\pi)^{-1/2}(1+\gamma_j^l)^{-1} \int_{-\infty}^{\infty} \exp[-\tfrac{1}{2}\{(1+\gamma_j^l)^{-2}(x-\phi_j^l)^2 - x^2\}]$$

$$\times \exp(-\tfrac{1}{2}x^2)dx$$

$$= (2\pi)^{-1/2}(1+\gamma_j^l)^{-1} \int_{-\infty}^{\infty} \exp[-\tfrac{1}{2}(1+\gamma_j^l)^{-2}(x-\phi_j^l)^2]dx = 1.$$

Therefore, if $n < \infty$, we can show inductively that

$$\tilde{\mathbf{E}}(N(t_j -) \prod_{l=2}^{n} N(t_j, l) \mid \mathcal{Y}(s)) = 1, \quad \text{for } s < t_j.$$

Let $\{\Lambda_j\}$ be a sequence of finite sets such that $\Lambda_1 \subset \Lambda_2 \subset \Lambda_3 \ldots$ and $\bigcup_{n=1}^{\infty} \Lambda_n = \Lambda$. Then, we have

$$\tilde{\mathbf{E}}\Big(\prod_{\substack{s<t_j\leq t\\ t_j\in\Lambda_n}} N(t_j -) \prod_{\substack{s<t_j<t\\ t_j\in\Lambda_n}} N^n(t_j +) | \mathcal{Y}(s)\Big) = 1 \quad s<t,$$

where

$$N^n(t_j +) \equiv \prod_{l=2}^{L_j\wedge n} N(t_j, l).$$

Let

$$[M_t^d]_n \equiv \prod_{\substack{t_j\leq t\\ t_j\in\Lambda_n}} N(t_j -) \prod_{\substack{t_j<t\\ t_j\in\Lambda_n}} N^n(t_j +),$$

then obviously, we have

$$\mathbf{E}\{[M_t^d]_n | \mathcal{Y}(s+)\} = [M_s^d]_n \quad \text{for} \quad s<t.$$

Now, we put $\mathcal{K} = \{[M_t^d]_n, n = 1, 2, \ldots,\}$, each belonging to $\mathrm{L}^1(\Omega, \mathcal{F}, P)$. $\mathcal{K}$ is uniformly integrable. In fact,

$$\begin{aligned}
\tilde{\mathbf{E}}([M_t^d]_n \log [M_t^d]_n) &= \sum_{t_j\leq t,\, t_j\in\Lambda_n} \tilde{\mathbf{E}}((\log N(t_j -))[M_{t_j}^d]_n) \\
&\quad + \sum_{t_j<t,\, t_j\in\Lambda_n} \sum_{l=2}^{L_j\wedge n} \tilde{\mathbf{E}}\Big((\log N(t_j, l))[M_{t_j}^d]_n \prod_{k=2}^{l} N(t_j, k)\Big) \\
&= \sum_{t_j\in\Lambda_n} \sum_{l=1}^{L_j\wedge n} \Big[\tfrac{1}{2}\{-\log(1+\gamma_j^l)^2 + (1+\gamma_j^l)^2 - 1\} \\
&\quad + \tilde{\mathbf{E}}((\phi_j^l)^2 \prod_{k=1}^{l-1} N(t_j, k))\Big] \\
&= \sum_{t_j\in\Lambda_n} \sum_{l=1}^{L_j\wedge n} [\tfrac{1}{2}\{-\log(1+\gamma_j^l)^2 + (1+\gamma_j^l)^2 - 1\} \\
&\quad + \tilde{\tilde{\mathbf{E}}}(\phi_j^l)^2],
\end{aligned}$$

which are bounded with respect to n by (6), where $\mathrm{d}\tilde{\tilde{P}}$ is a Gaussian measure defined by $\mathrm{d}\tilde{\tilde{P}} = [M_1^d]_n \,\mathrm{d}\tilde{P}$.

Therefore, $\lim_{n\to\infty} [M_t^d]_n = M_t^d$ is a martingale with respect to the system $(\mathcal{Y}(t+), \tilde{P})$. By the same way, M_t^d is a martingale with respect to the system $(\mathcal{Y}(t-), \tilde{P})$. Thus we have Lemma 3.

Now, define

$$M_t^c = \frac{M_t}{M_t^d} \quad \text{if} \quad M_t^d > 0 \quad \text{and} \quad M_t^c = 0 \quad \text{if} \quad M_t^d = 0.$$

By definition of M_t^d, M_t^c, Lemma 2 and a theorem of P. Lévy [theorem 2.8 (p. 55) in [10]] we have the following lemmas.

Lemma 4

$$\tilde{P}\left(\lim_{s\to t} M_s^c = M_t^c\right) = 1 \quad \textit{for each} \quad 0 \leq t \leq 1.$$

Lemma 5. *M_t^c is a martingale with respect to the system $(\mathcal{Y}(t), \tilde{P})$.*

Proof. If we put $[M_t^c]_n = M_t/[M_t^d]_n$ if $[M_t^d]_n > 0$, and $[M_t^c]_n = 0$ if $[M_t^d]_n = 0$, then we have

$$\begin{aligned}\tilde{\mathbf{E}}([M_t^c]_n) &= \mathbf{E}\left(\frac{1}{[M_t^d]_n}\right)\\ &= \mathbf{E}\left(\prod_{t_j\in\Lambda_n}\prod_{l=1}^{L_j\wedge n}(1+\gamma_j^l)\exp\{-\tfrac{1}{2}(B_{t_j}^l)^2-(1+\gamma_j^l)^2(B_{t_j}^l+\Phi_{t_j}^l)^2\}\right)\\ &= 1.\end{aligned}$$

By the same way as the proof of Lemma 3, we can see that $\{[M_t^c]_n\}$ is uniformly integrable. From it, it follows that $\lim_{n\to\infty}[M_t^c]_n = M_t^c$ is a martingale with respect to $(\mathcal{Y}(t), \tilde{P})$.

We shall seek a representation of M_t^c.

Lemma 6. *M_t^c is expressed in the form*

$$M_t^c = \exp\left\{\sum_{i=1}^N \int_0^t f_c^i(u,\omega)\mathrm{d}\tilde{B}^{(i)}(u) - \frac{1}{2}\sum_{i=1}^N \int_0^t (f_c^i(u,\omega))^2 \mathrm{d}m^i(\omega)\right\} \tag{7}$$

where $\{f_c^i(\omega)\}$ satisfies the following conditions,

(II.1) *(s, ω)-measurable,*
(II.2) *$\mathcal{Y}(t)$-measurable for each $t \in [0, 1]$,*
(II.3) $\mathbf{P}(\sum_{i=1}^N \int_0^1 (f_c^i(u,\omega))^2 \mathrm{d}m^i(u) < \infty) = 1$.

Proof. We need the following proposition.

Proposition 1 (Kunita–Watanabe) [9]). *Let $\mathcal{F}_t = \tilde{\mathcal{B}}(t)$ and and $(X_t, \mathcal{F}_t, 0 \leq t \leq 1)$ be a martingale with paths being continuous for each fixed t and with* $\sup_{0\leq t\leq 1} \tilde{\mathbf{E}}(|X_t|) < \infty$. *Then there exist $\mathcal{F}_t$-measurable process $\{f^i(t,\omega), 0 \leq t \leq 1\}$ such that*

$$\tilde{\mathbf{P}}\left(\sum_{i=1}^{N}\int_0^1 (f^i(s,\omega))^2 \mathrm{d}m^i(s) < \infty\right) = 1;$$

it holds

$$x_t = x_0 + \sum_{i=1}^{N}\int_0^t f^i(s,\omega)\mathrm{d}\tilde{B}^{(i)}(s).$$

Proof. The statement of Proposition 1 and Lemma 7 below are not described in the current context, but can be proved in analogous ways though it needs to be more carefully checked. The details of these proofs will be published elsewhere.

By making use of Proposition 1, M_t^c is expressed by

$$M_t^c = \sum_{i=1}^{N}\int_0^t g^i(s,\omega)\mathrm{d}B^{(i)}(s,\omega) + 1,$$

where

$$\mathbf{P}\left(\sum_{i=1}^{N}\int_0^t (g^i(s,\omega))^2 dm^i(s) < \infty\right) = 1.$$

Here, if we use Ito's formula on stochastic differentials, we have

$$M_t^c = \exp\left(\sum_{i=1}^{N}\int_0^t (M_s^c)^+ g^i(t,\omega)\mathrm{d}B^{(i)}(s,\omega)\right.$$

$$\left. - \tfrac{1}{2}\sum_{i=1}^{N}\int_0^t (M_s^c)^+ g^i(s,\omega)^2 \mathrm{d}m^i(s)\right).$$ [5]

If we put $f_c^i(s,\omega) = M_s^+ g^i(s,\omega)$, we have the representation (7), and we can see that $\{f_c^i(s,\omega)\}$ satisfies the conditions (II.1)–(II.3).

Lemma 7 (Givsanov [2]). *Let* $f_c^i(s,\omega)$ *be the functions obtained in Lemma* 6 *and let for* $i = 1, 2, \ldots, N$,

$$\tilde{\tilde{B}}^{(i)}(t) = \tilde{B}^{(i)}(t) - \int_0^t f_c^i(s,\omega)\mathrm{d}m^i(s). \tag{8}$$

Then $\{\tilde{\tilde{B}}^{(i)}(t)\}$ are Gaussian processes which enjoy the same law as $\{\tilde{B}^{(i)}(t)\}$ with respect to the measure P. So, we can identify $\{\tilde{\tilde{B}}^{(i)}(t)\}$ with $\{B^{(i)}(t)\}$.

[5] $M_s^+ = (M_s)^{-1}$ if $M_s > 0$. $M_s^+ = 0$ if $M_s = 0$.

Lemma 8

$$\tilde{\mathrm{E}}\left(\sum_{i=1}^{N}\int_0^1 (f_c^i(s,\omega))^2 \mathrm{d}m^i(s)\right)<\infty \tag{9}$$

Let $\mathcal{M}_t$ be the family of Gaussian random variables

$$\left\{\sum_{i=1}^{k_1} d_i\tilde{B}^{(d_i)}(s_i)+\sum_{j=1}^{k_2}\beta_j B_{t_{ij}}^{l_j},\ 0\leq s_i\leq t,\ 1\leq j_1,j_2,\ldots,j_{k_1}\leq N,\ t_{i_j}\leq t,\right.$$
$$\left. t_{i_j}\in\Lambda,\ 1\leq l_j\leq L_{i_j}\right\},$$

where $\alpha_1,\alpha_2,\ldots,\alpha_{k_1}$ and $\beta_1,\beta_2,\ldots,\beta_{k_2}$ are arbitrary constants.

Let $(\mathcal{M}_t)^P$ and $(\mathcal{M}_t)^{\tilde{P}}$ be the closure of $\mathcal{M}_t$ by L^2-norm with respect to the measure P and $\tilde{P}$, respectively. Then, we can see easily that $(\mathcal{M}_t)^P=(\mathcal{M}_t)^{\tilde{P}}$ and $(\mathcal{M}_t)^P$ consists of Gaussian random variables.

Lemma 9. *If we put*

$$F_t^i(\omega)=\int_0^t f_c^i(s,\omega)\mathrm{d}m^i(s),\ (i=1,2,\ldots,N),$$

then $F_t^i(\omega)\in(\mathcal{M}_t)^{\tilde{P}}$.

Lemma 10. *Let* η *be an element of* $(\mathcal{M}_t)^{\tilde{P}}$. *Then, there are measurable functions* $\{K_i(u), i=1,2,\ldots,N\}$ *and real numbers* $\{a_k^j\}$ *with*

$$\sum_{i=1}^{N}\int_0^t K_i^2(u)\mathrm{d}m^i(u)+\sum_j\sum_{k=1}^{L_j}(a_k^j)^2<\infty$$

such that

$$\eta-\sum_{j=1}^{N}\int_0^t K_j(u)\mathrm{d}\tilde{B}^{(j)}(u)+\sum_{t_j\leq t}\sum_{k=1}^{L_j}a_k^j\tilde{B}_{t_j}^k$$

Proof. The proof is similar to Theorem 5.6 in ([10] p. 189).

We now let Λ^i denote the (s,ω)-set

$$\left\{(s,\omega);\ \lim_{n\to\infty}(F_{s-1/n}^i)/(m^i(s)-m^i(s-1/n))\text{ does not exist or}\right.$$

$$\left.\lim_{n\to\infty}(F_s^i-F_{s-1/n}^i)/(m^i(s)-m^i(s-1/n)\neq f_c^i(s,\omega)\right\},$$

where we define $F_s^i=0$ for $s<0$. Then, $\bigcup_{i=1}^N\Lambda^i$ is (s,ω)-measurable and $\mu^i(\Lambda^i)=0$, where μ^i is the product measure of $\tilde{P}(\mathrm{d}\omega)$ and $m^i(\mathrm{d}s)$ on $[0,1]$. In fact, $m^i(\Lambda_\omega^i)=0$ with P-measure 1, where $\Lambda_\omega^i=\{s;(s,\omega)\in\Lambda^i\}$.

By Fubini's theorem, it follows that $\tilde{P}(\Lambda_s^i)=0$ except for the s-set of m^i-measure 0, where $\Lambda_s^i=\{\omega;(s,\omega)\in\Lambda^i\}$. Therefore,

$$\lim_{n\to\infty}(F_s^i-F_{s-1/n}^i)/(m^i(s)-m^i(s-1/n))=f_c^i(s,\omega)\quad\text{for}\quad\omega\notin\Lambda_s^i,$$

and $f^i(s,\omega)\in(\mathcal{M}_s)^P$ except for the s-set of m^i-measure 0.

Furthermore, by Lemma 10, we have the following representation

$$f_c^i(s,\omega)=\sum_{k=1}^{N}\int_0^s k_k^i(s,v)\mathrm{d}\tilde{B}^{(k)}(v)+\sum_{t_j\leq s}\sum_{k=1}^{L_j}a_k^{i,j}(s)\tilde{B}_{t_j}^k,$$

where

$$\sum_{k=1}^{N}\int_0^s (k_k^i(s,v))^2\mathrm{d}m^k(v)$$
$$+\sum_{t_j\leq s}\sum_{k=1}^{L_j}(a_k^{i,j}(s))^2<\infty\quad\text{for any}\quad 0\leq s\leq 1.$$

Here, we can choose $k_k^i(s,u)$ to be (s,u)-measurable. Put

$$\tilde{\Lambda}_s^i=\Big\{\omega;f_c^i(s,\omega)\neq\sum_{k=1}^{N}\int_0^s k_k^i(s,v)\mathrm{d}\tilde{B}^{(k)}(v)$$
$$+\sum_{t_j\leq s}\sum_{k=1}^{L_j}a_k^{i,j}(s)\tilde{B}_{t_j}^k\Big\}\cup\Lambda_s^i,$$

then $P(\tilde{\Lambda}_s^i)=0$ except for s-set of m^i-measure 0. Since

$$\tilde{\Lambda}^i=\Big\{(s,\omega);f_c^i(s,\omega)\neq\sum_{j=1}^{N}\int_0^s k_k^i(s,v)\mathrm{d}\tilde{B}^{(k)}(v)$$
$$+\sum_{t_j\leq s}\sum_{k=1}^{L_j}a_k^{i,j}(s)\tilde{B}_{t_j}^{(k)}\Big\}\cup\Lambda^i$$

is (s,ω)-measurable, it holds

$$f_c^i(s,\omega)=\sum_{j=1}^{N}\int_0^s k_k^i(s,v)\mathrm{d}\tilde{B}^{(k)}(v)+\sum_{t_j\leq s}\sum_{k=1}^{L_j}a_k^{i,j}(s)\tilde{B}_{t_j}^k$$
$$(i=1,2,\ldots,N),$$

for $s\notin\bigcup_{i=1}^N\tilde{\Lambda}_\omega^i=\bigcup_{i=1}^\infty\{s;(s,\omega)\in\tilde{\Lambda}^i\}$ with $\tilde{P}$-measure 1.

Consequently, we have, with $\tilde{P}$-measure 1,

$$B^{(i)}(t)=\tilde{B}^{(i)}(t)-\int_0^N\mathrm{d}m^i(s)\sum_{k=1}^{N}\int_0^s k_k^i(s,v)\mathrm{d}\tilde{B}^{(k)}(v)$$
$$-\int_0^t\mathrm{d}m^i(s)\sum_{t_j\leq s}\sum_{k=1}^{L_j}a_k^{i,j}(s)\tilde{B}_{t_j}^k. \tag{10}$$

By means of (9) in Lemma 8, we have

$$\tilde{\mathbf{E}}\Big(\Big(\int_0^t dm^i(s)\Big(\sum_{k=1}^N \int_0^s k_k^i(s,v)d\tilde{B}^{(k)}(v) + \sum_{t_j\le s}\sum_{k=1}^{L_j}(a_k^{i,j}(s)\tilde{B}_{t_j}^k)\Big)\Big)^2\Big) =$$
$$= \int_0^t dm^i(s)\Big(\sum_{k=1}^N \int_0^s (k_k^i(s,v))^2 dm^k(v) + \sum_{t_j\le s}\sum_{k=1}^{L_j}(a_k^{i,j}(s))^2\Big) < \infty. \tag{11}$$

Also, by means of Lemma 10, we have the following representation for $\phi_j^l(\omega)$ appeared in $B_{t_j}^l = (1+\gamma_j^l)^{-1}(\tilde{B}_{t_j}^l - \phi_j^l(\omega))$,

$$\phi_j^l(\omega) = \sum_{k=1}^N \int_0^{t_j} k_{j,k}^l(v)d\tilde{B}^{(k)}(v) + \sum_{t_m<t_j}\sum_{k=1}^{L_m} a_{j,k}^{l,m}\tilde{B}_{t_m}^k$$
$$+ \sum_{k=1}^{l-1} a_{j,k}^{l,j}\tilde{B}_{t_j}^k \tag{12}$$

Since $\sum_{t_j\in\Lambda}\sum_{l=1}^{L_j}\tilde{\mathbf{E}}((\phi_j^l(\omega))^2) < \infty$ by (5), we have,

$$\sum_{t_j\in\Lambda}\sum_{l=1}^{L_j}\Big(\sum_{k=1}^N \int_0^{t_j}(k_{j,k}^l(v))^2 dm^k(v) + \sum_{t_m<t_j}\sum_{k=1}^{L_m}(a_{j,k}^{l,m})^2$$
$$+ \sum_{k=1}^{l-1}(a_{j,k}^{l,j})^2)\Big) < \infty. \tag{13}$$

Let

$$\mathrm{L}^2 = \bigoplus\sum_{i=l}^N L^2(m^i)\bigoplus\sum_{t_j\in\Lambda}\sum_{l=1}^{L_j} L_j^l;$$

where

$$L^2(m^i) = \Big\{f;\, f \text{ is measurable and } \int_0^1 (f(x))^2 dm^i(x) < \infty\Big\},$$

$$L_j^1 = \{f_j^1(t);\, f_j^1(t) = \chi_{[t_j,1)}(t)\alpha,\ \alpha \text{ is a real number}\}$$

and

$$L_j^l = \{f_j^l(\cdot);\, f_j^l(t) = \chi_{(t_j,1]}(t)\alpha,\ \alpha \text{ is a real number}\},\ (l = 2, 3, \ldots, L_j).$$

An element $\mathfrak{f}$ of L^2 will be denoted by

$$\mathfrak{f} = \Big\{f^1(t), \ldots, f^N(t); f_1^1(t)\ldots f_1^{L_1}(t), f_2^1(t), \ldots, f_2^{L_2}(t), \ldots,\Big\},$$

where $f^i \in L^2(m^i)$, $f_j^l \in L_j^l$.

Lemma 11. *Let K be a linear transformation from L^2 into itself which is defined by $\mathfrak{g} = (I-K)\mathfrak{f}$, in component wise,*

$$g^i(t) = f^i(t) - \sum_{k=1}^N \int_0^t k_k^i(t,v)f^k(v)dm^k(v)$$
$$- \sum_{t_j\le t}\sum_{k=1}^{L_j} a_k^{i,j}(t)f_j^k(t) \qquad (i = 1, 2, \ldots, N),$$

$$g_j^l(t) = f_j^l(t) - \frac{1}{(1+\gamma_j^l)}\Big\{\gamma_j^l f_j^l(t) + \chi_{(t_j,1]}(t)\Big[\sum_{k=1}^{N}\int_0^{t_j} k_{j,k}^l(v) f^k(v)\mathrm{d}m^k(v)$$

$$+ \sum_{t_m<t_j}\sum_{k=1}^{L_m} a_{j,k}^{l,m} f_m^{(k)}(t_j)$$

$$+ \sum_{k=1}^{l-1} a_{j,k}^{l,j} f_j^{(k)}(t_j)\Big]\Big\} \quad (t_i \in \Lambda, l = 1, \ldots, L_j),$$

where for $l = 1$, we put $\chi_{[t_j,1)}(t)$ in place of $\chi_{(t_j,1]}(t)$.

Then, there exist a unique resolvent operator L *such that* $\mathfrak{f} = (I - L)\mathfrak{g}$; *in component wise,*

$$f^i(t) = g^i(t) - \sum_{k=1}^{N}\int_0^t l_k^i(t, v) g^k(v) m^k(\mathrm{d}v)$$

$$- \sum_{t_j \le t}\sum_{l=1}^{L_j} c_l^{i,j}(t) g_j^l(t) \qquad (i = 1, 2, \ldots, N)$$

$$f_j^l(t) = (1+\gamma_j^l)\Big\{g_j^l(t) - \chi_{(t_j,1]}(t) \times$$

$$\times \Big[\sum_{k=1}^{N}\int_0^{t_j} l_{j,k}^l g^k(t)\mathrm{d}m^k(t) + \sum_{t_m<t_j}\sum_{k=1}^{L_j} c_{j,k}^{l,m} g_m^k(t)$$

$$+ \sum_{k=1}^{l-1} c_{j,k}^{l,j} g_j^k(t)\Big]\Big\}$$

where for $l = 1$, we must replace $\chi_{(t_j,1]}(t)$ by $\chi_{[t_j,1)}(t)$.

Proof. Proof can be done in the same way as in case of K being a Volterra operator.

Corollary 1.

$$k_k^i(u, v) + l_k^i(u, v) - \int_v^u \sum_{l=1}^{N} l_l^i(u, w) k_k^l(v, w)\mathrm{d}m^k(w)$$

$$- \sum_{t_j \le u}\sum_{l=1}^{L_j} c_l^{i,j}(u)(1+\gamma_j^l)^{-1}\chi_{[0,t_j)}(v) k_{j,k}^l(v) = 0; \tag{14}$$

$$a_k^{i,j}(u) + \int_{t_j}^u \sum_{l=1}^{N} l_l^i(u, v) a_k^{l,j}(v)\mathrm{d}m^l(v) - c_k^{i,j}(u)(1+\gamma_j^k)^{-1}$$

$$+ \sum_{\{m;\, t_j<t_m\le u\}}\sum_{l=1}^{L_m} c_l^{i,m}(u)(1+\gamma_m^l)^{-1} a_{m,k}^{l,j} \tag{15}$$

$$+ \sum_{l=k+1}^{L_j} c_l^{i,j}(u)(1+\gamma_j^l)^{-1} a_{j,k}^{l,j} = 0;$$

$$
\begin{aligned}
&-(1+\gamma_j^l)^{-1}k_{j,k}^{\,l}(v)-l_k^{j,l}(v)+\sum_{h=1}^{N}\int_v^{t_j} l_{j,h}^{l}(u)k_k^{h}(u,v)\mathrm{d}m^h(u)\\
&\quad+\sum_{\{m;t_m<t_j\}}\sum_{h=1}^{L_j} c_{j,h}^{l,m}(1+\gamma_m^h)^{-1}k_{j,k}^{\,l}(v)\\
&\quad+\sum_{h=1}^{l-1} c_{j,h}^{l,j}(1+\gamma_j^h)^{-1}k_{j,k}^{\,h}(v)=0;
\end{aligned}
\tag{16}
$$

$$
\begin{aligned}
&-(1+\gamma_j^l)^{-1}a_{j,k}^{l,m}+\sum_{h=1}^{N}\int_{t_m}^{t_j} l_{j,h}^{l}(v)a_h^{j,m}(v)\mathrm{d}m^h(v)\\
&\quad-c_{j,k}^{l,m}(1+\gamma_m^k)^{-1}+\sum_{h=k+1}^{L_j} c_{j,k}^{l,m}(1+\gamma_m^k)^{-1}a_{m,k}^{h,m}\\
&\quad+\sum_{h=1}^{l-1} c_{j,h}^{l,j}(1+\gamma_j^h)^{-1}a_{j,k}^{h,m}=0;
\end{aligned}
\tag{17}
$$

$$
\begin{aligned}
&-(1+\gamma_j^l)^{-1}a_{j,k}^{l,j}-c_{j,k}^{l,j}(1+\gamma_j^k)^{-1}\\
&\quad+\sum_{h=k+1}^{l-1} c_{j,h}^{l,j}(1+\gamma_j^h)^{-1}a_{j,k}^{h,j}=0
\end{aligned}
\tag{18}
$$

$$(i=1,2,\ldots,N,\ k=1,\ldots,l-1),\ t_j\in\Lambda).$$

By using Corollary 1 of Lemma 11, we can obtain a representation of $\{\tilde{B}^{(i)}(t), i=1,2,\ldots,N;\ \tilde{B}_{t_j}^{l},\ t_j\in\Lambda, l=1,2,\ldots,L_j\}$.

Firstly, by (10) and (12), we have

$$
\begin{aligned}
&B^{(i)}(t)-\int_0^t \mathrm{d}m^i(u)\int_0^u\sum_{k=1}^{N} l_k^i(u,v)\mathrm{d}B^k(v)\\
&\qquad\int_0^t \mathrm{d}m^i(u)\Big(\sum_{t_j\leq u}\sum_{k=1}^{L_j} c_l^{i,j}(u)B_{t_j}^{l}\Big)\\
&\quad=\tilde{B}^{(i)}(t)-\int_0^t \mathrm{d}m^i(u)\int_0^u\sum_{k=1}^{N} k_k^i(u,v)\mathrm{d}\tilde{B}^{(k)}(v)\\
&\qquad-\int_0^t \mathrm{d}m^i(u)\Big(\sum_{t_j\leq u}\sum_{k=1}^{L_j} a_k^{i,j}(u)\tilde{B}_{t_j}^{k}\Big)\\
&\qquad-\int_0^t \mathrm{d}m^i(u)\int_0^u\sum_{k=1}^{N} l_k^i(u,v)\mathrm{d}\tilde{B}^{(k)}(v)\\
&\qquad+\int_0^t \mathrm{d}m^i(u)\int_0^u\sum_{k=1}^{N} l_k^i(u,v)\mathrm{d}m^{(k)}(v)\sum_{l=1}^{N}\int_0^v k_l^k(v,w)\mathrm{d}B^{(l)}(w)\\
&\qquad+\int_0^t \mathrm{d}m^i(u)\int_0^u\sum_{k=1}^{N} l_k^i(u,v)\mathrm{d}m^k(v)\sum_{t_j\leq v}\sum_{l=1}^{L_j} a_l^{k,j}(v)\tilde{B}_{t_j}^{l}
\end{aligned}
$$

$$-\int_0^t \mathrm{d}m^i(u)\Big(\sum_{t_j\leq u}\sum_{l=1}^{L_j} c_l^{i,j}(u)\Big)(1+\gamma_j^l)^{-1}\times$$

$$\times\Big[\tilde{B}_{t_j}^l-\sum_{k=1}^N\int_0^{t_j} k_{j,k}^l(v)\mathrm{d}\tilde{B}^{(k)}(v)-\sum_{t_m<t_j}\sum_{k=1}^{L_m} a_{j,k}^{l,m}\tilde{B}_{t_m}^k-\sum_{k=1}^{l-1} a_{j,k}^{l,j}\tilde{B}_{t_j}^k\Big].$$

By (14) of Corollary 1 of Lemma 11, we have

$$\int_0^t \mathrm{d}m^i(u)\sum_{k=1}^N\int_0^u \mathrm{d}\tilde{B}^{(k)}(v)\Big[k_k^i(u,v)+l_k^i(u,v)$$

$$-\int_v^u\sum_{l=1}^N l_l^i(u,w)k_k^l(w,v)\mathrm{d}m^k(w)$$

$$-\sum_{t_j\leq u}\sum_{l=1}^{L_j} c_l^{i,j}(u)(1+\gamma_j^l)^{-1}\chi_{[0,t_j)}(v)k_{j,k}^l(v)\Big]=0.$$

Also, by (15) we have

$$\int_0^t \mathrm{d}m^i(u)\Big(\sum_{t_j\leq u}\sum_{k=1}^{L_j} a_k^{i,j}(u)\tilde{B}_{t_j}^k+\int_0^u\sum_{k=1}^N l_k^i(u,v)\mathrm{d}m^{(k)}(v)\sum_{t_j\leq u}\sum_{k=1}^{L_j} a_l^{k,j}(v)\tilde{B}_{t_j}^l$$

$$-\sum_{\{j\,;\,t_j\leq u\}}\sum_{l=1}^{L_j} c_l^{i,j}(u)(1+\gamma_j^l)^{-1}\Big\{\tilde{B}_{t_j}^l-\sum_{\{m\,;\,t_m<t_j\}}\sum_{k=1}^{L_m} a_{j,k}^{l,m}\tilde{B}_{t_m}^k-\sum_{k=1}^{l-1} a_{j,k}^{l,j}\tilde{B}_{t_j}^k\Big\}\Big)$$

$$=\int_0^t \mathrm{d}m^i(u)\sum_{\{j\,;\,t_j\leq u\}}\sum_{k=1}^{L_j}\Big[a_k^{i,j}(u)+\int_{t_j}^u\sum_{l=1}^N l_l^i(u,v)a_k^{l,j}(v)\mathrm{d}m^l(v)$$

$$-c_k^{i,j}(u)(1+\gamma_j^k)^{-1}$$

$$+\sum_{\{m\,;\,t_j<t_m\leq u\}}\sum_{l=1}^{L_m} c_l^{i,m}(u)(1+\gamma_m^l)^{-1}a_{m,k}^{l,j}$$

$$+\sum_{l=k+1}^{L_j} c_l^{i,j}(u)(1+\gamma_j^l)^{-1}\Big]\tilde{B}_{t_j}^k=0.$$

Therefore, we have

$$\tilde{B}^{(i)}(t)=B^{(i)}(t)-\int_0^t \mathrm{d}m^i(u)\int_0^u\sum_{k=1}^N l_k^i(u,v)\mathrm{d}B^k(v)$$
$$-\int_0^t \mathrm{d}m^i(u)\Big(\sum_{t_j\leq u}\sum_{k=1}^{L_j} c_l^{i,j}(u)B_{t_j}^l\Big). \tag{19}$$

Again by (10) and (12), we have

$$B_{t_j}^l-\sum_{k=1}^N\int_0^{t_j} l_{j,k}^l(v)\mathrm{d}B^{(k)}(v)-\sum_{\{m\,;\,t_m<t_j\}}\sum_{k=1}^{L_j} c_{j,k}^{l,m}B_{t_m}^k$$

$$-\sum_{k=1}^{l-1} c_{j,k}^{l,j}B_{t_j}^k$$

$$= (1+\gamma_j^l)^{-1}\Big(\tilde{B}_{t_j}^l - \sum_{k=1}^{N}\int_0^{t_j} k_{j,k}^l(v)\mathrm{d}\tilde{B}^{(k)}(v)$$

$$- \sum_{\{m;t_m<t_j\}}\sum_{k=1}^{L_m} a_{j,k}^{l,m}\tilde{B}_{t_m}^k - \sum_{k=1}^{l-1} a_{j,k}^{l,j}\tilde{B}_{t_j}^k\Big)$$

$$- \sum_{k=1}^{N}\int_0^{t_j} l_{j,k}^l(v)\mathrm{d}\tilde{B}^{(k)}(v) +$$

$$+ \sum_{k=1}^{N}\int_0^{t_j} l_{j,k}^l(v)\mathrm{d}m^k(v)\sum_{k=1}^{N}\int_0^{v} k_h^k(v,u)\mathrm{d}\tilde{B}^{(h)}(u)$$

$$+ \sum_{k=1}^{N}\int_0^{t_j} l_{j,k}^l(v)\mathrm{d}m^k(v)\sum_{t_m\leq v}\sum_{h=1}^{L_m} a_h^{j,m}(v)\tilde{B}_{t_m}^h$$

$$- \sum_{\{m,t_m<t_j\}}\sum_{k=1}^{L_j} c_{j,k}^{l,m}(1+\gamma_m^k)^{-1}\tilde{B}_{t_m}^k$$

$$+ \sum_{\{m;t_m<t_j\}}\sum_{k=1}^{L_j} c_{j,k}^{l,m}(1+\gamma_m^k)^{-1}\sum_{h=1}^{N}\int_0^{t_j} k_{j,h}^l(v)\mathrm{d}\tilde{B}^{(h)}(v)$$

$$+ \sum_{\{m;t_m<t_j\}}\sum_{k=1}^{L_j} c_{j,k}^{l,m}(1+\gamma_m^k)^{-1}\sum_{h=1}^{k-1} a_{m,h}^{k,m}\tilde{B}_{t_m}^{(h)}$$

$$- \sum_{k=1}^{l-1} c_{j,k}^{l,j}(1+\gamma_j^k)^{-1}\tilde{B}_{t_j}^k$$

$$+ \sum_{k=1}^{l-1} c_{j,k}^{l,j}(1+\gamma_j^k)^{-1}\sum_{h=1}^{N}\int_0^{t_j} k_{j,h}^k(v)\mathrm{d}\tilde{B}^{(h)}(v)$$

$$+ \sum_{k=1}^{l-1} c_{j,k}^{l,j}(1+\gamma_j^k)^{-1}\sum_{t_m<t_j}\sum_{h=1}^{L_m} a_{j,h}^{k,m}\tilde{B}_{t_m}^h$$

$$+ \sum_{k=1}^{l-1} c_{j,k}^{l,j}(1+\gamma_j^k)^{-1}\sum_{h=1}^{k-1} a_{j,h}^{k,j}\tilde{B}_{t_j}^h$$

By (16), we have

$$\sum_{k=1}^{N}\int_0^{t_j}\Big(-(1+\gamma_j^l)^{-1}k_{j,k}^l(v) - l_{j,k}^l(v)$$

$$+ \sum_{h=1}^{N}\int_v^{t_j} l_{j,h}^l(u)k_k^h(u,v)\mathrm{d}m^h(u)$$

$$+ \sum_{\{m;t_m<t_j\}}\sum_{h=1}^{L_j} c_{j,h}^{l,m}(1+\gamma_m^h)^{-1}k_{j,k}^l(v)$$

$$+ \sum_{h=1}^{l-1} c_{j,h}^{l,j}(1+\gamma_j^h)^{-1}k_{j,k}^h(v)\Big)\mathrm{d}\tilde{B}^{(k)}(v) = 0.$$

By (17), we have

$$\sum_{\{m;t_m<t_j\}}\sum_{k=1}^{L_j}\Big[-(1+\gamma_j^l)^{-1}a_{j,k}^{l,m}+\sum_{h=1}^{N}\int_{t_m}^{t_j} l_{j,h}^{l}(v)a_{k}^{j,m}(v)\mathrm{d}m^h(v)$$

$$-c_{j,k}^{l,m}(1+\gamma_m^k)^{-1}+\sum_{h=k+1}^{L_j} c_{j,h}^{l,m}(1+\gamma_m^h)^{-1}a_{m,k}^{h,m}+$$

$$+\sum_{h=1}^{l-1} c_{j,h}^{l,j}(1+\gamma_j^h)^{-1}a_{j,k}^{h,m}\Big]\tilde{B}_{t_m}^{k}=0.$$

By (18), we have

$$\sum_{k=1}^{l-1}\Big(-(1+\gamma_j^l)^{-1}a_{j,k}^{l,j}-c_{j,k}^{l,j}(1+\gamma_j^k)^{-1}$$

$$+\sum_{h=k+1}^{l-1} c_{j,h}^{l,j}(1+\gamma_j^h)^{-1}a_{j,k}^{h,j}\Big)\tilde{B}_{t_j}^{k}=0.$$

Therefore, we have

$$\begin{aligned}(1+\gamma_j^l)^{-1}\tilde{B}_{t_j}^{1} &= B_{t_j}^{l}-\sum_{k=1}^{N}\int_0^{t_j} l_{j,k}^{l}(v)\mathrm{d}B^{(k)}(v)\\ &\quad -\sum_{\{m;t_m<t_j\}}\sum_{k=1}^{L_j} c_{j,k}^{l,m}B_{t_m}^{k}-\sum_{k=1}^{l-1} c_{j,k}^{l,j}B_{t_j}^{k}.\end{aligned} \tag{20}$$

By gathering (2), (19) and (20), we have (3). The other parts of the theorem are obvious.

References

[1] Fernique, X. (1971). Régularitẽ de processus Gaussiens, *Invent. Math.*, **12**, 304–320.

[2] Girsanov, I.V. (1960). On transforming a certain class of stochastic process by absolutely continuous substitution of measures, *Ther. Probability App.*, **5**, 285–301.

[3] Gohberg, I.C. and Krein, M.G. (1970). *Theory and Applications of Volterra operators in Hilbert spaces* (English Translation, American Mathematical Society, Providence, Rhode Island).

[4] Hida, T. (1960). Canonical representations of Gaussian processes and their applications, *Mem. Coll. Sci. Univ. Kyoto*, **33**, 109–155.

[5] Hitsuda, M. (1968). Representation of Gaussian processes equivalent to Wiener process, *Osaka J. Math*, **5**, 299–312.

[6] Hitsuda, M. (1968). Representation of Gaussian processes equivalent to Wiener process and its applications, *Proceedings of the first symposium of USSR–Japan on probability theory*.

[7] Kallianpur, G. (1975). Canonical representations of equivalent Gaussian processes. In: *Stochastic Processes and Related Topics*, Volume I (M.L. Puri, editor). Academic Press Inc., New York.

[8] Kallianpur, G. and Oodaira, H. (1973). Non-anticipative representations of equivalent Gaussian Processes, *Ann. Prob.*, **1**, 104–122.

[9] Kunita, H. and Watanabe, S. (1967). On square integrable martingales, *Nagoya Math. J.*, **30**, 209–245.

[10] Liptzer, P.C. and Shiryaev, A.N. (1974). *Statistics of Stochastic Processes*. Nauka, Moscow.

[11] Rozanov, Yu. A. (1968). *Infinite-dimensional Gaussian Distributions*, Nauka, Moscow (English Translations, Proceedings of the Steklov Mathematical Institute, 108 (1971) Amer. Math. Soc.).

P.R. Krishnaiah, ed., *Multivariate Analysis–IV*
© North-Holland Publishing Company (1977) 267–281

A STOCHASTIC EQUATION FOR THE OPTIMAL NON-LINEAR FILTER*

G. KALLIANPUR**
University of Minnesota, Minneapolis, Minn., U.S.A.

The basic stochastic differential equation for the optimal non-linear filter derived in the paper of Fujisaki–Kallianpur–Kunita is generalized so as to include applications to problems involving discontinuous control.

1. Introduction

In recent years, a great deal of attention has been devoted to the theory of non-linear filtering, in particular, to the problem of deriving a stochastic differential equation for the optimal filter. (See the Bibliography in [2].) A general form of such an equation when the noise in the observation process model is the Wiener process has been obtained in the paper of Fujisaki, Kallianpur and Kunita [2]. The latter work was motivated by applications to control theory in which the signal and observation processes are governed by an Ito stochastic differential equation or by a more general stochastic equation studied by Ito and Nisio (see [2]).

The aim of the present paper is to show that the approach to filtering theory adopted in [2] is not limited to this kind of application and to generalize the main result (Theorem 4.1) of [2]. The purpose of the extended theory is to make it applicable to problems involving discontinuous control. The need for such an extension was first pointed out to me by Professor C. Striebel to whom I wish to express my thanks for conversations on this topic.

The principal result of this paper was announced without proof in [3]. Applications to problems of linear discontinuous stochastic control will be considered in a later paper.

* This work was completed while the author was Visiting Professor at the Departments of Mathematics and System Science, University of California at Los Angeles.
** Work supported in part by NSF Grant No. MPS 74–21872.

2. Observation process model and the innovation process

The system or signal process $x_t(\omega)$ taking values in a complete metric space S and the observation process $z_t(\omega)$ $(t \in [0, T])$ are assumed given on some complete probability space $(\Omega, \mathcal{A}, P)$ and further related as follows.

$$z_t(\omega) = \int_0^t h_u(\omega)\,\mathrm{d}u + w_t(\omega), \tag{1}$$

where

$$w_t(\omega) \text{ is an } N\text{-dimensional standard Wiener process} \tag{2}$$

$$h_t(\omega) \text{ is a } (t, \omega) \text{ measurable } N\text{-dimensional process such that } \int_0^T \mathbf{E}|h_t|^2 \mathrm{d}t \text{ is finite.} \tag{3}$$

Let us introduce the following family of σ-fields.

$$\mathcal{G}_t = \sigma\{x_s, w_s, s \leq t\}, \qquad \mathcal{N}_t^T = \sigma\{w_v - w_u, t \leq u \leq v \leq T\} \tag{4}$$

and

$$\mathcal{F}_t = \sigma\{z_s, s \leq t\}.$$

It will be assumed that the σ-fields $\mathcal{F}_t$ and $\mathcal{G}_t$ are augmented by the addition of all P-null sets. In the model (1) the information about the signal process is carried by (h_t) by means of the measurability assumption:

$$\text{For each } t,\ h_t \text{ is } \mathcal{G}_t \text{ measurable, i.e. } (h_t) \text{ is } (\mathcal{G}_t)\text{-adapted.} \tag{5}$$

In order to take into account applications involving stochastic control we make the further assumption that for every t the σ-fields

$$\mathcal{G}_t \text{ and } \mathcal{N}_t^T \text{ are stochastically independent.} \tag{6}$$

Clearly (6) includes the case when the signal (x_t) and noise (w_t) are completely independent.

The derivation of the desired stochastic equation rests on two important results proved in Fujisaki–Kallianpur–Kunita [2]. We state them below without proof. From the assumptions made above on (h_t) it can be shown that one can work with a modification of the conditional expectation $\mathbf{E}(h_t \mid \mathcal{F}_t)$ which is jointly measurable and $(\mathcal{F}_t)$-adapted. This particular modification will be henceforth denoted by $\hat{h}_t$.

Let us now define the process (ν_t) by

$$\nu_t = z_t - \int_0^t \hat{h}_s \, \mathrm{d}s. \tag{7}$$

Proposition 1. *$(\nu_t, \mathscr{F}_t, P)$ is an N-dimensional Wiener martingale. Furthermore $\mathscr{F}_t$ and $\sigma\{\nu_v - \nu_u;\ t < u < v \leq T\}$ are independent.*

(ν_t) is called the *innovation process.*

Proposition 2 (A martingale representation theorem). *Under conditions* (1), (2), (3), (5) *and* (6) *every separable square integrable martingale $(Y_t, \mathscr{F}_t, P)$ is sample continuous and has the Ito stochastic integral representation*

$$Y_t - \mathbf{E}(Y_0) = \int_0^t (\Phi_s, \mathrm{d}\nu_s), \tag{8}$$

where

$$\int_0^T \mathbf{E}|\Phi_s|^2 \mathrm{d}s < \infty \tag{9}$$

and (Φ_s) is jointly measurable and adapted to $(\mathscr{F}_s)$.

We list here some auxiliary results from Meyer [5], Chapter VII which will be used in the lemma proved below.

An increasing process (A_t) $(0 \leq t < \infty)$,
is called *natural* if (and only if) (10)

$$\mathbf{E}\int_0^t Y_s \mathrm{d}A_s = \mathbf{E}\int_0^t Y_{s-} \mathrm{d}A_s$$

for every positive, right continuous, uniformly integrable martingale (Y_t).

Let X_t be a potential generated by a
natural increasing process (A_t). (11)

Then

$$\mathbf{E}(A_\infty^2) = \mathbf{E}\int_0^\infty (X_t + X_{t-})\mathrm{d}A_t$$

in the sense that either both sides are $+\infty$ or both finite and equal.

If (Y_t) is a positive right continuous martingale
and (A_t) an increasing process, then for each t (12)

$$\mathbf{E}(A_t Y_t) = \mathbf{E}\int_0^t Y_s \mathrm{d}A_s$$

Let (A_t), (B_t) be two increasing processes and $(\mathcal{F}_t)$ a right continuous family. Then

$$\mathbf{E}(A_t - A_s \mid \mathcal{F}_s) = \mathbf{E}(B_t - B_s \mid \mathcal{F}_s) \quad \forall s, t \ (s \leq t)$$

implies

$$\mathbf{E}\int_0^t Y_s \mathrm{d}A_s = \mathbf{E}\int_0^t Y_s \mathrm{d}B_s \tag{13}$$

if (Y_t) is an $(\mathcal{F}_t)$-predictable process.

Since the family $(\mathcal{F}_t)$ in (4) cannot be assumed right continuous we have to employ an argument based on the notion of the dual projection introduced by Dellacherie and Meyer ([1], [4]).

Lemma 1. *Let (A_t), $(0 \leq t \leq T)$ be an increasing process and (A_t^*) its dual projection relative to $(\mathcal{F}_t)$. Then*

$$\mathbf{E}(A_T)^2 < \infty \implies \mathbf{E}(A_T^*)^2 < \infty. \tag{14}$$

Proof. Let $X_t = \mathbf{E}(A_T^* - A_t^* \mid \mathcal{F}_{t+})$ and $Y_t = \mathbf{E}(A_T - A_t \mid \mathcal{F}_{t+})$ be the potentials generated by (A_t^*) and (A_t) respectively. We recall that (A_t^*) is natural and is the unique (right continuous) natural process determined by (A_t) via the decomposition of (Y_t). Furthermore, (A_t^*) is $(\mathcal{F}_t)$-adapted and predictable. Since $X_t = Y_t$ for each t a.s. we have

$$\mathbf{E}\int_0^T X_t \mathrm{d}A_t^* = \mathbf{E}\int_0^T Y_t \mathrm{d}A_t^* \leq \mathbf{E}\int_0^T M_t \mathrm{d}A_t^*, \tag{15}$$

where M_t is the right continuous version of the martingale $\mathbf{E}(A_T \mid \mathcal{F}_{t+})$. Since $M_t \geq 0$, from (12) applied to the right side of the inequality in (15) we obtain

$$\begin{aligned}\mathbf{E}\int_0^T X_t \mathrm{d}A_t^* &\leq \mathbf{E}(M_T A_T^*) \\ &= \mathbf{E}\{\mathbf{E}(A_T \mid \mathcal{F}_T) A_T^*\} = \mathbf{E}(A_T A_T^*).\end{aligned} \tag{16}$$

(The right-hand side in (16) could be infinite.) Writing $\mathcal{G}_t$ for $\mathcal{F}_{t+}$ we have (note that the left limit X_{t-} exists because we consider such versions for the martingale involved and because (A_t^*) is increasing)

$$X_{t-} = \lim_{n\to\infty} X_{t-1/n} = \lim_{n\to\infty} Y_{t-1/n} \quad \text{(a.s.)},$$

so that

$$\mathbf{E}\int_0^T X_{t-}\,\mathrm{d}A^*_t = \mathbf{E}\int_0^T \left(\lim_{n\to\infty} Y_{t-1/n}\right)\mathrm{d}A^*_t$$

$$= \mathbf{E}\int_0^T \lim_{n\to\infty} \mathbf{E}(A_T - A_{t-1/n} \mid \mathcal{G}_{t-1/n})\,\mathrm{d}A^*_t$$

$$\leq \mathbf{E}\int_0^T \lim_{n\to\infty} \mathbf{E}(A_T \mid \mathcal{G}_{t-1/n})\,\mathrm{d}A^*_t$$

$$= \mathbf{E}\int_0^T M_{t-}\,\mathrm{d}A^*_t, \qquad (M_t \text{ being the martingale introduced above})$$

$$= \mathbf{E}\int_0^T M_t\,\mathrm{d}A^*_t$$

by (10) since (A^*_t) is natural. Hence as in (15) and (16) we have

$\mathbf{E}\int_0^T X_{t-}\,\mathrm{d}A^*_t \leq \mathbf{E}(A_T A^*_T)$, and we obtain the inequality

$$\mathbf{E}(A^*_T)^2 \leq 2\mathbf{E}(A_T A^*_T). \tag{17}$$

Let us first consider the case when (A_t) is bounded:

$$A_T(\omega) \leq C < \infty \quad \text{for all } \omega. \tag{18}$$

Then from (17) $\mathbf{E}(A^{*2}_T) < \infty$ and hence

$\mathbf{E}(A^*_T)^2 \leq 2\{\mathbf{E}(A^2_T)\,\mathbf{E}(A^{*2}_T)\}^{1/2}$ which gives

$$\mathbf{E}(A^*_T)^2 \leq 4\mathbf{E}(A^2_T). \tag{19}$$

It remains to show that $\mathbf{E}(A^*_T)^2 < \infty$ in the general case (i.e., with no boundedness assumptions on A_T). For every positive integer n define

$$A^n_t = A_t \wedge n$$

and set

$$Y^n_t = \mathbf{E}(A^n_T - A^n_t \mid \mathcal{F}_{t+}).$$

It is easy to see that $Y^n_t \to Y_t$ a.s (for each t) and that for each n, Y^n_t is a potential of class (D) and so has a decomposition of the form $\mathbf{E}(A^{*n}_T - A^{*n}_t \mid \mathcal{F}_{t+})$ where (A^{*n}_t) is a predictable increasing process. Since A^n_T is bounded, by the case already considered we have

$$\mathbf{E}(A^{*n}_T)^2 \leq 4\mathbf{E}(A^n_T)^2 \leq 4\mathbf{E}(A^2_T) \quad \text{for all } n.$$

It follows that (A_T^{*n}) is a uniformly integrable sequence and also there exists a subsequence $(A_T^{*n_k})$ which converges weakly in $L^2(\Omega)$, i.e., in the weak or $\sigma(L^2, L^2)$-topology. Let us denote the limit by $\hat{A}_T$. Writing $X_t^n = \mathbf{E}(A_{Tn}^* - A_t^{*n} \mid \mathcal{F}_{t+})$ and $Y_t^n = \mathbf{E}(A_T^n - A_t^n \mid \mathcal{F}_{t+})$ we have a.s. for each t, $X_t^n = Y_t^n$ and since $\lim_{n\to\infty} Y_t^n = \mathbf{E}(A_T - A_t \mid \mathcal{F}_{t+})$ a.s. it follows that $\lim_{n\to\infty} X_t^n$ exists a.s and equals $\mathbf{E}(A_T^* - A_t^* \mid \mathcal{F}_{t+})$. In fact, by the uniform integrability of $(A_T^n - A_t^n)$ the above $\lim_{n\to\infty} Y_t^n$ (hence also $\lim_{n\to\infty} X_t^n$) exists in the L^1-sense. Since $X_t^n = \mathbf{E}(A_T^{*n} \mid \mathcal{F}_{t+}) - A_t^{*n}$ and since the operator of conditional expectation $\mathbf{E}(\cdot \mid \mathcal{F}_{t+})$ on $L^2(\Omega)$ is continuous with respect to the weak topology it follows that (using the subsequence (n_k)) the weak $\lim_{k\to\infty} A_t^{*n_k}$ exists which we denote by $\hat{A}_t$. Now for all $M \in \mathcal{F}_{t+}$ we have

$$\int_M Y_t^{n_k} \,\mathrm{d}P = \int_M \mathbf{E}(A_T^{*n_k} \mid \mathcal{F}_{t+}) \,\mathrm{d}P - \int_M A_t^{*n_k} \,\mathrm{d}P.$$

Making $n_k \to \infty$ we obtain

$$\int_M Y_t \,\mathrm{d}P = \int_M \mathbf{E}(\hat{A}_T \mid \mathcal{F}_{t+}) \mathrm{d}P - \int_M \hat{A}_t \,\mathrm{d}P.$$

Here we have used the L^1 convergence of the uniformly integrable sequence $\{Y_t^n\}$ and the weak convergence of $\mathbf{E}(A_T^{*n} \mid \mathcal{F}_{t+})$ and A_t^{*n}. It is easy to see that $\hat{A}_t$ is $\mathcal{F}_{t+}$-measurable and predictable. Hence we have

$$\int_M Y_t \,\mathrm{d}P = \int_M \mathbf{E}(\hat{A}_T - \hat{A}_t \mid \mathcal{F}_{t+}) \mathrm{d}P \quad (\forall M \in \mathcal{F}_{t+}).$$

Thus, a.s.

$$\mathbf{E}(\hat{A}_T - \hat{A}_t \mid \mathcal{F}_{t+}) = \mathbf{E}(A_T - A_t \mid \mathcal{F}_{t+}) = \mathbf{E}(A_T^* - A_t^* \mid \mathcal{F}_{t+})$$

from which it follows that

$$\hat{A}_t = A_t^* \quad \text{a.s.} \quad \text{for each } t.$$

Since $\hat{A}_t \in L^2(\Omega)$ for each t we obtain $\mathbf{E}(A_T^*)^2 = \mathbf{E}(\hat{A}_T^2) < \infty$. This proves the lemma.

3. A stochastic equation for the general non-linear filtering problem

Let f be a real measurable function on S (the state space of the signal process (x_t)) such that

$$\mathbf{E}[f(x_t)]^2 < \infty \quad \text{for all } t \text{ in } [0, T]. \tag{1}$$

The function f is said to belong to the class $\mathscr{D}$ if there exists a jointly measurable, real process $B_t[f](\omega)$ adapted to $(\mathscr{G}_t)$ and having the following properties.

Almost all trajectories of the process $(B_t[f])$ are right continuous and of bounded variation over the interval $[0, T]$ with $B_0[f] = 0$. (2)

$$\mathbf{E}(\operatorname{Var} B[f])^2 < \infty \tag{3}$$

where $\operatorname{Var} B[f](\omega)$ is the total variation of the trajectory $B_t[f](\omega)$ $(0 \leq t \leq T)$.

The process $M_t(f) \equiv f(x_t) - \mathbf{E}[f(x_0)] - B_t[f]$ is a $(\mathscr{G}_t, P)$ martingale. (4)

Note that from conditions (1) and (3) it follows that $(M_t(f))$ is a square integrable martingale.

Theorem 1. *Let the conditions of Section 2 and (1)–(4) hold. Then for every f in $\mathscr{D}$ there exists a jointly measurable process $(\bar{B}_t[f])$ adapted to the family $(\mathscr{F}_t)$ such that almost all its trajectories are right continuous and of bounded variation over the interval $[0, T]$ with $\bar{B}_0[f] = 0$. Furthermore,*

$$\mathbf{E}(\operatorname{Var} \bar{B}[f])^2 < \infty, \tag{5}$$

and the process

$$\bar{M}_t(f) \equiv \mathbf{E}[f(x_t) \mid \mathscr{F}_t] - \mathbf{E}[f(x_0) \mid \mathscr{F}_0] - \bar{B}_t[f] \tag{6}$$

is a square integrable $(\mathscr{F}_t, P)$ martingale.

Proof. Since f is fixed we shall suppress it for the time being and write B_t for $B_t[f]$, etc. The process $(\bar{B}_t)$ is nothing but the "dual projection" or "compensator" introduced earlier associated with an increasing integrable process. Write $B_t = U_t - V_t$ where (U_t), (V_t) are increasing processes with right continuous trajectories and such that $U_0 = V_0 = 0$. From (3) we also have $\mathbf{E}(U_T^2) < \infty$ and $\mathbf{E}(V_T^2) < \infty$. The process $\xi_t = \mathbf{E}(U_T - U_t \mid \mathscr{F}_{t+})$ is a positive supermartingale of class (D). Hence it has a Doob decomposition $Y_t \quad \bar{U}_t$ where (Y_t) is a martingale and $\bar{U}_t$ is a (uniquely determined) predictable, integrable increasing process adapted to $(\mathscr{F}_{t+})$ with $\bar{U}_0 = 0$. Hence for $s < t$, we have $\mathbf{E}(U_t - U_s \mid \mathscr{F}_{s+}) = \mathbf{E}(\bar{U}_t - \bar{U}_s \mid \mathscr{F}_{s+})$ which gives

$$\mathbf{E}(U_t - U_s \mid \mathcal{F}_s) = \mathbf{E}(\bar{U}_t - \bar{U}_s \mid \mathcal{F}_s). \tag{7}$$

The predictability of $(\bar{U}_t)$ also implies that $\bar{U}_t$ is actually $\mathcal{F}_t$-measurable. Define the process $(\bar{V}_t)$ in a similar fashion and write $\bar{B}_t = \bar{U}_t - \bar{V}_t$. It is clear that $\bar{B}_0 = 0$, almost all trajectories of $(\bar{B}_t)$ are right continuous and of bounded variation over $[0, T]$ and further that $\mathbf{E}(\operatorname{Var} \bar{B}) = \mathbf{E}(\bar{U}_T) + \mathbf{E}(\bar{V}_T) < \infty$. The square integrability of $\operatorname{Var} \bar{B}$ follows from Lemma 1. The fact that $(\bar{M}_t(f), \mathcal{F}_t, P)$ is a martingale follows easily. For $s < t$,

$$\begin{aligned}
&\mathbf{E}\{\bar{M}_t(f) - \bar{M}_s(f) \mid \mathcal{F}_s\} \\
&\quad = \mathbf{E}\{f(x_t) - f(x_s) \mid \mathcal{F}_s\} - \mathbf{E}(\bar{B}_t - \bar{B}_s \mid \mathcal{F}_s) \\
&\quad = \mathbf{E}\{M_t(f) - M_s(f) \mid \mathcal{F}_s\} + \mathbf{E}(B_t - B_s \mid \mathcal{F}_s) - \mathbf{E}(\bar{B}_t - \bar{B}_s \mid \mathcal{F}_s) \\
&\quad = 0 \qquad \text{from (4) and (7).}
\end{aligned}$$

Next we need the following two lemmas.

Lemma 2. *Let $(M_t(f), \mathcal{G}_t, P)$ be the square integrable martingale of* (4). *Then there exist unique sample continuous processes $\langle M(f), w^i\rangle$ $(i = 1, \ldots, N)$ adapted to $\mathcal{G}_t$ such that almost all sample functions are of bounded variation and $M_t(f) w_t^i - \langle M(f), w^i\rangle_t$ are $\mathcal{G}_t$-martingales. Furthermore each $\langle M(f), w^i\rangle_t$ has the following properties: It is absolutely continuous with respect to Lebesgue measure in $[0, T]$. There exists a modification of the Radon–Nikodym derivative which is (t, ω)-measurable and adapted to $(\mathcal{G}_t)$ and which we shall denote by $\tilde{D}_t^i f(\omega)$. Then using the vector notation $\tilde{D}_t f = (\tilde{D}_t^1 f, \ldots, \tilde{D}_t^N f)$,*

$$\langle M(f), w\rangle_t = \int_0^t \tilde{D}_s f \, \mathrm{d}s \quad \text{a.s.} \tag{8}$$

where

$$\int_0^T \mathbf{E} \mid \tilde{D}_s f \mid^2 \mathrm{d}s < \infty. \tag{9}$$

If the processes (x_t) and (w_t) are completely independent, then a.s.

$$\langle M(f), w\rangle_t = 0. \tag{10}$$

See [2] for the proof.

Lemma 3. *Let $Y(\omega) = \int_0^T f(s, \omega) \mathrm{d}\nu_s(\omega)$ where $\mathbf{E} Y^2 = \int_0^T \mathbf{E} f^2(s) \mathrm{d}s < \infty$ and let (B_t) be a stochastic process whose sample functions are of bounded variation in $[0, T]$ with $\operatorname{Var} B(\omega)$ denoting the total variation of $B_t(\omega)$. Write*

$$\gamma(\omega) = \operatorname{Var} B(\omega).$$

Let $\Pi_n = \{t_j^n\}$ *be a partition of* $[0, T]$. *Write*

$$\Delta_j Y = \int_{t_{j-1}}^{t_j} f(s)\,\mathrm{d}\nu_s \qquad \textit{and} \qquad \eta_n(\omega) = \max_{\Pi_n} |\Delta_j Y(\omega)|.$$

Let

$$\zeta_{\Pi_n} = \sum_{\Pi_n} \Delta_j B \Delta_j Y, \qquad \Delta_j B = B_{t_j} - B_{t_{j-1}}.$$

Suppose that $\mathbf{E}\gamma^2 < \infty$. *Then*

$$\lim_{|\Pi_n| \to 0} \mathbf{E}|\zeta_{\Pi_n}| = 0.$$

Proof. Let $Y_j = \int_0^{t_j} f(s)\,\mathrm{d}\nu_s$; $Y_0 = 0$ $(T_0 = 0)$, $t_n = T$, $Y_n = Y$. For $j \le n$,

$$\begin{aligned}
\mathbf{E}[Y_n \mid Y_1, \ldots, Y_j] &= \mathbf{E}\left[Y_j + \int_{t_j}^T f(s)\,\mathrm{d}\nu_s \mid Y_1, \ldots, Y_j \right] \\
&= Y_j + \mathbf{E}\left\{ \mathbf{E}\left[\left[\int_{tj}^T f(s)\,\mathrm{d}\nu_s \mid \mathcal{F}_{t_j} \right] \right| Y_1, \ldots, Y_j \right\} \\
&= Y_j. \qquad (\text{Here } \mathcal{F}_t = \sigma\{z_u,\ 0 \le u \le t\}.)
\end{aligned}$$

Hence from a well-known inequality

$$\mathbf{E}(\max |Y_j|)^2 \le 4\mathbf{E}|Y_n|^2 = 4 \int_0^T \mathbf{E}\{f^2(s)\}\,\mathrm{d}s = k, \text{ say.}$$

Thus

$$\eta_n = \max_j |\Delta_j Y| = \max_j |Y_j - Y_{j-1}| \le 2 \max_j |Y_j|$$

so that

$$\int_{\{\eta_n \ge A\}} \eta_n\,\mathrm{d}P \le \frac{\mathbf{E}(\eta_n^2)}{A} \le \frac{K}{A}, \qquad (K = 4k). \tag{11}$$

Next, let $\varepsilon > 0$ be arbitrary. Choose $C = C_\varepsilon$ positive such that

$$\int_{[\gamma^2 > C^2]} \gamma^2\,\mathrm{d}P < \frac{\varepsilon^2}{4K}.$$

Now

$$\begin{aligned}
|\zeta_{\Pi_n}| &\le \sum_{\Pi_n} |\Delta_j B|\,|\Delta_j Y| \\
&\le \max_j |\Delta_j Y| \cdot \sum_{\Pi_n} |\Delta_j B| \\
&\le \eta_n \sup_{\Pi} \left(\sum_{\Pi} |\Delta_j B| \right) \\
&= \eta_n \operatorname{Var}(B) = \eta_n \cdot \gamma.
\end{aligned}$$

Hence

$$\int_{[|\zeta_{\Pi_n}|\geq N]} |\zeta_{\Pi_n}|\,\mathrm{d}P \leq \int_{[\eta_n \geq N/C]\cap[\gamma\leq C]} \eta_n\,\gamma\,\mathrm{d}P + \int_{[\gamma>C]} \eta_n\gamma\,\mathrm{d}P. \tag{12}$$

The first term on RHS of (12)

$$\leq C\int_{[\eta_n\geq N/C]} \eta_n\,\mathrm{d}P < \varepsilon/2 \quad \text{for } N\geq N(\varepsilon), \text{ independent of } n.$$

This is done by taking $A = N/C$ in (11) and $N(\varepsilon)$ such that $KC_\varepsilon/N(\varepsilon) < \varepsilon/2$. The second term on RHS of (12)

$$= \int \eta_n\gamma I_{[\gamma>C]}\,\mathrm{d}P$$

$$\leq \left\{\mathbf{E}(\eta_n^2)\int_{[\gamma^2>C^2]} \gamma^2\mathrm{d}P\right\}^{1/2} \leq K^{1/2}\left(\frac{\varepsilon^2}{4K}\right)^{1/2} = \frac{\varepsilon}{2}.$$

Thus there exists $N(\varepsilon)$ such that $N\geq N(\varepsilon)$ implies

$$\int_{[|\zeta_{\Pi_n}|\geq N]} |\zeta_{\Pi_n}|\,\mathrm{d}P < \varepsilon \quad \text{for all } n,$$

i.e., $\{\zeta_{\Pi_n}\}$ is P-uniformly integrable. Since by assumption $\operatorname{Var} B < \infty$ a.s. P and

$$\eta_n = \max_{\Pi_n}\left|\int_{t_j}^{t_j'} f(s)\,\mathrm{d}\nu_s\right| \to 0 \quad (\text{a.s. } P) \quad \text{as } |\Pi_n|\to 0$$

if we choose $\int_0^t f(s)\,\mathrm{d}\nu_s$ to be sample continuous, we have $\zeta_{\Pi_n}\to 0$ a.s. P ($|\Pi_n|\to 0$). The uniform integrability of $\{\zeta_{\Pi_n}\}$ then implies $\mathbf{E}|\zeta_{\Pi_n}|\to 0$.

Looking back over our discussion so far what we have shown may be summarized as follows:

Proposition 3. *For $f\in\mathcal{D}$, let the process $(f(x_t))$ be a separable square integrable $(\mathcal{G}_t, P)$ semi-martingale with the form*

$$f(x_t) = f(x_0) + M_t(f) + B_t(f).$$

Then the conditional expectation process (writing $\mathbf{E}^t$ for $\mathbf{E}(\cdot\,|\,\mathcal{F}_t)$)$\{\mathbf{E}^t(f(x_t))\}$ is a square integrable $(\mathcal{F}_t, P)$ semi-martingale of the form

$$\mathbf{E}^t[f(x_t)] = \mathbf{E}[f(x_0)] + \int_0^t (\Phi_s, \mathrm{d}\nu_s) + \bar{B}_t(f).$$

We are now able to derive the basic result of this paper which describes the time evolution of the optimal non-linear filter. For this it is necessary to discover the particular form that Φ_s takes.

Theorem 2. *Assume conditions* (2.1)–(2.4). *If* $f \in \mathscr{D}$ *satisfies*

$$\int_0^T \mathbf{E}|f(x_t)h_t|^2 \, \mathrm{d}t < \infty,$$

then $\mathbf{E}^t[f(x_t)]$ *satisfies the stochastic equation*

$$\mathbf{E}^t[f(x_t)] = \mathbf{E}[f(x_0)] + \bar{B}_t(f) + \int_0^t ([\mathbf{E}^s(f(x_s)h_s) - \mathbf{E}^s(f(x_s))\mathbf{E}^s(h_s) + \mathbf{E}^s(\tilde{D}_s f)], \mathrm{d}\nu_s). \tag{13}$$

Proof. Equation (13) is equivalent to saying that $\bar{M}_t(f)$ equals the stochastic integral on the right-hand side which we denote by $M_t^*(f)$. The method of proof is to show $\mathbf{E}[\bar{M}_t(f)Y_t] = \mathbf{E}[M_t^*(f)Y_t]$ for all Y_t such that

$Y_t = \int_0^t (\Phi_s, \mathrm{d}\nu_s)$ (Φ_s bounded) since such Y_t are dense in $\mathrm{L}^2(\mathscr{F}_t, P)$

(up to constants) by virtue of Proposition 2. [Note that the Φ_s here is not to be confused with the same symbol in Proposition 3.] Since f will be fixed throughout the argument we suppress it and write B_t, M_t, $\bar{B}_t$, $\bar{M}_t$ and M_t^*. Also for ease of writing we shall consider the scalar case (i.e. $N = 1$). We calculate $\mathbf{E}(M_tY_t)$ and $\mathbf{E}[(\bar{M}_t - M_t)Y_t]$ separately. The details of each calculation are given in several steps and need no additional explanation.

$$\mathbf{E}(M_tY_t) = \mathbf{E}([f(x_t) - f(x_0) - B_t]Y_t) = \mathbf{E}\left[M_t\int_0^t \Phi_s \mathrm{d}w_s\right] + \mathbf{E}\left[f(x_t)\int_0^t g_s \mathrm{d}s\right] - \mathbf{E}\left[B_t\int_0^t g_s \mathrm{d}s\right] \tag{I}$$

where we set

$$g_s = \Phi_s(h_s - \hat{h}_s).$$

Consider the right side of I:

$$\text{1st term} = \mathbf{E}\left(\int_0^t \Phi_s \, \mathrm{d}\langle M, w\rangle_s\right)$$

$$= \mathbf{E}\int_0^t \Phi_s\tilde{D}_s f \, \mathrm{d}s \qquad \text{(by Lemma 2)}$$

$$= \mathbf{E}\left[Y_t\int_0^t \mathbf{E}^s(\tilde{D}_s f)\, \mathrm{d}\nu_s\right].$$

$$\text{2nd term} = \mathbf{E}\int_0^t f(x_t)g_s\,ds$$

$$= \int_0^t (\mathbf{E}\{[f(x_t) - f(x_s)]g_s\} + \mathbf{E}\{f(x_s)g_s\})ds)$$

$$= \int_0^t \mathbf{E}[f(x_s)g_s]\,ds + \int_0^t \mathbf{E}\{(M_t - M_s + B_t - B_s)g_s\}\,ds.$$

Now

$$\begin{aligned}\mathbf{E}\{(M_t - M_s + B_t - B_s)g_s\} &= \mathbf{E}(g_s\mathbf{E}\{M_t - M_s + B_t - B_s \mid \mathcal{G}_s\})\\ &= \mathbf{E}[g_s\mathbf{E}(B_t \mid \mathcal{G}_s)] - \mathbf{E}(g_sB_s)\\ &= \mathbf{E}(B_tg_s) - \mathbf{E}(B_sg_s).\end{aligned}$$

Hence

$$\mathbf{E}\int_0^t f(x_t)g_s\,ds = \mathbf{E}\int_0^t f(x_s)g_s\,ds + \mathbf{E}\Big(B_t\int_0^t g_s\,ds\Big) - \mathbf{E}\int_0^t B_sg_s\,ds,$$

and so

$$\mathbf{E}(M_tY_t) = \mathbf{E}\Big(Y_t\Big[\int_0^t \mathbf{E}^s(\tilde{D}_sf)d\nu_s\Big]\Big) + \mathbf{E}\Big(\int_0^t f(x_s)g_s\,ds\Big) - \mathbf{E}\Big(\int_0^t B_sg_s\,ds\Big).$$

The second term on the right equals

$$\mathbf{E}\int_0^t (\Phi_s\mathbf{E}^s[f(x_s)(h_s - \hat{h}_s)])d\nu_s = \mathbf{E}\Big(Y_t \cdot \int_0^t \mathbf{E}^s[f(x_s)(h_s - \hat{h}_s)]d\nu_s\Big).$$

Hence

$$\mathbf{E}(M_tY_t) = \mathbf{E}(M_t^*Y_t) - \mathbf{E}\Big(\int_0^t B_sg_s\,ds\Big). \tag{14}$$

Next,

$$\mathbf{E}[(\bar{M}_t - M_t)Y_t] = \mathbf{E}([\mathbf{E}^t(f(x_t)) - f(x_t) + \bar{B}_t - B_t]Y_t) \tag{II}$$

$$= \mathbf{E}[(\bar{B}_t - B_t)Y_t]. \tag{15}$$

Now introduce the following notation. $\Pi = \{t_j\}$ is a finite partition of $[0, t]$. Actually Π represents a sequence Π_n for which eventually $|\Pi_n| = \max_j (t_{j+1}^n - t_j^n) \to 0$ as $n \to \infty$. Set $\Delta_jB = B_{t_{j+1}} - B_{t_j}$ and similarly for $\Delta_j\bar{B}$, etc. We have

$$\mathbf{E}\{(B_t - \bar{B}_t)Y_t\} = \mathbf{E}\Big\{\sum_{\Pi}(\Delta_jB - \Delta_j\bar{B})Y_t\Big\} = \sum_{\Pi}\mathbf{E}\{(\Delta_jB - \Delta_j\bar{B})Y_{t_j}\}$$

$$+ \sum_{\Pi}\mathbf{E}\{(\Delta_jB - \Delta_j\bar{B})(Y_t - Y_{t_j})\}.$$

$$\text{The 1st term} = \sum_{\Pi}\mathbf{E}\{Y_{t_j}\mathbf{E}[(\Delta_jB - \Delta_j\bar{B}) \mid \mathcal{F}_{t_j}]\} = 0.$$

$$Y_t - Y_{t_j} = \int_{t_j}^{t} \Phi_s \mathrm{d}w_s + \int_{t_j}^{t} g_s \mathrm{d}s.$$

Thus

$$\begin{aligned}\mathbf{E}\{(B_t - \bar{B}_t)Y_t\} &= \sum_{\Pi} \mathbf{E}\left\{(\Delta_j B - \Delta_j \bar{B}) \int_{t_j}^{t_{j+1}} \Phi_s \mathrm{d}w_s\right\} \\ &\quad + \sum_{\Pi} \mathbf{E}\left\{(\Delta_j B - \Delta_j \bar{B}) \int_{t_{j+1}}^{t} \Phi_s \,\mathrm{d}w_s\right\} \\ &\quad + \sum_{\Pi} \mathbf{E}\left[(\Delta_j B - \Delta_j \bar{B}) \int_{t_j}^{t} g_s \,\mathrm{d}s\right].\end{aligned} \tag{16}$$

Consider the 2nd term on the right side of (16):

$$\begin{aligned}&\mathbf{E}\left[(\Delta_j B - \Delta_j \bar{B}) \int_{t_{j+1}}^{t} \Phi_s \,\mathrm{d}w_s\right] \\ &\quad = \mathbf{E}\left[(\Delta_j B - \Delta_j \bar{B}) \mathbf{E}\left(\int_{t_{j+1}}^{t} \Phi_s \,\mathrm{d}w_s \mid \mathcal{G}_{t_{j+1}}\right)\right] = 0\end{aligned}$$

since $\int_0^t (\Phi_s, \mathrm{d}w_s)$ is an L^2-$\mathcal{G}_t$ martingale.

In the 3rd term,

$$\begin{aligned}\mathbf{E}(\Delta_j B - \Delta_j \bar{B}) \int_{t_j}^{t} g_s \,\mathrm{d}s &= \mathbf{E}\left[(\Delta_j B - \Delta_j \bar{B}) \int_{t_j}^{t_{j+1}} g_s \,\mathrm{d}s\right] \\ &\quad + \mathbf{E}\left[\Delta_j B \int_{t_{j+1}}^{t} g_s \,\mathrm{d}s\right] - \mathbf{E}\left[\Delta_j \bar{B} \int_{t_{j+1}}^{t} g_s \,\mathrm{d}s\right]\end{aligned}$$

and

$$\begin{aligned}\mathbf{E}\left[\Delta_j \bar{B} \int_{t_{j+1}}^{t} g_s \,\mathrm{d}s\right] &= \mathbf{E} \int_{t_{j+1}}^{t} \mathbf{E}[\Delta_j \bar{B}(\Phi_s[h_s - \hat{h}_s]) \mid \mathcal{F}_s] \,\mathrm{d}s \\ &= \mathbf{E} \int_{t_{j+1}}^{t} \Delta_j \bar{B}(\Phi_s \mathbf{E}^{\mathrm{r}}(h_s - \hat{h}_s)) \mathrm{d}s = 0.\end{aligned}$$

Hence

$$\begin{aligned}&\mathbf{E}[(\bar{M}_t - M_t)Y_t] \\ &\quad = \mathbf{E}\left(\sum_{\Pi} (\Delta_j B - \Delta_j \bar{B}) \Delta_j Y\right) + \mathbf{E}\left\{\sum_{\Pi} (\Delta_j B) \int_{t_{j+1}}^{t} g_s \,\mathrm{d}s\right\}.\end{aligned}$$

The conditions $\mathbf{E}(\mathrm{Var}\, B)^2 < \infty$, $\mathbf{E}(\mathrm{Var}\, \bar{B})^2 < \infty$ (by Lemma 1) imply that $\mathbf{E}\,|\Sigma_{\Pi_n} (\Delta_j B)(\Delta_j Y)| \to 0$ and $\mathbf{E}\,|\Sigma_{\Pi_n} (\Delta_j \bar{B})(\Delta_j Y)| \to 0$. Next we show that

$$\mathbf{E}\left\{\sum_{\Pi_n} (\Delta_j B) \int_{t_{j+1}}^{t} g_s \mathrm{d}s\right\} \to \mathbf{E}\left(\int_0^t B_s g_s \,\mathrm{d}s\right).$$

For this write

$$D = \sum_{\Pi} \Delta_j B \int_{t_{j+1}}^{t} g_s \,\mathrm{d}s - \int_0^t B_s g_s \,\mathrm{d}s,$$

and let

$$a_k = \int_{t_k}^{t_{k+1}} g_s\, ds.$$

Then

$$\sum_{\Pi} (\Delta_j B)\{a_{j+1} + \ldots + a_{n-1}\}$$

$$= B_{t_1}(a_1 + \ldots + a_{n-1}) + [B_{t_2} - B_{t_1}](a_2 + \ldots + a_{n-1})$$

$$+ (B_{t_3} - B_{t_2})(a_3 + \ldots + a_{n-1}) + \ldots$$

$$+ (B_{t_{n-2}} - B_{t_{n-3}})(a_{n-2} + a_{n-1})$$

$$+ (B_{t_{n-1}} - B_{t_{n-2}})a_{n-1}$$

$$= B_{t_1}a_1 + B_{t_2}a_2 + \ldots + B_{t_{n-2}}a_{n-2} + B_{t_{n-1}}a_{n-1}.$$

So

$$D = \sum B_{t_j} \int_{t_j}^{t_{j+1}} g_s\, ds - \sum \int_{t_j}^{t_{j+1}} B_s g_s\, ds$$

$$= \sum \int_{t_j}^{t_{j+1}} (B_{t_j} - B_s) g_s\, ds.$$

Since for

$$s \in [t_j, t_{j+1}] \quad |B_{t_j} - B_s| = |U_{t_j} - U_s + V_s - V_{t_j}| \leq U_{t_{j+1}} - U_{t_j} + V_{t_{j+1}} - V_{t_j},$$

$$|D| \leq \left[\max_{\Pi_n} \int_{\Delta_j} |g_s|\, ds \right] \sum_{\Pi_n} \Delta_j |B|$$

$$\leq \max_{\Pi_n} \int_{\Delta_j} |g_s|\, ds\, (\operatorname{Var} B) \to 0 \text{ a.s.} \quad \text{as } |\Pi_n| \to 0.$$

Also $|D| \leq \int_0^T |g_s|\, ds\, (\operatorname{Var} B)$ and

$$\mathbf{E}\left[\int_0^T |g_s|\, ds\, (\operatorname{Var} B) \right] \leq \left\{ \mathbf{E} \int_0^T |g_s|^2\, ds\, T\, \mathbf{E}(\operatorname{Var} B)^2 \right\}^{1/2} < \infty.$$

Hence,

$$\mathbf{E}(D_{\Pi_n}) \to 0 \quad \text{as } |\Pi_n| \to 0 \qquad (\text{writing } D_{\Pi_n} \text{ for } D)$$

and so,

$$\mathbf{E}\left[\sum_{\Pi_n} (\Delta_j B) \int_{t_{j+1}}^{t} g_s\, ds \right] \to \mathbf{E} \int_0^t B_s g_s\, ds.$$

Thus finally we have

$$\mathbf{E}[(\bar{M}_t - M_t) Y_t] = \mathbf{E} \int_0^t B_s g_s\, ds \Big].$$

Comparing this with (14) we at once get

$$\mathbf{E}(\bar{M}_t Y_t) = \mathbf{E}(M_t^* Y_t)$$

concluding the proof of the theorem.

An important special case of the above theorem, suggested by applications in which signal and observation processes are governed by stochastic differential equations, is the following (Theorem 4.1 of [2]). We shall assume that the process $B_t[f](\omega)$ is given by $\int_0^t (\tilde{A}_s f)(\omega)\,\mathrm{d}s$ where $(\tilde{A}_t f)(\omega)$ is real-valued, jointly measurable, $\mathcal{G}_t$-adapted and such that

$$\int_0^T \mathbf{E}\,|\,\tilde{A}_t f\,|^2\,\mathrm{d}t < \infty.$$

In the present context we write $\mathcal{D}(\tilde{A})$ in place of $\mathcal{D}$.

Theorem 3. *Assume the conditions of Theorem* 1. *If* $f \in \mathcal{D}(\tilde{A})$ *satisfies*

$$\int_0^t \mathbf{E}\,|\,f(x_t)h_t\,|^2\,\mathrm{d}t < \infty$$

then $\mathbf{E}^t f(x_t)$ *satisfies the following stochastic differential equation*

$$\mathbf{E}^t[f(x_t)] = \mathbf{E}[f(x_0)] + \int_0^t \mathbf{E}^s[\tilde{A}_s f]\,\mathrm{d}s + \int_0^t (\mathbf{E}^s(f(x_s)h_s) - \mathbf{E}^s f(x_s)\mathbf{E}^s(h_s) + \mathbf{E}^s(\tilde{D}_s f), \mathrm{d}\nu_s). \tag{17}$$

Although Theorem 2 was established for a real-valued observation process (z_t), the extension to vector-valued process introduces no new difficulties.

References

[1] Dellacherie, C. (1972). Capacités et processus stochastiques, *Ergebnisse der Mathematik und ihrer Grenzgebiete*, Band **67**.

[2] Fujisaki, M., Kallianpur, G. and Kunita, H. (1972). Stochastic differential equations for the nonlinear filtering problem, *Osaka Journal of Mathematics*, **9**.

[3] Kallianpur, G. (1975). A general stochastic equation for the nonlinear filtering problem, *Optimization Techniques IFIP Technical Conference*, Springer-Verlag.

[4] Meyer, P.A. (1973). Sur un problème de filtration, Séminaire de Probabilités VII, Lecture Notes in Mathematics, Springer-Verlag.

[5] Meyer, P.A. (1966). *Probability and Potentials*, Blaisdell.

P.R. Krishnaiah, ed., *Multivariate Analysis–IV*

MULTIPLE TIME SERIES: DETERMINING THE ORDER OF APPROXIMATING AUTOREGRESSIVE SCHEMES*

Emanuel PARZEN

State University of New York at Buffalo, Amherst, N.Y., U.S.A.

One can distinguish three aims of the time series analysis of a finite sample $Y(t)$, $t = 1, 2, \ldots, T$ of a univariate or multivariate time series: (1) Spectral analysis, (2) Model identification, and (3) Prediction. In this paper we consider the case in which $Y(\circ)$ is a multiple time series which is stationary, normal, and zero mean. We describe an approach to the solution of these problems of time series analysis through a criterion called CAT (an abbreviation for criterion autoregressive transfer-function). CAT enables one to choose the order of an approximating autoregressive scheme which is "optimal" in the sense that its transfer function is a minimum overall mean square error estimator (called ARTFACT) of the infinite autoregressive transfer function (ARTF) of the filter which transforms the time series to its innovations (white noise). Algorithms for choosing the order of an ARTFACT (autoregressive transfer function approximation converging to the truth) enable one to carry out the approach to empirical multiple time series analysis introduced in Parzen (1969), in particular autoregressive spectral estimation of the spectral density matrix of a stationary multiple time series. Such estimators for univariate time series have been very successfully applied in geophysics [see Ulrych and Bishop (1975)] where they are called "maximum entropy spectral estimators." This paper provides a basis for an extension of these procedures to multiple time series.

1. Stationary multiple time series definitions

We consider a multiple time series $Y(t)$, $t = 1, 2, \ldots$ where $Y(t)$ is a random d-dimensional vector:

$$Y(t) = \begin{bmatrix} Y_1(t) \\ \cdots \\ Y_d(t) \end{bmatrix}. \tag{1.1}$$

We assume the time series $Y(t)$ is normal, zero mean, and covariance stationary with covariance matrix function $R(v)$. Then $\mathbf{E}[Y(t)] = 0$ and

$$R(v) = \mathbf{E}[Y(t)Y^*(t+v)], \quad v = 0, \pm 1, \pm 2, \ldots. \tag{1.2}$$

An asterisk * denotes the transpose of a matrix, and denotes the complex conjugate transpose if the components are complex valued.

* Research supported in part by the Office of Naval Research.

The (j, k)-th component of $R(v)$, denoted

$$R_{jk}(v) = \mathbf{E}[Y_j(t)Y_k(t+v)], \tag{1.3}$$

is assumed to be absolutely summable, $\Sigma_v |R_{jk}(v)| < \infty$; its Fourier transform

$$f_{jk}(\omega) = \frac{1}{2\pi} \sum_{v=-\infty}^{\infty} \mathrm{e}^{-iv\omega} R_{jk}(v) \tag{1.4}$$

is called the cross-spectral density function. The matrix

$$f(\omega) = \frac{1}{2\pi} \sum_{v=-\infty}^{\infty} \mathrm{e}^{-iv\omega} R(v) \tag{1.5}$$

is called the *spectral density matrix*; it provides a spectral representation of $R(v)$ since

$$R(v) = \int_{-\pi}^{\pi} \mathrm{e}^{iv\omega} f(\omega)\mathrm{d}\omega. \tag{1.6}$$

We define the inverse spectral density matrix by

$$\mathrm{fi}(\omega) = \frac{1}{4\pi^2} f^{-1*}(\omega). \tag{1.7}$$

We assume $\mathrm{fi}(\omega)$ to have integrable components; then its Fourier transform

$$\mathrm{Ri}(v) = \int_{-\pi}^{\pi} \mathrm{e}^{iv\omega}\, \mathrm{fi}(\omega)\mathrm{d}\omega \tag{1.8}$$

exists and is called the *covinverse* matrix (this concept was first named by Cleveland (1972) in the univariate case with the name "inverse autocorrelation function").

An important sufficient condition for the existence and integrability of the inverse spectral density matrix $\mathrm{fi}(\omega)$ is that the spectral density matrix $f(\omega)$ be bounded above and below in the sense that there exist positive finite constants λ_1 and λ_2 such that

$$\lambda_1 I \ll f(\omega) \ll \lambda_2 I, \qquad -\pi \leq \omega \leq \pi. \tag{1.9}$$

It has been shown [Masani (1966)] that the existence and integrability of $\mathrm{fi}(\omega)$ implies the existence of an infinite autoregressive representation for the time series $Y(\circ)$.

$$Y(t) + A_\infty(1)Y(t-1) + \ldots + A_\infty(m)Y(t-m) + \ldots = \varepsilon(t) \tag{1.10}$$

where $\varepsilon(t)$ is the one-step ahead prediction error given by

$$\varepsilon(t) = Y(t) - Y^{\mu}(t), \qquad Y^{\mu}(t) = \mathbf{E}[Y(t) \mid Y(t-1), Y(t-2), \ldots]. \quad (1.11)$$

In words, $Y^{\mu}(t)$ is the one-step ahead infinite memory minimum mean square error predictor of $Y(t)$ using an infinite memory $Y(t-j)$, $j = 1, 2, \ldots$. The prediction error $\varepsilon(t)$ is called the *innovation* at time t since it is that part of $Y(t)$ unpredictable from the past; we sometimes denote it by $Y^{\nu}(t)$. Its covariance matrix

$$K_{\infty} = \mathbf{E}[\varepsilon(t)\varepsilon^{*}(t)] \quad (1.12)$$

is called the innovation covariance matrix or the infinite memory prediction error covariance matrix. The innovation series $\varepsilon(\circ)$ is multiple white noise in the sense that

$$\mathbf{E}[\varepsilon(t)\varepsilon^{*}(t+v)] = 0 \quad \text{for} \quad v \neq 0. \quad (1.13)$$

In our approach to time series modeling, the time series problem is defined to be estimation of the filter (with transfer function denoted g_{∞}) which transforms the time series $Y(\circ)$ to the white noise innovation series $\varepsilon(\circ)$;

$$Y(\circ) \longrightarrow \boxed{g_{\infty}(\circ)} \longrightarrow \varepsilon(\circ).$$

Defining L to be the backward lag operator, $LY(t) = Y(t-1)$, we can write the filter as an operator $g_{\infty}(L)$,

$$g_{\infty}(L)Y(t) = \varepsilon(t). \quad (1.14)$$

Comparing this representation (1.14) with the infinite autoregressive representation (1.10) we can write the whitening filter transfer function $g_{\infty}(\circ)$ as the z-transform of the impulse response matrix sequence $\{A_{\infty}(m)\}$;

$$g_{\infty}(z) = I + A_{\infty}(1)z + \ldots + A_{\infty}(m)z^{m} + \ldots. \quad (1.15)$$

The rigorous mode of convergence of this infinite series will be clarified when we introduce the spectral Hilbert space of a multiple time series (in section 3).

We call $g_{\infty}(e^{i\omega})$, $-\pi \leq \omega \leq \pi$, the infinite autoregressive transfer function (ARTF). It provides an important representation of the spectral density matrix:

$$2\pi f(\omega) = g_{\infty}^{-1}(e^{i\omega})K_{\infty}g_{\infty}^{-1*}(e^{i\omega}). \quad (1.16)$$

Formula (1.16) has the following important consequences for spectral estimation; if one can find good estimators $\hat{K}_\infty$ of K_∞ and $\hat{g}_\infty(e^{i\omega})$ of $g_\infty(e^{i\omega})$ then a good estimator $\hat{f}(\omega)$ of $f(\omega)$ would be given by

$$2\pi\hat{f}(\omega) = \hat{g}_\infty^{-1}(e^{i\omega})\hat{K}_\infty\hat{g}_\infty^{-1*}(e^{i\omega}). \tag{1.17}$$

As shown in Parzen (1967), for many aspects of multiple time series modeling, knowledge (or estimation) of $\mathrm{fi}(\omega)$ is just as reasonable a starting point in finding relations among the variables as is a knowledge of $f(\omega)$. The estimator of $\mathrm{fi}(\omega)$ given by

$$2\pi\hat{\mathrm{fi}}(\omega) = \hat{g}_\infty^*(e^{i\omega})\hat{K}_\infty^{-1}\hat{g}_\infty(e^{i\omega}) \tag{1.18}$$

does not require matrix inversion at a large number of frequencies.

Our development of estimators of $g_\infty(e^{i\omega})$ will be guided by certain facts from prediction theory.

The infinite memory prediction error

$$\varepsilon(t) = Y^{\nu,\infty}(t) = Y(t) - \mathbf{E}[Y(t) \mid Y(t-k),\ k = 1, 2, \ldots] \tag{1.19}$$

is characterized by the normal equations

$$\mathbf{E}[\varepsilon(t)Y^*(t-k)] = 0, \qquad k = 1, 2, \ldots . \tag{1.20}$$

When $\varepsilon(t)$ can be represented as an infinite autoregression in past Y's, so that

$$\varepsilon(t) = \sum_{j=0}^{\infty} A_\infty(j)Y(t-j), \qquad A_\infty(0) = I,$$

the infinite autoregressive coefficients $A_\infty(j)$ can be found by solving an infinite set of linear equations

$$\sum_{j=0}^{\infty} A_\infty(j)R(j-k) = 0, \qquad k = 1, 2, \ldots, \tag{1.21}$$

$$= K_\infty, \qquad k = 0,$$

called the infinite memory matricial Yule Walker equations or Hopf–Wiener equations.

The m-memory prediction error, denoted

$$\varepsilon^{(m)}(t) = Y^{\nu,m}(t) = Y(t) - \mathbf{E}[Y(t) \mid Y(t-k),\ k = 1, \ldots, m], \tag{1.22}$$

can be represented as a finite linear combination

$$\varepsilon^{(m)}(t) = \sum_{j=0}^{m} A_m(j)Y(t-j), \qquad A_m(0) = I. \tag{1.23}$$

The memory m autoregressive coefficients $A_m(j)$ can be found by solving a finite set of linear equations

$$\sum_{j=0}^{m} A_m(j)R(j-k)=0, \qquad k=1,2,\ldots,m \tag{1.24}$$

called the memory m matricial Yule Walker equations. Then the memory m prediction error covariance matrix is given by

$$K_m = \mathbf{E}[\varepsilon^{(m)}(t)\varepsilon^{(m)*}(t)] = \sum_{j=0}^{m} A_m(j)R(j). \tag{1.25}$$

Introducing the memory m autoregressive transfer function, abbreviated ARTF(m),

$$g_m(z)=I+A_m(1)z+\ldots+A_m(m)z^m \tag{1.26}$$

we can represent the memory m prediction errors $\varepsilon^{(m)}(t)$ as the output of a filter with input $Y(t)$;

$$\varepsilon^{(m)}(t)=g_m(L)Y(t). \tag{1.27}$$

A zero mean covariance stationary multiple time series $Y(\circ)$ is called a joint autoregressive scheme of order m, denoted AR(m), if the infinite memory prediction error $\varepsilon(t)$ is in fact a finite linear combination of $Y(t-1),\ldots,Y(t-m)$ and the coefficient matrix of $Y(t-m)$ is non-singular; then $A_\infty(j)=A_m(j)$ and

$$\varepsilon(t)=Y(t)+A_m(1)Y(t-1)+\ldots+A_m(m)Y(t-m)=g_m(L)Y(t). \tag{1.28}$$

To clarify the relation between memory m predictors and finite order autoregressive schemes we have the following theorem.

Theorem. *Assume $A_m(m)$ is non-singular. Then $Y(\circ)$ is* AR(m) *if, and only if, $\varepsilon^{(m)}(\circ)$ is white noise.*

Proof. The "only if" part of the theorem holds by definition of AR(m). The "if" part is proved by induction. By definition of $\varepsilon^{(m)}(\circ)$, it satisfies the normal equations

$$\mathbf{E}[\varepsilon^{(m)}(t)Y^*(t-j)]=0, \qquad j=1,2,\ldots,m. \tag{1.29}$$

We desire to establish (1.29) for $j=m+1$ whence $\varepsilon^{(m)}(t)=\varepsilon^{(m+1)}(t)$. The assumption that $\varepsilon^{(m)}(\circ)$ is white noise implies in particular

$$\mathbf{E}[\varepsilon^{(m)}(t)\varepsilon^{(m)*}(t-1)]=0.$$

This covariance equals

$$\sum_{k=0}^{m} \mathbf{E}[\varepsilon^{(m)}(t)Y^*(t-1-k)]A_m^*(k) = \mathbf{E}[\varepsilon^{(m)}(t)Y^*(t-(m+1))]A_m^*(m).$$

One may now infer that (1.29) holds for $j = m+1$. Using $\mathbf{E}[\varepsilon^{(m)}(t)\varepsilon^{(m)*}(t-2)] = 0$, one can show that (1.29) holds for $j = m+2$, and so on.

2. Autoregressive order determination

One approach to modeling a multiple time series $Y(\circ)$ is to assume that there is a true order p such that the time series is AR(p). The problem of autoregressive order determination is then defined as follows: given a finite sample $Y(t)$, $t = 1, \ldots, T$, estimate the "true" order of the autoregressive scheme fitting the data. Our approach, first described in Parzen (1974), to modeling a multiple time series is based on the concept that the problem is to estimate the ARTF $g_\infty(\mathrm{e}^{\mathrm{i}\omega})$.

Given a sample $Y(t)$, $t = 1, 2, \ldots, T$ of a zero mean normal stationary multiple time series one forms:

(i) The sample spectral density matrix

$$f_T(\omega) = \frac{1}{2\pi T} \sum_{t=1}^{T} Y(t)\mathrm{e}^{\mathrm{i}\omega t} \sum_{t=1}^{T} Y^*(t)\mathrm{e}^{-\mathrm{i}\omega t}. \tag{2.1}$$

(ii) The sample covariance matrix

$$R_T(v) = \int_{-\pi}^{\pi} \mathrm{e}^{\mathrm{i}v\omega} f_T(\omega)\,\mathrm{d}\omega, \qquad v = 0, \quad \pm 1, \ldots, \tag{2.2}$$

which in practice is computed by the formula, for $v = 0, \pm 1, \ldots, \pm(T-1)$,

$$R_T(v) = \frac{2\pi}{2T} \sum_{j=0}^{2T-1} \exp\left(\mathrm{i}vk\frac{2\pi}{2T}\right) f_T\left(k\frac{2\pi}{2T}\right). \tag{2.3}$$

(iii) The sample autoregressive coefficients $\hat{A}_m(j)$ for $m = 1, 2, \ldots,$ and $j = 1, 2, \ldots, m$, by solving the sample matricial Yule Walker equations, defining $\hat{A}_m(0) = I$,

$$\sum_{j=0}^{m} \hat{A}_m(j) R_T(j-k) = 0, \qquad k = 1, 2, \ldots, m. \tag{2.4}$$

(iv) The sample memory m prediction error covariance matrix

$$\hat{K}_m = \sum_{j=0}^{m} \hat{A}_m(j) R_T(j). \tag{2.5}$$

(v) The sample ARTF(m) [memory m autoregressive transfer function]

$$\hat{g}_m(e^{i\omega}) = \sum_{j=0}^{m} \hat{A}_m(j)e^{ij\omega}. \tag{2.6}$$

(vi) The sample memory m autoregressive spectral density matrix estimator

$$\hat{f}_m(\omega) = (2\pi)^{-1}\hat{g}_m^{-1}(e^{i\omega})\hat{K}_m\hat{g}_m^{-1*}(e^{i\omega}). \tag{2.7}$$

When the time series $Y(\circ)$ is known to be AR(m), then $\hat{A}_m(j)$ and $\hat{K}_m$ can be shown to be *efficient* estimators of the parameters $A_m(j)$ and K_m of the joint autoregressive scheme [see Hannan (1970)]. In general, one can only show that $\hat{A}_m(j)$ and $\hat{K}_m$ are *consistent* estimators of the parameters of the memory m one step ahead prediction filter.

The problem remains: which order m to use (either for prediction or spectral estimation)? Hannan [(1970), p. 334] describes several older approaches to this problem. The aim of this paper is to describe a new approach, and contrast it with another new approach due to Akaike [see Akaike (1974) and Jones (1974)].

In the autoregressive order determination technique of Akaike, one assumes that there is a true order p such that the time series is AR(p). Fitting predictors of successive order $m = 1, 2, \ldots$, one estimates p by the value of m minimizing [Jones (1974), p. 895]

$$\text{AIC}(m) = \log\det\hat{K}_m + \frac{2d^2m}{T}. \tag{2.8}$$

This criterion is called Akaike's information criterion since it is a special case of the model fitting criterion:

AIC = − 2 log maximum likelihood + 2 (number of independently adjusted parameters within the model).

In the autoregressive order determination technique of Parzen, being introduced in this paper, one assumes that one is seeking to estimate the ARTF $g_\infty(e^{i\omega})$ by the sample ARTF(m)$\hat{g}_m(e^{i\omega})$, and chooses m to minimize a criterion, denoted CAT(m), which provides an estimate of the mean square overall error of $\hat{g}_m$ as an estimate of g_∞. We define

$$\text{CAT}(m) = \text{trace}\left\{\frac{d}{T}\sum_{j=1}^{m}\hat{\hat{K}}_j^{-1} - \hat{\hat{K}}_m^{-1}\right\}, \tag{2.9}$$

where

$$\hat{\hat{K}}_j = \frac{T}{T-jd}\hat{K}_j \tag{2.10}$$

is an unbiased estimator of K_j. We describe the theoretical basis of this criterion in the next section.

To compare CAT and AIC, one can show that AIC provides a larger bound for the quantity being estimated by CAT.

Theorem *For every order m, and for large T*

$$\frac{1}{d}\,\mathrm{CAT}(m) \leq -\exp\left\{\frac{-1}{d}\,\mathrm{AIC}(m)\right\}. \tag{2.11}$$

Proof. We use the asymptotically equivalent definition of AIC given by

$$\mathrm{AIC}(m) = -\log\det\hat{\hat{K}}_m^{-1} + \frac{d^2 m}{T}. \tag{2.12}$$

$$= -\log\det\hat{\hat{K}}_m^{-1} - d\log\left(1-\frac{dm}{T}\right).$$

The desired inequality is equivalent to (omitting the argument m)

$$\log d - \log(-\mathrm{CAT}) \leq \frac{1}{d}\,\mathrm{AIC}. \tag{2.13}$$

Since $\operatorname{tr} K_j^{-1}$ is an increasing function of j,

$$\mathrm{CAT}(m) \leq \left(\frac{dm}{T}-1\right)\operatorname{tr}\hat{\hat{K}}_m^{-1}, \qquad -\mathrm{CAT} \geq \left(1-\frac{dm}{T}\right)\operatorname{tr}\hat{\hat{K}}_m^{-1}.$$

We next use the fact: for any d by d positive definite matrix M

$$\log\operatorname{trace} M \geq \frac{1}{d}\log\det M + \log d. \tag{2.14}$$

Therefore

$$-\log d + \log(-\mathrm{CAT}) \geq \log\left(1-\frac{dm}{T}\right) + \frac{1}{d}\log\det\hat{\hat{K}}_m^{-1}.$$

The desired conclusion may now be inferred.

We have applied these criteria to many univariate time series, and have usually found AIC and CAT to achieve their minimum at the same value of m. We expect them to deviate somewhat more often in the multivariate case but still usually to yield the same determinations $\hat{m}$ of order m. Thus we have two criteria for order determination with quite different conceptual justifications but which yield similar answers in practice.

It is important to understand that the length $\hat{m}$ of the autoregressive scheme to be fitted to an observed time series is "long" only in theory. In practice, its order can be as small as 1. For the first order moving average MA(1), with dimension $d = 2$, $Y(t) = \varepsilon(t) + B\varepsilon(t-1)$, with $K = \mathbf{E}[\varepsilon(t)\varepsilon^*(t)] = I$; two cases of B:

$$B^{(1)} = \begin{bmatrix} 0.5 & 0 \\ 0.2 & -0.4 \end{bmatrix}, \qquad B^{(2)} = \begin{bmatrix} 0.9 & 0 \\ 0.2 & -0.8 \end{bmatrix};$$

sample sizes $T = 50, 100, 300$; and 25 trials (simulations) we found the following distribution of orders $\hat{m}$:

Sample size T	Case $B^{(1)}$	Case $B^{(2)}$
50	usually order 1	equally likely to be order 2 to 5
100	usually order 2	equally likely to be order 3 to 6
300	order 2 most likely, but often order 3, 4, or 5	equally likely to be order 7 to 9

3. Spectral Hilbert spaces of a multiple time series

The prediction theory of stationary time series can be developed in two distinct approaches, called the time domain approach and the frequency domain approach. The Box–Jenkins approach to time series modelling [see Box and Jenkins (1970)] adopts a time domain approach to fitting models which yield good predictions. In our approach we use the mathematical structure of the spectral-domain approach to prediction to justify and motivate finite memory autoregressive schemes as approximate models for stationary time series. More generally, they provide finite parametric approximations to any function which has the mathematical properties of spectral density functions and spectral density matrices.

To a d by d spectral density matrix $f(\omega)$ on $(-\pi, \pi)$ we define a Hilbert space $L_2(f)$ of d by d matrix-valued functions $\Phi(\omega) = \{\Phi_{jk}(\omega)\}$, with matrix-valued inner product

$$((\Phi, \Psi))_f = \int_{-\pi}^{\pi} \Phi(\omega) f(\omega) \Psi^*(\omega)\, d\omega \tag{3.1}$$

and complex-valued inner product

$$(\Phi, \Psi)_f = \operatorname{tr}((\Phi, \Psi))_f. \tag{3.2}$$

The space $L_2(f)$ of functions Φ such that

$$\| \Phi \|_f^2 = (\Phi, \Phi)_f < \infty \tag{3.3}$$

has been shown to be a Hilbert space [theorem of Rosenberg–Rosanov, quoted by Masani (1966), p. 362]. The power series representation of the ARTF $g_\infty(e^{i\omega})$ converges in $L_2(f)$, and $g_\infty(e^{i\omega})$ belongs to $L_2(f)$.

The problem of memory m one-step ahead prediction can be stated as an optimatization problem in $L_2(f)$ as follows. The prediction error $\eta(t)$ of a typical memory m predictor is of the form $\eta(t) = p(L)Y(t)$, where $p(z)$ is a polynomial of degree m with matrix coefficients C_j, $p(z) = C_0 + C_1 z + \ldots + C_m z^m$, with the constraint that $C_0 = I$. The prediction error covariance matrix can be expressed as the matrix-valued "norm" of p,

$$K_\eta = \mathbf{E}[\eta(t)\eta^*(t)] = ((p(e^{i\omega}), p(e^{i\omega})))_f. \tag{3.4}$$

Finding the optimal prediction error $\varepsilon^{(m)}(\circ)$ is equivalent to finding the minimum norm polynomial of degree m with constant term equal to the identity in the sense that the ARTF$(m)g_m(\circ)$ satisfies

$$K_m = ((g_m(e^{i\omega}), g_m(e^{i\omega})))_f \leqslant ((p(e^{i\omega}), p(e^{i\omega})))_f, \tag{3.5}$$

where $\leqslant$ denotes the ordering relation between non-negative definite matrices.

The matricial Yule Walker equations for the coefficients $A_m(j)$ of $g_m(z)$,

$$\begin{aligned} \sum_{j=0}^{m} A_m(j)R(j-k) &= 0, & k &= 1, \ldots, m, \\ &= K_m, & k &= 0 \end{aligned}$$

can be deduced from the projection theorem in Hilbert space

$$\begin{aligned} ((g_m(e^{i\omega}), e^{ik\omega}I))_f &= 0, & k &= 1, 2, \ldots, m, \\ &= K_m, & k &= 0. \end{aligned}$$

To obtain a theorem on the properties of the sample ARTF$(m)\hat{g}_m(e^{i\omega})$, we must modify our definitions of these transfer functions as follows:

$$\begin{aligned} \gamma_\infty(e^{i\omega}) &= K_\infty^{-1} g_\infty(e^{i\omega}), \\ \gamma_m(e^{i\omega}) &= K_m^{-1} g_m(e^{i\omega}), \\ \hat{\gamma}_m(e^{i\omega}) &= \hat{K}_m^{-1} \hat{g}_m(e^{i\omega}). \end{aligned} \tag{3.6}$$

As a measure of mean square overall estimation error we take

$$J_m = \mathbf{E}\,\|\,\hat{\gamma}_m(e^{i\omega}) - \gamma_\infty(e^{i\omega})\,\|_f^2 \tag{3.7}$$

$$= \mathbf{E}\,\mathrm{trace}\int_{-\pi}^{\pi} \{\hat{\gamma}_m(e^{i\omega}) - \gamma_\infty(e^{i\omega})\} f(\omega) \{\hat{\gamma}_m(e^{i\omega}) - \gamma_\infty(e^{i\omega})\}^* \, d\omega.$$

Using recursive relations from filtering theory [see Kailath (1974)] the following basic theorem can be proved.

Theorem. *For large T, approximately*

$$J_m = \mathrm{trace}\left\{\frac{d}{T}\sum_{j=1}^{m} K_j^{-1} + K_\infty^{-1} - K_m^{-1}\right\}. \tag{3.8}$$

This expression for mean square error is the sum of two terms, one representing *bias*:

$$((\gamma_m - \gamma_\infty, \gamma_m - \gamma_\infty))_f - K_\infty^{-1} - K_m^{-1}, \tag{3.9}$$

and the other representing the *variability* of the estimator $\hat{\gamma}_m$:

$$\mathbf{E}\,\|\,\hat{\gamma}_m - \gamma_m\,\|_f^2 = \mathrm{trace}\,\frac{d}{T}\sum_{j-1}^{m} K_j^{-1}. \tag{3.10}$$

Since K_∞^{-1} does not depend on m, to find the optimal order $\hat{m}$ it suffices to minimize $J'_m = J_m - \mathrm{tr}\,K_\infty^{-1}$.

Our motivation and justification for CAT(m) is that it is an estimator from a sample of size T of J'_m. The optimality of this estimation procedure remains to be investigated.

4. Sample spectral Hilbert spaces

The solution of the time series prediction problem requires a study of the spectral Hilbert space $L_2(f)$. The study of spectral Hilbert spaces turns out to be vital for time series modeling as well. Given a sample $Y(t)$, $t = 1, 2, \ldots, T$ and no knowledge of the spectral density matrix $f(\omega)$, the operations that prediction theory guides us to perform in $L_2(f)$ we will perform in practice in $L_2(f_T)$. We call $L_2(f_T)$ the sample spectral Hilbert space of the time series.

The relation between f_T and f, and therefore between $L_2(f)$ and $L_2(f_T)$, derives from the relation between the true covariance matrix function

$R(v)$ and the sample covariance matrix function $R_T(v)$. Under reasonable conditions, one can show that [see Hannan (1970), p. 210] for $v = 0, \pm 1, \dots$

$$\lim_{T\to\infty} R_T(v) = R(v) \quad \text{with probability one.} \tag{4.1}$$

It follows from the "continuity theorem of probability theory" that for any continuous matrix valued function $A(\omega)$

$$\lim_{T\to\infty} \int_{-\pi}^{\pi} \operatorname{tr} A(\omega) f_T(\omega)\mathrm{d}\omega = \int_{-\pi}^{\pi} \operatorname{tr} A(\omega) f(\omega)\mathrm{d}\omega.$$

Using these techniques one can prove the following theorem.

Theorem. *If* (4.1) *holds, then for any continuous matrix valued functions* Φ, Ψ

$$\lim_{T\to\infty} ((\Phi, \Psi))_{f_T} = ((\Phi, \Psi))_f \quad \textit{with probability one.}$$

Many other relations between $L_2(f_T)$ and $L_2(f)$ remain to be investigated.

5. Some practical considerations and open questions

Given a stationary multivariate time series $Y(t)$ of dimension $d = 2$, one can determine three orders: $\hat{m}_1$, the optimal order for the univariate series $Y_1(\circ)$; $\hat{m}_2$, the optimal order for the univariate series $Y_2(\circ)$; and $\hat{m}_{1,2}$, the optimal order for the joint series $Y(\circ)$.

It is our experience that the interpretation of $\hat{m}_{1,2}$ will not be satisfactory unless one computes it for the individually whitened and standardized series

$$\tilde{Y}_1(t) = \frac{1}{\hat{\sigma}_1} \hat{g}_{\hat{m}_1}(L) Y_1(t),$$

$$\tilde{Y}_2(t) = \frac{1}{\hat{\sigma}_2} \hat{g}_{\hat{m}_2}(L) Y_2(t).$$

These series were introduced in Parzen (1969) under the name of "individual innovations." The standard deviations $\hat{\sigma}_1$ and $\hat{\sigma}_2$ are of the series $\hat{g}_{\hat{m}_1}(L) Y_1(t)$ and $\hat{g}_{\hat{m}_2}(L) Y_2(t)$ respectively.

Theorem (Empirical). *Denote by* $\hat{\tilde{m}}_{1,2}$ *the order fitted to the multiple residual time series*

$$\tilde{Y}(t) = \begin{bmatrix} \tilde{Y}_1(t) \\ \tilde{Y}_2(t) \end{bmatrix}.$$

The three estimated orders $\hat{m}_1$, $\hat{m}_2$, and $\hat{\hat{m}}_{1,2}$ of approximating autoregressive schemes provide a quick and dirty summary of the correlation structure of the time series $Y_1(\circ)$ and $Y_2(\circ)$.

Any non-stationarity suspected to be present in a time series $Y_j(t)$ should be removed as part of the process of transformation to whitened residuals $\tilde{Y}_j(t)$.

Among the important open questions are the distribution theory of autoregressive cross-spectral density estimators; one approach to this investigation would be to extend the work of Berk (1974).

For some examples of autoregressive spectral estimation of stationary multiple time series, see Jones (1974).

References

Akaike, H. (1974). "A New Look at the Statistical Model Identification," *IEEE Trans. Auto. Control*, Vol. AC–**19**, pp. 716–723.

Berk, K. (1974). "Consistent autoregressive spectral estimates." *Annals Statistics*, Vol. **2**, pp. 489–502.

Box, G.E.P. and Jenkins, G.M. (1970). *Time Series Analysis, Forecasting, and Control*, San Francisco: Holden Day

Cleveland, W.S. (1972). "The inverse autocorrelations of a time series and their applications." *Technometrics*, Vol. **14**, pp. 277–293.

Hannan, E.J. (1970). *Multiple Time Series*, New York, Wiley.

Jones, R.H. (1974). "Identification and Autoregressive Spectrum Estimation." *IEEE Trans. Auto. Control*, Vol. AC–**19**, pp. 894–897.

Kailath, T. (1974). "A View of Three Decades of Filtering Theory, "*IEEE Trans. Inform. Theory*, Vol. IT–**20**, pp. 146–181.

Masani, P. (1966). "Recent Trends in Multivariate Prediction Theory," *Multivariate Analysis*, P.R. Krishnaiah, Ed., New York: Academic Press, pp. 351–382.

Parzen, E. (1967). "On Empirical Multiple Time Series Analysis." *Proc. Fifth Berkeley Symp. Math. Statist. Prob.*, Vol. **1**, pp. 305–340. Berkeley, California: Univ. California Press.

Parzen, E. (1969). "Multiple Time Series Modeling" in *Multivariate Analysis II*, P.R. Krishnaiah, Ed., New York: Academic Press, pp. 389–410.

Parzen, E. (1974). "Some Recent Advances in Time Series Analysis," *IEEE Trans. Auto. Control*, Vol. AC–**19**, pp. 723–730.

Ulrych, T.J. and Bishop, T.N. (1975). "Maximum Entropy Spectral Analysis and Autoregressive Decomposition, "*Reviews of Geophysics and Space Physics*, Vol. **13**, pp. 183–200.

P.R. Krishnaiah, ed., *Multivariate Analysis–IV*

ON THE SUPPORT OF GAUSSIAN PROBABILITY MEASURES ON LOCALLY CONVEX TOPOLOGICAL VECTOR SPACES

Balram S. RAJPUT
University of Tennessee, Knoxville, Tenn., U.S.A.

N.N. VAKHANIA
Tbilisi State University, Tbilisi, USSR,
and
University of Tennessee, Knoxville, Tenn., U.S.A.

Let μ be a Gaussian probability measure on a locally convex topological vector space E. Under the assumption of the existence of S_μ, the support of μ, and the mean of μ, it is shown that S_μ is the algebraic sum of the mean of μ and a closed subspace of E. Under an additional assumption on μ, this closed subspace is described in terms of the covariance operator and the reproducing kernel Hilbert space of μ. Two other results that provide sufficient conditions guaranteeing the existence of the mean of μ and the fact that the range of the covariance operator is contained in E are also proved.

1. Introduction

In studying the topological support (supp) of a Gaussian probability measure on a topological vector space (TVS), the main interest lies in investigating: (i) its algebraic structure and (ii) its possible description in terms of the parameters of the measure. Both of these aspects of the supp of Gaussian probability measures, satisfying certain regularity conditions, on various locally convex topological vector (LCTV) spaces have been extensively studied [2, 3, 6, 8–17, 21–24]. Among these the best known result, as regards to (i), asserts that the supp of a K-regular centered Gaussian probability measure on a LCTVS E is a (closed) subspace of E [2, 3]. Since the existence of the supp of a probability measure on any topological space is implied by τ-regularity and since τ-regularity is a weaker condition than K-regularity [20, p. 30], it is interesting to know whether the above result of [2, 3] can be proved assuming only the τ-regularity or better yet the existence of the supp of the measure[1]. In one

[1] It appears to be unknown that whether or not the existence of the support of a probability measure on an arbitrary topological space implies its τ-regularity. In the metric space setting, however, one can show that the answer to this question is affirmative.

of the main results (Theorem 1) of this paper, we give an affirmative answer to this question by showing that the above result does indeed hold under the sole assumption of the existence of the supp, thus providing the best possible result of this type.

In connection to (ii), we prove that, under certain conditions, the supp of a Gaussian probability measure on a LCTVS E is the algebraic sum of its mean and the closure of the range of its covariance operator (Theorem 2); an alternative, essentially equivalent, description of the supp in terms of the mean and the reproducing kernel Hilbert space (RKHS) of the measure is also given (Theorem 3, Corollary 1).

In addition to the above, we also give two other results (Propositions 1 and 2). In these we give sufficient conditions on a second order probability measure μ on a LCTVS E which guarantee the existence of the mean of μ and the fact that the range of the covariance operator of μ is contained in E. This second result along with Theorems 2 and 3 mentioned above extend the results, regarding the description of the support of Gaussian probability measures on different LCTV spaces, obtained by a number of authors using either the notion of the covariance operator [21, 23, 15, 17, 24] or the notion of the RKHS [9, 10, 11, 13, 12, 16, 14, 2, 3]. Related results [8, 6], which do not fall exactly in these two categories, can also be obtained from Theorem 1 (see, for example, [16, p. 302]).

Certain ideas from the theory of projective limits of topological spaces play a fundamental role in the proofs of the main results here; these ideas were used earlier in [16], where the supp of a K-regular Gaussian probability measure on complete separable LCTV spaces was considered. The organization of the rest of the paper is as follows: Section 2 is preliminary, and Sections 3 and 4, respectively, contain the statements and proofs of results.

2. Preliminaries

Unless stated otherwise the following conventions will remain fixed throughout the paper. In the following, we also list a few definitions and some known facts, which will be needed in the sequel.

All topological spaces considered here are assumed Hausdorff. If E is a topological space, then $\mathscr{B}_E$ (or just $\mathscr{B}$) will denote the σ-algebra generated by the class of open sets of E; further, if $A \subseteq E$, then $\bar{A}$ will denote the closure of A. By a probability measure on a topological space E, we will mean that it is defined on $\mathscr{B}_E$ (there is, however, an exception to this rule, see Definition 2.2).

We recall that a probability measure μ on a topological space E is called K-regular if $\mu(B) = \sup\{\mu(A): A \subseteq B,\ A \text{ compact}\}$, for every $B \in \mathscr{B}_E$, and τ-regular, if for every increasing net $\{U_\alpha\}$ of open sets of E we have $\lim \mu(U_\alpha) = \mu(\bigcup_\alpha U_\alpha)$. Now we give the definition of the supp of a probability measure.

Definition 2.1. Let μ be a probability measure on a topological space, then the smallest closed set with full measure (if it exists) is called the *supp* of μ.

It is clear that if $\text{supp}(\mu)$ (supp of μ) exists, then it is the intersection of all closed sets with μ-measure 1; further, it can be shown that it is also the set of those points of the space every open neighborhood (nbd) of which has positive μ-measure.

The underlying field for all vector spaces is the real field R. If E is a TVS then E^a and E^* will denote, respectively, the algebraic and topological dual of E. Now we proceed to give the definition of the mean and the covariance operator of a probability measure. Let E be a LCTVS and $\langle\,,\rangle$ the natural bilinear form between E and E^* and let $\mathscr{A}$ be the σ-algebra generated by the cylinder sets of E; and let μ be a probability measure on $(E, \mathscr{A})$. Then we say μ is of *first* or *second order* (in weak sense) according as $\int_E |\langle x, y\rangle| \mu(\mathrm{d}x) < \infty$ or $\int_E |\langle x, y\rangle|^2 \mu(\mathrm{d}x) < \infty$, for all $y \in E^*$.

Definition 2.2. Let E, E^* be as above and μ be of second order on $(E, \mathscr{A})$. Define $m_\mu(y) = \int_E \langle x, y\rangle \mu(\mathrm{d}x)$, for every $y \in E^*$, and $C_\mu(y_1, y_2) = \int_E \langle x, y_1\rangle\langle x, y_2\rangle \mu(\mathrm{d}x) - m_\mu(y_1)m_\mu(y_2)$, for every $y_1, y_2 \in E^*$; and K_μ be the map. $E^* \to E^{*a}$ defined by $(K_\mu y)(\cdot) = C_\mu(y, \cdot)$. It is easy to verify that $m_\mu \in E^{*a}$ and K_μ is a linear operator. The operator K_μ is called the *covariance operator* of μ; and if $m_\mu \in E$ (under the natural identification of E in E^{*a}), then m_μ is called the *mean* of μ (this definition of the covariance operator is formulated by Vakhania and Tarieladze in [25]). If the mean of μ exists, it will be denoted by m_μ, throughout.

It may be noted that the operator K_μ is symmetric and non-negative definite in the sense $(K_\mu y_1)(y_2) = (K_\mu y_2)(y_1)$ and $(K_\mu y_1)(y_1) \geq 0$, for all $y_1, y_2 \in E^*$; moreover, it can be shown easily that

$$|(K_\mu y_1)(y_2)|^2 \leq (K_\mu y_1)(y_1)(K_\mu y_2)(y_2), \tag{2.1}$$

for all $y_1, y_2 \in E^*$. Finally, we note that for the definition of the mean given above, one only needs that μ be of first order.

We will assume that the reader is familiar with the notions of Gaussian measures and Gaussian cylinder measures on TV spaces.

3. Statement and discussion of results

We begin with the following proposition, which itself is not a central result of this paper, but plays a fundamental role for the proofs of the main results and is also of independent interest.

Proposition 1. *Let I be a directed set and $\{E_\alpha, g_{\alpha\beta}, \alpha \leq \beta\}$ be a projective system of topological spaces relative to I and let $E = \varprojlim g_{\alpha\beta} E_\alpha$, the projective limit of E_α's [4, p. 41]. Let μ be a probability measure on E and $\mu_\alpha = \mu \circ g_\alpha^{-1}$, $\alpha \in I$, where g_α is the natural projection of E into E_α. Then if* $\operatorname{supp}(\mu)$ $(\equiv S_\mu)$ *exists, then so does* $\operatorname{supp}(\mu_\alpha)$, *for each $\alpha \in I$; further*

$$S_{\mu_\alpha} \equiv \operatorname{supp}(\mu_\alpha) = \overline{g_\alpha(S_\mu)} \tag{3.1}$$

and

$$S_\mu = \varprojlim \tilde{g}_{\alpha\beta} S_{\mu_\alpha}, \tag{3.2}$$

where $\tilde{g}_{\alpha\beta}$ is the restriction of $g_{\alpha\beta}$ to S_{μ_β}.

Using Proposition 1, we get quite easy proof of the fact that in certain LCTV spaces the supp of a centered Gaussian probability measure is a linear subspace. As we will see in Section 4, the proof of the analog of this result for general LCTV spaces is much more difficult.

Proposition 2. *Let E, F be two vector spaces in duality under the bilinear form $\langle , \rangle$ [19, p. 123] and let μ be a centered Gaussian probability measure on $(E, \mathscr{B}_E)$, where $\mathscr{B}_E$ is the σ-algebra generated by the $\sigma(E, F)$-open sets. If* $\operatorname{supp}(\mu)$ *exists, then it is a linear subspace of E.*

Now we are ready to state the main results of this paper.

Theorem 1.[2] *Let μ be a Gaussian probability measure on a LCTVS E. If* $\operatorname{supp}(\mu)$ *and the mean of μ exist, then*

[2] This result under τ-regularity assumption on μ has recently been announced (without proof) by H. Sato and A. Tortrat (1975) [C. R. Acad. Sc. Paris. Ser. A, 280, 909–911].

$$\text{supp}(\mu) = m_\mu + F,$$

where F is a closed subspace of E.

(Recall that throughout m_μ denotes the mean of μ, see Definition 2.2.)

Theorem 2. *Let μ and E be as in Theorem 1. If the* supp(μ) *and the mean of μ exist and if $K_\mu(E^*) \subseteq E$, where K_μ is the covariance operator of μ, then*

$$S_\mu \equiv \text{supp}(\mu) = m_\mu + \overline{K_\mu(E^*)}.$$

Theorem 3. *Let E and F be two LCTV spaces and u a continuous linear map from E into F. Let ν be a Gaussian cylinder probability measure on E such that if $y \in E^*$, $y \neq 0$ implies $\langle \cdot, y \rangle$ is a non-degenerate random variable and let μ be the cylinder measure $\nu \circ u^{-1}$ on F. If μ extends to $(F, \mathcal{B}_F)$ and if* supp(μ) *and the mean of μ exist, then*

$$\text{supp}(\mu) = m_\mu + \overline{u(E)}.$$

This immediately yields the following corollary.

Corollary 1. *Let H be a Hilbert space, E a LCTVS and u a continuous linear map from H into E. Let ν be the canonical Gaussian cylinder probability measure of H and μ be the cylinder measure $\nu \circ u^{-1}$ on E. If μ extends to $(E, \mathcal{B}_E)$ and if* supp(μ) *exists, then*

$$\text{supp}(\mu) = \overline{u(H)}.$$

Let μ be a Gaussian probability measure on a LCTVS E with covariance operator K_μ. The answer to the questions whether or not (1) the mean of μ always exists, (2) the range of K_μ is always contained in E, and (3) μ (assuming $m_\mu = 0$) can always be obtained as the image of the canonical Gaussian cylinder probability measure of its RKHS as in Corollary 1, is not known (to us anyway). Therefore, in order to apply the above results, it is interesting to find sufficient conditions on the measure μ and/or on the space E which provide affirmative answers to these questions. We now state two results in this direction.

Proposition 3. *Let E be a complete LCTVS and μ be a first order probability measure on $(E, \mathcal{B}_E)$. If* supp(μ) *exists and if, for some $\varepsilon > 0$,*

$$\inf_{y\in E^*} \mu\left\{x: |\langle x, y\rangle| \geq \varepsilon \left| \int_E \langle x, y\rangle \mu(\mathrm{d}x)\right|\right\} > 0, \tag{3.3}$$

then the mean of μ *exists.*

(Recall that $\langle , \rangle$ denotes the natural duality between E and E^*). This immediately yields the following corollary.

Corollary 2. *Let* μ *be a Gaussian probability measure on a complete LCTVS E. If* $\operatorname{supp}(\mu)$ *exists, then the mean of* μ *exists.*

Proposition 4. *Let* μ *be a centered Gaussian probability measure on a LCTVS. If there exists a sequence* $\{A_n\}$ *of* $\sigma(E, E^*)$*-compact, convex, circled sets of E such that* $\mu(A_n) \to 1$, *then* $K_\mu(E^*) \subseteq E$, *where* K_μ *is the covariance operator of* μ.

Remark 1. Under the same assumptions as in the above proposition, it is shown in [18] that the answer to Question 3 above is affirmative. Consequently, since any probability measure on a separable Fréchet space or dual of separable Fréchet space with weak or compact convergence topology satisfies the hypothesis of the above proposition (see, for example, [18]) and since any Gaussian probability measure on these spaces is K-regular and admits the mean value (see [18] and [24]), it follows that the results regarding the representation of the supp of Gaussian probability measures obtained in [6, 8–17, 21–23] and mentioned in Section 1 do indeed follow from our Theorems 1–3 and Proposition 4. Finally, we mention that our Theorems 1–3 also contain the results regarding the supp of Gaussian probability measures obtained in [2, 3, 16, 24].

4. Proof of results

We begin with a lemma which will be needed at several places in the proofs of results of Section 3. Its proof is routine and is therefore omitted.

Lemma 1. *Let E and F be two topological spaces and u a continuous map from E into F. Let* μ *be a probability measure on E and* $\nu = \mu \circ u^{-1}$. *Then if* $S_\mu \equiv \operatorname{supp}(\mu)$ *exists, so does* $\operatorname{supp}(\nu)$ *and*

$$S_\nu \equiv \operatorname{supp}(\nu) = \overline{u(S_\mu)}; \tag{4.1}$$

further, if E is a topological subspace of F and u is the injection map of E into F, then

$$S_\mu = E \cap S_\nu. \tag{4.2}$$

Now we are ready to prove Propositions 1 and 2.

Proof of Proposition 1. The existence of $\text{supp}(\mu_\alpha)$ and the proof of (3.1) is immediate from Lemma 1. The proof of (3.2) follows from (4.1), Corollary (ii) of [4, p. 49] and the fact that S_μ is a closed set of E.

Proof of Proposition 2. Let $\hat{E}$ be the completion of E relative to $\sigma(E, F)$; then $\hat{E}$ is algebraically and topologically isomorphic ($\cong$) to the topological product vector space $\Pi_{\lambda \in \Lambda} R_\lambda \equiv R^\Lambda$, for some index Λ (see [1, p. 38] and [5, p. 46]). Let Γ denote the family of all non-empty finite subsets of Λ ordered by inclusion, and $g_{\alpha\beta}$ denote the projection of R^β onto R^α, for $\alpha \leq \beta$; then $R^\Lambda \cong \varprojlim g_{\alpha\beta} R^\alpha$. It follows $\hat{E} \cong \varprojlim g_{\alpha\beta} R^\alpha$. For the purpose of the proof of this proposition, we can take $\hat{E} = \varprojlim g_{\alpha\beta} R^\alpha$.

Let i be the injection of E into $\hat{E}$ and $\nu = \mu \circ i^{-1}$. Then, clearly, ν is centered Gaussian on $\hat{E}$, and, by (3.2), $S_\nu = \varprojlim g_{\alpha\beta} S_{\nu_\alpha}$, where $S_{\nu_\alpha} \equiv \text{supp}(\nu_\alpha)$, $\nu_\alpha = \nu \circ g_\alpha^{-1}$ and g_α is the natural projection of $\hat{E}$ into R^α. But, since ν_α is clearly centered Gaussian on R^α, S_{ν_α} is a (closed) linear subspace of R^α. It follows that S_ν is a (closed) subspace of $\hat{E}$. Consequently, since by (4.2), $S_\mu = S_\nu \cap E$, we have the desired result.

For the proofs of some of the rest of the results stated in Section 3, we will need the following two lemmas, which we prove first.

Lemma 2. *Let E be a metric space with metric d and let μ be a probability measure on E. If* $\text{supp}(\mu)$ *exists, then it is separable.*

Proof. For $\varepsilon > 0$, let $A_\varepsilon = \{F \subset E: d(x, y) \geq \varepsilon, x, y \in F\}$. Since E contains at least two points (otherwise we are done), $A_{1/n} \neq \phi$, for all positive integers larger than, say, n_0. Now for every fixed $n > n_0$, every monotone increasing (ordered by inclusion) chain in $A_{1/n}$ has clearly a maximal element. It follows, from Zorn's Lemma, that $A_{1/n}$ has a maximal element, say, F_n. For every $\delta > 0$ and $x \in S_\mu \equiv \text{supp}(\mu)$ let $\Delta(x, \delta) = \{y \in S_\mu : d(x, y) < \delta\}$. If $n > n_0$, then, using maximality of F_n, we have $\bigcup_{x \in F_n} \Delta(x, 1/n) = S_\mu$; and using the facts that $F_n \in A_{1/n}$ and $\mu(\Delta(x, \delta)) > 0$, for all $x \in S_\mu$ and $\delta > 0$, we have that F_n is countable. Now it is clear that $F_0 = \bigcup_{n > n_0} F_n$ is a countable dense subset of S_μ. Completing the proof.

Lemma 3. *Let μ be a centered Gaussian probability measure on a Banach space E. If $S_\mu \equiv \operatorname{supp}(\mu)$ exists, then S_μ is a separable (closed) subspace of E.*

Proof. By Lemma 2, S_μ is a separable subset of E. Let F be the closure of the linear span of S_μ; then clearly F is a separable Banach subspace of E, and $\nu = \mu/F$ is centered Gaussian on $(F, \mathscr{B}_F)$. (There are a number of known proofs of the fact that the supp of a Gaussian probability measure on a separable Banach space is a subspace (see, for example, [9–11, 13, 14]). Recently, Rajput [18] has obtained a simple proof of the fact that the supp of a τ-regular symmetric stable measure with index $1 \leqslant \alpha \leqslant 2$ on certain TV spaces is a subspace, we present his proof for Gaussian case in separable Banach space setting).

Using the fact that characteristic functions determine measures uniquely in separable Banach spaces, we have $\nu^a * \nu^b = \gamma^{c(a,b)}$, where $c(a, b) = (a^2 + b^2)^{1/2}$, ν^d (for a fixed $d > 0$) is defined by $\nu^d(B) = \nu(d^{-1}B)$, for every $B \in \mathscr{B}_F$, and $*$ denotes the usual convolution. This along with the facts that $S_{\nu^d} \equiv \operatorname{supp}(\nu^d) = dS_\nu$, $(S_\nu \equiv \operatorname{supp}(\nu))$, for every $d > 0$, and $S_\nu = -S_\nu$ (since ν is symmetric) and Theorem 2.2.2 of [7, p. 38] imply

$$\overline{aS_\nu \pm bS_\nu} = c(a, b)S_\mu \tag{4.3}$$

for all $a > 0$, $b > 0$. Letting $a = b = 1/\sqrt{2}$ in (4.3), we conclude $\sqrt{2}S_\nu \subseteq S_\nu$, which in turn gives $2^n S_\nu \subseteq S_\nu$ for all $n = 1, 2, \ldots$. Further letting $a > 0$, $b > 0$ with $a^2 + b^2 = 1$ in (4.3), we get $dS_\nu \subseteq S_\nu$, for all $0 < d < 1$; consequently, since $0 \in S_\nu$ (again by (4.3)) and since S_ν is symmetric $dS_\nu \subseteq S_\nu$, for all $0 \leqslant |d| \leqslant 1$. Now let d be such that $|d| > 1$; choose a positive integer n so large that $|d|/2^n < 1$. Then $dS_\nu = 2^n((d/2^n)S_\nu) \subseteq 2^n S_\nu \subseteq S_\nu$. Thus $dS_\nu \subseteq S_\nu$, for all real d. This shows $dS_\nu = S_\nu$, for all real $d \neq 0$; consequently, since (by (4.3)) $S_\nu \supseteq aS_\nu + bS_\nu$ for any $a > 0$, $b > 0$ with $a^2 + b^2 = 1$, we have $S_\nu \supseteq S_\nu + S_\nu$. Completing the proof.

Now we are ready to give the proofs of Theorems 1, 2 and 3.

Proof of Theorem 1. Define a measure μ_0 on E by $\mu_0(B) = \mu(m_\mu + B)$, for every $B \in \mathscr{B}_E$. It is easy to verify that μ_0 is centered Gaussian on E, and by Lemma 1, $S_{\mu_0} \equiv \operatorname{supp}(\mu_0)$ exists. Further, clearly $S_\mu \equiv \operatorname{supp}(\mu) = m_\mu + S_{\mu_0}$. Thus, we need to show that S_{μ_0} is a (closed) subspace of E.

Let $\hat{E}$ be the completion of E and i be the natural injection of E into $\hat{E}$.

Let $\nu = \mu_0 \circ i^{-1}$. Clearly, ν is centered Gaussian on $\hat{E}$; and, by Lemma 1, $S_\nu \equiv \text{supp}(\nu)$ exists. Since $\hat{E}$ is a complete LCTVS, $\hat{E} = \varprojlim g_{\alpha\beta}E_\alpha$, the projective limit of a projective system $\{E_\alpha, g_{\alpha\beta}, \alpha \leq \beta\}$ of Banach spaces relative to some index set I [19, p. 53]. Let $\nu_\alpha = \nu \circ g_\alpha^{-1}$, $\alpha \in I$, where g_α is the natural projection of $\hat{E}$ into E_α. Then ν_α is, clearly, centered Gaussian on the Banach space E_α, for each $\alpha \in I$. Consequently, since by Lemma 1 $S_{\nu_\alpha} \equiv \text{supp}(\nu_\alpha)$ exists, we have from Lemma 3, that S_{ν_α} is a (closed) subspace of E_α. But from Proposition 1,

$$S_\nu = \varprojlim \tilde{g}_{\alpha\beta} S_{\nu_\alpha},$$

where $\tilde{g}_{\alpha\beta}$'s are the obvious maps. Thus S_ν is a subspace of $\hat{E}$. Therefore, since by (4.2), $S_{\mu_0} = E \cap S_\nu$, S_{μ_0} is a subspace of E. This completes the proof.

Proof of Theorem 2. As in the proof of Theorem 1, we assume without loss of generality that $m_\mu = 0$. Since both S_μ and $\overline{K_\mu(E^*)}$ are closed subspaces of E, it is sufficient to prove that $S_\mu^\perp = \overline{K_\mu(E^*)}^\perp$, where $A^\perp$ denotes the usual annihilator of A relative to the natural duality between E and E^*. Let $y \in \overline{K_\mu(E^*)}^\perp$, then y vanishes on $K_\mu(E^*)$. In particular, $(K_\mu y)(y) = \int_E |\langle x, y\rangle|^2 \mu(\mathrm{d}x) = 0$; consequently, $\mu\{x \in E: \langle x, y\rangle = 0\} = 1$. Thus, since S_μ is the supp(μ), $\{x \in E: \langle x, y\rangle = 0\} \supseteq S_\mu$; i.e., $y \in S_\mu^\perp$. This shows $\overline{K_\mu(E^*)}' \subseteq S_\mu^\perp$. To show the reserve inclusion, let $y \in S_\mu^\perp$, then $\langle x, y\rangle = 0$, for all $x \in S_\mu$. Therefore $\langle K_\mu y, y\rangle = 0$; consequently, using (2.1), we have $\langle K_\mu z, y\rangle = 0$, for all $z \in E^*$ (here we are using the obvious fact $(K_\mu z)(y) = \langle K_\mu z, y\rangle$). Therefore, by continuity of $\langle \cdot, y\rangle$, we have $\langle x, y\rangle = 0$, for all $x \in \overline{K_\mu(E^*)}$, i.e., $y \in \overline{K_\mu(E^*)}^\perp$. Completing the proof.

Proof of Theorem 3. Again we assume $m_\mu = 0$. Let $\Lambda = \overline{u(E)}$ and let $x_0 \in F \setminus \Lambda$. Choose $y \in F^*$ such that $\langle \cdot, y\rangle$ vanishes on Λ and $\langle x_0, y\rangle = 1$. Let $V = \{x \in F: |\langle x, y\rangle| > 0\}$, then V is an open cylinder set in F and $u^{-1}(V) = \emptyset$, since $\langle \cdot, y\rangle$ vanishes on Λ. It follows $\mu(V) = \nu(u^{-1}(V)) = 0$. Thus, since V is an open nbd of x_0, $x_0 \notin S_\mu$. Consequently, $S_\mu \subseteq \Lambda$. We now show that S_μ is indeed equal to Λ. If not, then there would exist an $x_0 \in \Lambda \setminus S_\mu$, $x_0 \neq 0$; and we can choose $z \in F^*$ such that $\langle x, z\rangle = 0$, for all $x \in S_\mu$ and $\langle x_0, z\rangle = 1$. Thus, on one hand $\mu\{x \in F: |\langle x, z\rangle| > 0\} = 0$, and on the other hand

$$\begin{aligned}\mu\{x \in F: |\langle x, z\rangle| > 0\} &= \nu \circ u^{-1}\{x \in F: |\langle x, z\rangle| > 0\}\\ &= \nu\{x \in E: |\langle u(x), z\rangle| > 0\} = 1,\end{aligned}$$

since the element $y \in E^*$ defined by $\langle x, y\rangle = \langle u(x), z\rangle$, $x \in E$, is not 0 (here we are using the same notation for the bilinear forms on E and F). This is a contradiction, and hence $\Lambda = S_\mu$.

It may be worth pointing out that for the proof of Theorem 2 one needs only the facts that m_μ and $\operatorname{supp}(\mu)$ exist, $\operatorname{supp}(\mu)$ is linear and $K_\mu(E^*) \subseteq E$ and no other special properties of Gaussian probability measures are needed; consequently, Theorem 2 is applicable for any second order probability measure satisfying these hypotheses. Similar remarks apply to Theorem 2.

For the proofs of Propositions 3 and 4, we find it convenient to write the value of $y \in E^*$ at $x \in E$ as $y(x)$ rather than as $\langle x, y\rangle$ (here E, of course, is a TVS). This change of notation will be used in the following:

Proof of Proposition 3. The proof of this result when E is a separable Banach space is obtained in [22]. Using this we first show that the result holds for arbitrary (not necessarily separable) Banach space. So we first assume that E is a Banach space and μ satisfies the hypotheses of the proposition.

Since $\operatorname{supp}(\mu)$ exists, by Lemma 2, it is a separable subset of E; and hence F, the closure of the linear span of $\operatorname{supp}(\mu)$, is a separable Banach subspace of E and $\mu(F) = 1$. We now show that $\nu = \mu/F$ on F satisfies the analog of (3.3) for the same $\varepsilon > 0$, as for μ on E. Let $y \in F^*$ and $\hat{y}$ be a continuous extension of y to E, then

$$\nu\left\{x \in F\colon |y(x)| \geq \varepsilon\left|\int_F y(x)\nu(\mathrm{d}x)\right|\right\} =$$

$$= \mu\left\{x \in F\colon |\hat{y}(x)| \geq \varepsilon\left|\int_F \hat{y}(x)\nu(\mathrm{d}x)\right|\right\}$$

$$= \mu\left\{x \in E\colon |\hat{y}(x)| \geq \varepsilon\left|\int_E \hat{y}(x)\mu(\mathrm{d}x)\right|\right\},$$

since $\mu(E\backslash F) = 0$. On the other hand if $\hat{y} \in E^*$ and $y = \hat{y}/F$, then the same argument as above shows

$$\nu\left\{x \in F\colon |y(x)| \geq \varepsilon\left|\int_F y(x)\nu(\mathrm{d}x)\right|\right\} =$$

$$= \mu\left\{x \in E\colon |\hat{y}(x)| \geq \varepsilon\left|\int_E \hat{y}(x)\mu(\mathrm{d}x)\right|\right\}$$

Consequently,

$$\inf_{y\in F^*} \nu\left\{x \in F: |y(x)| \geq \varepsilon \left| \int_F y(x)\nu(dx) \right| \right\} =$$

$$= \inf_{\hat{y}\in E^*} \mu\left\{x \in E: |\hat{y}(x)| \geq \varepsilon \left| \int_E \hat{y}(x)\mu(dx) \right| \right\} > 0.$$

Thus, since F is a separable Banach space, it follows from the result mentioned in the beginning of this proof that m_ν, the mean of ν, exists. It is easy now to see that m_ν is also the mean of μ.

Now let E and μ be as in the proposition. As in the proof of Theorem 1, we have $E = \varprojlim g_{\alpha\beta}E_\alpha$. Let $\mu_\alpha = \mu \circ g_\alpha^{-1}$, where g_α is the natural projection of E into E_α. It is clear that μ_α is a first order probability measure on the Banach space E_α, for each α. Since $\operatorname{supp}(\mu)$ exists, $S_{\mu_\alpha} = \operatorname{supp}(\mu_\alpha)$ exists for each α (Proposition 1). Now we show that μ_α satisfies the analog of (3.3). Let $y \in E_\alpha^*$ and let

$$\Lambda_y = \left\{x \in E_\alpha: |y(x)| \geq \varepsilon \left| \int_{E_\alpha} y(x)\mu_\alpha(dx) \right| \right\},$$

where $\varepsilon > 0$ is the same as in (3.3). By the change of variable formula we have

$$g_\alpha^{-1}(\Lambda_y) = \left\{x \in E: |y \circ g_\alpha(x)| \geq \varepsilon \left| \int_{E_\alpha} y(x)\mu_\alpha(dx) \right| \right\}$$

$$= \left\{x \in E: |y \circ g_\alpha(x)| \geq \varepsilon \left| \int_E y \circ g_\alpha(x)\mu(dx) \right| \right\}.$$

Consequently,

$$\mu_\alpha(\Lambda_y) = \mu\left\{x \in E: |y \circ g_\alpha(x)| \geq \varepsilon \left| \int_E y \circ g_\alpha(x)\mu(dx) \right| \right\},$$

therefore,

$$\inf_{y\in E_\alpha^*} \mu_\alpha(\Lambda_y) = \inf_{y\in E_\alpha^*} \mu\left\{x \in E: |y \circ g_\alpha(x)| \geq \varepsilon \left| \int_E y \circ g_\alpha(x)\mu(dx) \right| \right\}$$

$$\geq \inf_{y\in E^*} \mu\left\{x: |y(x)| \geq \varepsilon \left| \int_E y(x)\mu(dx) \right| \right\} > 0.$$

Thus, by the previous paragraph, m_{μ_α}, the mean of μ_α, exists, for each α. Let $m_\mu = \{m_{\mu_\alpha}\}$; then it is shown in [16, p. 289] that m_μ is the mean of μ.

Proof of Proposition 4. Let $\varepsilon > 0$ be given. Let A be a $\sigma(E, E^*)$-compact convex, circled set such that $\mu(A) \geq 1 - \varepsilon/3$; and let ψ denote the characteristic function of μ. Then, for any $y_1, y_2 \in E^*$, we have

$$|\psi(y_1)-\psi(y_2)| \leq \int_E |e^{iy_1(x)}[1-e^{iy(x)}]|\mu(dx)$$

$$\leq \int_A |1-e^{iy(x)}|\mu(dx) + \int_{A^c} |1-e^{iy(x)}|\mu(dx), \quad (4.4)$$

where $y = y_2 - y_1$ and A^c is the complement of A.

Let $V = \{y \in E^*: \sup_{x\in A}|y(x)| \leq \varepsilon/3\}$; then V is a nbd of 0 in E^* for the Mackey topology. Now it is clear from (4.4) that if $y_2 - y_1 \in V$, then $|\psi(y_1)-\psi(y_2)| < \varepsilon$. Consequently, ψ is continuous on E^* with Mackey topology. It is clear that $\psi(y) = e^{-1/2(K_\mu y)(y)}$, for every $y \in E^*$; it follows now, from (2.1), that $(K_\mu y)(\cdot)$ is continuous on E^* with Mackey topology. Consequently, since Mackey topology is consistent with the natural duality between E and E^*, it follows $K_\mu(y)(\cdot)$ belongs to E. Finishing the proof.

References

[1] Badrikian, A. (1970). *Séminare sur les Fonctions Alèatoires Linéaires et les Measures Cylindriques.* Springer-Verlag, New York.

[2] Badrikian, A., and Chevet, S. (1974). Questions liées à la théorie des espaces de Wiener. *Ann. Inst. Fourier, Grenoble*, **24**, 1–25.

[3] Badrikian, A., and Chevet, S. (1974). *Measures Cylindriques, Espaces de Wiener et Fonctions Aléatoires Gaussiennes.* Springer-Verlag, New York.

[4] Bourbaki, N. (1966). *Elements of Mathematics, General Topology*, Part I. Addison-Wesley, Massachusetts.

[5] Day, M.M. (1962). *Normed Linear Spaces*, 2nd. ed. Academic Press, New York.

[6] Garsia, A.M., Posner, E.C., and Rodemich, E.R. (1968). Some properties of measures on function spaces induced by Gaussian processes. *J. Math. Anal. Appl.*, **21**, 150–160.

[7] Grenander, U. (1963). *Probabilities on Algebraic Structures.* Wiley, New York.

[8] Ito, K. (1970). The topological support of Gauss measure on Hilbert space. *Nagoya Math. J.*, **38**, 181–183.

[9] Jain, N.C., and Kallianpur, G. (1970). A note on uniform convergence of stochastic processes. *Ann. Math. Statist.*, **41**, 1360–1362.

[10] Jain, N.C., and Kallianpur, G. (1970). Norm convergent expansions for Gaussian processes in Banach spaces. *Proc. Amer. Math. Soc.*, **25**, 890–895.

[11] Kallianpur, G. (1971). Abstract Wiener processes and their reproducing kernel Hilbert spaces. *Z. Wahrscheinlichkeitstheorie und Verw. Gebiete*, **17**, 113–123.

[12] Kallianpur, G., and Nadkarni, M. (1972). *Supports of Gaussian measures.* Proceedings of Sixth Berkeley Symposium on Mathematical Statistics and Probability, Berkeley, Univ. of Calif. Press, Vol. II.

[13] Kuelbs, J. (1971). Expansions of vectors in a Banach space related to Gaussian measures. *Proc. Amer. Math. Soc.*, **27**, 364–370.

[14] LePage, R.D. (1972). Note relating Bochner integrals and reproducing kernels to series expansion on a Gaussian Banach space. *Proc. Amer. Math. Soc.*, **32**, 285–288.

[15] Rajput, B.S. (1972). Gaussian measures on L_p spaces, $1 \leq p < \infty$. *J. Multivariate Anal.*, **2**, 382–403.

[16] Rajput, B.S. (1972). On Gaussian measures in certain locally convex spaces. *J. Multivariate Anal.*, **2**, 282–306.
[17] Rajput, B. S. (1972). The support of Gaussian measures on Banach spaces. *Theory Prob. and Appl.*, **17**, 775–782.
[18] Rajput, B.S. (1974). On the support of certain symmetric stable probability measures on TVS. *Proc. Am. Math. Soc.* (to appear).
[19] Schaefer, H.H. (1970). *Topological Vector Spaces*, 3rd. printing. Springer, New York.
[20] Schwartz, L. (1973). *Radon Measures on Arbitrary Topological Spaces and Cylindrical Measures.* Oxford University Press, London, Bombay.
[21] Vakhania, N.N. (1966). On non-degenerate probability distributions on l_p spaces, $1 \leq p < \infty$. *Theory Prob. and Appl.*, **11**, 524–528.
[22] Vakhania, N.N. (1971). Probability Distributions on Linear Spaces. Tbilisi.
[23] Vakhania, N.N. (1975). On the topological support of Gaussian measure in Banach space. *Nagoya Math. J.*, **57**, 59–63.
[24] Vakhania, N.N., and Nguen Zui Tien (1975). On the support of probability measures on topological vector spaces (preprint).
[25] Vakhania, N.N., and Tarieladze, V.T. (1975). On covariance operators of probability measures on dual systems (preprint).

P.R. Krishnaiah, ed., *Multivariate Analysis–IV*

INFERENCE IN STOCHASTIC PROCESSES–VI: TRANSLATES AND DENSITIES*

M.M. RAO

University of California, Riverside, Calif., U.S.A.

In the signal plus noise model $Y_t = f_t + X_t$ with $\{X_t, t \in T\}$ as noise governed by a probability measure P, if P_f is the induced measure of Y, then f is an admissible translate (signal) provided P_f is P-continuous. In this paper, if P (i.e., the noise) is not necessarily Gaussian and if M_P is the set of admissible signals, then sufficient conditions are presented for M_P to be a cone, a convex set, or a linear space when the signals are nonstochastic. Stochastic admissible signals are also discussed if the noise is Gaussian. Under certain further smoothness conditions, the likelihood ratios $\mathrm{d}P_f/\mathrm{d}P$ (f nonstochastic) are obtained for a class of noise processes. The work is an extension of some earlier results of T.S. Pitcher and A.V. Skorokhod.

1. Introduction

Let $X = \{X_t, t \in T\}$ be a real stochastic process on (Ω, Σ, P) where Ω has the linear structure (e.g., $\Omega = \mathbf{R}^T$ or a Hilbert space below). If f is a real function on T, called a *signal*, the process $Y = \{Y_t, t \in T\}$ defined by (+) $Y_t = f_t + X_t$ on the same (Ω, Σ) is termed an *output* with X as *noise*, in the context of a communication channel. Let P_f be the induced measure, by the probability P and the Y-process. Then f is an *admissible translate* of X if P_f is P-continuous. The class $M(X)$ (also denoted M_P) of admissible translates of X (or of P) is of importance in the present work. In fact, if f_1, f_2 are in M_P then from the communication design point of view it is natural to demand that every point on the chord joining f_1 and f_2 is also admissible so that M_P should be convex. In the case that X is Gaussian, f nonstochastic, not only the convexity property of M_P is true, but even this set is linear. However, if X is non-Gaussian, then M_P is not necessarily convex as the following example implies:

If P is Gaussian and f_0 is an inadmissible translate so that $P_f \perp P$ (mutually singular) by the Hájek–Feldman dichotomy Theorem (such f_0's exist in abundance), define a mixture measure Q on Σ as:

* Research was sponsored by the Air Force Aerospace Research Laboratories, Air Force Systems Command, USAF, Contract F33615–74–C–4009.

$$Q = \sum_{k=-\infty}^{\infty} \alpha_k P_{kf_0}, \qquad \alpha_k > 0, \qquad \sum_{k=-\infty}^{\infty} \alpha_k = 1. \tag{1}$$

Then Q is non-Gaussian, and $f_0 \in M_Q$, the set of admissible translates for Q. But one sees after a small computation that $\alpha f_0 \notin M_Q$ for $0 < \alpha < 1$. Note that if f_1, f_2 are in M_P, then $f_1 + f_2$ is also in M_P, as an easy consequence of the chain rule for derivatives of measures, so that for *any probability P on Σ, M_P is a semi-group under addition.*

The convexity property of M_P is important not only for the design problem mentioned above, but also for obtaining the likelihood ratios, or densities, $\mathrm{d}P_f/\mathrm{d}P$ in many cases. In fact, the class of output processes determined by $\{P_f, f \in M_P\}$ are related by certain semi-groups (or groups) of transformations from X if M_P is convex (or linear) and then the densities are tied up with infinitesimal generators of these transformations. This, in turn, implies that the densities are related to certain generalized exponential families. This observation emerges from the work of Pitcher [13]. For this reason, it is crucial to analyze the structure of M_P since *P will not be taken to be Gaussian in this paper.* Thus the next section contains some general results on M_P under various hypotheses including certain elementary considerations on stochastic signals. The main result there is based on Pitcher's work [12] and more particularly contains the proper (strengthened) hypothesis so that the main assertion of [12] obtains. In Section 3, under somewhat more stringent conditions than those of the preceding section, it is shown that M_P is linear and then the densities $\mathrm{d}P_f/\mathrm{d}P$ are computed. The main result here is a simplification (of the proof) and a generalization of some work of Skorokhod [18], based on the properties of Orlicz spaces. Some remarks on the multidimensional (or vector) processes are indicated in the last section.

The second order case of Section 2 was announced in [15]. If P were Gaussian, a more complete description is found in ([8, 10–11]). The difficulties in the general case are seen from [7] and [4] which contain a more detailed account of the results for certain general classes of processes. The approach of ([12, 13, 14]) and [18] and hence of the following sections is different from that of [4] and [7]; and both these are complementary in many respects.

2. Structure of M_P: General results

Since by definition $M_P \subset \mathbf{R}^T$, it can be described, for many classes of processes, as follows:

Theorem 1. (a) *Let X be a second order process with mean zero, $T = [a, b] \subset \mathbf{R}$, and continuous covariance r. Taking $\Omega = \mathbf{R}^T$, and thus (Ω, Σ, P) as the canonical representation of the process X, let $R: \mathrm{L}^2(T, \mathrm{d}t) \to \mathrm{L}^2(T, \mathrm{d}t)$ be the integral operator defined by $(Rg)(s) = \int_T r(s, t)g(t)\mathrm{d}t$ so that R is a positive definite Hermitian operator. Then for non-stochastic signals, $M_P \subset R^{1/2}(\mathrm{L}^2(T, \mathrm{d}t))$, i.e. $f \in M_P$ implies $f = R^{1/2}h$ for some $h \in \mathrm{L}^2(T, \mathrm{d}t)$ where $R^{1/2}$ is the positive square root of R in $\mathrm{L}^2(T, \mathrm{d}t)$. The inclusion may be strict.*

(b) *If in* (a) *X is moreover Gaussian but $f = \{f_t, t \in T\}$ is a measurable process independent of X and the $\{Y_t, t \in T\}$ is stochastically continuous, then $f \in M_P$ whenever $f_{(\cdot)}(\omega) \in R^{1/2}(\mathrm{L}^2(T, \mathrm{d}t))$ for a.a. (ω).* [*In this case $M_P = R^{1/2}(\mathrm{L}^2(T, \mathrm{d}t))$, a.e.*] *Also P_f is equivalent to P.* (*If f did not depend on ω* (*i.e. non-stochastic*) *this is a special case of the classical results, cf.* [11], [15].)

(c) *If $X = \{X_t, \mathcal{F}_t, t \in [a, b]\}$ is a standard Brownian motion and $f = \{f_t, \mathcal{F}_t, t \in [a, b]\}$ is an adapted process such that f_t is an integrated martingale, i.e. $f_t = \int_a^t Z_s \mathrm{d}s$ where $\{Z_s, \mathcal{F}_s, s \in [a, b]\}$ is a martingale* (Z_t *is a non-anticipated Brownian functional*) *then $f \in M(X) = M_P$. If moreover Z and X are independent then P_f and P are equivalent.* [$\mathcal{F}_{t-} = \sigma(\bigcup_{s<t} \mathcal{F}_s)$.]

(d) *If Ω is a Hilbert space with a countable base and Σ is its Borel σ field so that $P_f(A) = P(A - f)$, $A \in \Sigma$, $f \in \Omega$,* (*i.e. a non-stochastic translation*), *then $f \in M_P$ implies the existence of a positive definite trace class operator B on Ω such that $f = B^{1/2}h$ for some $h \in \Omega$* (*i.e. $M_P \subset B^{1/2}(\Omega)$*) *where $B^{1/2}$ is the positive square root of B.* [*As usual, $A - f = \{g - f: g \in A\} \subset \Omega$.*]

Proof. The proofs of (a) and (d) are essentially the same. The former is given in ([12], p. 539) and the latter based on [18] will be sketched (with a completion) here.

Thus for (d), since P is a measure on (Ω, Σ), the Fourier transform $C(\cdot)$ where $C(l) = \int_\Omega \mathrm{e}^{\mathrm{i}\langle l, \omega\rangle}\mathrm{d}P(\omega)$ is continuous for the 'topology of Sazonov', i.e. there exists a positive definite trace class operator B on the Hilbert space Ω (determined by P) such that

$$C(l) \to C(l_0) \quad \text{as} \quad \|B^{1/2}(l - l_0)\|^2 = \|l - l_0\|_B^2 = \langle B(l - l_0), (l - l_0)\rangle \to 0,$$

(cf. [17], Part II, Ch. II). If $f \in M_P$ and $C_f(\cdot)$ is the corresponding transform of P_f, and $\rho_f = \mathrm{d}P_f/\mathrm{d}P$, then

$$C_f(l) = \int_\Omega \mathrm{e}^{\mathrm{i}\langle l, \omega\rangle}\rho_f(\omega)\mathrm{d}P(\omega). \tag{2}$$

Thus, if $\rho_f^n = \rho_f$ for $\rho_f \leq n$, $= 0$ if not, then

$$\begin{aligned} |C_f(l) - 1| &\leq 2\,\mathrm{Re}\,(1 - C_f(l)) \\ &\leq 2n\,\mathrm{Re}\,(1 - C(l)) + 4\int_\Omega (\rho_f(\omega) - \rho_f^n(\omega))\,\mathrm{d}P, \end{aligned} \tag{3}$$

and the right side of (3) tends to zero if $\| l \|_B \to 0$ first and then $n \to \infty$ where B, associated to P, is given above.

Let $\Omega_1 = \{l : \| l \|_B \leq \delta\}$ where $\delta > 0$ is suitably chosen later, and let $\Omega_2 = \Omega - \Omega_1$. By (3) $C_f(l) = e^{i\langle l, f\rangle} C(l) \to 1$ as $\| l \|_B \to 0$ so that $\langle l, f\rangle \to 0$ also. For $\varepsilon > 0$, take the above δ $(= \delta_\varepsilon)$ such that $|\langle l, f\rangle| < \varepsilon$ if $\| l \|_B \leq \delta$. Let $\tilde{l} = l\delta / \| l \|_B$, $l \in \Omega_2$ so $\| \tilde{l} \|_B = \delta$ and $\tilde{l} \in \Omega_1$, $|\langle \tilde{l}, f\rangle| < \varepsilon$. Hence $|\langle l, f\rangle| < (\varepsilon/\delta) \| l \|_B$. Now replacing l by $B^{-1/2} l$ here (since on $B^{1/2}(\Omega)$, $B^{-1/2}$ exists) one has $|\langle B^{-1/2} l, f\rangle| < (\varepsilon/\delta) \| B^{-1/2} l \|_B$. This is $< \varepsilon$ if $l \in B^{1/2}(\Omega_1)$, and $< (\varepsilon/\delta) \| l \|_B$ if $l \in B^{1/2}(\Omega_2)$. Hence the linear functional $x_f^*(l) = \langle B^{-1/2} l, f\rangle = \langle l, B^{-1/2} f\rangle$ satisfies $\sup\{|x_f^*(l)| : \|(l)\|_B \leq 1\} \leq \max(\varepsilon, \varepsilon/\delta)$, and so by the Riesz representation there exists uniquely a $g \in \Omega$ such that $x_f^*(l) = \langle l, g\rangle$, $l \in \Omega$. Thus $g = B^{-1/2} f$ or $f = B^{1/2} g$ so that $M_P \subset B^{1/2}(\Omega)$.

Part (b) is essentially given in [9]. (See also [2], p. 694.)

To prove (c), first note that $\{Z_s, \mathcal{F}_s, s \in T\}$, $T = [a, b] \subset \mathbf{R}$ may be taken to be a right continuous martingale, replacing it by such a modification, since $\{f_t, t \in T\}$ is then unchanged. Hence $Z^* = \sup_t |Z_t|$ and $Z^{*2} = \sup_t |Z_t|^2$ are measurable. But $\{|Z_t|, \mathcal{F}_t, t \in T\}$ is a uniformly integrable submartingale (since T is a compact interval), so that by the maximal inequality ([6], p. 314), $\boldsymbol{P}[Z^* > \lambda] \leq (1/\lambda) E(|Z_b|)$, $\lambda > 0$; so $Z^* < \infty$ a.e., where $\boldsymbol{E}$ is the expectation operator. This implies also that $Z^{*2} < \infty$ a.e. Since $Z_t^2 \leq Z^{*2}$ a.e., it follows that $\int_T Z_t^2 \mathrm{d}t \leq Z^{*2}(b - a) < \infty$ a.e. By hypothesis, the process $\{Z_s, \mathcal{F}_s, s \in T\}$ is a non-anticipating Brownian functional. Hence by ([9], Theorem 1′) one deduces that $f \in M_P$. The last assertion is also a consequence of [9].

Remarks. 1. If $T = \mathbf{N}$, the natural numbers, then the hypothesis of (c) becomes (or weakened to): $X = \{X_n, n \in \mathbf{N}\}$ is a set of independent normalized Gaussian random variables and $f = \{f_n, \mathcal{F}_n, n \geq 1\}$ is such that (and this is the weakening) it is an $L^1(P)$-bounded martingale with X_n independent of $\mathcal{F}_n$, $n \geq 1$. Then by the remarkable result of Austin [1], $\Sigma_{n=1}^{\infty} g_n^2 < \infty$, a.e. where $g_n = f_n - f_{n-1}$ and this implies that $f \in M_P$ by ([9], Thm. 1). In the continuous parameter case, the somewhat involved condition was used above. Note that, if $\boldsymbol{E}(Z_s^2) < \infty$, $s \in T$, then $\boldsymbol{E}(Z^{*2}) < \infty$ by ([6], p. 317), and hence $\boldsymbol{E}(\int_T Z_t^2 \mathrm{d}t) < \infty$.

2. In both (a) and (d) $M_P \subset \Omega$ but generally $M_P \notin \Sigma$. If X is Gaussian, then the outer measure $P^*(M_P) = 0$, as is known for a long time (cf., e.g. [15]). In the present case, the size of M_P (i.e. the number $P^*(M_P)$) is not known, (cf. also [20]). Results related to certain aspects of the above theorem may be found in [2] and [5] as well as [16] and [10a].

For non-stochastic signals, conditions for M_P to be a cone or linear space can be obtained if the noise is a general second order process. The following is such a result whose conditions resemble those of the consistency of classical maximum likelihood estimators of a parameter. It was announced in [15], and is a strengthened version of a key result of [12].

Theorem 2. *Let $X = \{X_t, t \in T\}$, be a second order process with zero mean and continuous covariance r on $T = [a, b] \subset \mathbf{R}$. Let $\{\lambda_n\}$ and $\{\varphi_n\}$ be the eigenvalues and eigenfunctions of r and let $\{X_n, n \geq 1\}$ and $\{f_n, n \geq 1\}$ be defined by the Karhunen representation as:*

$$X_n = \lambda_n^{-1/2} \int_T X(t)\varphi_n(t)\,\mathrm{d}t,$$
$$f_n = \lambda_n^{1/2} \int_T f(t)\varphi_n(t)\,\mathrm{d}t, \quad f \in M_P = M(X), \tag{4}$$

so that $\{X_n\}_1^\infty \subset L^2(P)$ are orthonormal and $f = (f_1, f_2, \ldots) \in l^2$. Let $\{P_n\}$ be the n-dimensional distributions of $\{X_i, 1 \leq i \leq n\}$ with densities $\{p_n\}$, relative to the Lebesgue measure in $\mathbf{R}^n$. Suppose that $\{p_n, n \geq 1\}$ satisfies the following conditions:

(i) *$p_n > 0$ a.e., $n \geq 1$,*
(ii) *$\lim_{|t_j| \to \infty} p_n(t_1, \ldots, t_j, \ldots, t_n) = 0$, $1 \leq j \leq n$, for almost all t_i $(i \neq j)$,*
(iii) *$\partial p_n / \partial t_j$, $1 \leq j \leq n$ exists for each n, and*
(iv) *there exists an absolute constant K_0, such that*

$$\sum_{j=1}^{n} \int_{\mathbf{R}^n} \left(\frac{\partial \log p_n}{\partial t_j}\right)^2 \mathrm{d}P_n \leq K_0 < \infty, \qquad n \geq 1.$$

Then the set $M_1(X) = \{f \in M(X) : f = \{f_n, n \geq 1\} \in l^1\}$ is a positive cone. If, moreover $\sum_{n=1}^{\infty} \lambda_n^{-1/2} < \infty$, then $M_1(X) = M(X)$ and if p_n is symmetric about the origin of $\mathbf{R}^n$ for each n, then M_1 is linear.

Remark. If (iv) is weakened to "each term of the sum is uniformly bounded," then the result need not be valid, i.e., $M(X)$ is not a cone as M_Q of (1) shows. The strengthened hypothesis allows an adaption of the beautiful argument of [12], and it will be sketched here with appropriate changes. (*From now on only nonstochastic signals are considered.*)

Proof. By the Karhunen representation,

$$X(t) = \sum_{n=1}^{\infty} X_n \frac{\varphi_n(t)}{\sqrt{\lambda_n}}, \qquad t \in T, \tag{5}$$

the series converging in $L^2(P)$ for each t. Moreover, by Mercer's theorem

$$\sum_{n=1}^{\infty} |f_n|^2 = \int_T \int_T f(t)\overline{f(s)}r(s,t)\mathrm{d}s\,\mathrm{d}t < \infty, \tag{6}$$

so that $\boldsymbol{f} = \{f_1, f_2, \ldots\} \in l^2$.

Let $f \in M_1(X)$, $\boldsymbol{f} = \{f_1, f_2, \ldots\} \in l^1 \subset l^2$, and if $\alpha \geq 0$, define

$${}_\alpha Y_n^2(\omega) = p_n(X_1(\omega) - \alpha f_1, \ldots, X_n(\omega) - \alpha f_n)/p_n(X_1(\omega), \ldots, X_n(\omega)).$$

If $\mathcal{F}_n = \sigma(X_1, \ldots, X_n)$ then $\{{}_\alpha Y_n, \mathcal{F}_n, n \geq 1\}$ is a (positive) supermartingale on (Ω, Σ, P) and $\|{}_\alpha Y_n\|_2 = 1$. Hence ${}_\alpha Y_n \to {}_\alpha Y_\infty$ a.e. $[P]$ and in $L^2(P)$ by the standard martingale theory [6]. Let $V_n(\alpha)\colon L^2(P) \to L^2(P)$ be defined by

$$(V_n(\alpha)g)(X_1, \ldots, X_n) = {}_\alpha Y_n \cdot g(X_1 - \alpha f_1, \ldots, X_n - \alpha f_n), \tag{7}$$

where g is a bounded tame function, i.e. $g \circ \pi_n\colon \Omega \to \mathbf{R}$ and g depends on a "finite number of coordinates". It can be shown that $V_n(\alpha)h \to V(\alpha)h$ a.e. and in $L^2(P)$, for each $h \in \mathcal{M} = \bigcup_n L^\infty(\mathcal{F}_n)$, that for each n, $\{V_n(\alpha), \alpha \geq 0\}$ is a strongly continuous semi-group of isometries on $L^2(\mathcal{F}_n)$ and that the $V(\alpha)$ is defined as in (7) with ${}_\alpha Y_\infty$ in place of ${}_\alpha Y_n$. The proof of these facts may be found in [12] and [16], the latter has more detailed computations often omitted here. It then follows, from the work there, that $V(\alpha)$ is defined and is strongly measurable on all of $L^2(\mathcal{F}_\infty)$ where $\mathcal{F}_\infty = \sigma(\bigcup_n \mathcal{F}_n)$.

Suppose it is proved that $\{V(\alpha), \alpha \geq 0\}$ is also a strongly continuous semi-group of isometries on $L^2(\mathcal{F}_\infty)$. Since $V(1)$ is an isometry then for each $0 < \alpha < 1$ one has

$$\|h\|_2 = \|V(1)h\|_2 = \|V(\alpha)V(1-\alpha)h\|_2 \leq \|V(\alpha)h\|_2 = \|h\|_2, \quad h \in \mathcal{M},$$

and moreover for a tame h,

$$\begin{aligned}\int_\Omega (V(\alpha)1)^2 h(X_1, \ldots, X_n)(\omega)\,\mathrm{d}P(\omega) &= \\ &= \int_{\mathbf{R}^n} \frac{p_n(x_1 - \alpha f_1, \ldots, x_n - \alpha f_n)}{p_n(x_1, \ldots, x_n)} h(x_1, \ldots, x_n)\mathrm{d}P_n(x_1, \ldots, x_n) \\ &= \int_{\mathbf{R}^n} h(x_1 + \alpha f_1, \ldots, x_n + \alpha f_n)\mathrm{d}P_n(x_1, \ldots, x_n) \\ &= \int_\Omega h(X_1, \ldots, X_n)(\omega)\mathrm{d}P_{\alpha f}(\omega). \end{aligned} \tag{8}$$

Since such h's form a dense set in $L^2(\mathcal{F}_\infty)$, (8) implies that $(V(\alpha)1)^2 = \mathrm{d}P_{\alpha f}/\mathrm{d}P$ and $\alpha f \in M_1(X)$. Hence the result follows if the semi-group property of $\{V(\alpha), \alpha \geq 0\}$ stated above is established. This is the key step of the proof, and all the hypothesis of (ii)–(iv) is used for it, as will become clear from what follows.

Let $R_n(\lambda) = \int_0^\infty e^{-\lambda\alpha} V_n(\alpha) d\alpha$, $\lambda > 0$, and let R be similarly defined with $V(\cdot)$, as a strong integral. Since $\{V_n(\alpha), \alpha \geq 0\}$ is a strongly continuous semi-group, $\{R_n(\lambda), \lambda > 0\}$ is a resolvent (by the classical Hille–Yosida theorem) and the $R_n(\lambda) \to R(\lambda)$ strongly for $\lambda > 0$. Thus by the same theorem, it suffices to show that $\{R(\lambda), \lambda > 0\}$ is a resolvent on $L^2(\mathcal{F}_\infty)$ which then implies the desired result for $\{V(\alpha), \alpha \geq 0\}$. For this, by an important approximation theorem of Trotter [19], one needs to verify:

(1) $R(\lambda) - R(\lambda') = (\lambda' - \lambda) R(\lambda) R(\lambda')$ for $\lambda, \lambda' > 0$,

(2) $\| \lambda^n R(\lambda)^n \| \leq K_1 < \infty$ for all $n \geq 1$, $\lambda > 0$, and

(3) $\lim_{\lambda \to \infty} \lambda R(\lambda) = I$ strongly in $L^2(\mathcal{F}_\infty)$, where I is the identity operator.

Since $R_n(\lambda) \to R(\lambda)$ strongly, (1) follows. Also for $h \in \mathcal{M}$, using $R(\lambda)^n h = R(\lambda)(R^{n-1}(\lambda) h)$, a simple estimate on the integral shows that $\| R(\lambda)^n h \|_2 \leq \lambda^{-n} \| h \|_2$ and this implies (2) with $K_1 = 1$. To prove (3), let g be a tame function with uniformly bounded first order partial derivatives. Then

$$
\begin{aligned}
(\lambda R_n(\lambda) g)(X) &= \int_0^\infty {}_\alpha Y_n g(X_1 - \alpha_1 f_1, \ldots, X_n - \alpha f_n) \frac{d}{d\alpha}(-e^{-\lambda\alpha}) d\alpha \\
&= g(X) - \int_0^\infty e^{-\lambda\alpha} {}_\alpha Y_n \\
&\quad \times \left(\sum_{i=1}^n f_i \frac{\partial g}{\partial x_i}(X_1 - \alpha f_1, \ldots, X_n - \alpha f_n) \right) d\alpha \\
&\quad - \int_0^\infty e^{-\lambda\alpha} \left\{ \left[\left(\sum_{i=1}^n f_i \frac{\partial \sqrt{p_n}}{\partial x_i} \right) g \right] \right. \\
&\quad \left. \times (X_1 - \alpha f_1, \ldots, X_n - \alpha f_n) p_n(X)^{-1/2} \right\} d\alpha. \qquad (9)
\end{aligned}
$$

This is a strong integral for which integration by parts is known to be valid. Since $\| {}_\alpha Y_n \|_2 = 1$, $\partial g / \partial x_i$ is bounded, and $\Sigma_{i=1}^\infty |f_i| < \infty$, the first integral of (9) is $\leq C\lambda^{-1}$ ($C > 0$ is a generic constant *independent* of n). In the second integral the factor in $\{\}$ satisfies, by the Cauchy inequality,

$$
\begin{aligned}
\int_\Omega \{\}^2 dP_n &= \int_{\mathbf{R}^n} \left(\sum_{i=1}^n f_i \frac{\partial \sqrt{p_n}}{\partial x_i}(x_1 - \alpha f_1, \ldots, x_n - \alpha f_n) \right)^2 dx_1 \ldots dx_n \\
&\leq \sum_{i=1}^n f_i^2 \sum_{i=1}^n \int_{\mathbf{R}^n} \left(\frac{\partial \sqrt{p_n}}{\partial x_i}(x) \right)^2 dx_1, \ldots dx_n \\
&\leq K_0 \sum_{i=1}^\infty |f_i| < \infty, \quad \text{by (iv).}
\end{aligned}
$$

Hence the term $\|\{\}\|_2 \leqslant K_2 < \infty$, and so the second term is $\leqslant C\lambda^{-1}$. Thus $\|\lambda R(\lambda)g - g\|_2 = \lim_n \|\lambda R_n(\lambda)g - g\|_2 \leqslant C\lambda^{-1} \to 0$ as $\lambda \to \infty$. Since such g's form a dense set in $L^2(\mathcal{F}_\infty)$, the result follows.

It is clear that if $\Sigma_n \lambda_n^{-1} < \infty$, then $M_1(X) = M(X)$ and if also p_n is symmetric about the origin, then $\{V(-\alpha), \alpha \geqslant 0\}$ is a semi-group and thus $\{V(\alpha), \alpha \in \mathbf{R}\}$ is a group by the above proof. This implies that $M_1(X)$ is linear, and the theorem is proved.

3. Structure of M_P: Convexity and densities

In this section Ω will be a separable Hilbert space for (Ω, Σ, P) and M_P is the set of admissible translates, $M_P \subset \Omega$ (i.e. non-stochastic signals) with P as a Radon probability. The proposition here simplifies, explains and extends the result ([18], Theorem 7).

Let I be the class of all finite-dimensional subspaces of Ω. Since $G \subset \Omega$ of finite codimension implies $\Omega/G = G^\perp = \Omega \ominus G \in I$ (isometric equivalence) the set I is also identifiable as the class of all closed subspaces of Ω, of finite codimension. For α, β in I, let $\alpha < \beta$ if $\Omega_\alpha \subset \Omega_\beta$ so that I is directed. Let $\Pi_\alpha : \Omega \to \Omega_\alpha (\cong \mathbf{R}^\alpha)$, $\Pi_{\alpha\beta} : \Omega_\beta \to \Omega_\beta$ $(\alpha < \beta)$ be the coordinate projections. Then $\{\Omega_\alpha, \Pi_{\alpha\beta}, \alpha < \beta \text{ in } I\}$ is a projective system and the projective limit, $\varprojlim (\Omega_\alpha, \Pi_\alpha)$ exists and is homeomorphic with the adjoint space Ω^*, when endowed with its weak topology, and which under the present hypothesis may be identified with Ω (cf. [17], pp. 177–180). Moreover, $P^\alpha \circ \Pi_\alpha^{-1}$, the image of P^α on Ω_α is a Radon probability. Since Ω is separable, the projective limit of these probabilities $(P^\alpha, \Pi_\alpha, \alpha \in I)$ exists and may be identified with P (cf. [17]). In fact, one may take I as the finite subsets of $\mathbf{N}$ and $\Omega^* \cong \mathbf{R}^{\mathbf{N}}$, and the collection may be taken as $\{\mathbf{R}^n, n \geqslant 1\}$ resembling the reduction in Theorem 2. This (somewhat involved) identification must be recognized here because of the occurrence of several changes of spaces in the problem.

Let $\mathcal{B}_n$ be the Borel σ-algebra of $\mathbf{R}^\alpha$ (α having n-points) and $\mathcal{F}_n = \Pi_\alpha^{-1}(\mathcal{B}_n)$. Then $\mathcal{F}_n \subset \Sigma$ forms an increasing sequence and $P^\alpha = P \circ \Pi_\alpha^{-1} : \mathcal{B}_n \to [0, 1]$ can be identified with $P^\alpha = P_n \circ \Pi_\alpha^{-1}$, $P_n = P \mid \mathcal{F}_n$. Similarly if $Q = P_a$, $Q' = P_{ta}$, $t \geqslant 0$, $a \in M_P$, let $Q_n = Q \mid \mathcal{F}_n$, $Q'_n = Q' \mid \mathcal{F}_n$ and $Q^{t\alpha} = Q' \circ \Pi_\alpha^{-1}$. Then $Q' \perp P$ (mutually singular) if $Q'_n \perp P_n$ for some n. However, that Q'_n is P_n-continuous for all n need not imply the same for Q' (cf. e.g. [17]). For further analysis, *suppose that*:

(*) each Q^{tn} and P^n is absolutely continuous relative to the n-dimensional Lebesgue measure for all $t \geq 0$, so that also Q_n^t, P_n are dominated by a σ-finite μ_n $(= \text{Leb} \circ \Pi_n^{-1})$.

This is a restrictive assumption. [In fact, Skorokhod ([18], p. 564) has shown that (*) is satisfied whenever for an orthonormal basis $\{e_n, n \geq 1\} \subset \Omega$, $te_n \in M_P$ for all $t \geq 0$, $n \geq 1$. This assumption is not made here, but it indicates the restrictive nature of (*), which is stronger than the corresponding condition in Theorem 2.] If

$$f_n = \frac{\mathrm{d}P_n}{\mathrm{d}\mu_n}, \qquad g_n = \frac{\mathrm{d}Q_n^t}{\mathrm{d}\mu_n} \quad \text{and} \quad h_n = \frac{\mathrm{d}Q_n^t}{\mathrm{d}P_n} = \frac{g_n}{f_n}, \quad n \geq 1, \text{ a.e. } [\mu_n],$$

where Q_n^t is P_n-continuous, then $\{h_n, \mathcal{F}_n, n \geq 1\}$ is a martingale and $h_n \to h = \mathrm{d}Q^t/\mathrm{d}P$ a.e. iff $\{h_n, n \geq 1\}$ is uniformly integrable, uniformly in t in compact intervals of $\mathbf{R}^+$ (cf. [6]). The problem is to find "good" sufficient conditions for the above uniformities. If

$$\tilde{g}_n = \frac{\mathrm{d}Q^{tn}}{\mathrm{d}x}, \qquad \tilde{f}_n = \frac{\mathrm{d}P^n}{\mathrm{d}x}, \qquad \tilde{h}_n = \frac{\tilde{g}_n}{\tilde{f}_n}, \quad \text{then} \quad h_n = \tilde{h}_n \circ \Pi_n, \quad n \geq 1.$$

Observe that $\{\tilde{h}_n, n \geq 1\}$, defined on spaces of different dimensions, is *not a martingale*, and so the distinction between h_n, $\tilde{h}_n$ should be maintained. Also note that for $A \in \mathcal{B}_n$, since $A - y = \{x : x + y \in A\} \in \mathcal{B}_n$,

$$Q_n^t \circ \Pi_n^{-1}(A) = \int_A \tilde{g}_n(x)\mathrm{d}x = P_{ta} \circ \Pi_n^{-1}(A) = P(\Pi_n^{-1}(A) - ta)$$

$$= \int_{A - \Pi_n(ta)} \tilde{f}_n \mathrm{d}x = \int_A \tilde{f}_n(x - \Pi_n(ta))\mathrm{d}x. \tag{10}$$

Hence $\tilde{g}_n(x) = \tilde{f}_n(x - \Pi_n(ta))$ a.a. (x) and $h_n(\omega, a) = \tilde{h}_n(\Pi_n(\omega)) = \tilde{f}_n(\Pi_n\omega - t\Pi_n(a))/f_n(\Pi_n(\omega))$ a.e. $[\mu_n]$. Thus $f_n = \tilde{f}_n \circ \Pi_n$ and the domain of $\tilde{f}_n$, being $\mathbf{R}^n$, has the standard differential structure. A *smoothness assumption* will be made on $\tilde{f}_n$:

(**) the density $\tilde{f}_n$ is continuously differentiable in $\mathbf{R}^n$ for each $n \geq 1$ and that (for simplicity) $\tilde{f}_n > 0$, a.e. (Leb).

Thus the gradient

$$\nabla \tilde{f}_n = \left(\frac{\partial \tilde{f}_n}{\partial x_1}, \ldots, \frac{\partial \tilde{f}_n}{\partial x_n}\right)$$

exists and

$$((\nabla \tilde{f}_n)(x), y) = \frac{\mathrm{d}}{\mathrm{d}s} \tilde{f}_n(x + sy)\bigg|_{s=0},$$

using the scalar product notation of $\mathbf{R}^n$. If $h_n^*(\cdot, a)$, $\tilde{h}_n^*(\cdot, \Pi_n(a))$ are defined by

$$\tilde{h}_n^*(x, \Pi_n(a)) = h_n^*(\omega, a) = (\nabla \tilde{f}_n(\Pi_n\omega), \Pi_n a)/\tilde{f}_n(\Pi_n\omega),$$
$$x = \Pi_n(\omega), \qquad \omega \in \Omega, \qquad a \in M_P, \tag{11}$$

then it is easily seen that $\{h_n^*(\cdot, a), \mathcal{F}_n, n \geq 1\}$ is a martingale on (Ω, Σ, P). Moreover, $h_n^*(\omega, \cdot)$ is linear on Ω, and using (11) one gets:

$$\int_0^t h_n^*(\omega - sa, a)\mathrm{d}s = -\log h_n(\omega, a). \tag{12}$$

The problem thus is to find conditions on h_n^* so that $h_n \to h$ a.e. with the desired uniformities. This is given in the following result:

Theorem 3. *Let Φ be a twice differentiable non-negative symmetric convex function on $\mathbf{R}$ with $\Phi(0) = 0$, $|x\Phi''(x)| \leq C < \infty$, $x \geq 0$ and $\Phi'(x) \uparrow \infty$ where the primes denote differentiation. If $\Psi(x) = \sup\{|x|y - \Phi(y)\colon y \geq 0\}$, called the complementary function of Φ, suppose that (in addition to (*) and (**) above)*

$$\sup_n \int_\Omega \Psi(\beta h_n^*(\omega, a))\mathrm{d}P(\omega) \leq C_1 < \infty, \qquad a \in M_P, \tag{13}$$

for some $\beta > 0$. Then $ta \in M_P$ for all $t \in \mathbf{R}$ (i.e. M_P is linear) and moreover, $h_n^* \to h^*$ a.e. and for all $t \in \mathbf{R}$; the density $h(\cdot, ta) = \mathrm{d}Q'/\mathrm{d}P$ is given by

$$h(\omega, ta) = \exp\left\{-\int_0^t h^*(\omega - sa, a)\mathrm{d}s\right\} \text{ a.e. } [P], \tag{14}$$

so that P_{ta} is equivalent to P.

Remark. If $\Phi(x) = |x|\log^+|x|$, so that $\Psi(x) = \mathrm{e}^{|x|} - |x| - 1$, the result reduces to that given in [18]. Also if $\Psi(x) = \mathrm{e}^{x^2} - 1$, then the corresponding Φ (which cannot be expressed in a closed form) satisfies the other hypotheses if (13) is again assumed. Similarly other convex Φ's are possible, as seen from the theory of Orlicz spaces. It is also interesting that the conditions yield the density (14) to be from a generalized exponential family.

Proof. Let $a \in M_P$, $t \in \boldsymbol{R}$ and consider the integral

$$I_n(t) = \int_\Omega \Phi''\left(\frac{f_n(\omega)}{f_n(\omega + ta)}\right)\mathrm{d}P_n(\omega). \tag{15}$$

The hypothesis implies that $I_n(\cdot)$ is differentiable and that the derivative I_n' is given by

$$I_n'(t) = -\int_\Omega \Phi''\left(\frac{f_n(\omega)}{f_n(\omega + ta)}\right) h_n^*(\omega + ta, a) \frac{f_n(\omega)}{f_n(\omega + ta)} \mathrm{d}P_n(\omega).$$

Using the growth conditions on Φ'', and the Young's inequality, this is simplified to:

$$\begin{aligned}
I_n'(t) &\leq \frac{C}{\beta}\int_{\mathbf{R}^n} \Phi\left(\frac{\tilde{f}_n(x)}{\tilde{f}_n(x + t\Pi_n(a))}\right)\tilde{f}_n(x + t\Pi_n(a))\mathrm{d}x \\
&\quad + \frac{C}{\beta}\int_{\mathbf{R}^n} \Psi(\beta\tilde{h}_n^*(x + t\Pi_n(a), \Pi_n(a))\cdot \tilde{f}_n(x + t\Pi_n(a))\mathrm{d}x \\
&\leq \frac{C}{\beta}\int_{\mathbf{R}^n} \frac{\tilde{f}_n(x)}{\tilde{f}_n(x + t\Pi_n(\omega))}\Phi'\left(\frac{\tilde{f}_n(x)}{\tilde{f}_n(x + t\Pi_n(a))}\right)\tilde{f}_n(x + t\Pi_n(a))\mathrm{d}x \qquad (16) \\
&\quad + \frac{C}{\beta}\int_\Omega \Psi(\beta h_n^*(\omega, a))\mathrm{d}P_n(\omega), \quad (\text{since } \Phi(x) \leq |x\Phi'(x)|), \\
&= \frac{C}{\beta} I_n(t) + \frac{C}{\beta} C_1, \quad \text{by (13)}.
\end{aligned}$$

Integrating (16) and using the fact that $I_n(0) = \Phi'(1)$, one has

$$I_n(t) \leq \frac{C}{\beta}\int_0^t I_n(s)\,\mathrm{d}s + C_2 + \Phi'(1), \tag{17}$$

where $C_2 = CC_1/\beta$. However it is well known that a measurable and integrable (on compacts) function $0 \leq g(t) \leq a + b\int_0^t g(s)\mathrm{d}s$, $b \geq 0$, must satisfy $g(t) \leq a e^{bt}$, so that (17) implies:

$$I_n(t) \leq \max\left(\frac{C}{\beta}, C_2\right)\exp[(\Phi'(1) + 1)t]. \tag{18}$$

Using again $\Phi(x) \leq |x\Phi'(x)|$, one has with (15),

$$\int_\Omega \Phi\left(\frac{f_n(\omega - ta)}{f_n(\omega)}\right)\mathrm{d}P_n(\omega) \leq \int_\Omega \Phi'\left(\frac{f_n(\omega)}{f_n(\omega + ta)}\right)\mathrm{d}P_n(\omega) = I_n(t). \tag{19}$$

Thus (18) and (19) imply for t in compact intervals:

$$\sup_n \int_\Omega \Phi\left(\frac{f_n(\omega - ta)}{f_n(\omega)}\right)\mathrm{d}P_n(\omega) < \infty. \tag{20}$$

Since $\Phi'(x) \uparrow \infty$, by the de la Vallée Poussin criterion $\{h_n(\cdot, a), n \geq 1\}$ is uniformly integrable uniformly in t in compact intervals of $\mathbf{R}$. Since this is also a martingale sequence, it converges a.e. (and in $L^1(P)$) to $h(\omega, ta) = dP_{ta}/dP(\omega)$, a.e., by the classical Andersen–Jessen–Doob theorem. Thus $ta \in M_P$ for t in compacts of $\mathbf{R}$ and hence (by the σ-compactness of $\mathbf{R}$) for all $t \in \mathbf{R}$. So M_P is linear.

It is well-known that $\Phi'(t) \uparrow \infty$ implies that $\Psi(\cdot)$ is continuous, and since $\{h_n^*(\cdot, a), \mathcal{F}_n, n \geq 1\} \subset L^\Psi(P)$ and is in a bounded part (by (13)), it follows again that $h_n^*(\cdot, a) \to h^*(\cdot, a)$ a.e. and in $L^1(P)$, $h^*(\cdot, a) \in L^\Psi(P)$. This implies, with the preceding paragraph and (12),

$$h(\omega, ta) = \lim_{n\to\infty} h_n(\omega, a) = \exp\left\{-\lim_n \int_0^t h_n^*(\omega - sa, a)\mathrm{d}s\right\}, \text{ a.e.} \quad (21)$$

It remains to show that (14) obtains from (21).

By the above $h_n^*(\omega - sa, a) \to h^*(\omega - sa, a)$, a.a. (ω) and all s in a compact interval. Let N be the P-null set such that this convergence holds for $\omega \in \Omega - N$. Using the exponential form, it is easy to see that $\lim_n \int_a^b h_n^*(\omega - sa, a)\mathrm{d}s$ exists for all $-\infty < a < b < \infty$, and by considering differences of these intervals it is also true that $\lim_n \int_A h_n^*(\omega - sa, a)\mathrm{d}s$ exists for all $\omega \in \Omega - N$ and Borel sets $A \subset T$, a compact interval. Let $\omega \in \Omega - N$ be arbitrarily fixed and define $\nu_{n,\omega}(A) = \int_A h_n^*(\omega - sa, a)\mathrm{d}s$. Then $\{\nu_{n,\omega}(\cdot), n \geq 1\}$ is a set of bounded σ-additive functions on the Borel σ-field of T, which converges set-wise. Hence by the classical Vitali–Hahn–Saks theorem $\nu_\omega(A) = \lim_n \nu_{n,\omega}(A)$ exists and $\nu_\omega(\cdot)$ is a signed measure which is absolutely continuous relative to the Lebesgue measure since each $\nu_{n,\omega}$ is, and that $\{\nu_{n,\omega}, n \geq 1\}$ is uniformly (in n) absolutely continuous for the Lebesgue measure. So given $\varepsilon > 0$, there is a $\delta_\varepsilon > 0$ such that $\mathrm{Leb}(B) < \delta_\varepsilon$ implies $|\nu_{n,\omega}(B)| < \varepsilon$ uniformly in n. Consequently, by a theorem of Bartle ([3], p. 348) the limit can be taken inside the integral in (21) and one has

$$\lim_n \int_A h_n^*(\omega - sa, a)\,\mathrm{d}s = \int_A h^*(\omega - sa, a)\,\mathrm{d}s, \quad \omega \in \Omega - N. \quad (22)$$

Thus (21) and (22) imply (14), and this completes the proof.

4. Some remarks

The preceding considerations were for scalar processes. The key reduction in Theorem 2 is the Karhunen representation. A multidimensional extension of this representation (due to W.L. Root and his colleagues) is known, and with it the multivariate extension of Theorem 2 is not difficult. On the other hand, only the Hilbert space structure of Ω is essentially used for Theorem 3. Thus this result also admits a multidimensional extension without much trouble. However, the notational complexity and some aspects of both these extensions are not entirely trivial. The details will be omitted here.

In case the signal and noise are independent Gaussian processes, Parzen [11] has given necessary and sufficient conditions for $\{f_t(\omega), t \in T\}$ to belong to M_P for almost all $\omega \in \Omega$, and moreover he gave an interesting derivation of the density using the tensor products of Aronszajn space theory. His results admit an easy extension to multidimensional signals and noises. On the other hand the general case, with a deep analysis considered by Pitcher [13], should be investigated in multidimensions and it will then be necessary and useful to find some algorithms to apply the results for practical problems. All these questions remain for future research.

References

[1] Austin, D.G. (1966). A sample function property of martingales. *Ann. Math. Statist.* **37**, 1396–1397.

[2] Baker, C.R. (1973). On equivalence of probability measures. *Ann. Prob.* **1**, 690–698.

[3] Bartle, R.G. (1956). A bilinear vector integral. *Studia Math.* **15**, 337–352.

[4] Briggs, V.D. (1975). Densities for infinitely divisible random processes. *J. Multivariate Anal.* **5**, 178–205.

[5] Cambanis, S. (1975). The relationship between the sure and stochastic signals in noise. *SIAM J. Appl. Math.* (to appear).

[6] Doob, J.L. (1953). *Stochastic Processes.* Wiley and Sons, New York.

[7] Gikhman, I.I. and Skorokhod, A.V. (1966). On densities of probability measures on functional spaces. *Uspekhi Math. Nauk* **21** (No. 6) 83–152.

[8] Girsanov, I.V. (1960) On transforming a certain class of stochastic processes by absolutely continuous substitution of measures. *Theor. Prob. Appl.* **5**, 285–301.

[9] Kadota, T.T. and Shepp, L.A. (1970). Conditions for absolute continuity between a certain pair of probability measures. *Z. Wahrs.* **16**, 250–260.

[10] Kailath, T. (1971). The structure of Radon–Nikodýn derivatives with respect to Wiener and related measures. *Ann. Math. Statist.* **42**, 1054–1067.

[10a] Kallianpur, G. (1959). A problem in optimum filtering with finite data. *Ann. Math. Statist.*, **30**, 659–669.

[11] Parzen, E. (1963). Probability density functionals and reproducing kernel Hilbert spaces. *Proc. Symp. on Time Series Analysis*, 155–169. Wiley & Sons, New York.

[12] Pitcher, T.S. (1963). The admissible mean values of a stochastic process. *Trans. Amer. Math. Soc.* **108**, 538–546.

[13] Pitcher, T.S. (1964). Parameter estimation for stochastic processes. *Acta Math.* **112**, 1–40.

[14] Pitcher, T.S. (1965). The behavior of likelihood ratios of stochastic processes. *Ann. Math. Statist.* **36**, 529–534.

[15] Rao, M.M. (1975). Inference in stochastic processes–V: Admissible means. *Sankhyā*, Ser. A (to appear).

[16] Rao, M.M. (1974). Admissible means and likelihood ratios of time series. Technical Report, ARL TR 74–0137.

[17] Schwartz, L. (1973). *Radon Measures on Arbitrary Topological Spaces and Cylindrical Measures.* Tata Institute, Bombay.

[18] Skorokhod, A.V. (1970). On admissible translates of measures in Hilbert space. *Theor. Prob. Appl.* **15**, 557–580.

[19] Trotter, H.F. (1958). Approximations of semi-groups of operators. *Pacific J. Math.* **8**, 887–919.

[20] Zinn, J. (1974). Zero–one laws for non-Gaussian measures. *Proc. Amer. Math. Soc.* **44**, 179–185.

P.R. Krishnaiah, ed., *Multivariate Analysis–IV*
© North-Holland Publishing Company (1977) 325–337

APPLICATION OF SEMI-INVARIANTS TO ASYMPTOTIC ANALYSIS OF DISTRIBUTIONS OF RANDOM PROCESSES

V.A. STATULEVIČIUS

Academy of Sciences of the Lithuanian SSR, Vilnius, U.S.S.R.

A method of semi-invariants is proposed for the investigation of asymptotical behaviour of distributions of sums of dependent random variables and some statistics for dependent observations.

1. Estimates for semi-invariants. Limit theorems under RMT regularity conditions

Let us consider a real random process $X_t, t \in T$ with a continuous $T = (-\infty, +\infty)$ or discrete $T = \{0, \pm 1, \pm 2, \ldots\}$ time. The process X_t is said [1, 2] to belong to the class $T^{(k)}$ if $\mathbf{E}|X_t|^k \leq C_k < \infty$ for all $t \in T$ and to the class $S^{(k)}$, if $X_t \in T^{(k)}$ and $\mathbf{E}X_{t_1} \ldots X_{t_l} = \mathbf{E}X_{t_1+\tau} \ldots X_{t_l+\tau}$ for all $1 \leq l \leq k$, $t_i \in T$, $t_i + \tau \in T$.

If $X_t \in T^{(k)}$, then there exists a correlation function

$$s_X^{(k)}(t_1, \ldots, t_k) = \Gamma\{X_{t_1}, \ldots, X_{t_k}\}$$

of the k-th order of the random process X_t. Here

$$\Gamma\{X_{t_1}, \ldots, X_{t_k}\} = \frac{1}{\mathrm{i}^k} \frac{\partial^k}{\partial u_1 \ldots \partial u_k} \ln \mathbf{E} \exp\left\{\mathrm{i} \sum_{j=1}^{k} u_j X_{t_j}\right\}\Bigg|_{u_1 \ldots u_k = 0}$$

is a sample semi-invariant of the random vector $(X_{t_1}, \ldots, X_{t_k})$.

Lemma 1 ([3]). *If $X_t \in T^{(k)}$, $t_1 < t_2 < \ldots < t_k$, then*

$$\Gamma\{X_{t_1}, \ldots, X_{t_k}\} = \sum_{\nu=1}^{k} (-1)^{\nu-1} \sum_{\cup_{p=1}^{\nu} I_p = I} N_\nu(I_1, \ldots, I_\nu) \prod_{p=1}^{\nu} \widehat{\mathbf{E}}(I_p). \quad (1)$$

Here $\Sigma_{\cup_{p=1}^{\nu} I_p = I}$ denotes the summation over all the partitions of the set $I = \{1, 2, \ldots, k\}$ into subsets $I_p \subset I$. If

$$I_p = \{i_1, i_2, \ldots, i_m\}$$

then

$$\mathbf{E}(I_p) = \mathbf{E}X_{t_{i_1}} \ldots X_{t_{i_m}}$$

and

$$\widehat{\mathbf{E}}(I_p)=\widehat{\mathbf{E}}X_{t_{i_1}}\dots X_{t_{i_m}}=\mathbf{E}X_{t_{i_1}}\widehat{X_{t_{i_2}}\dots\widehat{X_{t_{i_{m-1}}}\widehat{X}_{t_{i_m}}}}.$$

The sign "$\frown$" over a random variable means that it is centered, i.e.

$$\widehat{X}=X-\mathbf{E}X.$$

The integers $N_\nu(I_1,\dots,I_\nu)\geq 0$ depend only on the partition $I_1,\dots,I_\nu$ and, if $N_\nu(I_1,\dots,I_\nu)>0$, then

$$\sum_{p=1}^{\nu}\max_{i,j\in I_p}(t_j-t_i)\geq\max_{1\leq i,j\leq k}(t_j-t_i).$$

Moreover, there exists an absolute constant H_0 such that

$$\sum_{\nu=1}^{k}\sum_{\cup_{p=1}^{\nu}I_p=I}N_\nu(I_1,\dots,I_\nu)\leq H_0^k.$$

$\Gamma\{X_{t_1},\dots,X_{t_k}\}$ can be expressed in terms of the moments $\mathbf{E}(I_p)$:

$$\Gamma\{X_{t_1},\dots,X_{t_k}\}=\sum_{\nu=1}^{k}(-1)^{\nu-1}(\nu-1)!\sum_{\cup_{p=1}^{\nu}I_p=I}\prod_{p=1}^{\nu}\mathbf{E}(I_p),\tag{2}$$

but this formula is not helpful for the investigation of $\Gamma\{X_{t_1},\dots,X_{t_k}\}$ since the order of the summands in (2) is higher than that of $\Gamma\{X_{t_1},\dots,X_{t_k}\}$. Some of these terms must be combined to reduce these orders. This is done in formula (1).

If ξ_t, $t\in T$, $t\geq 0$ is a Markov process with initial distribution $P(A)$ and transition probabilities $P_{s,t}(x,A)$ and $X_t=g_t(\xi_t)$ where $g_t(x)$ is a measurable x function, then

$$\widehat{\mathbf{E}}X_{t_1}\dots X_{t_k}=$$

$$=\underbrace{\int\dots\int}_{k}g_{t_1}(x_1)P_{t_1}(\mathrm{d}x_1)\prod_{j=2}^{k}g_{t_j}(x_j)(P_{t_{j-1}t_j}(x_{j-1},\mathrm{d}x_j)-P_{t_j}(\mathrm{d}x_j)),$$

where $t_1<t_2<\dots<t_k$ and $P_t(A)=\mathbf{P}\{\xi_t\in A\}$. Therefore if

$$\sup_{x,A}|P_{s,t}(x,A)-P_t(A)|=\varphi(s,t),$$

$$\sup_x\int|g_t(y)||P_{s,t}(x,\mathrm{d}y)-P_t(\mathrm{d}y)|=\rho(s,t)$$

then

$$|\widehat{\mathbf{E}}X_{t_1}\dots X_{t_k}| \leqslant$$

$$\leqslant 2\varphi^{\frac{1}{2}}(t_1, t_2)\sqrt{\mathbf{E}X_{t_1}^2\mathbf{E}X_{t_2}^2}\prod_{j=2}^{k}\rho(t_{j-1}, t_j). \tag{3}$$

Analogous estimates hold not only for a process X_t related to a Markov process ξ_t but also for a process X_t satisfying a Markov type regularity (RMT) condition. Let T be discrete and $\mathscr{F}_s^t = \sigma\{X_u, s \leqslant u \leqslant t\}$. Assume $s < t$ and let

$$\gamma(s, t) = \sup_{A\in\mathscr{F}_t^t, B\in\mathscr{F}_{s+1}^{t-1}, C\in\mathscr{F}_s^s} |\mathbf{P}\{A \mid B, C\} - \mathbf{P}\{A \mid B\}|,$$

$$\rho_1(s, t) = \sup_{B\in\mathscr{F}_{s+1}^{t-1}, C\in\mathscr{F}_s^s} \int |x|\,|\mathbf{P}\{\mathrm{d}x \mid B, C\} - \mathbf{P}\{\mathrm{d}x \mid B\}|.$$

The estimate analogous to (3) holds:

Theorem 1. *If $X_t \in T^{(k)}$, then*

$$|\widehat{\mathbf{E}}X_{t_1}\dots X_{t_k}| \leqslant H_1^k\varphi^{\frac{1}{2}}(t_1, t_2)\sqrt{\mathbf{E}X_{t_1}^2\mathbf{E}X_{t_2}^2}\max\prod_{j=2}^{m}\rho_1(t_{i_{j-1}}, t_{i_j}), \tag{4}$$

$$|\widehat{\mathbf{E}}X_{t_1}\dots X_{t_k}| \leqslant$$

$$\leqslant H_2^k\left(\max\prod_{j=1}^{m}\gamma(t_{i_{j-1}}, t_{i_j})\right)^{1/p_1}\prod_{j=1}^{k}(\mathbf{E}|X_{t_j}|^{p_j})^{1/p_j}, \tag{5}$$

for any integers $p_j > 0$ with $1/p_1 + \dots + 1/p_k = 1$.

Here H_1 and H_2 are absolute constants, $t_{i_0} < t_{i_1} < \dots < t_{i_m}$ a subset of the numbers $t_1, \dots, t_k$, $t_{i_0} = t_1$, $t_{i_1} = t_2, \dots, t_{i_m} = t_k$. "max" is taken over all such subsets.

If $\sup_{s<t}\gamma(s, t) \to 0 (t - s \to \infty)$, then we say that the RMT condition is satisfied. When this condition is satisfied, it is not difficult, with the help of estimates (4), (5) and Lemma 1, to investigate asymptotic properties of the distribution $\mathbf{P}\{Z_n < x\}$ of the normed sum

$$Z_n = \frac{S_n - ES_n}{B_n}, \qquad S_n = X_1 + X_2 + \dots + X_n, \qquad B_n^2 = DS_n$$

of random variables $X_1, \dots, X_n$ because the semi-invariant of k-th order

$$\Gamma_k\{S_n\} = \Gamma\{\underbrace{S_n, \dots, S_n}_{k}\}$$

of a sum S_n satisfies the equation

$$\Gamma_k\{S_n\} = \sum_{1\leq t_1,\ldots,t_k\leq n} \Gamma\{X_{t_1},\ldots,X_{t_k}\}. \tag{6}$$

For instance, the following theorem holds.

Theorem 2. *If* $X_t \in T^{(\infty)}$, $\mathbf{E}X_t = 0$, $t = 1,\ldots,n$,

$$\sup_{1\leq s,t\leq n} \gamma(s,t) \leq C_1 \mathrm{e}^{-\gamma_n(t-s)}, \qquad \gamma_n > 0,$$

$$\mathbf{E}\{|X_j^p|\} \leq p!\,K^p \cdot DX_j, \qquad j = 1,2,\ldots,n \tag{7}$$

for all $p \geq 3$, *then for* $0 \leq x \leq \mathrm{o}(\Delta_n)$,

$$\Delta_n = \frac{\gamma_n B_n}{\max\limits_{1\leq j\leq n} \sqrt{DX_j}}$$

the equation of large deviations

$$\frac{\mathbf{P}\{Z_n > x\}}{1-\Phi(x)} = \mathrm{e}^{(x^3/\Delta_n)\lambda_n(x/\Delta_n)}\left(1+\mathrm{O}\left(\frac{1+x}{\Delta_n}\right)\right) \tag{7'}$$

holds.

One can get acquainted with the results in the field of limit theorems for $\mathbf{P}\{Z_n < x\}$ in [4]. A theorem analogous to Theorem 2 was proved there but only with a condition similar to (7) imposed on the conditional moments.

2. Estimation of the spectral distribution function and spectral density

The method of semi-invariants is useful for the investigation of asymptotic properties of statistics of random processes.

Let X_t, $t = 0, \pm 1, \pm 2, \ldots$ be a real stationary Gaussian sequence with mean 0 and spectral density $f(\lambda)$. $-\pi < \lambda \leq \pi$. Let us denote a spectral mean with weight function ψ by

$$A = A(\psi) = \int_{-\pi}^{\pi} \psi(\lambda) f(\lambda)\,\mathrm{d}\lambda.$$

The known estimates of the variables A and $f(\lambda)$, constructed from the sample $X_1,\ldots,X_N$ of size N, are

$$A_N = A_N(\psi) = \int_{-\pi}^{\pi} \psi(\lambda) I_N(\lambda) d\lambda$$

and

$$f_N(\lambda) = \int_{-\pi}^{\pi} \varphi_N(\tau - \lambda) I_N(\tau) d\tau,$$

where

$$I_N(\lambda) = \frac{1}{2\pi N} \left| \sum_{t=1}^{N} X_t \, e^{-it\lambda} \right|^2$$

is the so-called periodogram, φ_N is an arbitrary kernel, i.e. a family of functions $\{\varphi_N, N = 1, 2, \dots\}$, satisfying the conditions

(1) $\forall N \int_{-\pi}^{\pi} \varphi_N(\lambda) d\lambda = 1$,

(2) $\sup_N \int_{-\pi}^{\pi} |\varphi_N(\lambda)| d\lambda < \infty$,

(3) $\forall \delta > 0 \lim_{N \to \infty} \int_{(-\pi, \pi] \setminus [-\delta, \delta]} |\varphi_N(\lambda)| d\lambda = 0$.

We shall construct the kernel φ_N in the following manner:

$$\varphi_N(\lambda) = \begin{cases} \dfrac{1}{\varepsilon_N} \varphi\left(\dfrac{\lambda}{\varepsilon_N}\right), & -\pi\varepsilon_N < \lambda \leq \pi\varepsilon_N, \\ 0, \ \lambda \in (-\pi, \pi] \setminus (-\pi\varepsilon_N, \pi\varepsilon_N], \end{cases}$$

where φ is some bounded function, satisfying the condition

$$\int_{-\pi}^{\pi} \varphi(\lambda) d\lambda = 1,$$

$\{\varepsilon_N, N = 1, 2, \dots\}$ is a numerical sequence such that $0 < \varepsilon_N \leq 1$ and $\varepsilon_N \to 0$, $N\varepsilon_N \to \infty$ when $N \to \infty$.

We shall find exponential estimates of probabilities

$$\mathbf{P}\{|A_N - \mathbf{E}A_N| \geq x\}, \qquad \mathbf{P}\{|A_N - A| \geq x + K_N\},$$

$$\mathbf{P}\{|f_N(\lambda) - \mathbf{E} f_N(\lambda)| \geq x\} \qquad \mathbf{P}\{|f_N(\lambda) - f(\lambda)| \geq x + G_N(\lambda)\},$$

where $x \geq 0$, $\lambda \in (-\pi, \pi]$, $K_N = |A - \mathbf{E}A_N|$, $G_N(\lambda) = |f(\lambda) - \mathbf{E} f_N(\lambda)|$, as well as estimates of the rates at which K_N and $G_N(\lambda)$ tend to 0 as $N \to \infty$.

Let

$$\kappa(t, g) = \int_{-\pi}^{\pi} g(x) e^{-itx} dx, \qquad t \in \mathbf{R}^1.$$

In practice it is more convenient to write the estimates A_N, $f_N(\lambda)$ in the form

$$A_N = \frac{1}{2\pi} \sum_{k=-(N-1)}^{N-1} \kappa(k, \psi) C_N(k, 0)$$

and

$$f_N(\lambda) = \frac{1}{2\pi} \sum_{k=-(N-1)}^{N-1} \kappa(k\varepsilon_N, \varphi) C_N(k, 0) \mathrm{e}^{-ik\lambda},$$

where

$$C_N(k, 0) = \begin{cases} \frac{1}{N} \sum_{1 \le s, t \le N} X_s X_t, & |k| < N, \\ 0, & |k| \ge N \end{cases}$$

is the estimate of the correlation function $\Gamma\{X_k, X_0\}$.

In fact, be applying Parseval's equation (if $\psi \in L_2[-\pi, \pi]$) we have

$$A_N = \int_{-\pi}^{\pi} \psi(\lambda) I_N(\lambda) \mathrm{d}\lambda = \frac{1}{2\pi} \sum_{k=-\infty}^{\infty} \kappa(k, \psi) \overline{\kappa(k, \bar{I}_N)}.$$

But $\overline{\kappa(k, \bar{I}_N)} = C_N(k, 0)$.

Taking into account that $\kappa(k, \varphi_N(\cdot - \lambda)) = \kappa(k\varepsilon_N, \varphi)\mathrm{e}^{-ik\lambda}$ we get the relation for $f_N(\lambda)$.

Let

$$C_{h,N}(g) = \sum_{k=-(N-1)}^{N-1} |\kappa(kh, g)| h,$$

$$\gamma_{h,N}(g) = \sup_{-(N-1) \le k \le N-1} |\kappa(kh, g)|, \qquad h > 0,$$

$$C_N(g) = C_{1,N}(g), \qquad \gamma_N(g) = \gamma_{1,N}(g).$$

Theorem 3 ([7]). *For any $x \ge 0$ we have*

$$\begin{aligned} &\mathbf{P}\{|A_N - \mathbf{E}A_N| \ge x\} \le \\ &\le 2 \exp\left\{C_1 C_2 N\left[-\ln\left(1 - \frac{x}{C_2 + x}\right) - \frac{x}{C_2 + x}\right] - \frac{C_1 x^2 N}{C_2 + x}\right\} \\ &\le 2 \exp\left\{-\frac{C_1 x^2 N}{2(C_2 + x)}\right\}, \qquad N = 1, 2, \ldots, \end{aligned}$$

where

$$C_1 = \frac{\pi}{C_N(\psi) C_N(f)}, \qquad C_2 = \frac{1}{2\pi} \gamma_N(\psi) C_N(f).$$

Theorem 4 ([7]). *For any $\lambda \in (-\pi, \pi]$ and $x \ge 0$ we have*

$$\mathbf{P}\{|f_N(\lambda)-\mathbf{E}f_N(\lambda)|\geqslant x\}\leqslant$$

$$2\exp\left\{C_1'C_2'N\varepsilon_N\left[-\ln\left(1-\frac{x}{C_2'+x}\right)-\frac{x}{C_2'+x}\right]-\frac{C_1'x^2}{C_2'+x}N\varepsilon_N\right\}$$

$$\leqslant 2\exp\left\{-\frac{C_1'x^2N\varepsilon_N}{2(C_2'+x)}\right\},\qquad N=1,2,\dots,$$

where

$$C_1'=\frac{\pi}{C_{\varepsilon_N,N}(\varphi)C_N(f)},\qquad C_2'=\frac{1}{2\pi}\gamma_{\varepsilon_N,N}(\varphi)C_N(f).$$

If $f(\lambda)$ and the weight function ψ, which define the constants C_1 and C_2 in Theorem 3, are absolutely continuous and their derivatives belong to the space $L_2[-\pi,\pi]$, then $\inf C_1>0$,

$$\sup_N C_2<\left|\int_{-\pi}^{\pi}g(t)\mathrm{d}t\right|+\left(\frac{2\pi^3}{3}\right)^{\frac{1}{2}}\cdot\|g'\|_{L_2}<\infty.$$

If in Theorem 4 the spectral function $f(\lambda)$ and the weight function ψ are absolutely continuous, their derivatives belong to the space $L_2[-\pi,\pi]$, and if $\varphi(-\pi)=\varphi(\pi)=0$, then

$$\inf_N C_1'>0,\qquad \sup_N C_2'<\infty.$$

The estimates of rates, at which biases $K_N=|\mathbf{E}A_N-A|$ and $G_N(\lambda)=|\mathrm{E}f_N(\lambda)-f(\lambda)|$ converge to 0, are given in Theorems 5 and 6.

Let

$$\|g\|_{L_p}=\left(\int_{-\pi}^{\pi}|g(\lambda)|^p\,\mathrm{d}\lambda\right)^{1/p},\qquad p\in[1,\infty),$$

$$\|g\|_{L_\infty}=\operatorname{ess}\sup_{\lambda\in(-\pi,\pi)}|g(\lambda)|,$$

$$g_x=g(\cdot+x),\qquad x\in\mathbf{R}^1.$$

$\mathrm{Lip}_H(\alpha,q)$ is the space of all functions satisfying an integral Lipschitz condition with exponents $\alpha\in(0,1]$, $q\in[1,\infty]$ and the constant H, i.e. the space of all functions g, satisfying the condition

$$\forall x\in(-\pi,\pi]\,\|g_x-g\|_{L_q}\leqslant H|x|^\alpha. \tag{8}$$

By virtue of the periodicity of the function g inequality (8) is also satisfied for all $x\in\mathbf{R}^1$.

Theorem 5 ([7]). *If the spectral density belongs to* L_p *and if the weight function* ψ *belongs to* $\mathrm{Lip}_H(\alpha, q)$, *where* $p \in [1,\infty]$, $1/p + 1/q = 1$, $\alpha \in (0,1]$, *then*

$$\forall N\, K_N \leqslant \pi H \|f\|_{L_p} \cdot \begin{cases} \left[1 - \dfrac{1+\alpha}{(N\pi)^{1-\alpha}2^{\alpha}}\right] \dfrac{2^{\alpha}}{N^{\alpha}(1-\alpha^2)}, & \alpha < 1, \\ (\frac{1}{2} + \ln \frac{1}{2}\pi + \ln N) \cdot 1/N, & \alpha = 1. \end{cases}$$

If the periodically extended weight function ψ is r times differentiable $r \in \{1,2\}$ at every point of the real line, then

$$\forall N\, K_N \leqslant \pi \|f\|_{L_1} \|\psi^{(r)}\|_{L_\infty} \cdot \begin{cases} (\frac{1}{2} + \ln \frac{1}{2}\pi + \ln N) \cdot 1/N, & r = 1, \\ \dfrac{\pi}{4N}, & r = 2. \end{cases}$$

If the periodically extended weight function ψ is differentiable at every point of the real line, and if $\psi' \in \mathrm{Lip}_H(\alpha, \infty)$, then

$$\forall N\, K_N \leqslant \pi H \|f\|_{L_1} \cdot \begin{cases} \left[1 - \dfrac{2^{1+\alpha}}{(\pi N)^{\alpha}(2+\alpha)}\right] \cdot \dfrac{\pi^{\alpha}}{\alpha N}, & \alpha < 1 \\ \dfrac{\pi}{4N}, & \alpha = 1. \end{cases}$$

Theorem 6 ([7]). *If the periodically extended spectral density* f *is* r *times differentiable* $r \in \{1, 2, \ldots\}$ *at every point of the real line, and if the weight function* φ *satisfies the condition*

$$\int_{-\pi}^{\pi} z^k \varphi(z) \mathrm{d}z = 0, \qquad k = 1, 2, \ldots, r-1,$$

then for all $\lambda \in (-\pi, \pi]$ *and* $N = 1, 2, \ldots$ *we have*

$$G_N(\lambda) \leqslant \frac{\|f^{(r)}\|_{L_\infty}}{r!} \int_{-\pi}^{\pi} |z|^r |\varphi(z)| \mathrm{d}z \cdot \varepsilon_N^r$$

$$+ \begin{cases} \|f'\|_{L_\infty} \|\varphi\|_{L_1} (\frac{1}{2} + \ln \frac{1}{2}\pi + \ln N) \cdot \pi/N, & r = 1, \\ \|f''\|_{L_\infty} \|\varphi\|_{L_1} \cdot \pi^2/4N, & r \geqslant 2. \end{cases}$$

If the periodically extended spectral density f *is* r *times differentiable* $r \in \{0, 1, 2, \ldots\}$ *at every point of the real line, if* $f^{(r)} \in \mathrm{Lip}_H(\alpha, \infty)$ *and if the weight function* φ *satisfies the condition*

$$\int_{-\pi}^{\pi} z^k \varphi(z) \mathrm{d}z = 0, \qquad k = 1, 2, \ldots, r,$$

then for all $\lambda \in (-\pi, \pi]$ *and* $N = 1, 2, \ldots$

$$G_N(\lambda) \leqslant \frac{H}{r!}\int_{-\pi}^{\pi} |z|^{r+\alpha}|\varphi(z)|\,dz \cdot \varepsilon_N^{r+\alpha} +$$

$$+\pi\|\varphi\|_{L_1}\begin{cases} H\left[1-\dfrac{1+\alpha}{(N\pi)^{1-\alpha}2^\alpha}\right]\cdot\dfrac{2^\alpha}{N^\alpha(1-\alpha^2)}, & r=0, \quad \alpha\in(0,1),\\ H(\tfrac{1}{2}+\ln\tfrac{1}{2}\pi+\ln N)\cdot 1/N, & r=0, \quad \alpha=1,\\ H\left[1-\dfrac{2^{1+\alpha}}{(N\pi)^\alpha(2+\alpha)}\right]\dfrac{\pi^\alpha}{\alpha N}, & r=1, \quad \alpha\in(0,1),\\ H\cdot\pi/2N, & r=1, \quad \alpha=1,\\ \|f''\|_{L_\infty}\cdot\pi/4N, & r\geqslant 2.\end{cases}$$

The proofs of Theorems 3, 4 mainly rest on the estimate

$$|\Gamma_k(A_N)| \leqslant \frac{(k-1)!}{2\pi^k}\,\gamma_N(\psi)C_N^k(f)\frac{1}{N^{k-1}}. \tag{9}$$

It also holds, if in the right part of the inequality we exchange the functions ψ and f.

Since $\kappa(k,\varphi_N(\cdot-\lambda)) = \alpha(k\varepsilon_N,\varphi)e^{-i\lambda k}$, $\gamma_N(\varphi_N(\cdot-\lambda)) = \gamma_{\varepsilon_N,N}(\varphi)$ and

$$C_N(\varphi_N(\cdot-\lambda)) = \frac{1}{\varepsilon_N}C_{\varepsilon_N,N}(\psi),$$

then from (9) it follows that

$$|\Gamma_k(f_N(\lambda))| \leqslant \frac{(k-1)!}{2\pi^k}\,\gamma_{\varepsilon_N,N}(\varphi)C_{\varepsilon_N,N}^{k-1}(\varphi)$$

$$\times C_N^k(f)\frac{1}{(N\varepsilon_N)^{k-1}}.$$

Further, if the distribution of a random variable Z with $\mathbf{E}Z=0$, $DZ=1$ and finite moments satisfies the inequality

$$|\Gamma_k\{Z\}| \leqslant \frac{(k-1)!\,H}{\Delta^{k-2}}, \qquad k=3,4,\ldots,$$

then for all $x\geqslant 0$ we have

$$\mathbf{P}\{|Z|\geqslant x\}\leqslant$$

$$\leqslant 2\exp\left\{H\Delta^2\left[\ln\left(1-\frac{x}{H\Delta+x}-\frac{x}{H\Delta+x}\right)\right]-\frac{x^2\Delta}{H\Delta+x}\right\}$$

$$\leqslant 2\exp\left\{-\frac{x^2\Delta}{2(H\Delta+x)}\right\}.$$

And if the inequality

$$|\Gamma_k\{Z\}| \leq \frac{k!\,H}{\Delta^{k-2}}, \qquad k = 3, 4, \ldots$$

is satisfied then for $\mathbf{P}\{Z > x\}$ the equation of large deviations (7′) holds for all $0 \leq x < \delta_H \Delta$, where $\delta_H > 0$ depends only on H.

These results can also be generalized for the case of non-Gaussian random processes $X_t \in S^{(\infty)}$, if, for example, the regularity condition

$$\sum_{t_1, \ldots, t_k} |\Gamma\{X_{t_1}, \ldots, X_{t_{k-1}}, X_0\}| \leq k!\, H_3 H_4^{k-2} \tag{10}$$

is satisfied.

Let X_t be a strictly stationary random sequence with $\mathbf{E}X_t \equiv 0$ and $\mathbf{E}|X_t^p| < \infty$ for all $p \geq 1$. Let $F(\lambda) = \int_0^\lambda f(\lambda)\mathrm{d}\lambda$ be a spectral function of the process X_t with bounded density $\sup_\lambda f(\lambda) \leq L$. Let

$$\hat{F}_N(\lambda) = \int_0^\lambda I_N(\lambda)\mathrm{d}\lambda$$

be an estimate for $F(\lambda)$.

Theorem 7 ([4]). *If* (10) *is satisfied, then*

$$\sup_x \left| \mathbf{P}\left\{\sqrt{N} \sup_{0 \leq \lambda \leq \pi} |\hat{F}_N(\lambda) - F(\lambda)| \leq \sigma_N \cdot x\right\} - \mathbf{P}\left\{\sup_{0 \leq t \leq 1} |W(t)| \leq x\right\}\right| \leq C_{L, H_3, H_4} \frac{\ln(1+N)}{\sigma_N \cdot \sqrt{N}},$$

where $W(t)$ *is a Gaussian process,*

$$\sigma_N^2 = D\sqrt{N}[F_N(\pi) - F(\pi)] \to 2\pi \int_0^\pi f^2(\lambda)\mathrm{d}\lambda + 2\pi \int_0^\pi \int_0^\pi f(\alpha, -\alpha, \beta, -\beta)\mathrm{d}\alpha\,\mathrm{d}\beta, \quad N \to \infty.$$

An estimate of this type for the Gaussian sequence was obtained by T. Arak [15].

3. Limit theorems under strong mixing conditions

Let

(SM) $$\alpha(s, t) = \sup_{A \in \mathcal{F}_t^\infty,\, B \in \mathcal{F}_{-\infty}^s} |P(AB) - P(A)P(B)|,$$

$$(U,S,M)\quad \varphi(s,t)=\sup_{\substack{A\in\mathscr{F}_t^\infty,\,P(A)>0,\\ B\in\mathscr{F}_{-\infty}^s}}|P(A\mid B)-P(A)|.$$

As is known, it is impossible to obtain estimates analogous to (4), (5) with the help of $\alpha(s,t)$ or $\varphi(s,t)$.

The following statements are true:

There exist positive absolute constants H_5, H_6 such that

$$|\widehat{\mathbf{E}}X_{t_1}\dots X_{t_k}|\le H_5^k\left(\alpha\left(\max_{1\le j<k}|t_{j+1}-t_j|\right)\right)^{\delta/(2+\delta)}\max_{\substack{t,m\\ m\le(2+\delta)k}}\mathbf{E}|X_t|^m,\quad \delta>0,\tag{11}$$

$$|\widehat{\mathbf{E}}X_{t_1}\dots X_{t_k}|\le H_6^k\left(\varphi\left(\max_{1\le j<k}|t_{j+1}-t_j|\right)\right)^{1-1/p}\max_{\substack{t,m\\ m\le p\cdot k}}\mathbf{E}|X_t|^m,\quad p>2,\tag{12}$$

where

$$\alpha(\tau)=\sup_t\alpha(t,t+\tau),\qquad \varphi(\tau)=\sup_t\varphi(t,t+\tau).$$

There exist examples of processes X_t [9], [10], [12] such that estimates for them of the type (11), (12) are accurate with respect to $\alpha(\cdot)$ and $\varphi(\cdot)$ both for $\widehat{\mathbf{E}}X_{t_1}\dots X_{t_k}$ and $\Gamma\{X_{t_1},\dots,X_{t_k}\}$. In this case for the investigation of asymptotic behaviour one has to make use of a formula analogous to (1) though not for the logarithmic derivatives of the chacteristic function $\mathbf{E}\exp\{-\mathrm{i}\Sigma u_kX_{t_k}\}$ at the point $u=0$ but in a neighbourhood of zero. Let $X_t\in T^{(k)}$,

$$\Gamma\{X_{t_1},\dots,X_{t_k};u\}=$$

$$=\frac{1}{\mathrm{i}^k}\frac{\partial^k}{\partial u_{t_1}\dots\partial u_{t_k}}\ln\mathbf{E}\exp\left\{\mathrm{i}\sum_{j=1}^n u_jX_j\right\}\bigg|_{u_1=\dots=u_n=u},$$

$$\Gamma_k\{S_n;u\}=\frac{1}{\mathrm{i}^k}\frac{\mathrm{d}^k}{\mathrm{d}u^k}\ln\mathbf{E}\exp\{\mathrm{i}uS_n\},$$

then

$$\Gamma_k\{S_n;u\}=\sum_{1\le t_1,\dots,t_k\le n}\Gamma\{X_{t_1},\dots,X_{t_k};u\},\ \Gamma\{X_{t_1},\dots,X_{t_k};u\}=$$

$$=\sum_{\nu=1}^k(-1)^{\nu-1}\sum_{\cup_{p=1}^\nu I_p=I}N_\nu(I_1,\dots,I_\nu)\prod_{p=1}^\nu\hat{\mathbf{E}}(I_p;u),\tag{13}$$

where

$$\hat{\mathbf{E}}(I_p;u)=\frac{\hat{\mathbf{E}}\,\mathrm{e}^{\mathrm{i}uS_n}X_{t_{i1}}\dots X_{t_{im}}}{\mathbf{E}\,\mathrm{e}^{\mathrm{i}uS_n}},\qquad I_p=\{i,\dots,i_m\},$$

$$\hat{\mathbf{E}}\,\mathrm{e}^{\mathrm{i}uS_n}X_{t_{i_1}}\dots X_{t_{i_m}}=\mathbf{E}\,\mathrm{e}^{\mathrm{i}uS_n}X_{t_{i_1}}\widehat{X_{t_{i_2}}\dots\widehat{X_{t_{i_{m-1}}}X_{t_{i_m}}}}$$

and sign " $\wedge$ " over X_t means the following:

$$\hat{X}_t = X_t - \frac{\mathbf{E}\, e^{iuS_n} X_t}{\mathbf{E}\, e^{iuS_n}}\,.$$

We shall mention a theorem.

Theorem 8. *If* $\mathbf{E}|X_t|^3 < \infty$, $\mathbf{E}X_t = 0$, $t = 1, 2, \ldots, n$, $\varphi(\tau) \leq e^{-\alpha_n \tau}$, $\alpha_n > 0$, *then an absolute constant* C_3 *exists such that*

$$\sup_x |\mathbf{P}\{Z_n < x\} - \Phi(x)| \leq C_3 L_{3n} \ln(1 + L_{3n}^{-1}),$$

where

$$L_{3n} = \frac{\sum_{j=1}^{n} \mathbf{E}|X_j|^3}{\alpha_n^2 B_n^3}, \qquad B_n^2 = DS_n.$$

In order to prove Theorems 2, 8 and employ the relations (1), (13) we use the following representation

$$S_n = \sum_{j=1}^{N} I_j,$$

where

$$I_j = \zeta_{l_j} + S_{l_{j-1} l_j} - \zeta_{l_{j-1}}, \qquad j = 1, \ldots, N-1,$$

$$I_N = \mathbf{E}(S_n \mid \mathcal{F}_0^0), \qquad \zeta_k = \mathbf{E}(S_n \mid \mathcal{F}_1^k),$$

$$S_{kl} = \sum_{j=k+1}^{l} X_j, \qquad \mathcal{F}_k^l = \sigma\{X_t, k \leq t \leq l\},$$

$$0 = l_0 < l_1 < \ldots < l_{N-1} = n$$

is any partition of the interval $[0, n]$ by the help of integers. Then

$$DS_n = \sum_{j=1}^{N} Dy_j.$$

Usually we have $l_j - l_{j-1} \asymp 1/\gamma_n$ in Theorem 2 and $l_j - l_{j-1} \asymp 1/\alpha_n$ in Theorem 8.

The method of semi-invariants is very helpful for the investigation of asymptotic properties of quadratic and multilinear forms, constructed from the sequence of independent or weakly dependent random variables tending to the normal law.

References

[1] Leonov, V.P. and Širjaev, A.N. (1959). Sur le calcul des semi-invariants, *Teor. Verojatnost. i Primenen.*, **4**, 342–355. English translation, *Theor. Probability Appl.*, **4** (1959), 319–329. *MR* # 5058.

[2] Ширяев, А. Н. (1960). Некоторые вопросы спектральной теории старших моментов, Теория вероятн. и ее примен., 3, 5, 293–313.

[3] Statulevičius, V. (1970). Limit theorems for random functions, I, *Litovsk. Mat. Sb.*, **10**, 583–592. (Russian) *MR* **43** # 2765.

[4] Statulevičius, V. (1974). Limit theorems for dependent random variables under various regularity conditions, *Proceedings of the International Congress of Mathematicians*, Vancouver, Canada.

[5] Statulevičius, V. (1966). On large deviations, *Z. Wahrscheinlichkeitstheorie und Verw. Gebiete*, **6**, 133–144. *MR* **36** # 4612.

[6] Statulevičius, V. (1969). Limit theorems for sums of random variables that are connected in a Markov chain. I, II, III, *Litovsk. Mat. Sb.*, **9**, 345–362, 635–672; ibid. **10**, (1970), 161–169. (Russian) *MR* **42** # 1188.

[7] Бенткус, Р., Рудзкис, Р., Статулявичус, В. (1975). Экспоненциальные неравенства для оценок спектра стационарной гауссовской последовательности, Литовский мат. сб., 3, 15.

[8] Журбенко, И. Г. (1974). Об оценке спектральной функции стационарного процесса, ДАН СССР, 5, 214, 1005–1008.

[9] Журбенко, И. Г., Зуев, Н. М. (1975). Оценки старших спектральных плотностей стационарных процессов, удовлетворяющих условию Крамера, с переимснованием "по Розенблату", Литовский мат. сб., 1, 15.

[10] Зуев, Н. М. (1973). Об оценках смешанных семиинвариантов случайных процессов, Математические заметки, 4, 13, 581–586.

[11] Stein, Ch. (1973). A bound for error in the normal approximation to the distribution of a sum of dependent random variables, *Proc. Sixth Berkeley Sympos. on Math. Statist. and Probability*, Univ. of California Press, Berkeley, Calif., pp. 583–602.

[12] Žurbenko, I.G. (1971). Strong estimates for fixed semi-invariants of random processes, *Sibirsk. Mat. Ž.* **13**, 293–308 = *Siberian Math. J.*, **13**, (1971), 202–213. *MR* # 6141.

[13] Žurbenko, I.G. (1972). Spectral semi-invariant of stationary processes with sufficiently strong mixing, *Teor. Verojatnost. i Primenen.* **17**, 150–152 = *Theor. Probability Appl.* **17** (1972), 151–152. *MR* **45** # 7806.

[14] Phillipp, W. (1969). The remainder in the central limit theorem of mixing stochastic processes, *Ann. Math. Statist.* **40**, 601–609. *MR* **39** # 4912.

[15] Arak, T. (1971). On the rate of convergence of the distribution of the maximum deviation of an estimate of the spectrum of a Gaussian random sequence, *Dokl. Akad. Nauk SSSR* **201**, 1019–1021 = *Soviet Math. Dokl.*, **12**, (1971), 1747–1750. *MR* **45** # 2787.

PART IV

INFORMATION AND CONTROL THEORY

P.R. Krishnaiah, ed., *Multivariate Analysis–IV*

STOCHASTIC CONTROL OF SYSTEMS GOVERNED BY PARTIAL DIFFERENTIAL EQUATIONS

A.V. BALAKRISHNAN*

A theory of stochastic control (Linear Quadratic Regulator) of systems governed by partial differential equations featuring both control and random disturbance on the boundary, exploiting the theory of semigroups of linear operators and using white noise theory in place of Wiener Process theory.

1. Introduction

Although the potential applications are many—from flexible aircraft to water quality to energy systems—control theory for partial differential equations, and more specifically, Stochastic Control Theory is still far less understood than the similar theory for lumped parameter systems. This is due no doubt to the far greater variety and complexity of the basic theory of partial differential equations. In particular relatively little is known about stochastic processes distributed in time and space. Obviously any practical application has to be preceded by a considerable body of theoretical knowledge supplemented by simulation.

In this paper we consider a class of stochastic control problems with control on the boundary. Such a problem has no analogue in ordinary differential equations. Our treatment of the theory is also novel in two additional aspects compared with the current literature, [1,2] for example (where additional references may be found), namely in the use of the theory of semigroups of operators, and in the use of 'white noise' in contrast to Wiener process, especially in modelling errors on observed data. The full significance of the latter does not appear however until we consider non-linear operations [3,4]. In the linear case, as here, the results are the same.

We begin with a loose, 'physical' statement of the problem we shall consider, postponing precision to the abstract version to keep the otherwise cumbersome notation in check.

* Research supported in part under grant no. 73–2492, Applied Mathematics Division, AFOSR, USAF.

Let $\mathscr{D}$ denote a bounded domain in the Euclidean space E^n. Points in Euclidean space will be denoted generally by ξ, with components ξ_i. The stochastic control problem is characterized in the first instance by an observed process $y(t, \xi)$, $0 < t$, $\xi \in \mathscr{D}$, t denoting time. (The domain $\mathscr{D}$ may degenerate to a finite number of points.) We assume that $y(t, \cdot)$ represents the response of the system under consideration, subject only to the unavoidable additive error term of thermodynamic origin:

$$y(t, \xi) = s(t, \xi) + n(t, \xi), \tag{1.1}$$

where $n(t, \xi)$ is the error or 'noise' term, and $s(t, \xi)$ is itself composed of a response to the input, as well as the response to an unknown disturbance modelled as a stochastic process, arising in the following way: let τ denote a formal partial differential operator (to be specified more fully later), Ω a bounded domain in E^m, with boundary Γ. Then

$$s(t, \cdot) = Cf(t, \cdot) \tag{1.2}$$

where C is a linear transformation, and

$$\frac{\partial f}{\partial t} f(t, \xi) = \tau f(t, \xi) + v_1(t, \xi) + n_1(t, \zeta), \qquad 0 < t, \quad \xi \in \Omega \tag{1.3}$$

$$f(t, \cdot)|_\Gamma = v_2(t, \cdot) + n_2(t, \cdot) \qquad f(0, \xi) \text{ specified.}$$

Here $v_1(t, \cdot)$, $v_2(t, \cdot)$ represent distributed and boundary inputs respectively; similarly $n_1(t, \xi)$ is 'distributed' noise (stochastic process in t and ξ), and $n_2(t, \cdot)$ is a stochastic process on the boundary Γ. The inputs $v_1(t, \cdot)$, $v_2(t, \cdot)$ may themselves be realized as linear transformations on 'controls':

$$v_1(t, \cdot) = B_1 u_1(t, \cdot) \qquad v_2(t, \cdot) = B_2 u_2(t, \cdot).$$

The optimization problem is to minimize

$$\int_0^T \mathbf{E}(\| g(t, \cdot)\|^2) \mathrm{d}t + \lambda \int_0^T \mathbf{E}(\| u(t, \cdot)\|^2) \mathrm{d}t \tag{1.4}$$

where $\mathbf{E}$ denotes expectation,

$$g(t, \cdot) = Rf(t, \cdot), \tag{1.5}$$

R being a linear transformation, $\|\cdot\|$ denoting the desired norm in the space variable. The second term, with λ a positive constant, accounts for the constraints placed on the controls:

$$u(t, \cdot) = \begin{pmatrix} u_1(t, \cdot) \\ u_2(t, \cdot) \end{pmatrix}.$$

The control at time t will naturally be based on the observation up to time t. (We assume that T is finite. Additional conditions are required if T is infinite.)

2. Semigroup formulation

Before we proceed to the complete abstract formulation of the problem, we consider first a simplified, non-stochastic version to see how the semigroup theoretic version is obtained. Thus let

$$\frac{\partial f}{\partial t}(t,\xi) = \tau f(t,\xi) + v_1(t,\xi),\ 0 < t,\ \xi \in \Omega$$
$$f(t,\cdot)|_{\Gamma = v_2(t,\cdot)}, \qquad f(0,\xi) \ \text{given} \tag{2.1}$$

Assume that $f(0,\xi)$ is in $L_2(\Omega)$.

Here and in what follows we shall take τ to be a second order strongly elliptic operator:

$$\tau f = \sum_1^m \sum_1^m a_{ij}(\xi)\frac{\partial^2 f}{\partial\xi_i\partial\xi_j} + \sum_1^m a_i(\xi)\frac{\partial f}{\partial \xi_i} \quad \text{in } \Omega, \tag{2.2}$$

where the coefficients are continuous in the closure of Ω, and real, and the quadratic form

$$\sum_1^m \sum_1^m a_{ij}(\xi)\chi_i\chi_j \geqslant \beta \sum_1^m \chi_i^2,\ \beta > 0 \quad \text{for all } \xi \text{ in } \Omega.$$

Let A denote its 'zero boundary value' restriction; more specifically let it denote the smallest closed extension in $L_2(\Omega)$ of τ restricted to $C_0^\infty(\Omega)$. Then (see [5]), A generates a strongly continuous, analytic semigroup $S(t)$ over $L_2(\Omega)$. We assume next that the boundary Γ is smooth enough so that the Dirichlet problem

$$f = 0 \quad \text{in } \Omega, \qquad f|_{\Gamma = g} \tag{2.3}$$

where g is in $L_2(\Gamma)$ has a unique solution given by:

$$f = Dg$$

where, furthermore D is a linear bounded transformation mapping $L_2(\Gamma)$ into $L_2(\Omega)$. Assuming now that

$$v_2(t,\cdot)\in L_2((0,T);\ L_2(\Gamma)),$$

$$v_1(t,\cdot)\in L_2((0,T);\ L_2(\Omega)),$$

it can be shown that the unique (generalized) solution of (2.1) with the necessary continuity properties is given by:

$$x(t)=S(t)x(0)+\int_0^t S(t-s)v_1(s,\cdot)ds-\int_0^1 AS(t-s)Dv_2(s,\cdot)ds \quad \text{a.e.}$$

$$0\leqslant t\leqslant T.\ldots \tag{2.4}$$

The details may be found in [6]. It is important to note that the third term is defined only a.e., where we have exploited the analyticity of the semigroup. Washburn [7] has shown that

$$\|AS(t)D\|=O(t^{-3/4})$$

although we shall not need to use this estimate here. Analyticity alone only yields

$$\|AS(t)\|=O(t^{-1})$$

3. White noise and weak random variables

Before we proceed to the stochastic control problem, let us review briefly the notion of white noise and related estimation theory for weak random variables. See [6] for a more detailed treatment. Let H denote a real separable Hilbert space and let

$$W=L_2((0,T):H)$$

By white noise we mean the weak random variable in W with characteristic function:

$$C(h)=\exp -\tfrac{1}{2}[h,h], \qquad h\in W.$$

Thus if $\omega(\cdot)$ denotes the elements in W,

$$\int_0^T [\omega(s),h(s)]ds$$

defines a Gaussian random variable with mean zero and variance $[h,h]$. More generally, for any g, h in W,

$$\mathbf{E}([\omega,h][\omega,g])=[h,g].$$

Thus ω is a weak random variable with zero mean and covariance operator equal to the identity. Let L denote a linear bounded transformation mapping W into another separable Hilbert space H_0. Then $L\omega$ defines a weak random variable in H_0 with covariance operator LL^*. Let ξ, η denote two weak random variables with finite second moment and let us denote the covariances by $R_{\xi\xi}$ and $R_{\eta\eta}$ respectively and cross covariance $R_{\xi\eta}$. Then the variational problem: minimise for each h:

$$\mathbf{E}([\xi - L\eta, h]^2)$$

over the class of linear bounded transformations, has, assuming that R_η has a bounded inverse, the solution:

$$L_0 = R_{\xi\eta}\, R_{\eta\eta}^{-1} \tag{3.1}$$

We define the conditional expectation by:

$$\mathbf{E}(\xi \mid \eta) = L_0 \tag{3.2}$$

Thus defined we note an important property: Let M denote any bounded linear operator then

$$\mathbf{E}(M\xi \mid \eta) = ML_0\eta \tag{3.4}$$

We also need a slight generalization of this result. Suppose A is a closed linear operator (possibly unbounded), with dense domain, and suppose that

$$\xi \in (\text{domain of } A) \tag{3.4}$$

and that $A\xi$ has finite second moment. Then L_0 maps into the domain of A, and

$$\mathbf{E}(A\xi \mid \eta) = AL_0\eta \tag{3.5}$$

This follows from the easily verified fact that $R_{\xi\eta}$ maps into the domain of A.

4. Main results

We begin with an abstract formulation of the problem. Thus let H_n denote a real separable Hilbert space, and let

$$W_n = L_2((0, T); H_n)$$

Let $\omega(\cdot)$ denote white noise in W_n, μ the corresponding Gauss measure on W_n, and

$$\mathbf{E}(e^{i[\omega,h]}) = \exp - [h, h]/2.$$

Let F denote a bounded linear operator mapping H_n into $L_2(\Omega)$, L a bounded linear operator on H_n into $L_2(\Gamma)$ and similarly G bounded linear on H_n into $L_2(\mathscr{D})$ such that

$$FG^* = 0; \qquad LG^* = 0; \qquad GG^* = \text{Identity} \tag{4.1}$$

Let H_u denote a real separable Hilbert space, and let

$$W_u = L_2((0, T); H_u).$$

Then the abstract version of (1.3) can be written as

$$\begin{aligned} x(t, \omega) = S(t)x(0) &+ \int_0^t S(t-s)B_1 u(s)\mathrm{d}s \\ &+ (-1)\int_0^t AS(t-s)DB_2 u(s)\mathrm{d}s \\ &+ \int_0^t S(t-s)F\omega(s)\mathrm{d}s + (-1)\int_0^t AS(t-s)DL\omega(s)\mathrm{d}s, \end{aligned} \tag{4.2}$$

$$y(t, \omega) = Cx(t, \omega) + G\omega(t), \qquad 0 \leq t \leq T, \tag{4.3}$$

where B_1, B_2 are linear bounded transformations on H_u into $L_2(\Omega)$, and $L_2(\Gamma)$ respectively, and C is a linear bounded transformation into $L_2(\mathscr{D})$, which is restricted so that C^* maps into the domain of A^* with the consequence that

$$CAx = (A^*C^*)^*x, \qquad x \in (\text{domain of } A) \tag{4.4}$$

can be extended to be linear bounded. This is a technical restriction (requiring a degree of smoothness on the observation operation) and can perhaps be removed. The control $u(t)$ is required to be a linear transformation on the input of the form:

$$u(t) = \int_0^t K(t, s)\, y(s, \omega)\mathrm{d}s \tag{4.5}$$

and because of the dependence of ω, we shall write $u(t, \omega)$ from now on. Another technical requirement is that the mapping defined by (4.5) representable as

$$u(\cdot, \omega) = K\omega$$

defined on W_n into W_u be Hilbert–Schmidt. This restriction is put in so that

$$\mathbf{E}([u(\cdot,\omega),u(\cdot,\omega)]) = TrKK^* < \infty$$

where the inner product is in W_u. The operator R in (1.5) is, for a similar reason, required to be Hilbert–Schmidt, and furthermore such that

$$(\text{Range of } R^*) \subset (\text{Domain of } A^*) \tag{4.6}$$

and

$$A^*R^* \text{ is also Hilbert–Schmidt.} \tag{4.7}$$

The optimal control problem can then be stated as choosing controls of the form (4.5) so as to minimize:

$$\mathbf{E}([Rx(\cdot,\omega),Rx(\cdot,\omega)]) + \lambda\mathbf{E}([u(\cdot,\omega),u(\cdot,\omega)]). \tag{4.8}$$

Without loss of generality we shall take $\lambda = 1$ in what follows.

The main step in the solution is to note that the response to the boundary input can be expressed:

$$\int_0^t AS(t-s)DB_2u(s)ds = A\int_0^t S(t-s)DB_2u(s)ds \quad \text{a.e.} \quad 0<t<T,$$

and hence we can rewrite (4.2), (4.3) as:

$$\dot{z}(t,\omega) = A\,z(t,\omega) + B_1\,u(t,\omega) + F\omega(t),$$

$$\dot{v}(t,\omega) = A\,v(t,\omega) - D\,B_2\,u(t,\omega) - DL\omega(t),$$

$z(0) = x(0)$ given; $v(0) = 0$;

$$y(t,\omega) = Cz(t,\omega) + (A^*C^*)^*v(t,\omega) + G\omega(t),$$

where A^*C^* is a bounded operator, and the problem is to minimize:

$$\mathbf{E}(\|(A^*R^*)^*v(\cdot,\omega)\|^2) + \mathbf{E}([u(\cdot,\omega),u(\cdot,\omega)])$$

with $u(\cdot,\omega)$ determine in the form (4.5). Again (A^*R^*) is bounded and Hilbert–Schmidt, and hence writing

$$Z(t,\omega) = \begin{pmatrix} z(t,\omega) \\ v(t,\omega) \end{pmatrix},$$

and noting that

$$\mathscr{A} = \begin{vmatrix} A & 0 \\ 0 & A \end{vmatrix}$$

is again an infinitesimal generator over the product space $L_2(\Omega) \times L_2(\Omega)$, we can reduce the problem to the semigroup formulation given in [8] (See also [6]). Thus we have:

$$\dot{Z}(t, \omega) = \mathscr{A}Z(t, \omega) + \mathscr{B}u(t, \omega) + \mathscr{F}\omega(t),$$

$$y(t, \omega) = \mathscr{C}Z(t, \omega) + G\omega(t),$$

where

$$\mathscr{B} = \begin{bmatrix} B_1 \\ -DB_2 \end{bmatrix}; \qquad \mathscr{F} = \begin{bmatrix} F \\ -DL \end{bmatrix}; \qquad \mathscr{C} = [C(A^*C^*)^*]$$

and the cost functional to be minimized, becomes in terms of $Z(t, \omega)$:

$$\mathbf{E}([QZ(\cdot, \omega), QZ(\cdot, \omega)]) + \mathbf{E}([u(\cdot, \omega), u(\cdot, \omega)]),$$

where

$$Q = [0 \quad (A^*R^*)^*]$$

The optimal control solution then is given by (taking $x(0) = 0$ for simplicity, cf. [8]):

$$u_0(t, \omega) = -\mathscr{B}^*P_c(t)\hat{Z}(t, \omega),$$

$$\dot{\hat{Z}}(t, \omega) = \mathscr{A}\hat{Z}(t, \omega) + P_f(t)\mathscr{C}^*(y(t, \omega) - \mathscr{C}\hat{Z}(t, \omega)) + \mathscr{B}u_0(t, \omega)$$

where $P_f(t)$ is the unique solution of (x, y in domain of $\mathscr{A}^*$)

$$\frac{\mathrm{d}}{\mathrm{d}t}[P_f(t)x, y] = [P_f(t)\mathscr{A}^*x, y] + [P_f(t)x, \mathscr{A}^*y] + [\mathscr{F}\mathscr{F}^*x, y] - [P_f(t)\mathscr{C}^*\mathscr{C}P_f(t)x, y]$$

$$P_f(0) = 0,$$

and $P_c(t)$ is the unique solution of (x, y in domain of $\mathscr{A}$)

$$\frac{\mathrm{d}}{\mathrm{d}t}[P_c(t)x, y] + [P_c(t)x, \mathscr{A}y] + [P_c(t)\mathscr{A}x, y] + [Qx, Qy] - [\mathscr{B}^*P_c(t)x, \mathscr{B}^*P_c(t)y] = 0$$

$$P_c(T) = 0.$$

Further reduction is clearly possible in terms of the originally given operators. For the case where $F = 0$, $B_1 = 0$ See [9].

References

[1] Bensoussan, A. (1970). *Filtrage Optimal des Systèmes Linears*, Dunod, Paris.
[2] Dawson, D.A. (1975). Stochastic Evolution Equation and Related Measure Processes, *Journal of Multivariate Analysis*, March.
[3] Balakrishnan, A.V. (1975). 'Stochastic Bilinear Partial Differential Equations,' in *Variable Structure Systems with Applications to Economics and Biology*, Lecture Notes in Economics and Mathematical Systems, No. 111, Springer-Verlag.
[4] Balakrishnan, A.V. (1974). White Noise versus Wiener Process Models in Filtering and Control, *Proceedings of the IFAC Symposium on Stochastic Control*, Budapest.
[5] Dunford, N. and Schwartz, J.T. (1963). *Linear Operators*, Part II, Interscience Publishers.
[6] Balakrishnan, A.V. (1976). *Applied Functional Analysis*, Springer-Verlag (to be published).
[7] Washburn, D. (1974). A semigroup approach to time-optimal boundary control of diffusion processes, Dissertation, UCLA.
[8] Balakrishnan, A.V. (1974). Stochastic Optimization Theory in Hilbert Spaces, *Journal of Applied Mathematics and Optimization*.
[9] Balakrishnan, A.V. (1975). Identification and Stochastic Control of a Class of Distributed Systems with Boundary Noise, in Lecture Notes in Economics and Mathematical Systems, no. 107, Springer-Verlag.

P.R. Krishnaiah, ed., *Multivariate Analysis–IV*

EQUILIBRIUM PROPERTIES OF ARBITRARILY INTERCONNECTED QUEUEING NETWORKS*

Frederick J. BEUTLER
Benjamin MELAMED
and
Bernard P. ZEIGLER
The University of Michigan, Ann Arbor, Mich., U.S.A.

Considered are properties of arbitrarily interconnected queueing networks of single server nodes with exponential service and Poisson exogenous inputs. Customers departing node i are instantaneously routed to node j with probability p_{ij}, or depart the system with probability q_i.

The queue length vector process is formulated axiomatically through a stochastic integral equation characterizing this process as a Markov jump process. The associated Kolmogorov forward equation then leads to an infinite set of birth-and-death equations.

Solvability of the traffic equation and relative traffic intensity below unity are shown to be necessary and sufficient for equilibrium solutions to exist. A novel stochastic integral description of the departure process is crucial in the proof.

Conditions for the existence and uniqueness of the solution of the traffic equation are found. When these conditions are met, there is a unique equilibrium (of vector queue length probabilities) which is asymptotically attained.

The expectation of the service time for a customer entering the network is finite iff the traffic equation possesses a unique solution. The defect in the service time distribution is related to the probability that a customer remains in the network. Its generating function is rational, leading to a direct computation of the service time distribution.

The expected number of visits to each server by an arbitrary customer is discussed. The results are generally similar to those described in the preceding paragraph.

1. Introduction

In this paper, we shall investigate properties of queueing systems with arbitrary interconnections. The distinguishing feature of such networks is the random routing of customers, suggesting in particular that a given

* This research was partially supported by the Office of Naval Research under contracts N00014–67–A–0181–0047 (NR 042–296), N00014–75–C–0492 (NR 042–296) and the Air Force Office of Scientific Research under Grant AFOSR–76–2903.

customer encounters the same server more than once in his passage through the system. This feedback property engenders stochastic dependencies in which the input stream to a server depends on that server's previous behavior, the actions of other servers in the system, and on various exogenous customer inputs.

Because the server inputs in feedback queueing systems consist (in general) of superpositions of dependent streams, the analysis of these networks appears insuperably difficult. Nevertheless, the richness of applications to computer systems, communication networks and maintenance stations has motivated researchers [8], [9], [7], [1], [10] to attempt a derivation of relevant properties. In particular, the equilibrium waiting line length probabilities are known [8] (under certain assumptions), and this classical result[1] has led to more recent variations on the same theme.

Since the equilibrium waiting line probabilities are also the starting point for much of our study, we shall begin by summarizing the work of Jackson [8] in this regard. We assume m nodes, each consisting of a single server. A customer leaving node i is immediately and independently routed to node j with probability p_{ij}; since no losses are permitted, the customer departs the system from node i with probability $q_i \stackrel{d}{=} 1 - \Sigma_1^m p_{ij}$. The matrix formed by the p_{ij} is denoted by P, and is called the *switching matrix*. Further, the service at each node is exponential, the rate at node i being σ_i; it is convenient to express this parameter set by the row vector $\sigma \stackrel{d}{=} (\sigma_1, \ldots, \sigma_m)$. In like manner, we write $\alpha \stackrel{d}{=} (\alpha_1, \ldots, \alpha_m)$ for the intensities α_i of the input streams, with the additional proviso that these streams are independent Poisson.

The above hypotheses enabled Jackson to write a set of birth-and-death differential equations describing the probability of the state as a function of time. If ${}_iZ_t$ is the waiting line length (including any customer in service) at node i and time t, the *state* is the row vector $Z_t \stackrel{d}{=} ({}_1Z_t, \ldots, {}_mZ_t)$, and the probability row vector $q(t)$ is defined componentwise by

$$[q(t)]_n \stackrel{d}{=} \mathbf{P}(Z_t = n) \tag{1.1}$$

where $n = (n_1, \ldots, n_m)$. Then the birth-and-death equations become

[1] Actually, various generalizations are possible [8], [9], but we shall not pursue them here.

$$\dot{q}(t) = q(t)A, \tag{1.2}$$

and Jackson showed that $q(t) \equiv q_0$ is a solution if we take[2]

$$(q_0)_n \overset{d}{=} \prod_1^m (1-\rho_i)\rho^{n_i}. \tag{1.3}$$

The ρ_i appearing in (1.3) is defined by

$$\rho_i \overset{d}{=} \frac{\delta_i}{\sigma_i} \tag{1.4}$$

where δ is the m-dimensional row vector solution of the *traffic equation*

$$\delta = \alpha + \delta P. \tag{1.5}$$

Intuitively, (1.5) is a balance or conservation equation [1], in which δ_i is interpreted as the output (input) traffic intensity when the system is in equilibrium. It is thus plausible that (1.5) must hold if equilibrium is to be maintained. Similarly, Jackson requires

$$\rho_i < 1 \quad \text{for all } i, \tag{1.6}$$

which is needed to make sense of (1.3), and also corresponds to the usual queueing equilibrium demand that the relative traffic intensity be less than unity.

However, a deeper understanding of probabilistic equilibrium is desirable. We should ask, for example, if the q_0 of (1.3) is unique, and whether its existence implies that

$$q(t) \to q_0 \tag{1.7}$$

for any initial probability vector $q(0)$. Next, we ought to study the traffic equation (1.5) with regard to existence and uniqueness of a solution δ satisfying $\delta \geq 0$,[3] relate this solution to the input intensity α and the switching matrix P, and validate (if true) the heuristic notion that δ is the traffic intensity vector. Finally, we might consider whether the solubility of (1.5) together with the inequality (1.6) constitute necessary conditions for an equilibrium.

[2] It is remarkable that the ${}_iZ_t$ are mutually independent in equilibrium, and that the distribution of each ${}_iZ_t$ is precisely the same as for an M/M/1 queue with input rate δ_i and service rate σ_i.

[3] A non-negative vector (or matrix) is consistently defined as one all of whose components are non-negative.

In what follows, we shall resolve each of the posed questions. To establish a sound foundation, we establish an axiomatic formulation for the feedback queueing network; the waiting lines are then modeled by a Markov jump process of which the Kolmogorov forward equation has well-behaved solutions. A proof of the necessity of (1.5) and (1.6) for equilibrium (1.7) is then possible. To sharpen the necessity result, and to demonstrate sufficiency of (1.5) and (1.6) for convergence to equilibrium (1.7), additional background is needed. Accordingly, the switching matrix P is decomposed in a way which leads naturally to the analysis of the solution(s) for the traffic equation (1.5). Thereafter, we can complete the discussion of necessary and sufficient conditions for convergence to equilibrium (1.7).

Since total service times for a customer entering the system are closely related to the traffic equation, we shall be able to make use of the previous material to obtain a rational generating function. The defect in the corresponding probability vector for the service time appears explicitly in terms of the switching matrix P. If there is no defect, the expected service time is finite, and appears as a function of δ.

It is found that the total number of visits to service nodes is likewise related to the traffic equation. In fact, the expectation of this random variable (when this expectation is finite) solves a normalized form of the traffic equation (1.5).

Thus, it is seen that the traffic equation plays a central role, not only with reference to equilibrium line length probabilities, but also in relation to service times and number of services rendered.

2. The feedback queueing network model

The purpose of this section is to set forth the assumptions underlying the remainder of this paper. Most of these have already been stated in the introductory section, and we shall adhere to the notation introduced there. The service discipline is irrelevant to our needs, but we shall have to specify arrivals, departures and routings.

The arrivals to the respective service nodes are mutually independent Poisson streams, whose intensities are described by the row m-vector $\alpha \geqslant 0$. It is occasionally convenient to normalize this vector to the probability vector r, in which

$$r_i \stackrel{d}{=} \alpha_i \left(\sum \alpha_j \right)^{-1} \tag{2.1}$$

is the probability that an arriving customer enters the system via node i.

Upon completion of service at the ith node, a customer is routed (independently of all other routings) to the jth node with probability p_{ij}. This is best expressed by the mutually independent zero-one random variables $\{_{ij}X_k\}$, $k = 1, 2, \ldots$. We require $\mathbf{P}[_{ij}X_k = 1] = p_{ij}$ and $\Sigma_{j=0}^{m}\, _{ij}X_k = 1$ almost surely (a.s.) for each k. Then, if $_iD_t$ is defined as the number of departures from node i in $(0, t]$, the number of arrivals at node j from node i during this same period is

$$_{ij}A_t^e \stackrel{d}{=} \sum_{k=1}^{_iD_t} {_{ij}X_k} \tag{2.2}$$

by definition. The total number of endogenous arrivals at node j is taken to be the sum of the arrivals from all the server nodes, or

$$_jA_t^e \stackrel{d}{=} \sum_{i=1}^{m} {_{ij}A_t^e} \tag{2.3}$$

There are, in addition, the exogenous arrivals to j as already specified; their number in $(0, t]$ is denoted $_jA_t^\circ$.

Next, we must hypothesize the service mechanism and indicate how it leads to departures. To this end, let $_iY_t$ be a mutually independent collection of right continuous Poisson counting processes with respective parameters σ_i.[4] Now take

$$_iB_t \stackrel{d}{=} \begin{cases} 1 & \text{if } _iZ_{t-} \geq 1, \\ 0 & \text{if } _iZ_{t-} = 0. \end{cases} \tag{2.4}$$

With the above notation, departures $_iD_t$ from i in $(0, t]$ are expressed by the stochastic integral formulation

$$_iD_t \stackrel{d}{=} \int_0^t {_iB_s}\, \mathrm{d}_iY_s, \tag{2.5}$$

in which $_iY_s$ differs from a martingale only by $\sigma_i s$, so that the integral can be defined in the sense of [4], IX.5. Although (2.5) is directly identifiable as exponential service, the stochastic integral formulation has certain advantages which become evident later.

[4] It should be mentioned that the exogenous arrivals A°, the service process Y and the routing variables are also mutually independent.

Finally, a stochastic version of continuity expresses customer arrivals to and departures from the network. The relevant equation is

$$Z_t = Z_s + A^{\mathrm{o}}_{(s,t]} + A^{\mathrm{e}}_{(s,t]} - D_{(s,t]}. \tag{2.6}$$

From (2.6), one deduces that Z_t is a Markov jump process with denumerable state space and stationary transition probabilities. Indeed, the integral equation (2.6) for Z_t is a special case of a more general form which admits a unique solution consisting of a jump Markov process ([11], Chapter 4). The intuitive key to the Markov property lies in the Poisson nature of A^{o} and Y, and the independence of their jumps from the present and past state of the system.

When the system is in state n at time y, the probability of no services and no exogenous arivals (and hence no changes in state) over a period of duration u is $\mathrm{e}^{-c_n u}$, where

$$c_n = \sum_1^m \alpha_i + \sum_1^m \sigma_i,$$

the second sum being taken only over those indices for which $n_i > 0$. Here (2.7) is once more the consequence of the mutual independence and Poisson character of A^{o} and Y.

Because Z_t is a jump Markov process as described above, its transition probabilities must satisfy ([6], p. 458)

$$P_{nn'}(t) = \delta_{nn'}\mathrm{e}^{-c_n t} + \sum_k \int_0^t c_n \mathrm{e}^{-c_n s} q_{nk} P_{kn'}(t-s)\mathrm{d}s. \tag{2.8}$$

In (2.8) above, the c_n are given by (2.7), and the q_{nk} are easily calculated. For instance, if $n'_i = n_i - 1 \geq 0$, $n'_j = n_j + 1$ and $n'_r = n_r$ for the other $m-2$ indices, $q_{nn'} = (c_n)^{-1}\sigma_i p_{ij}$.

Of course, (2.8) may be rewritten as a Kolmogorov backward equation. Because the c_n are bounded, both the backward and forward Kolmogorov differential equations exhibit the same (unique) solution matrix $P(t)$ such that $\Sigma_n P_{n'n}(t) = 1$ for each n' and $t > 0$ (cf. [6], p. 462).[5] Moreover, the forward equation treats the transition matrix $P(t)$ as an operator on row probability vectors; hence, a row probability vector must satisfy

[5] The existence and uniqueness of the probability solution to the forward equation holds under far more general conditions than the boundedness of the c_n. This suggests that the methods applied here are valid under a variety of assumptions on the arrival and service mechanism. In some instances, invariant probability vectors can be found for such models [9].

$$\dot{q}(t) = q(t)A \tag{2.9}$$

in which $q(t) = q(0)P(t)$.

Equation (2.9) — which is identical with the birth-and-death equations used by Jackson — is used later to show that

$$q(t) \to q_0 \tag{2.10}$$

for any initial probability vector $q(0)$, provided that q_0 is a probability vector satisfying $q_0A = \theta$, and that the recurrent states of Z_t constitute an irreducible set. In fact, q_0 is then an invariant probability vector, because taking $q(0) = q_0$ in (2.9) yields $q(t) = q_0$ for every t. A standard Markov chain result ([6], p. 466) now implies (pointwise) convergence (2.10).

Theorem 2.1. *Let there exists an invariant probability vector q_0, and let $\mathbf{P}(Z_0 = n) = (q_0)_n$. If δ is defined by*

$$\delta \stackrel{d}{=} \mathbf{E}[D_{(t,t+1]}], \tag{2.11}$$

the traffic equation

$$\delta = \alpha + \delta P \tag{2.12}$$

is valid. Furthermore, each ρ_i (recall $\rho_i \stackrel{d}{=} \delta_i/\sigma_i$) is subject to the inequality

$$\rho_i \leq 1. \tag{2.13}$$

Proof. We begin by showing (2.13). From the expectation properties of martingale-type stochastic integrals ([4], IX.5) and the definitions (2.5) and (2.11) of D_t and δ respectively,

$$\delta_i = \sigma_i \int_t^{t+1} \mathbf{E}({}_iB_s)\,\mathrm{d}s. \tag{2.14}$$

But B_s has the same distribution for all s, since this is true for Z_s according to our hypotheses. Therefore, (2.14) becomes

$$\delta_i = \sigma_i P({}_iZ_t > 0); \tag{2.15}$$

this is equivalent to the inequality (2.13).

We turn to the proof of (2.12), applying generating function methods to the relation $q_0A = \theta$. Take

$$\phi(z_1, \ldots, z_m) \stackrel{d}{=} \sum_{\substack{n_i=0 \\ i=1,\ldots,m}}^{\infty} (q_0)_n \prod_1^m z_i^{n_i}, \tag{2.16}$$

with $n = (n_1, \ldots, n_m)$ as the generating function of Z_t. After some algebraic simplification, the relations among the various components of q_0 are expressed in terms of the generating function (hereafter abbreviated g.f.) by

$$\begin{aligned} &\sum_{i=1}^m \alpha_i (z_i - 1)\phi(z_1, \ldots, z_m) \\ &\quad + \sum_{i=1}^m [\phi(z_i, \ldots, z_m) - \phi(z_1, \ldots, 0_i, \ldots, z_m)]\sigma_i q_i\left[\frac{1}{z_i} - 1\right] \\ &\quad + \sum_{i=1}^m \sum_{\substack{j=1 \\ j\neq i}}^m [\phi(z_1, \ldots, z_m) - \phi(z_1, \ldots, 0_j, \ldots, z_m)]\sigma_j p_{ji}\left[\frac{z_i}{z_j} - 1\right] = 0 \end{aligned} \tag{2.17}$$

for $0 < |z_i| \leq 1$.

By setting $z_k = 1$ in (2.16) for any k, ϕ is reduced to a lower order g.f. Accordingly, choose $z_k = 1$ for all $k \neq 1$; this choice leads to a verification of (2.12) for its first component. First, setting these $z_k = 1$ simplifies (2.17) to

$$\begin{aligned} &\left\{\alpha_1\phi(z_1) + \sum_{i=2}^m [\phi(z_1) - \phi(z_1, 0_i)]\sigma_i p_{i1}\right\}(z_1 - 1) \\ &\quad + [\phi(z_1) - \phi(0_1)]\sigma_1(1 - p_{11})(z_1^{-1} - 1) = 0, \end{aligned} \tag{2.18}$$

where (for example)

$$\phi(z_1, 0_i) \stackrel{d}{=} \mathbf{P}[\{{}_1Z_t = k\} \cap \{{}_iZ_t = 0\}]z_1^k. \tag{2.19}$$

Let us now write Q for ${}_1Z_t$, Q_i for ${}_iZ_t$, and adopt the convention $\mathbf{P}(Q = -1) = 0$. Equating powers of z_1 in (2.18) thus yields

$$\begin{aligned} &\alpha_1[\mathbf{P}(Q = n - 1) - \mathbf{P}(Q = n)] \\ &\quad + \sum_{i=2}^m [\mathbf{P}(Q = n - 1, Q_i > 0) - \mathbf{P}(Q = n, Q_i > 0]\sigma_i p_{i1} \\ &\quad + [\mathbf{P}(Q = n + 1) - \mathbf{P}(Q = n) + \delta_{n0}\mathbf{P}(Q = 0)]\sigma_1(1 - p_{11}) = 0 \end{aligned} \tag{2.20}$$

for $n = 0, 1, 2, \ldots$. Next, add the equations (2.20) together for $n = 0, 1, \ldots, k$ and then again for $k = 0, 1, 2, \ldots$. The result is

$$\alpha_1 + \sum_{i=2}^m \mathbf{P}(Q_i > 0)\sigma_i p_{i1} - \mathbf{P}(Q > 0)\sigma_1(1 - p_{11}) = 0. \tag{2.21}$$

Upon substitution of (2.15) in (2.21), one obtains the first component of the traffic equation (2.12). The arguments for equality of the remaining components in (2.12) are identical to the above.

In general, (2.13) need not be a strict inequality, so that Theorem 2.1 cannot be strengthened in this regard. Indeed, if $p_{ii} = 1$, ρ_i may be either equal to or less than unity. There is no such ambiguity if $p_{ii} < 1$, for then we have

Corollary 2.1. *If* $p_{ii} < 1$, *then* $\rho_i < 1$.

Proof. If the assertion of the corollary is false for $i = 1$, $\mathbf{P}(Q = k) = 0$ for every $k \geq 0$, thereby contradicting the assumption of a stationary probability vector. To show $\mathbf{P}(Q = k) = 0$, we proceed by induction. Except in the trivial case $\delta_1 = 0$, $\mathbf{P}(Q = 0) = 0$ is implied by (2.15) if $\delta_i = \sigma_i$. Furthermore, if $\mathbf{P}(Q = k) = 0$ is true for $k = 0, \ldots, n$ we obtain $\mathbf{P}(Q = n + 1) = 0$ by (2.20).

3. Decomposition of the switching matrix

We have seen from the preceding section that the traffic equation is closely connected with equilibrium for the queueing network. In turn, the characteristics of the traffic equation (1.5) depend directly on the switching matrix P. Viewed in a different context, the non-zero elements of P represent the queueing network topology, and the values of these entries determine the probabilities with which customers pursue alternative paths through the network.

Since P is a substochastic matrix, it may be extended to a transition matrix $\tilde{P}$ by the adjunction of an absorbing (exit) node 0:

$$\tilde{P} = \left[\begin{array}{c|ccc} 1 & 0 & \cdots & 0 \\ \hline p_{10} & & & \\ \vdots & & P & \\ p_{m0} & & & \end{array}\right] \tag{3.1}$$

The entries p_{i0} must satisfy $p_{i0} = 1 - \Sigma p_{ij} = q_i$, which is to say that p_{i0} is the probability that a customer departs the system immediately upon receiving service at node i.

Although we have no present direct interest in Markov chains associated with $\tilde{P}$, we shall nevertheless adopt the usual Markov chain conventions and notation (cf. [2], Part I). For instance, $p_{ij}^{(n)}$ is the usual n-step transition probability, and $i \rightsquigarrow j$ signifies that node j is accessible from node i.

In queueing theory, it is customary to classify the server nodes as *open* or *closed*; a node i closed if there is no possiblity that a customer at i eventually leaves the system. If a node is not closed, it is perforce open. A translation of this concept into the language of Markov chains is simply: a node i is open if $i \rightsquigarrow 0$, and closed if $i \not\rightsquigarrow 0$. However, a somewhat different (and new) notion is more appropriate to the analysis of queueing networks.

Definition 3.1. A node i is *completely open* if

$$p_{i0}^{(n)} \to 1. \tag{3.2}$$

The set of completely open nodes will be called T_0, and its complement in P is designated C.

It is easy to verify (using standard Markov chain theory — see [2]) each of the following:

$$\text{If } i \in T_0 \text{ and } i \rightsquigarrow j \text{ for } j \in P, \text{ we have } j \in T_0. \tag{3.3}$$

T_0 is a collection of equivalence classes of transient nodes (i.e., states). (3.4)

If C is non-empty, there exists a recurrent class $C_r \subset C$ (3.5)

For each $i \in C$, there exists a $j \in C_r$ such that $i \rightsquigarrow j$. (3.6)

The presence of a non-void C corresponds to a positive probability that a customer at certain service nodes is trapped in the network. A network for which T_0 is empty is usually called closed [10]. At the other extreme, a system with all open nodes actually consists only of completely open nodes; here C is void, and such a system is termed open.

From (3.3) and (3.4), one obtains immediately a decomposition of P useful in the impending discussion of the solution(s) of the traffic equation.

Lemma 3.1. *The matrix P can be partitioned as follows*:

$$P = \left[\begin{array}{c|c} T_0 & \theta \\ \hline R & C \end{array}\right] \tag{3.7}$$

4. The traffic equation — solution properties

We next study the existence and uniqueness of the solutions to the traffic equation $\delta = \alpha + \delta P$. We restrict consideration to $\alpha \geqslant 0$, and regard δ as a solution only if $\delta \geqslant 0$. By virtue of the decompositon (3.7), the traffic equation can equivalently be written as the two equations

$$_1\delta = {}_1\alpha + {}_1\delta T_0 + {}_2\delta R \tag{4.1a}$$

and

$$_2\delta = {}_2\alpha + {}_2\delta C \tag{4.1b}$$

in which δ and α are partitioned in accordance with $\delta = ({}_1\delta \mid {}_2\delta)$ and $\alpha = ({}_1\alpha \mid {}_2\alpha)$, respectively. When written in the form (4.1a) and (4.1b), the solution properties of the traffic equation can be definitively expressed by

Theorem 4.1. *Solutions exist to the traffic equation iff* $_2\alpha = \theta$. *The solution is unique iff* C *is empty; however, the solution* $_1\delta$ *is unique in any event, and is furnished by*

$$_1\delta = {}_1\alpha \sum_{n=0}^{\infty} (T_0)^n. \tag{4.2}$$

Proof. If $_2\alpha = \theta$ and C is non-void, (4.1b) possesses solutions satisfying $_2\delta = {}_2\delta C$, with $\delta_i \geqslant 0$ for $i \in C_r$ and $\delta_i = 0$ for $i \in (C - C_r)$ (the nodes in $C - C_r$ are transient).[6] But C_r is closed, so that $p_{ij} = 0$ if $i \in C_r$ and $j \in T_0$. Therefore, $\delta_i p_{ij} = 0$ for each $i \in C$, $j \in T_0$, and this is the same as $_2\delta R = \theta$.

We have just shown that $_2\alpha = \theta$ reduces (4.1a) to

$$_1\delta(I - T_0) = {}_1\alpha. \tag{4.3}$$

From (3.4), the series on the right of (4.2) must converge, so that we find by direct computation that

$$(I - T_0)^{-1} = \sum_{n=0}^{\infty} (T_0)^n. \tag{4.4}$$

This proves (4.2), and also demonstrates the uniqueness of $_1\delta$ whenever $_2\alpha = \theta$. If C is empty, $_1\delta = \delta$ is obvious, so that the traffic equation then has a unique solution.

[6] It is assumed that the reader is familiar with the elementary properties of discrete parameter Markov chains. Consequently, these properties will be used throughout without reference or specific justification.

Finally, consider a ${}_2\alpha$ with $\alpha_i > 0$, $i \in C$. Let $j \in C_r$ be a node such that $i \rightsquigarrow j$; for these indices we then have $\Sigma_{n=0}^{\infty} p_{ij}^{(n)} = \infty$. Moreover, substituting repeatedly in (4.1b) yields

$$_2\delta = {}_2\alpha \sum_0^k C^n + {}_2\delta C^{k+1} > {}_2\alpha \sum_0^k C^n \tag{4.5}$$

from which we obtain a fortiori

$$\delta_j > \alpha_i \sum_{n=0}^k p_{ij}^{(n)} \quad \text{for every } k. \tag{4.6}$$

This shows that no bounded solution can exist for (4.1b) whenever ${}_2\alpha \neq \theta$.

Theorem 4.1 leads one to suspect that the simplest and most elegant results are to be obtained for an open network, i.e. one in which C is empty. In fact, it is easy to show that Z_t is a reducible Markov process if C is non-empty, for it follows from the stochastic continuity equation (2.6) that $\Sigma_{i \in C, i} Z_t$ is almost surely non-decreasing. For non-empty C and ${}_2\alpha = \theta$ one can also exhibit a family of invariant probabilities q_0, in which the probabilities associated with the coordinates of C_r have been calculated by Gordon and Newell [7] for closed networks with a fixed number of customers.

On the other hand, if C is void, we can state and prove

Lemma 4.1. *If C is empty, the recurrent states of the Markov process Z_t constitute an irreducible set.*

Remark. It is not true that Z_t is an irreducible process; there may be transient states, as is the case when $\alpha_i = 0$ and $\Sigma_k p_{ki} = 0$ for some i.

Proof. We prove the somewhat stronger statement that $n \rightsquigarrow \theta$ for every $n = (n_1, \ldots, n_m)$. For this it suffices to prove $n' \rightsquigarrow n$, where $n' = (n_1, \ldots, n_i + 1, \ldots, n_m)$; $n \rightsquigarrow \theta$ then follows by induction.

For each i, there is a sequence of nodes $j_1, \ldots, j_k$ with $k \leq m-1$ such that $p_{ij_1} p_{j_1 j_2} \cdots p_{j_k 0} > 0$. Then, for every $t_0 > 0$ there exists a $\beta > 0$ (which does not depend on n' or n) for which

$$P_{n'n}(t) \geq \beta p_{ij_1} p_{j_1 j_2} \cdots p_{j_k 0}, \qquad t \in [t_0, \infty). \tag{4.7}$$

To verify (4.7), we return to the integral equation (2.8). This equation obviously implies $P_{nn}(t) \geq \exp(-c_n t)$. Substituting this P_{nn} back into (2.8) yields in general for any t_0 and states a, b: there is a $\beta' > 0$ such that for any $t \in [t_0 m^{-1}, \infty)$

$$P_{ab}(t) \geq \beta' q_{ab}. \tag{4.8}$$

Because each c_n (in this instance, c_a) is bounded from above and below β' does not depend on a. Of course, the right side is non-zero only if a and b are neighboring lattice points for which the corresponding $p_{ij} > 0$. The right side of (4.8) can therefore be expressed in terms of this p_{ij}, and the other constants absorbed in β'. If we now apply the Chapman–Kolmogorov equations repeatedly to the modified form of (4.8) and once more redefine β, we finally obtain (4.7).

With the aid of Lemma 4.1 above we can use the Markov character of Z_t and the knowledge of Jackson's equilibrium probability vector to present sufficient condition for the convergence to the equilibrium probability.

Theorem 4.2. *Assume that the queueing network is open (i.e., C is empty). Let the solution δ to the traffic equation* (1.5) *be such that each ρ_i [see* (1.4) *for the definition] satisfies* (1.6), *that is, $\rho_i < 1$. Then the probability vector $q(t)$ converges pointwise according to*

$$q(t) \to q_0, \tag{4.9}$$

where the probability vector q_0 is given by (1.3).

Proof. ρ_i and q_0 are well defined, since the solution δ to (1.5) exists and is unique from Theorem 4.1. As the argument following (2.10) demonstrates, q_0 is an invariant probability vector for the Markov process Z_t. Finally, (4.9) is true because the recurrent states of Z_t constitute an irreducible set by the preceding lemma ([6], p. 466).

Primarily for emphasis, we reiterate the conditions which enable us to assert the equilibrium probabilistic behavior of Z_t.

Theorem 4.3. *If an invariant probability vector q_0 exists, the following are equivalent:*

(1) *q_0 is unique,*
(2) *the network is open,*
(3) *C is empty,*
(4) *the traffic equation has a unique solution.*

Proof. Everything stated in this theorem has already been proved.

5. Service times

The total service time for customers entering the system is, perhaps surprisingly, closely connected with the traffic equation and its solution(s). For this reason, we can study the service time by the methods introduced in the preceding sections.

Our approach makes use of generating functions (g.f.), which in this case are Laplace-Stieltjes transforms of possibly defective probability distributions. We shall call $f_i(s)$ the g.f. of the total service time of a customer, given that he enters the queueing system by node i. Further, f will be the column vector with entries f_i, $i = 1, 2, \ldots, m$, and q is the column vector whose components q_i are the exit probabilities (also designated p_{i0}) from the various server nodes. With this notation, we can state and prove

Theorem 5.1. *For* $s > 0$, f *satisfies*

$$f(s) = Q^{-1}(s)q \tag{5.1}$$

where Q *is defined by its entries*

$$[Q(s)]_{ij} = (s\sigma_i^{-1} + 1)\delta_{ij} - p_{ij}. \tag{5.2}$$

The f_i *may be defective g.f.; if* S_j *is the service time of a customer entering the system through node* j, *we have*

$$\mathbf{P}(S_j < \infty) = \lim_n p_{j0}^{(n)}. \tag{5.3}$$

The right side of (5.3) *can be found from*

$$\lim_n p_{j0}^{(n)} = \lim_{s \to 0+} f_j(s) \tag{5.4}$$

or alternatively as the minimal positive solutions u_i *of*

$$u = q + Pu. \tag{5.5}$$

Proof. A customer at node i receives service with g.f. v_i, and either exits (probability q_i) or is routed to node j with probability p_{ij}. In the latter event, his additional time of service, starting at j, has g.f. f_j. We are therefore led to

$$f_i(s) = \left[q_i + \sum_j p_{ij} f_j(s)\right] v_i(s), \tag{5.6}$$

which takes advantage of the presumed independence of service times. We have consistently assumed $v_i(s) = \sigma_i(s + \sigma_i)^{-1}$, since this corresponds to exponential service. Substituting v_i into (5.6), and introducing vector notation gives

$$(sI + \sigma)f(s) = \sigma q + \sigma P f(s) \tag{5.7}$$

in which σ is defined as the diagonal matrix with entries σ_i. The same equation can be rewritten once more as $Q(s)f(s) = q$, where

$$Q(s) = (s\sigma^{-1} + I) - P \tag{5.8}$$

as in (5.2).

Clearly, (5.1) is true if $Q(s)$ is invertible for every $s > 0$. In the first place, $(s\sigma^{-1} + I)$ is invertible. Respective to the vector norm $\|a\| = \max |a_j|$ we have $\|(s\sigma^{-1} + I)^{-1}\| < 1$ and $\|P\| \leq 1$. This means $\|(s\sigma^{-1} + I)^{-1}P\| < 1$, whence $Q(s)$ has the desired inverse. The proof is thus complete.

It remains to demonstrate that the f provided by (5.1) is in fact a g.f., and to compute the defect in f. For this purpose, consider the recursive system $f^0(s) = \theta$,

$$(sI + \sigma)f^{n+1}(s) = \sigma q + \sigma P f^n(s). \tag{5.9}$$

For each $s \geq 0$, the following properties of the f^n are readily verified by induction:

$$f_i^n(s) \leq f_i^{n+1}(s), \tag{5.10}$$

$$0 \leq f_i^n(s) \leq 1, \tag{5.11}$$

$$f_i^n \text{ is a possibly defective g.f.} \tag{5.12}$$

In addition, one sees that $f^n(s) \uparrow f^n(0)$ as $s \to 0+$.

Since the f^n are monotone in n, their limit f^∞ exists. If we then take limits on n in (5.9), we find f^∞ solves (5.7) for every $s \geq 0$. Moreover f^∞, being the limit of a sequence of g.f., is itself a g.f. ([6], XIII.1, Theorem 2). Actually, the uniqueness of the solution $f(s) = Q^{-1}(s)q$ guarantees that $f(s) = f^\infty(s)$ for all $s > 0$. This means that f as given by (5.1) is also a g.f., whose defect is the difference between unity and $\lim_{s\to 0+} f(s)$. To compute the latter, we make use of the interchange of monotone limits to write

$$\lim_{s\to 0+} f(s) = \lim_{s\to 0+} \lim_{n} f^n(s) = \lim_{n} \lim_{s\to 0+} f^n(s) = \lim_{n} f^n(0). \tag{5.13}$$

For notational convenience, call $f^n(0) \stackrel{\mathrm{d}}{=} u^n$. Then consider (5.9) with $s = 0$. We obtain the recursive equations

$$u^{n+1} = q + Pu^n, \qquad u^0 = \theta. \tag{5.14}$$

Again, $\{u^n\}$ is a non-decreasing sequence whose limit we call u. Limits on n in (5.14) identifies this u as solution of (5.5); also, induction shows u to be the minimal positive solution of (5.5). In view of (5.13), and since u is merely $\lim_n f^n(0)$, we obtain at once

$$\mathbf{P}(S_i < \infty) = u_i. \tag{5.15}$$

The proof of the theorem is completed by showing that

$$u_i = \lim_n p_{i0}^{(n)}. \tag{5.16}$$

Let us expand (5.5) by components, writing p_{i0} for q_i. This gives

$$u_i = \sum_{j=1}^{m} p_{ij}u_j + p_{i0}. \tag{5.17}$$

It is a standard result in Markov chain theory that the minimal non-negative solution of (5.17) is precisely the probability of eventually entering state 0, given i as the initial state ([5], XV.8, Theorem 2). In other words, the u_i of (5.5) statisfies (5.17).

Theorem 5.1 permits us to characterize the service times S_j according to nature of the service node j. This simple classification follows from definition (3.2) and properties (3.5) and (3.6). More specifically, we have

Corollary 5.1. $\mathbf{P}(S_j < \infty)$ *is zero for* $j \in C_r$, *unity for* $j \in T_0$, *and* $0 < \mathbf{P}(S_j < \infty) < 1$ *for* $j \in (C - C_r)$.

It is of interest to note that f is of relatively simple form, so that the probability density functions for the service times can be easily found; in fact, these densities are at worst of mixed exponential type. This fact is a consequence of

Corollary 5.2. *The g.f.* f_i *are rational functions whose denominators are polynomials of degree no higher than* m.

Proof. As (5.8) indicates, the determinant of $Q(s)$ is a polynomial of degree m in s. This determinant appears in the denominator of each entry of $Q^{-1}(s)$ and, by (5.1), in the denominator of each $f_i(s)$.

Corollary 5.2 and the equation form (5.7) suggest that $f(s)$ coincides with the transform description of linear dynamical systems.[7] The theory of such systems features various realizations interpretable as networks of servers with exponential services and random routings, all these networks having the same service g.f. $f(s)$. Unfortunately, the routing "probabilities" may be negative and the service density functions complex. Thus, the dynamical systems analogue is appealing but does not appear promising for queueing network analysis. It has also been proposed [10] that any queueing network can be simplified to delete the feedback interconnections without changing the f_i; again, there is no evidence that the scheme would only involve routings having non-negative probabilities.

The final topic of this section is the expected service time for an arbitrary customer entering the queueing network. The service time under consideration will be denoted by S_a. It is evident that the expectation is

$$\mathbf{E}(S_a) = \sum_{i=1}^{m} \mathbf{E}(S_i) r_i \tag{5.18}$$

where r_i is the probability that an arbitrary customer enters the network by node i, as defined by (2.1). From (5.18) and Corollary 5.1 it follows at once that $\mathbf{E}(S_a) = \infty$ unless $\alpha_i = 0$ for every $i \in C$; this is precisely the same condition required in Theorem 4.1 for the existence of solutions to the traffic equation.

Conversely, if $\alpha_i = 0$ for all $i \in C$, $\mathbf{E}(S_a) < \infty$. Furthermore, $\mathbf{E}(S_a)$ appears in terms of ${}_1\delta$ [cf. (4.1a)], which is uniquely defined. To simplify the pertinent result and its proof we shall consider below only the more restrictive case of $P = T_0$.

Theorem 5.2. *Let C be empty. Then*

$$\mathbf{E}(S_a) = \left(\sum_{1}^{m} \alpha_j\right)^{-1} \delta\sigma^{-1} e \tag{5.19}$$

in which δ and σ are respectively the row vector and diagonal matrix introduced earlier, and e is the column vector whose entries are each unity.

Proof. We will have from (5.18) and the moment property of g.f.

$$\mathbf{E}(S_a) = -\lim_{s \to 0+} rf'(s), \tag{5.20}$$

[7] See [3], especially Chapter 4, for the pertinent dynamical systems background.

the prime indicating a derivative. Now differentiating (5.1) and substituting back for $f(s)$ yields

$$-f'(s) = Q^{-1}(s)Q'(s)f(s). \tag{5.21}$$

As $s \to 0+$ $f(s)$ tends toward unity, since f cannot be defective. The limit of $Q'(s)$ is σ^{-1}, since from (5.8) $Q'(s) = \sigma^{-1}$ identically. Finally, we assert that $Q^{-1}(s)$ has $\Sigma_0^\infty P^n$ for its limit as $s \to 0+$. To derive the limit of $Q^{-1}(s)$, we observe $[Q(s)\sigma]^{-1} = R_s$, where R_λ is defined as the resolvent operator for $(I - P)\sigma$ at the value λ. The origin belongs to the resolvent set, with $R_0 = \sigma^{-1}\Sigma_0^\infty P^n$. Hence $R_s \to R_0$, or $\sigma^{-1}Q^{-1}(s) \to \sigma^{-1}Q^{-1}(0)$; since $Q^{-1}(0) = \Sigma_0^\infty P_n^n$, we have the desired result.

Combining (5.20) and (5.21) and letting $s \to 0+$ yields

$$\mathbf{E}(S_a) = r\left(\sum_0^\infty P^n\right)\sigma^{-1}e. \tag{5.22}$$

The ultimate result (5.19) now follows from (5.22) substituting the solution for δ from (4.2), and noting the relationship between r and α.

Remark. $\mathbf{E}(S_a)$ is given by (5.19) also for non-empty C providing we take ${}_2\delta = \theta$.

6. Expected number of visits to server nodes

Yet another queueing network statistic directly connected with the traffic equation is the expected number of services furnished a customer entering the system. For this parameter, as well as others concerned solely with routing, it suffices to study the discrete parameter Markov chain with transition matrix $\tilde{P}$. In particular, if N_{ij} is the expected number of visits to server node j, given that the customer enters the system at node i, we find that

$$N_{ij} = \sum_n p_{ij}^{(n)}; \tag{6.1}$$

this well-known result is derived by representing the number of visits by the sum of counting random variables, and taking the indicated conditional expectation.

We recall from (2.1) that r_i denotes the probability of a customer entering the queueing system via node i. This suggests we can find the (unconditional) expectation N_j of the number of visits to node j by an arbitrary customer. If we take N to be the row vector with components $N_1, \ldots, N_m$ we obtain from (6.1) above

$$N = \sum_0^\infty rP^n. \tag{6.2}$$

We shall now relate (6.2) to the traffic equation $\delta = \alpha + \delta P$.

Theorem 6.2. *For specified $\alpha \geqslant 0$, $N < \infty$ iff the traffic equation has a solution. When N is finite, it satisfies the normalized version of the traffic equation*

$$N = r + NP \tag{6.3}$$

so that

$$N = \left(\sum_1^m \alpha_i \right)^{-1} \delta. \tag{6.4}$$

Here δ is taken as the solution to the traffic equation with ${}_2\delta = \theta$ [cf. (4.1b)].

Proof. We determine first the condition under which N is finite. For any $j \in T_0$, $\Sigma p_{ij}^{(n)} < \infty$ since j is perforce transient; thus, $N_j < \infty$. On the other hand, for $i \in C$ there exists a $j \in C_r$ such that $i \rightsquigarrow j$ and hence $\Sigma p_{ij}^{(n)} = \infty$. This means $r_i > 0$ for any $i \in C$ implies $N_j = \infty$ for some j. In terms of the partition of Theorem 4.1, the finiteness of N is therefore equivalent to ${}_2\alpha = \theta$, which is the same necessary and sufficient condition for the existence of solution(s) to the traffic equation.

If now ${}_2\alpha = \theta$ (i.e., ${}_2r = \theta$), substitution of the partitioned form of r and P [see (3.7)] in (6.2) implies

$${}_1N = {}_1r \sum_0^\infty (T_0)^n, \qquad {}_2N = \theta. \tag{6.5}$$

It can be directly verified that the N of (6.5) solves (6.3). We need only note that (6.3) reduces to

$${}_1N = {}_1r + {}_1NT_0, \tag{6.6}$$

so that the desired result is obtained by applying (4.4). The truth of (6.4) is ascertained from (6.5) also; we compare (6.5) with (4.2) for the relation between ${}_1N$ and ${}_1\delta$, and observe that ${}_2\delta = \theta$ leads to the correct ${}_2N$.

We close with two remarks. First, the expectation of the total number of services provided an arbitrary customer is readily computed from N. In fact, if e is the column m-vector whose entries are all ones, the scalar Ne represents this expectation.

Secondly, the generating function for the number of services could have been employed to calculate the above expectation, and also to obtain other statistics on visits. The pertinent generating function is defined by

$$F_{ij}(z) = \sum_{n=0}^{\infty} \mathbf{P}[X_n = j \mid X_0 = i] z^n \tag{6.7}$$

where X_k is the node position of a customer after the kth service. Since

$$\mathbf{P}[X_n = j \mid X_0 = i] = \delta_{n0}\delta_{ij} + (1 - \delta_{n0}) \sum_{k=1}^{m} \mathbf{P}[X_{n-1} = k \mid X_0 = i] p_{kj} \tag{6.8}$$

the matrix form of the g.f. (6.7) must satisfy

$$F(z) = I + zF(z)P \tag{6.9}$$

for $|z| < 1$. Within the unit disc, therefore,

$$F(z) = (I - zP)^{-1} = \sum_{n=0}^{\infty} z^n P^n. \tag{6.10}$$

The last equation makes it possible to secure the expectations N_{ij} and N by suitable passage to the limit $z \to 1$.

References

[1] Chandy, K., Howard, J., Keller, T. and Towsly, D. (1973). Local Balance, Robustness, Poisson Departures and the Product Function in Queueing Networks, Technical Report TR-15, Department of Computer Science, University of Texas at Austin.

[2] Chung, K.L. (1960). *Markov Chains with Stationary Transition Probabilities*, Springer-Verlag, Berlin.

[3] Desoer, C.A. (1970). *Notes for a Second Course on Linear Systems*, Van Nostrand Reinhold, New York.

[4] Doob, J.L. (1953). *Stochastic Processes*, Wiley, New York.

[5] Feller, W. (1968). *An Introduction to Probability Theory and Its Applications*, Volume I (3rd ed.), Wiley, New York.

[6] Feller, W. (1966). *An Introduction to Probability Theory and Its Applications*, Volume II, Wiley, New York.

[7] Gordon, W.J. and Newell, G.F. (1967). Closed queueing systems with exponential servers, *Operations Research*, **15**, 254–265.

[8] Jackson, J.R. (1957). Networks of waiting lines, *Operations Research*, **5**, 518–521.

[9] Jackson, J.R. (1963). Jobshop-like queueing systems, *Management Science*, **10**, 131–142.

[10] Muntz, R.R. and Baskett, F. (1972). Open, Closed and Mixed Networks of Queues with Different Classes of Customers, Technical Report No. 33, Digital Systems Laboratory, Stanford University.

[11] Skorokhod, A.V. (1965). *Studies in the Theory of Random Processes*, Addison-Wesley, Reading, Mass.

P.R. Krishnaiah, ed., *Multivariate Analysis–IV*
© North-Holland Publishing Company (1977) 371–385

AN INTRODUCTION TO QUANTUM ESTIMATION THEORY

Carl W. HELSTROM

University of California, San Diego, La Jolla, Calif., U.S.A.

A Bayesian theory of the optimum estimation of parameters of a quantum-mechanical density operator is described. Estimators are formulated as probability-operator measures (p.o.m.) on the Hilbert space of the quantum system. Operator equations for the p.o.m. minimizing the average cost of estimation are presented. Quantum counterparts of the Cramér–Rao inequality exist. An estimation-theoretical version of the uncertainty relation of quantum mechanics is derived. Other estimation problems that have been analyzed by means of this theory are mentioned with references to the literature.

1. Quantum observation

Quantum mechanics offers a framework for generalizing the concepts of statistical hypothesis testing and the statistical estimation of parameters. The need for generalization arose in studying the detection of optical signals and the estimation of their parameters in optical communications, radar, and astronomy. The receiver of an optical signal admits light through an aperture and focuses it by means of lenses or mirrors onto a detector such as a photoelectric cell or a photographic plate. In detection the presence or absence of the signal being sought is deduced from the observed response of that final photosensitive element; in parameter estimation that response is appropriately analyzed. Whatever is thus done to the incident light is basically a processing of the electromagnetic field at the aperture of the receiver. In order to discover the ultimate limitations on the detectability of an optical signal or on the accuracy with which its parameters can be estimated, one must seek the optimum processing of the aperture field. It is subject to the laws of quantum mechanics, which constrict the totality of observations that can be made upon it. Ordinary hypothesis-testing theory and estimation theory cannot cope with the quantum-mechanical constraints on the observability of an optical field, and it is necessary to generalize them. The applications to optics have been reviewed elsewhere [1, 2]; we shall not discuss them here, but treat instead an elementary estimation problem pertinent to quantum mechanics itself. The new theory has been extended and placed on a rigorous mathematical basis by A.S. Holevo [3]. These papers [1–3] supply references to the previous literature.

In nonrelativistic quantum mechanics a system is characterized by a vector in a Hilbert space $\mathscr{H}$. The elements of this space, denoted by $|\psi\rangle$, $|\varphi\rangle$, and so on, are called *kets* in Dirac's formulation. Associated with them are adjoint elements, or linear functionals, $\langle\psi|$, $\langle\varphi|$, and so on, known as *bras* and defined through the scalar product between any two elements, which is the complex number $\langle\varphi|\psi\rangle$;

$$\langle\psi|\varphi\rangle = \langle\varphi|\psi\rangle^*.$$

The norm of an element $|\psi\rangle$ is $\langle\psi|\psi\rangle^{\frac{1}{2}}$. Kets standing for physical states of a system have unit norm. All the usual properties of a Hilbert space are assumed [4].

The vectors in the space $\mathscr{H}$ are transformed by linear operators. With the dynamical variables of a physical system, such as the coordinates and momenta of its constituent particles, are associated certain self-adjoint linear operators, which play a role in describing the temporal evolution of the system. In the course of time, the state vector $|\psi(t)\rangle$ undergoes a continuous unitary transformation,

$$|\psi(t)\rangle = U(t, t_0)|\psi(t_0)\rangle,$$

$$U(t_0, t_0) = \mathbf{1}, \qquad U^+(t, t_0) = [U(t, t_0)]^{-1} = U(t_0, t),$$

where $\mathbf{1}$ is the identity operator in $\mathscr{H}$. The unitary operator $U(t, t_0)$ obeys the Schrödinger equation

$$\mathrm{i}\frac{\partial}{\partial t}U(t, t_0) = (H/\hbar)U(t, t_0),$$

where H is a self-adjoint operator corresponding to the energy of the system, and $\hbar = h/2\pi$ is Planck's constant. The key to understanding the behavior of a quantum system is finding the proper Hamiltonian operator, the way to whose discovery is shown by analogies with classical mechanics and electromagnetism, but whose correct specification usually requires considerable physical insight.

Quantum systems are for the most part microscopic, and thought must be given to how information about them is conveyed into the gross macroscopic world in which our perceptions operate. Some instrument must interact with the system in such a way that the instrument is finally in one of a number of states or conditions distinguishable by measurements with classical-physical apparatus such as meters or counters, and furthermore that final state of the instrument must depend on the state of the quantum system. The readings of the meters or counters provide a set of numbers $(z_1, z_2, \ldots, z_m)$ that serve as the outcome of the observation. Any such instrument consists of a huge number of atoms and molecules, whose initial condition and whose detailed behavior throughout the interaction with the quantum system cannot be specified with precision. The final condition of the instrument as evidenced by the meter readings can be related to the initial state of the quantum system only in a statistical fashion, and the outcomes $(z_1, z_2, \ldots, z_m)$ must be treated as random variables.

Let the outcomes of an observation be represented by a point $z = (z_1, z_2, \ldots, z_m)$ in an m-dimensional outcome space Z. Suppose at first that limitations of range and accuracy in the meters or counters render the set of possible outcomes countable or even finite. The probability p_k of the k-th set of outcomes z_k is postulated to be the quadratic form $p_k = \langle\psi \mid F_k \mid \psi\rangle$ when the quantum system was in the state $\mid\psi\rangle$ immediately before the observation, with F_k an appropriate self-adjoint linear operator. Because the probability p_k must be non-negative, the operator F_k must be nonnegative definite, $F_k \geqslant \mathbf{0}$. Furthermore, as the probabilities p_k must sum to 1, the set of operators $\{F_k\}$ characterizing the observation must form a generalized resolution of the identity,

$$\sum_k F_k = \mathbf{1}.$$

They are not necessarily projection operators.

More generally, if we allow the outcome space Z to be the m-dimensional continuum $\mathbf{R}_m$, observation of the quantum system by a given instrument is equivalent to a mapping of regions $\Delta \subset Z$ into a set of non-negative definite self-adjoint operators $F(\Delta)$ having the following properties:

(a) The empty set $\emptyset$ maps into the zero operator in $\mathscr{H}$,

$$\emptyset \rightarrow F(\emptyset) = \mathbf{0}, \tag{1.1}$$

(b) The entire space Z maps into the identity operator,

$$Z \to F(Z) = \mathbf{1}, \tag{1.2}$$

(c) Countable sets of disjoint regions map into the sums of the corresponding operators,

$$\begin{gathered}\Delta_1 + \Delta_2 + \ldots + \Delta_k + \ldots \to F(\Delta_1 + \Delta_2 + \ldots + \Delta_k + \ldots) = \\ = F(\Delta_1) + F(\Delta_2) + \ldots + F(\Delta_k) + \ldots, \\ \Delta_1 \cap \Delta_2 \cap \ldots \cap \Delta_k \cap \ldots = \emptyset.\end{gathered} \tag{1.3}$$

The set of operators $\{F(\Delta)\}$ is called a *probability-operator measure* (p.o.m.). The probability that the outcome z of the observation lies in the region $\Delta \subset Z$ is then postulated as

$$\mathbf{P}(z \in \Delta) = \langle \psi | F(\Delta) | \psi \rangle. \tag{1.4}$$

The instrument is said to 'apply' the p.o.m. $\{F(\Delta)\}$ to the quantum system. The pertinent operators $F(\Delta)$ must be determined by analyzing the interaction between the system and the observing instrument.

The preparation of the quantum system may entail uncertainties about its exact state, which may be one of a number of states $|\varphi_j\rangle$, $j = 1, 2, \ldots$, with probabilities ν_j of occurrence; these states $|\varphi_j\rangle$ are not necessarily orthogonal. The probability that the outcome z lies in the region Δ is then

$$\mathbf{P}(z \in \Delta) = \sum_j \nu_j \langle \varphi_j | F(\Delta) | \varphi_j \rangle, \qquad \sum_j \nu_j = 1, \tag{1.5}$$

which can be written

$$\begin{aligned}\mathbf{P}(z \in \Delta) &= \mathrm{Tr}\left[\sum_j \nu_j |\varphi_j\rangle\langle\varphi_j| F(\Delta)\right] \\ &= \mathrm{Tr}\,[\rho F(\Delta)],\end{aligned} \tag{1.6}$$

where 'Tr' stands for the trace of the succeeding operator, and

$$\rho = \sum_j \nu_j |\varphi_j\rangle\langle\varphi_j| \tag{1.7}$$

is known as the *density operator* of what is now a statistical ensemble of quantum systems.

The density operator ρ is self-adjoint and non-negative definite. Its trace is

$$\mathrm{Tr}\,\rho = \sum_j \nu_j = 1. \tag{1.8}$$

It therefore possesses non-negative eigenvalues w_k and orthonormal eigenvectors $|\psi_k\rangle$,

$$\rho\,|\psi_k\rangle = w_k\,|\psi_k\rangle,$$
$$\rho = \sum_k w_k\,|\psi_k\rangle\langle\psi_k|, \qquad w_k \geq 0, \qquad \sum_k w_k = 1. \tag{1.9}$$

Comparison with (1.7) shows that the representation of a density operator as a mixture of states is not unique, unless ρ is a projector onto a single ket $|\psi\rangle$,

$$\rho = |\psi\rangle\langle\psi|, \qquad \langle\psi\,|\,\psi\rangle = 1,$$

in which case the ensemble is composed of systems in a *pure* state $|\psi\rangle$. Otherwise the ensemble or its constituent systems are said to be in a *mixed* state. The set of states of an ensemble is convex, and the pure states form its boundary.

Compatible propositions about a quantum system—or more accurately, about the results of observing it—were placed by von Neumann [5] into correspondence with a set of projection operators forming a resolution of the identity in $\mathcal{H}$. In particular, the spectral theorem associates with a self-adjoint operator a resolution of the identity into orthogonal projectors onto its eigenvectors. An instrument applying to the system such an orthogonal resolution of the identity is said to *measure* the dynamical variable represented by the operator. Recent studies by Davies and Lewis [6] and Benioff [7, 8] have shown that a formulation of quantum measurement in terms of orthogonal projectors is unnecessarily restrictive and that it must be replaced by one in terms of a p.o.m. or 'generalized resolution of the identity'. Von Neumann's formulation can be retrieved, however, by invoking a theorem of Naĭmark's [9] as interpreted in this context by Holevo [3]. It is only necessary to adjoin to the original quantum system S an ancillary system A whose states span a Hilbert space $\mathcal{H}_A$ of sufficient dimensionality. In the product space $\mathcal{H} \otimes \mathcal{H}_A$ of states of the combination there exists a resolution of the identity $\mathbf{1} \otimes \mathbf{1}_A$ into orthogonal projectors in one-to-one correspondence with the elements $F(\Delta)$ of the given p.o.m. on $\mathcal{H}$. If the ancillary system A is in an appropriately selected pure state, an instrument applying this orthogonal p.o.m. to the combination of S and A yields outcomes in Z having the same probability measure as governs application of the original p.o.m. to the system S alone.

2. Estimation theory

Let the density operator of a quantum system depend on certain parameters $\theta_1, \theta_2, \ldots, \theta_m$, which we represent as a point $\boldsymbol{\theta}$ in an m-dimensional parameter space Θ, $\rho = \rho(\boldsymbol{\theta})$. These parameters might specify, for example, the location of a system, its velocity, or the angle through which it has been turned. In quantum electromagnetism, the amplitude of a coherent light signal, its arrival time, and its frequency are parameters of the state of the electromagnetic field at the aperture of a receiver of the signal. The coordinates of a star are parameters of the density operator of the field that its light creates at the aperture of an observing telescope. In a particular situation some or all of such parameters may be unknown, and it may be necessary to estimate them by an observation of the quantum system.

The observation must be accomplished by means of some macroscopic instrument that yields numbers z as outcomes; these would be the readings of its meters or counters. Upon these numbers certain calculations may be performed in order to obtain the estimates $(\hat{\theta}_1, \hat{\theta}_2, \ldots, \hat{\theta}_m) = \hat{\boldsymbol{\theta}}$ of the parameters $\boldsymbol{\theta}$. These calculations could be carried out by a computer that reads the meters automatically and prints out its estimates afterward, and the computer might just as well be treated as part of the observing instrument. The whole process is one grand observation of the kind described in Section 1; the outcome space Z is now the parameter space Θ; and the estimation corresponds to applying to the system some p.o.m. representing a mapping of regions Δ in the parameter space Θ into nonnegative definite self-adjoint operators $F(\Delta)$ subject to the rules in (1.1–.3). The conditional probability that the estimate $\hat{\boldsymbol{\theta}}$ lies in a region $\Delta \subset \Theta$ is then

$$\mathbf{P}(\hat{\boldsymbol{\theta}} \in \Delta \mid \boldsymbol{\theta}) = \mathrm{Tr}\,[\rho(\boldsymbol{\theta})\, F(\Delta)] \tag{2.1}$$

when the true values of the parameters are $\boldsymbol{\theta}$.

With these conditional probabilities a Bayesian estimation theory can be constructed as a generalization of ordinary estimation theory. To this end it is convenient to define infinitesimal elements $\mathrm{d}F(\boldsymbol{\theta})$ of the p.o.m. at each point $\boldsymbol{\theta}$ in such a way that

$$F(\Delta) = \int_{\Delta} \mathrm{d}F(\boldsymbol{\theta}) \tag{2.2}$$

for all regions $\Delta \subset \Theta$, and in particular

$$\int_{\Theta} \mathrm{d}F(\boldsymbol{\theta}) = \mathbf{1}. \tag{2.3}$$

Then for the continuous parameters $\boldsymbol{\theta}$ with which we exclusively concern ourselves here, we can define the conditional probability density function (p.d.f.) of the estimates as

$$q(\hat{\boldsymbol{\theta}} \mid \boldsymbol{\theta})\, \mathrm{d}^m\hat{\boldsymbol{\theta}} = \mathrm{Tr}\,[\rho(\boldsymbol{\theta})\, \mathrm{d}F(\hat{\boldsymbol{\theta}})], \tag{2.4}$$

where $\mathrm{d}^m\hat{\boldsymbol{\theta}}$ is the volume element in Θ. The p.o.m. $\{F(\Delta)\}$ serves as a quantum estimator of the parameters $\boldsymbol{\theta}$.

Let the prior p.d.f. of the estimanda be $z(\boldsymbol{\theta})$,

$$\int_{\Theta} z(\boldsymbol{\theta})\, \mathrm{d}^m\boldsymbol{\theta} = 1, \tag{2.5}$$

and let the costs of errors be specified by a cost function $C(\hat{\boldsymbol{\theta}}, \boldsymbol{\theta})$. Typical cost functions are the quadratic,

$$C(\hat{\boldsymbol{\theta}}, \boldsymbol{\theta}) = \sum_{i=1}^{m} \sum_{j=1}^{m} G_{ij}\, (\hat{\theta}_i - \theta_i)(\hat{\theta}_j - \theta_j), \tag{2.6}$$

with $\| G_{ij} \|$ a positive-definite matrix, and the delta-function

$$C(\hat{\boldsymbol{\theta}}, \boldsymbol{\theta}) = -\prod_{i=1}^{m} \delta(\hat{\theta}_i - \theta_i). \tag{2.7}$$

In ordinary estimation theory, the former leads to the conditional-mean, the latter to the maximum-likelihood estimator.

The average cost of applying an estimator specified by the p.o.m. $\{F(\Delta)\}$ is

$$\begin{aligned} \bar{C} &= \mathrm{Tr} \int_{\Theta} \int_{\Theta} z(\boldsymbol{\theta})\, C(\hat{\boldsymbol{\theta}}, \boldsymbol{\theta})\, \rho(\boldsymbol{\theta})\, \mathrm{d}F(\hat{\boldsymbol{\theta}})\, \mathrm{d}^m\boldsymbol{\theta} \\ &= \mathrm{Tr} \int_{\Theta} W(\hat{\boldsymbol{\theta}})\, \mathrm{d}F(\hat{\boldsymbol{\theta}}) \end{aligned} \tag{2.8}$$

with

$$W(\hat{\boldsymbol{\theta}}) = \int_{\Theta} z(\boldsymbol{\theta})\, C(\hat{\boldsymbol{\theta}}, \boldsymbol{\theta})\, \rho(\boldsymbol{\theta})\, \mathrm{d}^m\boldsymbol{\theta} \tag{2.9}$$

a known self-adjoint 'risk' operator. The p.o.m. for which the cost $\bar{C}$ is minimum is sought. Holevo [3] has shown that it must satisfy the operator equations

$$[W(\hat{\boldsymbol{\theta}}) - \Upsilon]\mathrm{d}F(\hat{\boldsymbol{\theta}}) = \mathbf{0}, \tag{2.10}$$

$$W(\hat{\boldsymbol{\theta}}) - \Upsilon \geq \mathbf{0}, \tag{2.11}$$

where Υ is a self-adjoint operator given by

$$\Upsilon = \int_{\Theta} W(\hat{\boldsymbol{\theta}})\,\mathrm{d}F(\hat{\boldsymbol{\theta}}) = \int_{\Theta} \mathrm{d}F(\hat{\boldsymbol{\theta}})\,W(\hat{\boldsymbol{\theta}}). \tag{2.12}$$

This operator Υ can be thought of as a Lagrange multiplier taking account of the self-adjoint constraint (2.3), and we call it the Lagrange operator. The minimum average cost is

$$\bar{C}_{\min} = \mathrm{Tr}\,\Upsilon \tag{2.13}$$

from (2.8) and (2.3).

Let $\{F'(\Delta)\}$ be some other p.o.m. satisfying (2.3) and incurring an average cost $\bar{C}'$. Then by (2.8) and (2.13)

$$\bar{C}' - \bar{C}_{\min} = \mathrm{Tr} \int_{\Theta} [W(\hat{\boldsymbol{\theta}}) - \Upsilon]\mathrm{d}F'(\hat{\boldsymbol{\theta}}).$$

As the trace of the product of two non-negative definite self-adjoint operators is non-negative,

$$\bar{C}' - \bar{C}_{\min} \geq 0$$

by (2.11). The first of the conditions (2.10) makes this an equality for the optimum p.o.m. and follows in fact from (2.3), (2.11), (2.12), and the non-negative definiteness of $\mathrm{d}F(\hat{\boldsymbol{\theta}})$.

When the density operators $\rho(\boldsymbol{\theta})$ commute for all possible pairs of parameter sets $\boldsymbol{\theta}$, they possess a common resolution of the identity into commuting projection operators as in (1.9),

$$\rho(\boldsymbol{\theta}) = \sum_k w_k(\boldsymbol{\theta}) |\psi_k\rangle\langle\psi_k|. \tag{2.14}$$

The optimal estimator of $\boldsymbol{\theta}$ involves first observing the system with an instrument applying the orthogonal p.o.m. composed of projectors $|\psi_k\rangle\langle\psi_k|$ onto the common eigenvectors of the density operators, and then analyzing the outcome by the methods of conventional estimation theory, which thus manifests itself as a special case of the quantum-mechanical theory.

Inequalities of the Cramér–Rao type are to be found in quantum estimation theory as well. The conditional expected value of an estimate is defined by

$$\bar{\theta}_j = \mathbf{E}(\hat{\theta}_j \mid \boldsymbol{\theta}) = \operatorname{Tr} \int_{\Theta} \hat{\theta}_j \rho(\boldsymbol{\theta}) \mathrm{d}F(\hat{\boldsymbol{\theta}}), \tag{2.15}$$

and the conditional covariance of two estimates is

$$B_{ij} = \{\hat{\theta}_i, \hat{\theta}_j\} = \operatorname{Tr} \int_{\Theta} (\hat{\theta}_i - \bar{\theta}_i)(\hat{\theta}_j - \bar{\theta}_j)\rho(\boldsymbol{\theta}) \mathrm{d}F(\hat{\boldsymbol{\theta}}); \tag{2.16}$$

the matrix of these covariances is denoted by $\boldsymbol{B}$. Now define the right-logarithmic derivative (r.l.d.) operators L_k by

$$\frac{\partial \rho}{\partial \theta_k} = \rho L_k, \tag{2.17}$$

and the symmetrized logarithmic derivative (s.l.d.) operators $\mathscr{L}_k$ by

$$\frac{\partial \rho}{\partial \theta_k} = \tfrac{1}{2}(\rho \mathscr{L}_k + \mathscr{L}_k \rho). \tag{2.18}$$

(For some density operators ρ, and in particular for those representing pure states, the r.l.d. operators may not exist.) Define furthermore the matrices $\mathbf{A}$ and $\mathscr{A}$ as those whose elements are

$$A_{ij} = \operatorname{Tr} \rho L_i L_j^+ = \operatorname{Tr}\left(\frac{\partial \rho}{\partial \theta_i} L_j^+\right), \tag{2.19}$$

$$\mathscr{A}_{ij} = \tfrac{1}{2} \operatorname{Tr}[\rho(\mathscr{L}_i \mathscr{L}_j + \mathscr{L}_j \mathscr{L}_i)] = \operatorname{Tr}\left[\frac{\partial \rho}{\partial \theta_i} \mathscr{L}_j\right]. \tag{2.20}$$

Then for unbiased estimators, $\bar{\theta}_i \equiv \theta_i$, the quantum-mechanical Cramér–Rao inequalities correspond to the matrix inequalities

$$\boldsymbol{B} \geqslant \mathbf{A}^{-1}, \qquad \boldsymbol{B} \geqslant \mathscr{A}^{-1} \tag{2.21}$$

[10–12]. In particular, the variance of an unbiased estimator is bounded by the diagonal elements of the inverse matrices,

$$\operatorname{Var} \hat{\theta}_i \geqslant (\mathbf{A}^{-1})_{ii}, \qquad \operatorname{Var} \hat{\theta}_i \geqslant (\mathscr{A}^{-1})_{ii}. \tag{2.22}$$

If a single parameter θ is unknown and to be estimated, the s.l.d. inequality yields the superior lower bound;

$$\operatorname{Var} \hat{\theta} \geqslant [\operatorname{Tr}(\rho \mathscr{L}^2)]^{-1} = \left[\operatorname{Tr}\left(\frac{\partial \rho}{\partial \theta} \mathscr{L}\right)\right]^{-1}, \qquad \frac{\partial \rho}{\partial \theta} = \tfrac{1}{2}(\rho \mathscr{L} + \mathscr{L} \rho). \tag{2.23}$$

With multiple estimanda sometimes the one, sometimes the other inequality is superior [13].

3. The uncertainty principle

Estimation theory furnishes counterparts to the quantum-mechanical uncertainty principle for what we term *displacement parameters.* The standard uncertainty principle applies to an ensemble of systems known to be in state ρ. On one half of the ensemble the dynamical variable corresponding to a self-adjoint linear operator A_1 is measured by applying an orthogonal resolution of the identity into projectors onto its eigenvectors. On the systems in the other half, the dynamical variable corresponding to a second self-adjoint linear operator A_2 is measured in the same sense. The dispersions ΔA_1 and ΔA_2 of the outcomes are defined by

$$\Delta A_i^2 = \operatorname{Tr}[\rho(A_i - \bar{A}_i \mathbf{1})^2], \qquad \bar{A}_i = \operatorname{Tr}(\rho A_i), \qquad i = 1, 2. \tag{3.1}$$

The uncertainty principle states that

$$\Delta A_1 \Delta A_2 \geqslant \tfrac{1}{2} |\operatorname{Tr} \rho [A_1, A_2]|, \tag{3.2}$$

where $[A_1, A_2] = A_1 A_2 - A_2 A_1$ is the commutator of the two operators [14]. The principle is most familiar as applied to the position and momentum operators. There are, however, classical physical variables, such as an angle of rotation or the phase of an oscillation, for which corresponding quantum-mechanical operators do not exist, and setting up relevant uncertainty relations becomes much more complicated [15]. An enlightening alternative viewpoint considers these not as dynamical variables, but as parameters of the density operator and as estimanda.

Suppose that an apparatus prepares a quantum system in the state ρ_0. Let both be displaced by an amount θ. All other conditions being the same, the apparatus then prepares the system instead in the state

$$\rho(\theta) = e^{-iN\theta} \rho_0 e^{iN\theta}, \tag{3.3}$$

where N is a self-adjoint linear operator termed the infinitesimal generator of the displacement group. If the displacement is a translation along the x-axis, $N = P/\hbar$, where P is the operator for the x-component of the momentum of the system, and $\hbar$ is again Planck's constant. If θ is an angle through which the system and apparatus have been turned about the z-axis, $N = L_z/\hbar$, where $L_z = XP_y - YP_x$ is the operator for the z-

component of angular momentum. If θ represents the phase of a coherent signal in a quantum oscillator, $N = a^{\dagger}a$, where a and $a^{\dagger}$ are the annihilation and creation operators for photons, whose commutator is

$$[a, a^{\dagger}] = aa^{\dagger} - a^{\dagger}a = \mathbf{1}, \tag{3.4}$$

$a^{\dagger}a$ being the operator corresponding to the number of photons in the oscillator. If θ is the time at which some distinctive event occurs in the evolution of the system, $N = H/\hbar$, where H is the Hamiltonian or energy operator. We call θ in general a displacement parameter and investigate how accurately it can be estimated by observation of the quantum system.

If the parameter θ is an angle of rotation or an oscillator phase, the density operator $\rho(\theta)$ is periodic in θ with period 2π, and we restrict our analysis to estimators that are also periodic, $\mathrm{d}F(\hat{\theta}) = \mathrm{d}F(\hat{\theta} + 2\pi)$. We define the expected value of an estimate $\hat{\theta}$ as

$$\bar{\theta} = \mathbf{E}(\hat{\theta} \mid \theta) = \mathrm{Tr} \int_{\theta-\pi}^{\theta+\pi} \hat{\theta}\rho(\theta)\,\mathrm{d}F(\hat{\theta}) \tag{3.5}$$

and the mean square error $\mathscr{E}$ as

$$\mathscr{E} = \mathbf{E}[(\hat{\theta} - \theta)^2 \mid \theta] = \mathrm{Tr} \int_{\theta-\pi}^{\theta+\pi} (\hat{\theta} - \theta)^2 \rho(\theta)\,\mathrm{d}F(\hat{\theta}), \tag{3.6}$$

adopting these definitions because the error in the estimate $\hat{\theta}$ should be considered as the smaller of $|\hat{\theta} - \theta|$ and $|2\pi - \hat{\theta} + \theta|$ when both θ and $\hat{\theta}$ lie in $(-\pi, \pi)$.

Employing the same technique as in deriving the Cramér–Rao inequalities [11], we differentiate (3.5) with respect to θ,

$$\begin{aligned} \frac{\mathrm{d}\bar{\theta}}{\mathrm{d}\theta} &= \mathrm{Tr} \int_{\theta-\pi}^{\theta+\pi} \hat{\theta} \frac{\partial \rho}{\partial \theta} \mathrm{d}F(\hat{\theta}) + (\theta + \pi)\mathrm{Tr}\,\rho(\theta) \frac{\mathrm{d}F(\theta + \pi)}{\mathrm{d}\theta} \\ &\quad - (\theta - \pi)\,\mathrm{Tr}\,\rho(\theta) \frac{\mathrm{d}F(\theta - \pi)}{\mathrm{d}\theta} \\ &= \mathrm{Tr} \int_{\theta-\pi}^{\theta+\pi} (\hat{\theta} - \theta) \frac{\partial \rho}{\partial \theta} \mathrm{d}F(\hat{\theta}) + 2\pi q(\theta + \pi \mid \theta) \end{aligned} \tag{3.7}$$

by (2.4) and

$$\mathrm{Tr} \int_{\theta-\pi}^{\theta+\pi} \rho(\theta)\,\mathrm{d}F(\hat{\theta}) = 1, \qquad \mathrm{Tr} \int_{\theta-\pi}^{\theta+\pi} \frac{\partial \rho}{\partial \theta} \mathrm{d}F(\hat{\theta}) = 0, \tag{3.8}$$

which follow from (1.8) and (2.3). Now by (3.3)

$$\frac{\partial \rho}{\partial \theta} = \mathrm{i}(\rho N - N\rho),$$

whence

$$\mathrm{Tr\,i}\int_{\theta-\pi}^{\theta+\pi} (\hat{\theta}-\theta)(\rho N - N\rho)\,\mathrm{d}F(\hat{\theta}) = -2\,\mathrm{Im\,Tr}\int_{\theta-\pi}^{\theta+\pi} (\hat{\theta}-\theta)\,\rho N\,\mathrm{d}F(\hat{\theta})$$

$$= -2\mathrm{Im\,Tr}\int_{\theta-\pi}^{\theta+\pi} (\hat{\theta}-\theta)\rho(N-\bar{N}\mathbf{1})\,\mathrm{d}F(\hat{\theta})$$

$$= \frac{\mathrm{d}\bar{\theta}}{\mathrm{d}\theta} - 2\pi\, q(\theta+\pi \mid \theta) \stackrel{\mathrm{d}}{=} m(\theta), \tag{3.9}$$

with the expectation $\bar{N}$ defined as in (3.1). The introduction of $\bar{N}$ changes the trace of the integral only by a real number. By virtue of the Schwarz inequality for traces, $\mathrm{d}F(\hat{\theta})$ being non-negative definite [11],

$$\tfrac{1}{4}[m(\theta)]^2$$

$$\leqslant \left| \mathrm{Tr}\int_{\theta-\pi}^{\theta+\pi} (\hat{\theta}-\theta)\,\rho(N-\bar{N}\mathbf{1})\,\mathrm{d}F(\hat{\theta}) \right|^2$$

$$\leqslant \left[\mathrm{Tr}\int_{\theta-\pi}^{\theta+\pi} (\hat{\theta}-\theta)^2\rho\,\mathrm{d}F(\hat{\theta})\right]\left[\mathrm{Tr}\int_{\theta-\pi}^{\theta+\pi} \rho(N-\bar{N}\mathbf{1})\mathrm{d}F(\hat{\theta})(N-\bar{N}\mathbf{1})\right]$$

$$= \mathscr{E}\,\mathrm{Tr}[\rho(N-\bar{N}\mathbf{1})^2],$$

by (3.6) and (2.3). Thus we obtain for the mean square error the lower bound

$$\mathscr{E} = \mathbf{E}[(\hat{\theta}-\theta)^2 \mid \theta] \geqslant \frac{[m(\theta)]^2}{4\Delta N^2},$$

where as in (3.1) ΔN is the dispersion of the infinitesimal generator of the displacement group.

If the parameter θ is not periodic, the integrals in (3.5) and (3.6) are simply taken over the one-dimensional continuum of its values, and the term $2\pi q(\theta+\pi \mid \theta)$ is omitted from $m(\theta)$ in (3.9); for an unbiased estimate $m(\theta) = 1$. Even for a periodic parameter, the term $2\pi q(\theta+\pi \mid \theta)$ will be small whenever the width of the posterior p.d.f. $q(\hat{\theta} \mid \theta)$ about the true value θ is much less than 2π. Under these circumstances the inequality

$$\mathrm{Var}\,\hat{\theta} \geqslant \tfrac{1}{4}(\Delta N)^{-2}$$

sets a lower bound to the variance of an unbiased estimate of a displacement parameter.

Thus we obtain an estimation-theoretical version of the uncertainty principle by treating the displacement θ not as an observable represented by an operator, but as a parameter of the density operator of the system. If for instance θ is the time of an event in its evolution,

$$\mathrm{Var}\,\hat{\theta} \geq \hbar^2/4(\Delta H)^2,$$

where ΔH is the dispersion of the energy of the system. The difficulties attending time-energy uncertainty relations of the usual kind (3.2) have been described by Allcock [16].

In a recent paper [17] the maximum-likelihood estimator of a periodic displacement parameter θ, such as an angle of rotation or a phase, was derived for a system in a pure state $|\psi(\theta)\rangle$ by solving the optimization equations (2.10–.12). It was assumed that the prior p.d.f. of θ is uniform over $(0, 2\pi)$. The same estimator minimizes the average cost of error as assessed by the cost function

$$C(\hat{\theta}, \theta) = 4\sin^2[\tfrac{1}{2}(\hat{\theta} - \theta)],$$

which for small errors is approximately $(\hat{\theta} - \theta)^2$. The minimum average cost is

$$C_{\min} = \sum_k (|\langle k \mid \psi(\theta)\rangle| - |\langle k-1 \mid \psi(\theta)\rangle|)^2,$$

where $|k\rangle$ is an eigenvector of the infinitesimal generator N, whose eigenvalues k are integers or odd multiples of $\frac{1}{2}$, $N|k\rangle = k|k\rangle$. This minimum cost is the smaller, the broader the distribution of the components $|\langle k \mid \psi(\theta)\rangle|$ among the eigenstates $|k\rangle$. We encounter here a second form of the uncertainty principle for a quantum displacement parameter.

4. Remarks

Several papers have been devoted to the quantum-mechanical counterpart of the optimum estimation of the mean values of a Gaussian probability density function. The normal modes of the electromagnetic field of an ideal receiver of a coherent optical signal are described by a Gaussian density operator in Glauber's P-representation [18] when the input to the receiver contains thermal background radiation; the mean values of the complex modal amplitudes correspond to the signal and are treated as estimanda. Their optimum estimators have been found for a variety of cost functions under the assumption that their prior p.d.f. has a Gaussian form, and maximum-likelihood estimators have also been determined [3, 19–21]. The p.o.m. figuring in these estimators is based upon the overcompleteness relation for the coherent states of the harmonic oscillator [18]; it can also be applied to the simultaneous estimation of the position and velocity of a quantum system [22]. By means of the quantum-mechanical Cramér–Rao inequality it has been possible to set lower bounds under the variances of estimates of the arrival time and frequency of a coherent optical signal at an ideal receiver [13].

A number of problems remain. Supposing that one has solved the optimization equations (2.10–2.12) and found the best estimator $\{F(\Delta)\}$ of a particular set of parameters, how can an instrument be designed that will effectively apply it to the quantum system under observation? It is not even known in quantum mechanics whether an instrument exists to measure an arbitrary self-adjoint linear operator in the conventional sense. There is no way of characterizing mathematically the set of instruments that contemporary technology can construct and apply to quantum systems; the analyst must work with the broadest conceivable class of instruments, those applying probability-operator measures as defined in Section 1.

Computational methods for solving the optimization equations of quantum hypothesis-testing and estimation theories seem not to exist. It has been necessary to guess the correct solutions on the basis of symmetry or other features of the problem, and then to verify that they satisfy the equations. Only the simplest problems can be treated in this way. One might start by representing the density operators of Section 2, at least approximately, by finite square matrices, and by discretizing the parameter space. To the writer's knowledge, there are no methods for solving such operator equations even in finite matrix form. Accounting for the constraints of non-negative definiteness seems to present particular difficulty.

General properties of conventional estimators, such as asymptotic normality of maximum-likelihood estimators, might instructively be examined to see whether they hold for the corresponding quantum estimators. The quantum counterparts of other types of error bounds, such as those associated with the names of Bhattacharyya [23] and Barankin [24], might be sought. Mathematical statisticians will doubtless perceive still other aspects of conventional estimation theory whose generalization to the quantum theory are worthy of investigation.

References

[1] Helstrom, C.W., Liu, J., and Gordon, J. (1970). Quantum mechanical communication theory. *Proc. IEEE* **58**, 1578–1598.

[2] Helstrom, C.W. (1972). Quantum detection theory. *Progress in Optics*, E. Wolf, ed. North-Holland Publishing Co., Amsterdam. **10**, 289–369.

[3] Holevo, A.S. (1973). Statistical decision theory for quantum systems. *J. Multivariate Anal.* **3**, 337–394.

[4] Dirac, P.A.M. (1958). *Quantum Mechanics*. Oxford University Press, Oxford, Eng. (4th edition).

[5] Von Neumann, J. (1932). *Mathematische Grundlagen der Quantenmechanik*, Springer Verlag, Berlin. Translated by R.T. Beyer, *Mathematical Foundations of Quantum Mechanics*, Princeton University Press, Princeton, N.J. (1955).

[6] Davies, E.B. and Lewis, J.T. (1970). An operational approach to quantum probability. *Comm. Math. Phys.* **17**, 239–260.

[7] Benioff, P.A. (1972). Operator valued measures in quantum mechanics: finite and infinite processes. *J. Math. Phys.* **13**, 231–242.

[8] Benioff, P.A. (1972). Decision procedures in quantum mechanics. *J. Math. Phys.* **13**, 908–915.

[9] Naĭmark, M.A. (1940). Spectral functions of a symmetric operator (in Russian, English summary). *Izv. Acad. Nauk USSR*, ser. mat. **4**, 277–318.

[10] Helstrom, C.W. (1968). The minimum variance of estimates in quantum signal detection. *Trans. IEEE IT*–**14**, 234–242.

[11] Helstrom, C.W. (1973). Cramér Rao inequalities for operator valued measures in quantum mechanics. *Int. J. Theoret. Phys.* **8**, 361–376.

[12] Yuen, H.P. and Lax, M. (1973). Multiple-parameter quantum estimation and measurement of nonselfadjoint observables. *Trans. IEEE IT*–**19**, 740–750.

[13] Helstrom, C.W. and Kennedy, R.S. (1974). Noncommuting observables in quantum detection and estimation theory. *Trans. IEEE IT*–**20**, 16–24.

[14] Robertson, H.P. (1929). The uncertainty principle. *Phys. Rev.* **34**, 163–164.

[15] Carruthers, P. and Nieto, M.M. (1968). Phase and angle variables in quantum mechanics. *Rev. Mod. Phys.* **40**, 411–440.

[16] Allcock, G.R. (1969). The time of arrival in quantum mechanics. *Annals of Physics* **53**, 253–285.

[17] Helstrom, C.W. (1974). Estimation of a displacement parameter of a quantum system. *Int. J. Theoret. Phys.* **11**, 357–378.

[18] Glauber, R.J. (1963). Coherent and incoherent states of the radiation field. *Phys. Rev.* **131**, 2766–2788.

[19] Holevo, A.S. (1973). Optimal quantum measurements. *Teoret. i Matemat. Fizika* **17**, 319–326.

[20] Helstrom, C.W. (1974). Quantum Bayes estimation of the amplitude of a coherent signal. *Trans. IEEE IT*–**20**, 374–376.

[21] Belavkin, V.P. (1974). Optimal linear randomized filtration of quantum boson signals. *Problems of Control and Information Theory* **3**, 47–62.

[22] Helstrom, C.W. (1974). 'Simultaneous measurement' from the standpoint of quantum estimation theory. *Found. Phys.* **4**, 453–463.

[23] Bhattacharyya, A. (1946–48). On some analogues of the amount of information and their use in statistical estimation. *Sankhyā, Indian J. Statistics* **8**, 1–14, 201–218, 315–328.

[24] Barankin, E.W. (1949). Locally best unbiased estimates. *Ann. Math. Stat.* **20**, 477–501.

P.R. Krishnaiah, ed., *Multivariate Analysis–IV*
© North-Holland Publishing Company (1977) 387–406

A SCATTERING THEORY FRAMEWORK FOR FAST LEAST-SQUARES ALGORITHMS*

T. KAILATH and L. LJUNG**
Stanford University, Stanford, Calif.

In scattering theory the Riccati equation arises in a natural family of equations evolving forwards as well as backwards in time. We show how this framework allows an interesting derivation of the fast Chandrasekhar-type equations for linear least-squares filtering of processes generated by a time-invariant, finite-dimensional linear system driven by white noise. The processes are not required to be stationary. The same ideas can be used to obtain Levinson- and Cholesky-type differential equations for the impulse responses of the whitening filter and the innovations representation of such processes. The scattering framework brings out clearly both the significance of the time-invariance of the parameters of the underlying finite-dimensional system and of the associated family of nonstationary processes. For stationary processes, it also becomes clear that the assumption of finite-dimensionality is unnecessary, but the proper extension of the nonstationary class of processes raises some interesting questions.

1. Introduction

It is well known that the linear least-squares filtering problem for stochastic process can be solved via an associated Wiener–Hopf equation or, if a state space model for the process is available, via a certain Riccati equation to obtain the Kalman filter. The Kalman filter solution is numerically quite efficient, in particular if the dimension of the state vector is small. For high-order systems, however, solution of the Riccati equation requires long computation time, and may even be unfeasible. For such cases, Kailath [4] showed that under certain conditions it is possible to solve directly for the Kalman gain matrix, with much less computational effort. This is done with the so-called fast Chandrasekhar-equations for the gain matrix. These Chandrasekhar equations were first introduced in astrophysics to solve certain Wiener–Hopf equations with "displacement" kernels corresponding in the estimation context to covariances of station-

* This work was supported in part by the Air Force Office of Scientific Research, AF Systems Command, under contract AF 44–620–74–C–0068, the Joint Services Electronics Program under contract N–00014–67–A–0112–0044, and the Industrial Affiliates Program at the Information Systems Laboratory, Stanford University.
** Ljung is now with the Dept. of Electrical Engineering, Linköping University, Linköping, Sweden.

ary processes. However, an important feature of the derivation in [4], is that it showed how similar results could be obtained for a large class of nonstationary processes, namely those obtainable by driving finite-dimensional (but not necessarily stable) time-invariant linear systems with white noise and random initial conditions. This natural extension of the class of stationary processes seems to be good model of many physical problems, e.g., in speech analysis [2], and in earthquake models [7].

In this paper we shall further develop some of the properties of this class of processes. We begin by rederiving the Chandrasekhar-type equations of [4] in a new way using some simple but fundamental ideas from scattering theory (à la Redheffer [15], [16], and others). In this theory, the Riccati equation arises naturally in certain sets of backwards as well as forwards evolution equations for a semi-group of reflection and transmission operators. For time-invariant systems the equations in the two directions must be the same except for the sign, and we shall show that the resulting ("Stokes") identities lead naturally to the general Chandrasekhar-type equations. We see in this process the close Janus-like relationship between the Chandrasekhar- and Riccati-equation solutions to the estimation problem. The invariance ideas we use in this paper are quite simple and differ from the shift-invariance ideas earlier [18]–[19]—the present use of invariance is in a sense a natural completion of the ideas originally introduced by Ambartzumian and Chandrasekhar and further developed by Redheffer [16], Preisendorfer [13], Bellman, Wing, Kalaba and their colleagues [1].

We further exploit the interplay between forwards and backwards equations to develop what we call Levinson-type equations for the impulse response of the optimum whitening filter for the above-mentioned class of non-stationary processes. Previous results in this problem have all been for stationary processes, and we shall show how the interplay between forwards and backwards equations provides certain key ideas (esp. that of what we call "adjoint" covariance functions) that enable the extension to the general class of nonstationary processes of [4]. Similar results are obtained for what we call Cholesky-type equations for determining the innovations representations of such processes.

We should emphasize that though we have couched our presentation in terms of stochastic processes and covariance functions, this is not an essential restriction and most of our results hold also for "nonsymmetric" Riccati- and other equations as arise, for example, in studying general two-point linear boundary-value problems.

2. Some facts from scattering theory

Scattering theory is a way of analyzing how certain propagating entities (like neutrons or electromagnetic waves) travel through a medium and interact with it. The analysis is based on certain operators called transmission and reflection operators and the interesting fact is that Riccati equations arise naturally in this analysis (see, e.g., [15], [16]). Since Riccati equations also arise in least-squares estimation equations, we have the possibility of interpreting certain estimation equations in terms of a scattering model. The advantages of such an identification will soon be made clear.

We shall only need some simple facts from the theory, which can be developed in a self-contained way [16]. Consider entities $u_+(\tau)$ and $u_-(s)$ incident upon an "obstacle", say a layer extending from τ to s of some medium, and let $u_-(\tau)$ and $u_+(s)$ be emergent entities. We assume there exist four operators a, r, ρ, α such that

$$\begin{bmatrix} u_+(s) \\ u_-(\tau) \end{bmatrix} = \begin{bmatrix} a_0(\tau, s) & \rho_0(\tau, s) \\ r_0(\tau, s) & \alpha_0(\tau, s) \end{bmatrix} \begin{bmatrix} u_+(\tau) \\ u_-(s) \end{bmatrix} \tag{1}$$

The matrix of the 4 operators will be written $S_0(\tau, s)$ and called the *scattering matrix*. If we now have another layer extending from s to t, so that (see Fig. 1)

$$\begin{bmatrix} u_+(t) \\ u_-(s) \end{bmatrix} = S_0(s, t) \begin{bmatrix} u_+(s) \\ u_-(t) \end{bmatrix},$$

then it is easy to see by algebraic manipulation (or by tracing rays through the figure) that

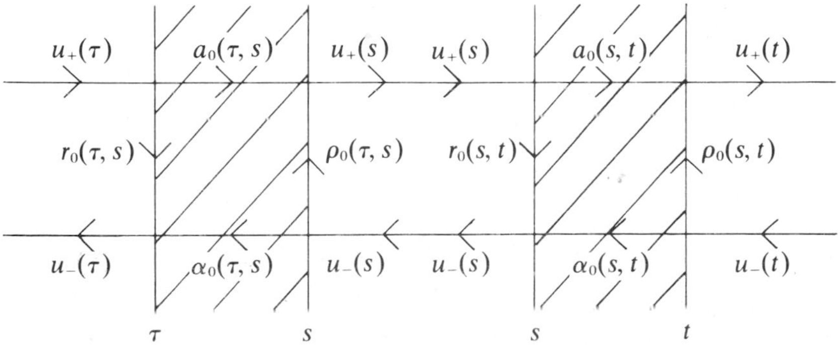

Fig. 1. To illustrate propagation through two adjacent layers (shown apart for clarity).

$$\begin{bmatrix} u_+(t) \\ u_-(\tau) \end{bmatrix} = S_0(\tau, t) \begin{bmatrix} u_+(\tau) \\ u_-(t) \end{bmatrix}$$

where $S_0(\tau, t)$ is obtained by the composition rule

$S_0(\tau, t) =$

$$= \begin{bmatrix} a_0(s,t)[I-\rho_0(\tau,s)r_0(s,t)]^{-1}a_0(\tau,s) & \rho_0(s,t)+a_0(s,t)\rho_0(\tau,s)\times \\ & \times[I-r_0(s,t)\rho_0(\tau,s)]^{-1}\alpha_0(s,t) \\ r_0(\tau,s)+\alpha_0(\tau,s)r_0(s,t)\times & \alpha_0(\tau,s)[I-r_0(s,t)\rho_0(\tau,s)]^{-1}\times \\ \times[I-\rho_0(\tau,s)r_0(s,t)]^{-1}a_0(s,t) & \times\,\alpha_0(s,t) \end{bmatrix}$$

$$= S_0(\tau, s) * S_0(s, t) \tag{2}$$

which is known as Redheffer's ''star-product'' composition rule.

Now we can, if we wish, ignore the underlying physical problem of propagating entities and layers and concentrate on the fact that what we have is a semi-group of operators closed under a particular composition rule. We can, therefore, develop a calculus of such operators by examining the effect of infinitesimal changes. Thus suppose that

$$S_0(s, s+\Delta s) = S_0(s, s) + M(s)\Delta s + o(\Delta s) \tag{3}$$

where

$$S_0(s, s) = I$$

and

$$M(s) = \begin{bmatrix} f(s) & g(s) \\ h(s) & e(s) \end{bmatrix}.$$

Clearly,

$$M(s) = \lim_{\Delta s \to 0} \frac{S_0(s, s+\Delta s) - I}{\Delta s}$$

is the infinitesimal generator of S. Now note that

$$S_0(\tau, s+\Delta s) = S_0(\tau, s) * S_0(s, s+\Delta s) \tag{4}$$

$$= o(\Delta s) + S_0(s, \tau) + \Delta s \times$$

$$\begin{bmatrix} [f(s)+\rho_0(\tau,s)h(s)]a_0(\tau,s) & g(s)+f(s)\rho_0(\tau,s) \\ & +\rho_0(\tau,s)e(s)+\rho_0(\tau,s)h(s)\rho_0(\tau,s) \\ \alpha_0(\tau,s)h(s)a_0(\tau,s) & \alpha_0(\tau,s)[e(s)+h(s)\rho_0(\tau,s)] \end{bmatrix} \tag{5}$$

From this we readily obtain the differential equations

$$\frac{\mathrm{d}}{\mathrm{d}s} S_0(\tau, s) = \begin{bmatrix} \frac{\mathrm{d}}{\mathrm{d}s} a_0(\tau, s) & \frac{\mathrm{d}}{\mathrm{d}s} \rho_0(\tau, s) \\ \frac{\mathrm{d}}{\mathrm{d}s} r_0(\tau, s) & \frac{\mathrm{d}}{\mathrm{d}s} \alpha_0(\tau, s) \end{bmatrix}$$

$$= \begin{bmatrix} [f(s) + \rho_0(\tau, s)h(s)]a_0(\tau, s) & g(s) + f(s)\rho_0(\tau, s) + \rho_0(\tau, s)e(s) \\ & + \rho_0(\tau, s)h(s)\rho_0(\tau, s) \\ \alpha_0(\tau, s)h(s)a_0(\tau, s) & \alpha_0(\tau, s)[e(s) + h(s)\rho_0(\tau, s)] \end{bmatrix}$$

$$S_0(\tau, \tau) = I = \begin{pmatrix} I & 0 \\ 0 & I \end{pmatrix} \tag{6}$$

Now note that the reflection coefficient ρ_0 obeys a Riccati equation.

As is well known in circuit theory, a particular advantage of the scattering formulation is the ease with which initial conditions can be changed. Thus note that

$$S(\tau, s; \Pi)) = \Pi * S_0(\tau, s) \tag{7}$$

obeys the same differential equations as S_0, but with $S(\tau, \tau; \Pi) = \Pi$.

Another fundamental aspect of scattering theory is the equal emphasis on "forwards" and "backwards" solutions, since we could equally well extend a given medium to the left or to the right. In this connection, however, the particular matrix $S_0(\tau, t)$ is especially useful since its "backwards" derivative is much easier to find than that of $S(\tau, t; \Pi)$. Thus note that we can write

$$\begin{aligned} S_0(\sigma - \mathrm{d}\sigma, t) &= S_0(\sigma - \mathrm{d}\sigma, \sigma) * S_0(\sigma, t) \\ &= [I + M(\sigma)d\sigma + \mathrm{o}(\mathrm{d}\sigma)] * S_0(\sigma, t) \end{aligned} \tag{8}$$

which leads to

$$-\frac{\mathrm{d}}{\mathrm{d}\sigma} S_0(\sigma, t) =$$

$$\begin{bmatrix} a_0(\sigma, t)[f(\sigma) + g(\sigma)r_0(\sigma, t)] & a_0(\sigma, t)g(\sigma)\alpha_0(\sigma, t) \\ h(\sigma) + e(\sigma)r_0(\sigma, t) + r_0(\sigma, t)f(\sigma) & [e(\sigma) + r_0(\sigma, t)g(\sigma)]\alpha_0(\sigma, t) \\ + r_0(\sigma, t)g(\sigma)r_0(\sigma, t) & \end{bmatrix}$$

$$S_0(t, t) = I. \tag{9}$$

Clearly there are certain interesting "dualities" between (6) and (9).

However, our main point now is that $\Pi * S_0(\sigma, t)$ does *not* obey the same backwards equation (9), as $S_0(\sigma, t)$. The proper equation can nevertheless be calculated as follows, where (for later application) we also explicitly let Π be a function of σ. We can write

$$\begin{aligned} S(\sigma-\mathrm{d}\sigma, t; \Pi) &= \Pi(\sigma-\mathrm{d}\sigma) * S_0(\sigma-\mathrm{d}\sigma, t) \\ &= \Pi(\sigma-\mathrm{d}\sigma) * S_0(\sigma-\mathrm{d}\sigma, \sigma) * S_0(\sigma, t) \\ &= \Pi(\sigma-\mathrm{d}\sigma) * (I + M(\sigma)\mathrm{d}\sigma \\ &\quad + \mathrm{o}(\mathrm{d}\sigma)) * \Pi^{-1}(\sigma) * \Pi(\sigma) * S_0(\sigma, t) \\ &= (I + \mathcal{M}_\Pi(\sigma)\mathrm{d}\sigma) * S(\sigma, t; \Pi) \end{aligned} \tag{10}$$

where $\mathcal{M}_\Pi(\sigma)$ is defined implicitly by

$$\begin{aligned} &\Pi(\sigma-\mathrm{d}\sigma) * (I + M(\sigma)\mathrm{d}\sigma + \mathrm{o}(\mathrm{d}\sigma)) * \Pi^{-1}(\sigma) \\ &\quad = I + \mathcal{M}_\Pi(\sigma)\mathrm{d}\sigma + \mathrm{o}(\mathrm{d}\sigma). \end{aligned} \tag{11}$$

It follows by comparing (8) and (11) that $S(\sigma, t; \Pi)$ obeys a differential equation in σ that has the same structure as (9), where however the entries f, g, h and e of $M(\sigma)$ are replaced by the corresponding entries of $\mathcal{M}_\Pi(\sigma)$.

This concludes our presentation of the simple facts we shall need from scattering theory. It is clear that the above results can be derived without any reference to star-products or scattering theory (as done, e.g., by Reid [17]), but we have found the above language to be mnemonic, suggestive and simple.

3. Preliminary formulas

The estimation problems that we shall consider in this paper will be connected with the following "state-space" model,

$$\begin{aligned} &\frac{\mathrm{d}}{\mathrm{d}t}x(t) = F(t)x(t) + u(t), \qquad t \geqslant \tau, \\ &z(t) = H(t)x(t), \\ &y(t) = z(t) + v(t), \end{aligned} \tag{12}$$

where the matrices $F(\cdot)$, $H(\cdot)$ are assumed to be known and $v(\cdot)$, $u(\cdot)$, $x(\tau)$ are random with known means and covariances,

$$\begin{aligned} &\mathbf{E}x(\tau) = 0 = \mathbf{E}u(s) = \mathbf{E}v(s), \qquad \mathbf{E}u(t)u'(s) = Q(t)\delta(t-s), \\ &\mathbf{E}x(\tau)x'(\tau) = \Pi, \qquad \mathbf{E}v(t)v'(s) = I\delta(t-s), \\ &\mathbf{E}u(t)v'(s) = \mathbf{E}x(\tau)u'(s) = 0. \end{aligned} \tag{13}$$

For future reference, we note that the covariance function of the process $z(\cdot)$ can be written as

$$\mathbf{E}z(s)z'(t) = K_z(s,t) = \begin{cases} H(s)\phi_F(s,t)\Pi(t)H'(t) & \text{for } s \geq t, \\ H(s)\Pi(s)\phi'_F(t,s)H'(t) & \text{for } t \geq s, \end{cases} \tag{14}$$

where $\Pi(\cdot)$ is defined by

$$\frac{\mathrm{d}}{\mathrm{d}s}\Pi(s) = F(s)\Pi(s) + \Pi(s)F'(s) + Q(s); \qquad \Pi(\tau) = \Pi \tag{15}$$

and $\phi_F(\cdot,\cdot)$ is the state-transition matrix of $F(\cdot)$, viz.

$$\frac{\mathrm{d}}{\mathrm{d}t}\phi_F(t,s) = F(t)\phi_F(t,s); \qquad \phi(s,s) = I.$$

We shall also sometimes use the cross-covariance function

$$\mathbf{E}z(s)x'(t) = K_{zx}(s,t) = \begin{cases} \Pi(s)\phi_F(s,t)\Pi(t) & \text{for } s \geq t, \\ H(s)\Pi(s)\phi'_F(t,s) & \text{for } t \geq s. \end{cases} \tag{16}$$

The problem is to find the best linear estimate, denoted by $\hat{z}(t)$, of $z(t)$ given the observations $\{y(s), \tau \leq s \leq t\}$. The solutions to this problem are well known. If the process $z(\cdot)$ is given by (12)–(13), it is perhaps most natural to use the Kalman filter solution:

$$\begin{aligned} &\hat{z}(t) = H(t)\hat{x}(t), \\ &\frac{\mathrm{d}}{\mathrm{d}t}\hat{x}(t) = [F(t) - K(t,\tau)H(t)]\hat{x}(t) + K(t,\tau)y(t); \ \hat{x}(\tau) = 0, \\ &K(t,\tau) = P(t,\tau)H'(t), \end{aligned} \tag{17}$$

and $P(t,\tau)$ satisfies the nonlinear Riccati equation

$$\begin{aligned} \frac{\mathrm{d}}{\mathrm{d}t}P(t,\tau) = {} & F(t)P(t,\tau) + P(t,\tau)F'(t) + Q(t) \\ & - P(t,\tau)H'(t)H(t)P(t,\tau); \\ P(\tau,\tau) = {} & \Pi. \end{aligned} \tag{18}$$

If no state model is available for $z(\cdot)$, or $K_z(s,t)$, we can of course calculate $\hat{z}(\cdot)$ as

$$\hat{z}(t) = \int_\tau^t h(t,s;\tau)y(s)\mathrm{d}s \tag{19}$$

where (by the condition that $z(t) - \hat{z}(t)$ is orthogonal to $y(s)$, $\tau \leq s \leq t$, we see that) $h(t,s;\tau)$ satisfies the Wiener–Hopf equation

$$h(t,s;\tau)+\int_\tau^t h(t,\sigma;\tau)K_z(\sigma,s)\mathrm{d}\sigma = K_z(t,s); \qquad \tau\leqslant s\leqslant t. \tag{20}$$

Clearly, if a state model as in (13)–(14) is available, we can write

$$h(t,s;\tau)=H(t)\phi(t,s;\tau)K(s;\tau), \tag{21}$$

where $\phi(t,s;\tau)$ is the state-transition matrix of $[F(t)-K(t,\tau)H(t)]$, i.e.,

$$\frac{\mathrm{d}}{\mathrm{d}t}\phi(t,s;\tau)=[F(t)-K(t,\tau)H(t)]\phi(t,s;\tau); \qquad \phi(s,s;\tau)=I.$$

The identity (21) will be very useful later, but for the present we note that the filtering Riccati equation (18) can be identified with the one arising in the scattering formulation (cf. (6)), if we set

$$f(s)=e'(s)=F(s); \qquad g(s)=Q(s), \qquad h(s)=-H'(s)H(s).$$

Then we see from equation (6) with initial condition

$$S(\tau,\tau)=\bar{\Pi}=\begin{bmatrix} I & \Pi \\ 0 & I\end{bmatrix} \tag{22}$$

that in addition to the relation $\rho(\tau,t)=P(t,\tau)$, we will also have $a(\tau,t)=\alpha'(\tau,t)=\phi(t,\tau;\tau)$. Let us also denote $r(\tau,t)=W(t,\tau)$, and for easy reference, let us repeat equation (6) in this new notation,

$$\frac{\mathrm{d}}{\mathrm{d}s}\phi(s,\tau;\tau)=[F(s)-P(s,\tau)H'(s)H(s)]\phi(s,\tau;\tau); \quad \phi(\tau,\tau;\tau)=I, \tag{23a}$$

$$\frac{\mathrm{d}}{\mathrm{d}s}P(s,\tau)=F(s)P(s,\tau)+P(s,\tau)F'(s)+Q(s)-P(s,\tau)H'(s)H(s)P(s,\tau);$$

$$P(\tau,\tau)=\Pi, \tag{23b}$$

$$\frac{\mathrm{d}}{\mathrm{d}s}W(s,\tau)=-\phi'(s,\tau;\tau)H'(s)H(s)\phi(s,\tau;\tau); \qquad W(\tau,\tau)=0. \tag{23c}$$

An important feature of the connection to the scattering framework is, as noted earlier, that we also immediately have a set of backwards equations (9), which is, with $\bar{\Pi}$ as in (22) and in the present notation:

$$-\frac{\mathrm{d}}{\mathrm{d}\sigma}\phi(t,\sigma;\sigma)=\phi(t,\sigma;\sigma)[\tilde{F}(\sigma)+\tilde{Q}(\sigma)W(t,\sigma)]; \qquad \phi(t,t;t)=I, \tag{24a}$$

$$-\frac{\mathrm{d}}{\mathrm{d}\sigma}P(t,\sigma)=\phi(t,\sigma;\sigma)\tilde{Q}(\sigma)\phi'(t,\sigma;\sigma); \qquad P(t,t)=\Pi, \tag{24b}$$

$$-\frac{\mathrm{d}}{\mathrm{d}\sigma}W(t,\sigma)=\tilde{F}'(\sigma)W(t,\sigma)+W(t,\sigma)\tilde{F}(\sigma)-\tilde{H}'(\sigma)\tilde{H}(\sigma)$$

$$+W(t,\sigma)\tilde{Q}(\sigma)W(t,\sigma); \qquad W(t,t)=0 \tag{24c}$$

where (cf. (11)),

$$\begin{aligned}\tilde{F}(\sigma) &= F(\sigma) - \Pi H'(\sigma)H(\sigma),\\ \tilde{Q}(\sigma) &= F(\sigma)\Pi + \Pi F'(\sigma) + Q(\sigma) - \Pi H'(\sigma)H(\sigma)\Pi,\\ \tilde{H}(\sigma) &= H(\sigma).\end{aligned} \tag{25}$$

4. The Chandrasekhar-type equations

Considering the state-space model (12)–(13) and the Kalman filter solution (17), we see that knowledge of the gain function $K(t, \tau)$ suffices to determine $\hat{x}(t)$ and that no direct knowledge of $P(t, \tau)$ is necessary. Since $K(t, \tau) = P(t, \tau)H'(t)$ often has lower dimensions than $P(t, \tau)$ it would be desirable to find differential equations directly for $K(t, \tau)$. Such a set is provided by certain Chandrasekhar-type equations derived by Kailath [4] for the case of time-invariant state-space models:

$$\frac{\mathrm{d}}{\mathrm{d}t}K(t, \tau) = L(t, \tau)\Sigma L'(t, \tau)H'; \qquad K(\tau, \tau) = \Pi H', \tag{26a}$$

$$\frac{\mathrm{d}}{\mathrm{d}t}L(t, \tau) = [F - K(t, \tau)H]L(t, \tau); \qquad L(\tau, \tau) = L_0, \tag{26b}$$

where L_0 and Σ are $n \times \alpha$ and $\alpha \times \alpha$ matrices determined by the factorization

$$F\Pi + \Pi F' + Q - \Pi H'H\Pi = L_0\Sigma L_0', \tag{26c}$$

where α is the rank of the LHS of (26c) and Σ is its signature (or inertia) matrix. [Recall also that Π is the value of $P(t, \tau)$ at $t = \tau$.]

Now from the formulas (23) and (24) we can obtain a set of such equations also for time-variant models, but as we shall see the equations are useless because there is no way of computing the solution except when the model is time-invariant.

To see the issues involved, note that we could get an equation for $K(t, \tau) = P(t, \tau)H'(t)$ by multiplying equation (24b) on the right by $H'(t)$, giving

$$-\frac{\mathrm{d}}{\mathrm{d}\sigma}K(t, \sigma) = \phi(t, \sigma; \sigma)\tilde{Q}(\sigma)\phi'(t, \sigma; \sigma)H'(t).$$

However, this does not seem to help, since another Riccati equation (24c) has to be solved in order to calculate $\phi(t, \sigma; \sigma)$ as a function of σ (cf.

(24a)). This difficulty can be alleviated by using (23a) instead, and finding $\phi(\cdot,\tau;\tau)$ via

$$\frac{\mathrm{d}}{\mathrm{d}s}\phi(s,\tau;\tau)=[F(s)-K(s,\tau)H(s)]\phi(s,\tau;\tau), \qquad \phi(\tau,\tau;\tau)=I.$$

Now introduce a decomposition of $\tilde{Q}(\sigma)$ (cf. (25) and (26c))

$$\tilde{Q}(\sigma)=\tilde{G}(\sigma)\Sigma(\sigma)\tilde{G}'(\sigma), \tag{27}$$

where the number of columns in $\tilde{G}(\sigma)$ equals the rank of $\tilde{Q}(\sigma)$ and where $\Sigma(\sigma)$ is the signature matrix of $\tilde{Q}(\sigma)$. Also define

$$L(s,r)=\phi(s,r;r)\tilde{G}(r).$$

The reason for doing all this is that now (23a) and (24b) can be written (note that τ and t are the two endpoints),

$$\frac{\mathrm{d}}{\mathrm{d}s}L(s,\tau)=[F(s)-K(s,\tau)H(s)]L(s,\tau); \qquad L(\tau,\tau)=\tilde{G}(\tau), \tag{28a}$$

$$-\frac{\mathrm{d}}{\mathrm{d}\sigma}K(t,\sigma)=L(t,\sigma)\Sigma(\sigma)L'(t,\sigma)H'(t); \qquad K(t,t)=\Pi H'(t) \tag{28b}$$

which may be recognized (cf. (26)) as a set of Chandrasekhar-type equations for $K(\cdot,\cdot)$ in the general time-variant case.

Unfortunately, however, (28) cannot be solved since the time arguments do not "fit". That is, the intermediate values of $K(s,\tau)$ required to solve (28a) are not provided by (28b) and vice versa. This is a fundamental difficulty in the general time-variant case.

However, the impasse can be resolved in the *time-invariant* case by noting that if $F(\cdot)$, $Q(\cdot)$ and $H(\cdot)$ are constant (time-invariant), then

$$-\frac{\mathrm{d}}{\mathrm{d}\sigma}K(t,\sigma)=\frac{\mathrm{d}}{\mathrm{d}t}K(t,\sigma) \quad \text{and} \quad K(t,t)=K(\tau,\tau). \tag{29}$$

[In the scattering formulation the interpretation of (29) is that, for a *homogeneous* (i.e. with constant $f(\cdot)$, $e(\cdot)$, $g(\cdot)$, $h(\cdot)$) medium, the result is the same whether we extend the medium to the right or to the left. Such observations were apparently first used by Stokes in studying light-transmission through a pile of plates [23]; the identities obtained by equating the right-hand sides of (23) and (24) have been called Stokes identities by Bellman, Kalaba and Wing [24].]

Then (28b) can be replaced by

$$\frac{\mathrm{d}}{\mathrm{d}s} K(s,\tau) = L(s,\tau)\Sigma L'(s,\tau)H'; \quad K(\tau,\tau) = \Pi H' \tag{30}$$

and now eqs. (28a) and (30) are exactly the equations (26), originally given in Kailath [4].

Though the technique used in [4] is still perhaps more direct, the present derivation gives more insight into the significance of the assumption of a time invariant model. It also extends immediately to the discrete-time case, where the method of [4] is now less obvious and requires the introduction of more quantities [12]; all this becomes much more natural in the scattering framework [3].

5. The Levinson-type equations

The Levinson algorithm [8], [20] is an efficient scheme for computation of the impulse response of the optimal filter, [$h(t,s)$ in (19)] for stationary discrete-time problems. It is directly specified by knowledge of the covariance function for $z(\cdot)$ and no underlying state-space model is required. However, the derivation relies heavily upon the assumption of stationary, i.e., the fact that $K_z(s,t) = K_z(s-t) = K_z'(t-s), s > t$. Some continuous time analogs of this equation have been reported ([9], [14], [19]) for stationary processes $z(\cdot)$.

Here we shall present a Levinson-type algorithm for non-stationary processes of a certain class, namely for those that admit a state-space model with time-invariant parameters (precisely the same class as arose in the Chandrasekhar-type equations in Section 4).

The impulse response $h(t,s;\tau)$ can be readily calculated knowing the gain function $K(t,\tau)$ of the optimum filter (cf. eq. (21)),

$$h(t,s;\tau) = H(t)\phi(t,s;\tau)K(s,\tau).$$

Therefore it would seem that we could easily obtain differential equations for $h(t,s;\tau)$ by using the Riccati equation formulas (23) or (24), or in the time-invariant case the Chandrasekhar equations (28a), (30). Unfortunately, this is not as simple as it might seem, as shown by the following calculation where we assume that H is time-invariant,

$$\frac{\partial}{\partial t} h(t,s;\tau) = H\frac{\partial}{\partial t}\phi(t,s;\tau)K(s,\tau) = H[F - P(t,\tau)H'H]\phi(t,s;\tau)P(s,\tau)H'.$$

Our goal would be to express this quantity in terms of $h(t,s;\tau)$ and possibly some auxiliary quantity. However, it is not obvious how to account for the effect of the $n \times n$ matrix F in terms of lower dimensional functions. It will turn out that it is more suitable to consider

$$\frac{\partial}{\partial t} h(t, t-s; \tau) = \frac{\partial}{\partial x} h(x, y; \tau)\Bigg|_{\substack{x=t \\ y=t-s}} + \frac{\partial}{\partial y} h(x, y; \tau)\Bigg|_{\substack{x=t \\ y=t-s}}$$

This expression can be evaluated making use of the equations (23) and (24). However the algebra involved is quite laborious, and we shall instead derive the expression through a more intuitive, though somewhat indirect, route.

It is well known that the discrete-time Levinson algorithm is based on a certain interplay between forwards and backwards estimation filters. Therefore, before plunging into detailed direct calculations, it will be helpful to briefly review some recent results on backwards filtering [10], [11], which turn out to yield certain key identities ((35) and (37)) that will essentially trivialize the calculations.

5.1. *Backwards models*

Consider the state-space model

$$-\frac{\mathrm{d}}{\mathrm{d}\sigma} x_B(\sigma) = F_B(\sigma) x_B(\sigma) + u_B(\sigma), \qquad \tau \leqslant \sigma \leqslant t,$$
$$z_B(\sigma) = H(\sigma) x_B(\sigma) \tag{31}$$

with

$$\mathbf{E} x_B(t) = \mathbf{E} u_B(\sigma) = 0; \qquad \mathbf{E} u_B(\sigma) u_B'(\lambda) = Q(\sigma)\delta(\sigma - \lambda);$$
$$\mathbf{E} x_B(t) x_B'(t) = \Pi(t); \qquad \mathbf{E} u_B(\sigma) x'(t) = 0 \tag{32}$$

and

$$F_B(\sigma) = F(\sigma) - \Pi^{-1}(\sigma) Q(\sigma), \tag{33}$$

where $\Pi(s)$ is given by (16).

As shown in [11], this model gives the same state covariance function as (13), that is

$$\mathbf{E} x_B(s) x_B'(t) = \mathbf{E} x(s) x'(t) \tag{34}$$

Moreover, this model is also such that the state $x_B(\cdot)$ is Markov in *decreasing* time. Consequently the linear estimation problem can be solved using either (31) or (13). Suppose that $x(\tau)$ is to be estimated using the data $y(s), \tau \leq s \leq t$. Using the "forwards" model (13), this is a smoothing problem and it can be solved, for example, by using the innovations formula [6]. However, if the backwards model is used instead, we have a filtering problem and the Kalman filter approach can be used. By equating the coefficients used to weight the data $y(s)$ in the formulas obtained by these two interpretations, we can obtain an important identity [10, Section IX],

$$\psi(\tau, s; t)P_B(s, t) = \\ = \Pi\left[-\int_s^t \phi'(\sigma, \tau; \tau)H'(\sigma)H(\sigma)\phi(\sigma, s; \tau)\mathrm{d}\sigma P(s, \tau) + \phi'(s, \tau; \tau)\right] \quad (35)$$

where $\phi(t, s; \tau)$ is the state-transition matrix for $[F(t) - P(t, \tau)H'(t)H(t)]$, with $\phi(s, s; \tau) = I$, $P(s, \tau)$ is as defined before in (23b), $P_B(s, t)$ is the solution of

$$-\frac{\mathrm{d}}{\mathrm{d}\sigma}P_B(\sigma, t) = F_B(\sigma)P_B(\sigma, t) + P_B(\sigma, t)F_B'(\sigma) + Q(\sigma) \\ - P_B(\sigma, t)H'(\sigma)H(\sigma)P_B(\sigma, t); \quad (36) \\ P_B(t, t) = \Pi(t)$$

and $\psi(\tau, s; t)$ is the state transition matrix for

$$[F_B(\tau) - P_B(\tau, t)H'(\tau)H(\tau)], \text{ with } \psi(s, s; t) = I.$$

Similarly, we can obtain the identity [10],

$$\phi(t, s; \tau)P(s, \tau) = \\ = \Pi(t)\left[-\int_\tau^s \psi'(\sigma, t; t)H'(\sigma)H(\sigma)\psi(\sigma, s; t)\mathrm{d}\sigma P_B(s, t) + \psi'(s, t; t)\right] \quad (37)$$

It should also be clear from (35) and (36) that

$$H(\tau)\psi(\tau, s; t)P_B(s, t)H'(s) = h_B(\tau, s; t) \quad (38)$$

is the filter for finding the estimate of $z(\tau)$ from $y(s)$, $\tau \leq s \leq t$, i.e.,

$$\hat{z}(\tau) = \int_\tau^t h_B(\tau, s; t)y(s)\mathrm{d}s.$$

Notice the similarity of (38) and (22).

The details of these calculations can be found in [10], where also a direct algebraic proof of (35) and (37) is provided. We only remark here that the relation (35) is in fact a version of the Krein–Siegert–Bellman resolvent identity for the Wiener–Hopf equation, see, e.g. [5].

5.2. *Differential equations for $h(t, s; \tau)$*

We can now return to our objective of finding a closed set of numerically convenient recursive equations for the optimal filter $h(t, s; \tau)$. In terms of a state space realization (13) for the covariance $K_z(s, t)$ (given by (15)), $h(t, s; \tau)$ is defined by (22). Similarly, we have the backwards filter $h_B(\tau, s; t)$ given by (38). Using (35) and (37), we readily obtain the equations

$$\begin{aligned}\frac{\partial}{\partial t} h_B(\tau, s; t) &= \frac{\partial}{\partial t} H(\tau)\psi(\tau, s; t)P_B(s, t)H'(s)\\ &= -H(\tau)\Pi\phi'(t, \tau; \tau)H'(t)H(t)\phi(t, s; \tau)P(s, \tau)H'(s)\\ &= -h_B(\tau, t; t)h(t, s; \tau)\end{aligned} \tag{39}$$

and

$$\begin{aligned}\frac{\partial}{\partial \tau} h(t, s; \tau) &= \frac{\partial}{\partial \tau} H(t)\phi(t, s; \tau)P(s, \tau)H'(s)\\ &= H(t)\Pi(t)\psi'(\tau, t; t)H'(\tau)H(\tau)\psi(\tau, s; t)P_B(s, t)H'(s)\\ &= h_B(\tau, t; t)'h_B(\tau, s; t)\end{aligned} \tag{40}$$

Now, eqs. (39) and (40) are of "Levinson-type" [8], [20] and are valid for general, time-varying systems. They are, however, not feasible to solve, due to the difference in time-arguments. If the process $z(\cdot)$ was stationary, then obviously

$$\frac{\partial}{\partial \tau} h(t, t-s; \tau) = -\frac{\partial}{\partial t} h(t, t-s; \tau) \tag{41}$$

and using (41) and (39) would give a feasible set of equations, which were in [9], [14], [18] (though without discussion of initial conditions—see below). In the case of a *non-stationary process* $z(\cdot)$, with a *time-invariant model* (13), a relation like (41) still holds, provided that we take into account that in the evaluation of $(\partial/\partial\tau)h(t, t-s; \tau)$ the "initial" value $\Pi(\tau)$ also is changing. [This is in contrast to the case in Section 4 (cf. (29)), where the initial condition Π was held fixed as τ varied.] In order to find the derivative with respect to t, this change in Π must be accounted for. Thus, in symbolic notation, we will have

$$\frac{\partial}{\partial t} h(t, t-s; \tau) = -\frac{\partial}{\partial \tau} h(t, t-s; \tau) + \frac{\mathrm{d}}{\mathrm{d}\tau}\Pi(\tau) \times \frac{\partial}{\partial \Pi} h(t, t-s; \tau). \tag{42}$$

Evaluation of this expression gives

$$\frac{\partial}{\partial t} h(t, t-s; \tau) = -H\Pi(t)\psi'(\tau, t; t)\tilde{H}'\Sigma\tilde{H}\psi(\tau, s; t)P_B(s, t)H', \tag{43}$$

where

$$\tilde{H}' = \Pi^{-1}\tilde{G} \tag{44}$$

and $\tilde{G}$ and Σ defined (25) and (27). Since the model is time-invariant, these matrices are not functions of time. Note that $\tilde{H}$ is of size $\alpha \times n$. Next we introduce the function

$$\tilde{h}_B(\tau, s; t) = \tilde{H}\psi(\tau, s; t)P_B(s, t)H', \tag{45}$$

which can be interpreted as the optimal filter for estimating $\tilde{H}x(\tau)$ from the measurements $y(s); \tau \leq s \leq t$,

$$\tilde{H}\hat{x}(\tau) = \int_\tau^t \tilde{h}_B(\tau, s; t)y(s)\mathrm{d}s.$$

Consequently, it also obeys the Wiener–Hopf equation

$$\tilde{h}_B(\tau, s; t) + \int_\tau^t \tilde{h}_B(\tau, \sigma; t)K_z(\sigma, s)\mathrm{d}\sigma = \tilde{K}_z'(s, \tau) \tag{46}$$

where

$$\tilde{K}_z(s, \tau) - K_{zx}(s, \tau)\tilde{H}' = H\phi_F(s, \tau)\Pi(\tau)\tilde{H}', \qquad \tau \leq s. \tag{47}$$

Then, for nonstationary processes with time-invariant state space models, we can obtain (using (22), (39), (43) and (45)) the generalized Levinson equations,

$$\frac{\partial}{\partial t} h(t, t-s; \tau) = -\tilde{h}_B(\tau, t; t)'\Sigma\tilde{h}_B(\tau, t-s; t);$$
$$\tau \leq t \leq T; \qquad 0 \leq s \leq t-\tau, \tag{48}$$

$$\frac{\partial}{\partial t}\tilde{h}_B(\tau, s; t) = -\tilde{h}_B(\tau, t; t)h(t, s; \tau); \qquad \tau \leq s \leq t \leq T. \tag{49}$$

Equations (48), (49) express a certain relationship between the functions $h(t, s; \tau)$ and $\tilde{h}_B(\tau, s; t)$. It should be noted that they do not contain the covariance function of $z(\cdot)$, and that hence such information must be introduced via the boundary conditions. We shall see that this can be done using the Wiener–Hopf equations (21) and (46), or rather, their derivatives with respect to s, viz.,

$$\frac{\partial}{\partial s} h(t, s; \tau) = -\int_\tau^t h(t, \sigma; \tau) \frac{\partial}{\partial s} K_z(\sigma, s) \mathrm{d}\sigma + \frac{\partial}{\partial s} K_z(t, s), \tag{50}$$

$$\frac{\partial}{\partial s} \tilde{h}_B(\tau, s; t) = -\int_\tau^t \tilde{h}_B(\tau, \sigma; t) \frac{\partial}{\partial s} K_z(\sigma, s) \mathrm{d}\sigma + \frac{\partial}{\partial s} \tilde{K}_z'(s, \tau). \tag{51}$$

A simple numerical scheme for solving (48), (49) is outlined in [25]. It should be noticed that the computational efficiency largely depends on α, the rank of $\tilde{Q} = \dot{P}(\tau)$ (cf., (25)), since it determines the dimension of the auxiliary function $\tilde{h}_B$.

If the process is stationary, then $\tilde{H} = H$, $\tilde{K}_z = K_z$ and $h(t, \tau; \tau) = \tilde{h}_B(\tau, t; t) = h_B(\tau, t; t)$, which reduces the required computations.

6. Cholesky factorization

By Cholesky factorization of the covariance function $K_z(t, s)$; $\tau \leq s$, $t \leq T$ is meant the following: Find two functions

$$\begin{aligned} &k(t, s; \tau); \qquad k(t, s; \tau) = 0 \quad \text{for } t < s, \\ &k^*(t, s; \tau) = k'(s, t; \tau) \end{aligned} \tag{52}$$

such that

$$K_z(t, s) = k(t, s; \tau) + k^*(t, s; \tau) + \int_\tau^T k(t, \sigma; \tau) k^*(\sigma, s; \tau) \mathrm{d}\sigma \tag{53}$$

or, symbolically,

$$I + K = (I + k)(I + k^*). \tag{54}$$

The term Cholesky factorization comes from the discrete-time analog, where (54) is just the decomposition of the covariance matrix into an upper and a lower triangular matrix.

The Cholesky factor $k(t, s; \tau)$ is of interest, because it gives the innovations representation of the process $y(\cdot)$ specified by (13). More specifically, if the innovations $\nu(\cdot)$ of $y(\cdot)$ were known, we could find $y(\cdot)$ as

$$y = (I + k)\nu \quad \text{or} \quad y(t) = \nu(t) + \int_\tau^t k(t, s; \tau) \nu(s) \mathrm{d}s. \tag{55}$$

Now, since $\nu(t) = y(t) - H(t)\hat{x}(t)$, with $\hat{x}(t)$ defined by (17), we can write (17) as

$$\frac{\mathrm{d}}{\mathrm{d}t}\hat{x}(t) = F(t)\hat{x}(t) + K(t, \tau)\nu(t),$$
$$y(t) = H(t)\hat{x}(t) + \nu(t) \tag{56}$$

from which we obtain, cf. (55),

$$k(t, s; \tau) = H(t)\phi_F(t, s)K(s, \tau). \tag{57}$$

In view of the discussions in the previous sections, to find a set of (low rank) equations for $k(t, s; \tau)$ we assume immediately that F, Q and H are time-invariant, and proceed as follows. Take

$$\begin{aligned}\frac{\partial}{\partial t}k(t, t-s; \tau) &= H\phi_F(t, t-s)\frac{\partial}{\partial t}P(t-s, \tau)H' \\ &= -H\phi_F(t, t-s)\phi(t-s, \tau; \tau)\Pi\tilde{H}'\Sigma\tilde{H}\Pi\phi'(t-s, \tau; \tau)H'\end{aligned} \tag{58}$$

where the second equality follows from the backwards equation for P, (24b), (27) and (44). Now introduce the function

$$\tilde{k}(t, s; \tau) = H\phi_F(t, s)\phi(s, \tau; \tau)\Pi\tilde{H}' \tag{59}$$

which has dimensions $p \times \alpha$ (where p is the dimension of $z(\cdot)$). Then calculate,using (23a) (the forwards equation for ϕ)

$$\begin{aligned}\frac{\partial}{\partial s}\tilde{k}(t, s; \tau) &= H\phi_F(t, s)[-F(s) + F(s) - P(s, \tau)H'H]\phi(s, \tau; \tau)\Pi\tilde{H}' \\ &= -k(t, s; \tau)\tilde{k}(s, s; \tau).\end{aligned} \tag{60}$$

Collecting (58), (59) and (60), we obtain the Cholesky equations

$$\begin{aligned}&\frac{\partial}{\partial t}k(t, t-s; \tau) = -\tilde{k}(t, t-; \tau)\Sigma\tilde{k}'(t-s, t-s; \tau), \\ &\frac{\partial}{\partial s}\tilde{k}(t, t-s; \tau) = k(t, t-s; \tau)\tilde{k}(t-s, t-s; \tau), \\ &\tau \leqslant t \leqslant T, \qquad 0 \leqslant s \leqslant t-\tau.\end{aligned} \tag{61}$$

In the stationary case, these equations are the continuous-time analog of the corresponding discrete-time equations given by Rissanen [21] and by Morf [22], who has also studied the discrete-time non-stationary case. The initial conditions for (61) are given by

$$k(t, \tau; \tau) = K_z(t, \tau), \qquad \tilde{k}(t, \tau; \tau) = \tilde{K}_z(t, \tau) \tag{62}$$

where $K_z(t, \tau)$ and $\tilde{K}_z(t, \tau)$ (defined by (17)) are given.

Again, note that if $z(\cdot)$ is stationary, then $\tilde{K}_z(t, \tau) = K_z(t, \tau)$ and all necessary information is provided by the covariance function $K_z(t, \tau)$.

We may also point out the following connections with the Levinson equations (48),

$$\tilde{h}_B(\tau, t; t) = \tilde{k}(t, t; \tau)', \qquad h(t, t; \tau) = k(t, t; \tau). \tag{63}$$

In fact, the functions h and k are quite closely related to each other by

$$(I + k)^{-1} = I - h$$

or

$$k(t, s; \tau) - h(t, s; \tau) - \int_s^t k(t, r; \tau) h(r, s; \tau) \mathrm{d}r = 0. \tag{64}$$

Indeed, it is clear that once either of h and k has been found, the other can be found from (6) [by letting $s : t \to \tau$] more easily than from the Levinson or Cholesky equation. In particular the Levinson algorithm can be derived from the Cholesky factorization and vice versa, so that we do not need two independent derivations (see, e.g., [25]). Computational schemes for the Cholesky equations are similar to, and in fact somewhat simpler than, those for the Levinson equations (see [25]).

7. Conclusions

It is a well known fact (see, e.g. [20]), that the original Levinson-algorithm for discrete-time stationary processes is closely related to forwards and backwards estimation filters. It has also been demonstrated, [9], [18], how the fast Chandrasekhar-type algorithms for stationary processes can be derived using backwards innovations.

We have here used a framework to handle the forwards and backwards equations of estimation theory, that arises quite naturally in scattering theory. With this tool we have given an alternative derivation of the Chandrasekhar algorithm, and also derived fast Levinson-type and Cholesky-factorization algorithms for processes that admit a time invariant, finite-dimensional state-space model. Furthermore we have introduced the assumption of time-invariance as late as possible in the derivation, in order to emphasize its fundamental importance. For the Levinson- and the Cholesky-algorithm, we have also demonstrated how an assumption of stationarity will substantially simplify the result.

Our derivations emphasize and exploit the close relationships between these different "fast" algorithms, and a further merit of the approach is that it directly extends to the discrete-time case [3]. In this paper, we have exploited the assumed existence of an underlying, finite-dimensional, state-space model of the process [or a corresponding factorization of the covariance function as in (14)]. It should be clear that without much difference in the derivations we could have introduced structure information in other forms, say, as an ARMA-representation of the process, or as a pattern for the sucessive derivatives of the covariance function. The only point in which the structure information plays a role in the Levinson- and Cholesky-agorithms is in the definition of the "adjoint" covariance function $\tilde{K}_z(t, \tau)$. It is interesting to note that this function can be directly defined through the expression

$$\frac{\partial}{\partial t} K_z(t, s) + \frac{\partial}{\partial s} K_z(t, s) = K_z(t, \tau) K_z'(s, \tau) + \tilde{K}_z(t, \tau) \Sigma \tilde{K}_z'(s, \tau)$$

without reference to state space quantities. This fact and other generalizations are pursued in [25] and [26].

References

[1] Bellman, R., and Wing, G.M. (1975). *An Introduction to Invariant Imbedding*. Wiley, New York.

[2] Makhoul, J. (1975). Linear prediction: A tutorial review. *Proc. IEEE*, **63**, 561–580.

[3] Friedlander, B., Kailath, T., and Ljung, L. (1976). Scattering theory and linear least squares estimation, Part II: Discrete-time Problems. *J. Franklin Inst.*, 301, 71–82.

[4] Kailath, T. (1973). Some new algorithms for recursive estimation in constant linear systems. *IEEE Trans. Info. Theory*, IT–**19**, 750–760.

[5] Kailath, T. (1974). A view of three decades of linear filtering theory. *IEEE Trans. Info. Theory*, IT–**20**, 146–181.

[6] Kailath, T., and Frost, P. (1968). An innovations approach to least-squares estimation, Part II: Linear smoothing in additive white noise. *IEEE Trans. Autom. Control*, AC–**13**, 655–660.

[7] Caughey, T., and Stumpf, H. (1961). Transient response of a dynamic system under random excitation. *ASME J. Appl. Mech.*, **28**, 563–566.

[8] Levinson, N. (1947). The Wiener RMS (Root-Mean Square) error criterion in filter design and prediction *J. Math. Phys.*, **25**, 261–278.

[9] Lindquist, A. (1974). Optimal filtering of continuous-time stationary processes by means of the backward innovation process. *SIAM J. Control*, **12**, No. 6, 747–755.

[10] Ljung, L., and Kailath, T. (1976). A unified approach to smoothing formulas. *Automatica*, **12**, 147–157.

[11] Ljung, L., and Kailath, T. (1976). Backwards Markovian models for second-order stochastic processes. *IEEE Trans. Info. Theory*, IT–22, July.

[12] Morf, M., Sidhu, G.S., and Kailath, T. (1974). Some new algorithms for recursive estimation in constant, linear, discrete-time systems. *IEEE Trans. Autom. Control*, AC–**19**, 315–323.

[13] Preisendorfer, R.W. (1965). *Radiative Transfer in Discrete Spaces.* Pergamon Press, Oxford.

[14] Pusey, L.C., and Baggeroer, A.B. (1974). Role of the stationarity equation of least-squares linear filtering in spectral estimation and wave propagation. MIT Quarterly Progress Report, No 114, July 1974. See also Pusey, L.C. (1975). Ph.D. dissertation to appear, MIT.

[15] Redheffer, R. (1961). Difference equations and functional equations in transmission-line theory. Ch 12 in *Modern Mathematics for the Engineer,* second series (Beckenbach, E.F., Ed.), McGraw-Hill, New York.

[16] Redheffer, R. (1962). On the relation of transmission-line theory to scattering and transfer. *J. Math. Phys.*, Vol XLI, 1–41.

[17] Reid, W.T. (1972). *Riccati Differential Equations.* Academic Press, N.Y.

[18] Sidhu, G.S. and Kailath, T. (1974). The shift-invariance approach to continuous-time fast estimation algorithms. *Proc.* 1974 *IEEE Decision and Control Conference,* Phoenix, Arizona, 839–845.

[19] Sidhu, G.S. (1975). Ph.D. dissertation, Dept. of Electrical Engineering, Stanford University.

[20] Wiggins, R.A., and Robinson, E.A. (1965). Recursive solution to the multichannel filtering problem. *J. Geophys. Res.*, **70**, 1885–1891.

[21] Rissanen, J. (1973). Algorithms for triangular decomposition of block Hankel and Toeplitz matrices with application to factorizing positive matrix polynomials. *Math. Comput.* **27**, 147–154.

[22] Morf, M. (1974). Ph.D. dissertation, Dept. of Electrical Engineering, Stanford University.

[23] Stokes, G.C. (1883), *Collected Papers,* Vol 2, Cambridge University Press, 91–93.

[24] Bellman, R., Kalaba, R., and Wing, J. (1960). Invariant imbedding and mathematical physics-I: Particle processes. *J. Math. Phys.*, **1**, 280–308.

[25] Kailath, T., and Ljung, L. (1975). A new approach to the determination of Fredholm resolvents of non-displacement kernels. Submitted for publication.

[26] Kailath, T., and Ljung, L. (1975). Levinson- and Chandrasekhar-type equations for a general estimation process. Submitted for publication.

P.R. Krishnaiah, ed., *Multivariate Analysis–IV*

A NEW APPROACH TO SCATTERING PROBLEMS IN RANDOM MEDIA*

David MIDDLETON
New York, N.Y., U.S.A.

A new approach to scattering of radiation from random media and interfaces is outlined. The basic elements of the method are: (1), development of the radiating and spatial properties of a differential element of the scattering region Λ_M (M = a volume or a surface) into a hierarchy of independent random "point" processes $dN^{(k)}$, $k \geqslant 0$; (2), use of the appropriate Langevin equation, where the (moving) scattering elements, represented by scattering operators $\hat{g}_M *$ act like new sources whose statistical properties determine in detail the statistical character of the various components of the "point-" process hierarchy, and whose detailed physical properties are embodied in a time-varying weighting or filter operator $\hat{h}_M *$ itself in general dependent on spatial location; (3), determination of the basic scattered wavefield and received waveforms using standard iterated integral operator techniques, now, however, extended to include the time-variable character of the scattering operators, which stems in turn from the general random motion of the scattering elements. Among the advantages of the new method is the elimination, by transfer to the *statistical* properties of the scatter process (embodied in the $dN^{(k)}$), particularly of the distributed boundary conditions associated with the random media interfaces, as well as within the volume itself. The result (for interfaces) is a direct development of the scattered field (and received waveforms) into a hierarchy of physically distinguishable and measurable classes ($k \geqslant 0$) of inhomogeneity. In addition, the method provides an explicit mechanism for handling *intermode* interactions, e.g. multiple scatter between the volume and an interface. The new approach is concisely illustrated by an example from underwater acoustics [1], involving scattering from the random, moving ocean surface.

1. Introduction

The purpose of this paper is to describe in a concise fashion a new approach to the scattering of radiation from random moving media and interfaces. In so doing, we shall perforce use various basic and specialized

* Based on work supported primarily by the Office of Naval Research, Codes 412, 222, under Contract N00014–70–C–0198. A concise account of the method (however, excluding the case of strong continuum scatter in volumes) is also given in Part I, Ref. [1], with detailed applications in Part II.

techniques and concepts of multivariate analysis (MVA), in order to obtain the desired statistical results which are among the goals of scattering theory. In particular, since we are dealing here with scattered radiation, our choice of statistical methods is controlled by the underlying physics of the problem in question. This is embodied, (1); in the Langevin equation which governs the propagation of the incident and scattered fields in the medium under study and at bounding interfaces, when appropriate; and (2); in the interaction of the incident radiation with whatever scattering element is involved. Statistical mechanisms also necessarily appear in the (stochastic) Langevin equation (which is an ensemble of dynamical equations containing random elements), and in the spatio-temporal properties of the scattering elements themselves. Our desired results, of course, are various statistics of the scattered field before and after reception. Of principal interest here, and in many physical applications, are the mean, mean square (intensity), covariances and associated spectral forms. Higher order moments and probability density functions (pdf's) and distributions (PDF's) may also be sought.

A vast body of literature on scattering exists, which we shall not attempt to cite here; (for this, however, we refer the reader to Middleton [1], for mainly recent acoustical applications, e.g. underwater sound and a large number of pertinent references). What is new in our present approach (and in Ref. [1], except for (vi) below), is:

(i) The development of the radiation (and spatial) properties of a differential element $d\Lambda_M$ of the scattering region Λ_M (a volume, e.g., $M = V$, or a surface, $M = S$ or B (bottom) into a hierarchy of *independent* random point processes $dN^{(k)}$, $(k \geq 0)$, specifically a linear sum, $dN = \Sigma_k \, dN^{(k)}$, where the $dN^{(k)}$ form a set of increasing statistical complexity.

(ii) That the statistical properties of the $dN^{(k)}$ embody the statistical description, which are the only *physically observable* quantities, of the incident field, scattering element, and the resulting random, emitted radiation.

(iii) Explicit introduction of the *dynamic doppler effects** produced by the moving scattering elements.

(iv) The incorporation of the development (i) into the governing Langevin equations, where it is seen that the scattering elements $\{d\Lambda_M\}$ act like new (moving) sources with weight and dynamic responses (representing "cross-sections", range delays, doppler, and frequency selectivity), whose

* Reference [1], Appendix A.2. See also footnote 9.

statistical properties determine in detail the statistical character of the various components of the point-process hierarchy $[\mathrm{d}N^{(k)}, k \geqslant 0]$.

(v) The elimination of the generally intractable distributed boundary conditions (b.c.'s) (for surfaces and *bounded* volumes), by transfer of their conditioning effects to the *statistical properties* of the hierarchy $[\mathrm{d}N^{(k)}, k \geqslant 0]$: in effect, distributed b.c.'s are replaced by "point" b.c.'s, for which local solutions are now exactly obtainable.

(vi) An explicit formalism for handling *intermode* interactions, e.g. multiple scatter between surfaces and volumes, for example.

(vii) That the distinction between scattering classes (various orders k) can be obtained quantitatively by experiment [cf. Sections V, VII, Part II, Ref. [1], for specific examples], so that our formal development of a scattering element [(i) above] can be explicitly associated with physically real effects.*

In the present paper we shall touch only upon the highlights of the method, illustrating it with the results of an example taken from underwater acoustics [1]. This example is concerned specifically with scattering from the random, moving ocean surface, where among the many quantities of interest to be obtained from the scattered field are the point- and directional multivariate statistics of (sea) surface elevation $\zeta(\boldsymbol{r}, t)$ and its (vertical) velocity $\dot{\zeta}(\boldsymbol{r}, t)$. Techniques of multivariate analysis (MVA) play an obviously direct rôle in the analysis, as the example and preceeding text below indicate.

Accordingly, the next Section (2) presents a general model of a typical differential scattering element $\mathrm{d}\Lambda_M$, with some general assumptions regarding the physical environments we shall consider; Section 3 is devoted to moment properties of the point processes $\{\mathrm{d}N^{(k)}\}$. Section 4 treats a number of Langevin equations with conditions on the $\{\mathrm{d}N^{(k)}\}$ appropriate to various scattering régimes, involving different degrees of coupling between scattering elements, the scattered field, and different scattering *modes*, e.g. surface scatter vs. volume scatter, for instance. Section 5 provides illustrative results of the above-cited example of acoustic surface scatter, and Section 6 concludes with a brief summary and critique.

* This is true for scattering from interfaces (air/sea, water/bottom, for example) when the intermode interactions are negligible. However, when the coupling between the (total) scattered field and a scattering element $\mathrm{d}\Lambda_M$ is not ignorable, an exact resolution of the effects of individual classes ($k > 0$) of inhomogeneity does not appear possible. See the comments at end of Sec. 4.

2. Model of a differential scattering element

With each differential element $d\Lambda_M(\boldsymbol{\lambda})$ of the scattering medium (volume, interface, or combinations thereof) we associate a differential stochastic operator of the form

$$d\hat{g}_{T\text{-scat}}(\tau, t \mid Z)_M\{\ \} \equiv A_M \hat{h}_M(\tau, t \mid Z)\{\ \} dN(Z)_M, \tag{2.1}$$

where $Z = [\boldsymbol{\lambda}, \boldsymbol{r}, \boldsymbol{\rho}, \boldsymbol{\theta}]$ denotes four sets of random (vector) variables; (we also shall use $Z \equiv Z_\lambda \times Z_r \times Z_\rho \times Z_\theta = \boldsymbol{\lambda} \times \boldsymbol{r} \times \boldsymbol{\rho} \times \boldsymbol{\theta}$, to represent the spaces over which these random variables are defined; the distinction will be clear from the context subsequently). Here $\hat{h}_M$ is a stochastic operator, which may be linear or nonlinear, and which embodies the *deterministic* radiative structure of the scattering element, located in $d\Lambda_M(\boldsymbol{\lambda})$. [In our formalism, we find it convenient to introduce *all* statistical properties of the scattering element by means of the weighting point-process $dN(Z)$. Thus, as we shall see in Section 3 following, for the Z in $\hat{h}_M$ one has appropriate pdf's in $dN(Z)$, cf. (3.9)–(3.11); also, A_M is a constant, often set equal to unity.][1] In our present notation (which follows that of Ref. [1]) we have $\boldsymbol{\lambda}(= \boldsymbol{R}/c_0)$, a spatial vector (measured in seconds, with c_0 = av. velocity of propagation in the medium in question) which locates the scattering element Λ_M vis-à-vis a given (source) coördinate system, in the manner of Fig. (2.1). Similarly, $\boldsymbol{r}$, $\boldsymbol{\rho}$ are spatial vectors, associated with possible statistical variations of $\hat{h}_M$ in space, e.g. $\boldsymbol{r}$ for magnitude and $\boldsymbol{\rho}$ for changes in the *form* of $\hat{h}_M$, analogous to the roles of τ and t in $\hat{h}_M$, which respectively represent a filtering action, or "memory", and possible time-variability of $\hat{h}_M$ (cf. eq. (10), Ref. [1], Part 1).

For many of the important physical situations the scattering mechanism is linear (which implies a *linear* Langevin equation, cf. (4.1) et seq. below). Then $\hat{h}_M(= h_M *)$ in (2.1) is a linear, time-varying stochastic filter weighting function, in the terminology of statistical communication theory (SCT). Various specific structures for h_M have been devised, for different physical situations. For example, for the (scalar) Basic "Point" Scatter Model (BPSM) where the scattering element (S), source (T), and receiver (R) are in motion vis-à-vis a fixed medium, one has[2]

[1] Ref. [1]; Part I, Section II, up to A.

[2] Ref. [1], part II, Section IA. Also, Eq. (A.2–17a) applied to Eq. (1), therein. Useful approximate forms (2.2) are developed in the text of this reference, cf. Appendix A.2, and Sections IA, B, C, Part II. See also footnote 9.

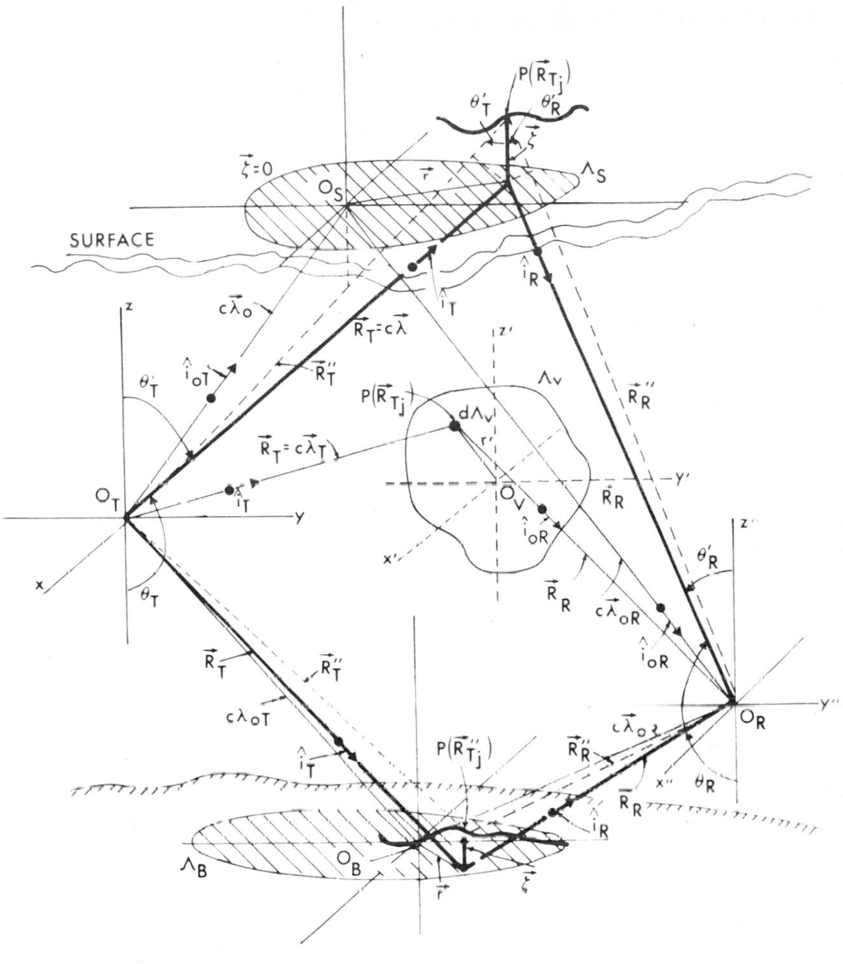

Fig. 2.1. Schema of scattering elements $d\Lambda_j$ at $P(\vec{R}_{Tj})$ on the surface and bottom at elevations $\zeta(\vec{r}, t_0)$, and in the volume, for the ensonified and viewed regions Λ_S, Λ_B, Λ_V.

$$h_M \doteq \gamma_0(\boldsymbol{\lambda}, \boldsymbol{r}, \boldsymbol{\rho})_M\, \delta\left(\tau - T(\boldsymbol{\lambda}) + \int_0^{t-\tau} \varepsilon(t')\mathrm{d}t'\right), \tag{2.2}$$

where [for the small-doppler, far-field conditions usually postulated, $\varepsilon(t')$ $(= [v_{TS}(t') + v_{ST}(t') + v_{RS}(t') + v_{SR}(t')]/c_0)$ is the sum of the relative (normalized) *doppler velocities* between T, R and S]. Usually, ε contains both a random and a deterministic component. Here γ_0 is a *cross-section*, which is a measure of the scattering strength of the scattering element $d\Lambda_M$, which is located (at any time t) at $\tau c_0 = c_0[T(\boldsymbol{\lambda}) - \int_0^{t-\tau} \varepsilon(t')\mathrm{d}t']$, and $T(\boldsymbol{\lambda})$ is a (random) path travel time from source-to-scattering element-to-receiver. For this model, a typical scatterer introduces a (stationary) phase modulation ($\sim T(\boldsymbol{\lambda})$) or an equivalent frequency modulation of the incident radiation, because of random location and motion, generally. Other structures for h_M have also been suggested, such as resonant filters, to account for bubbles or fish in underwater acoustic applications, for example [2]. For scattering of electromagnetic waves, γ_{0M} in (2.2), and therefore h_M, become second-rank tensors (or dyadics), coupling the incident and emitted vector fields [3].

3. Statistical properties of the $dN(\boldsymbol{Z})_M$: moments

We begin by developing $dN(\boldsymbol{Z})_M$ for domain (M) in question into the aforementioned hierarchy [cf. (i), Section 1 above] of *independent* random point processes [4], $dN_M^{(k)}$, according to

$$dN(\boldsymbol{Z})_M = \sum_{k=0}^{\infty} dN^{(k)}(\boldsymbol{Z})_M, \tag{3.1}$$

which embodies the statistical properties of the scattering element $d\Lambda_M$ [cf. (2.1) and comments]. This is not a unique development for $d\Lambda_M$, until all (existing) moments *and* the pdf's of dN_M of all orders, are specified. This, in turn, may be achieved when the physical model itself, whose scattering properties (including boundary conditions, cf. (iv), Section 1.) are contained in the $\{dN^{(k)}\}$, is (statistically) given, by experiment and analysis. In practise, we shall usually be content with a partial spcification, based on the lower order moments, (e.g., 1st and 2nd order).

Now, what the development (3.1) says, effectively, is that the common scattering domain Λ_M contains sets of independent scattering elements, which represent different orders (k) of inhomogeneities, i.e. different degrees of coupling, in amounts determined by the weights $dN^{(k)}$. The statistical independence derives at once from this categorizing by definition: elements which are actually excited and coupled in "pairs" $(k = 2)$ are always distinct from those actually excited which are coupled as "triples" $(k = 3)$, etc. The important feature of the $dN^{(k)}$ processes (apart from their postulated independence) is that these $dN^{(k)}$ are sets of scattering elements ("points" or continua) which are *actually interacting*, at any given instant, with the incident field, as distinct from the totality $dN_T^{(k)}$ of scattering elements *available* for such interactions, e.g. $dN^{(k)} \leqslant dN_T^{(k)}$. [In this fashion, shadowing effects are automatically included in the various moments, etc., of the $dN^{(k)}$.]

Fig. (3.1) shows this hierarchy of scattering elements, in a schema where the scattering domain Λ_M is resolved into a superposition of identically sized domains Λ_M, containing, respectively, the following orders of scattering elements:

(i) $k = 0$: identified with a *coherent* (e.g., specular) *component*;

(ii) $k = 1$: a purely random component (as far as radiative properties only are concerned, cf. [5]). [These elements can be shown to obey poisson statistics, cf. Ref. [1], Part I, Sec. II, C, D.]

(iii) $k \geqslant 2$: multiple, or correlated scatter terms, which involve pairs, triples, etc., of coupled scattering elements, whether these be discrete or ultimately continuous.[3]

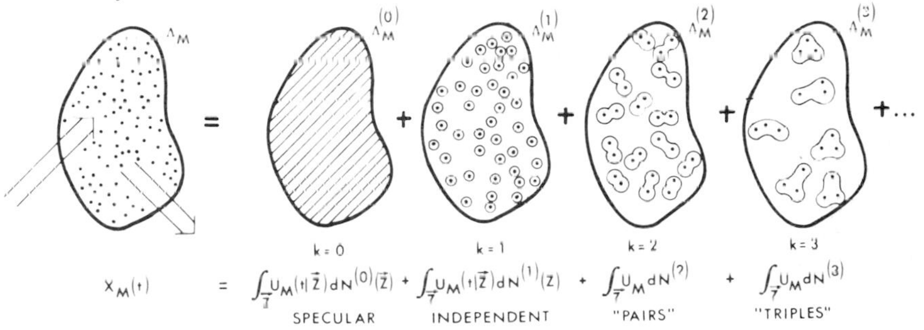

Fig. 3.1. Scattering domain Λ_M resolved into a hierarchy of independent scatter processes (and received scattered processes; received scattered process $X_M(t)$, [eqs. 4.14b, c] with scattering elements not interacting with the scattered field [cf. Section 4, eqs. (4.6a, b), Cases I, II].

This division of the acting scattering elements in k independent classes of inhomogeneities accordingly employs all available scattering elements *which are activated at any given instant*, numbers (or weights) of which will vary from moment to moment in response to the dynamics of the inhomogeneous medium in question. [The expansion of the differential scattering element as the operator (2.1), with (3.1), into "orthogonal" (≡ independent) processes (or operators) may be regarded as formally analogous to the familiar Karhunen–Loève expansion of a random process into (wide-sense) independent, orthogonal components[4], except that here the point processes $dN^{(k)}$ play the rôle of the (deterministic) eigenfunctions of the process covariance.]

[3] Ref. [1], (ii) following eq. (11).

[4] See, for example, Section 8.2 of Ref. [6].

Let us focus now on the *radiation events*, e.g., scattering, of the subset $dN(\boldsymbol{\lambda} \mid \boldsymbol{Z}_s)_M$ of $dN(\boldsymbol{Z}_\lambda \ (=\boldsymbol{\lambda}) \times \boldsymbol{Z}_s)_M = dN(\boldsymbol{Z})_M$, where for the moment the $\boldsymbol{Z}_s$ $(\equiv \boldsymbol{r} \times \boldsymbol{\rho} \times \boldsymbol{Z}_\theta)$ are regarded as specified. Later, cf. (3.8) et seq., we shall include the random spatially-dependent and parametric properties associated with $\boldsymbol{Z}_s$. We have the following moment properties of the $dN^{(k)}$:

(1) $k = 0$:

$$\langle dN_M^{(0)}(\boldsymbol{\lambda} \,\|\, \boldsymbol{Z}_s)_M \rangle_R = R_M^{(0)}(\boldsymbol{\lambda})d\boldsymbol{\lambda} \quad (\geqslant 0); \tag{3.2}$$

$$\langle dN^{(k)} \rangle_R = 0, \qquad k \geqslant 1, \tag{3.2a}$$

where $\langle\ \rangle_R$ denotes the average with respect to the radiation events. Next, we have for these moment properties

(2) $\langle dN_1^{(k_1)} \dots dN_n^{(k_n)} \rangle_R =$ products of averages over equal superscript indexes $k_1, \dots, k_n$; $(k \leqslant n,\ n \geqslant 1)$, without duplications. (3.3)

This property simply reflects the fact that the radiation events in different classes $(k_i \neq k_j)$ are independent, cf. (3.1) and ff. Also, we have

(3) $\langle dN_1^{(k)} \dots dN_n^{(k)} \rangle_R = 0, \qquad k > n \quad (\geqslant 1).$ (3.4)

This condition expresses the fact that each unit (of class k) acts like a single radiating aperture (cf. Fig. (3.1)). However, when $k \leqslant n$, eq. (3.4) becomes

(4) $\langle dN_1^{(k)} \dots dN_n^{(k)} \rangle_R = \sum$ (averages over all combinations of $dN^{(k)}$'s, $k \leqslant n$, taken in all different groups of subindexes, without duplication). (3.5)

With a little effort we can also show that

$$(5)\quad \langle dN_1 \dots dN_n \rangle_R = \sum_{k=0}^{n} \langle dN_1^{(k)} \dots dN_n^{(k)} \rangle_R . \tag{3.6}$$

Some examples of the "rules" (1)–(5) are:

(i) $\langle dN_1^{(0)} dN_2^{(2)} dN_3^{(2)} \rangle_R = \langle dN_1^{(0)} \rangle_R \langle dN_2^{(2)} dN_3^{(2)} \rangle_R$, cf. (2), eq. (3.3); (3.7a)

(ii) $\langle dN_1^{(0)} dN_2^{(0)} dN_3^{(0)} \rangle_R = \langle dN_1^{(0)} \rangle_R \langle dN_2^{(0)} \rangle_R \langle dN_3^{(0)} \rangle_R$, cf. (4), eq. (3.5);

$$\begin{aligned}\langle dN_1^{(2)} dN_2^{(2)} dN_3^{(2)} \rangle_R &= \langle dN_1^{(2)} dN_2^{(2)} dN_3^{(2)} \rangle_{R\text{-pair}} + \langle dN_1^{(2)} \rangle_R \langle dN_2^{(2)} dN_3^{(2)} \rangle_R \\ &\quad + \langle dN_1^{(2)} dN_2^{(2)} \rangle_R \langle dN_3^{(2)} \rangle_R + \langle dN_1^{(2)} dN_3^{(2)} \rangle \langle dN_2^{(2)} \rangle_R \\ &= \langle dN_1^{(2)} dN_2^{(2)} dN_3^{(2)} \rangle_{R\text{-pair}} \quad \text{[cf. 3.7c], by} \\ &\quad \text{(3.2a), (3.4), (3.5);}\end{aligned} \tag{3.7b}$$

$$\text{(iii)}\quad \langle dN_1 dN_2\rangle_R = \langle dN_1^{(0)}\rangle_R \langle dN_2^{(0)}\rangle_R + \langle dN_1^{(1)} dN_2^{(1)}\rangle_{R\text{-singles}} + \langle dN_1^{(2)} dN_2^{(2)}\rangle_{R\text{-pairs}}$$
$$= \{R_M^{(0)}(\boldsymbol{\lambda}_1)R_M^{(0)}(\boldsymbol{\lambda}_2) + R_M^{(1)}(\boldsymbol{\lambda}_1)\delta(\boldsymbol{\lambda}_2 - \boldsymbol{\lambda}_1) + R_M^{(2)}(\boldsymbol{\lambda}_1, \boldsymbol{\lambda}_2)\} d\boldsymbol{\lambda}_1 d\boldsymbol{\lambda}_2; \tag{3.7c}$$

$$\text{(iv)}\quad \langle dN_1 dN_2 dN_3\rangle_R = \langle dN_1^{(0)} dN_2^{(0)} dN_3^{(0)}\rangle_{\text{coh.}} + \langle dN_1^{(1)} dN_2^{(1)} dN_3^{(1)}\rangle_{\text{indep.}} + \langle dN_1^{(2)} dN_2^{(2)} dN_3^{(2)}\rangle_{\text{pairs}} + \langle dN_1^{(3)} dN_2^{(3)} dN_3^{(3)}\rangle_{\text{triples}}$$
$$= \{R_1^{(0)}(\boldsymbol{\lambda}_1)R_1^{(0)}(\boldsymbol{\lambda}_2)R_1^{(0)}(\boldsymbol{\lambda}_3) + 3R_M^{(1)}(\boldsymbol{\lambda}_1)\delta(\boldsymbol{\lambda}_3 - \boldsymbol{\lambda}_1)\delta(\boldsymbol{\lambda}_2 - \boldsymbol{\lambda}_1) + 3R_M^{(2)}(\boldsymbol{\lambda}_1, \boldsymbol{\lambda}_2)\delta(\boldsymbol{\lambda}_3 - \boldsymbol{\lambda}_2) + R_M^{(3)}(\boldsymbol{\lambda}_1, \boldsymbol{\lambda}_2, \boldsymbol{\lambda}_3)\} d\boldsymbol{\lambda}_1 d\boldsymbol{\lambda}_2 d\boldsymbol{\lambda}_3,$$

by (1)–(4) above. (3.7d)

Next, we account for the statistical dependence of the $dN, dN^{(k)}$ on possible random parameters ($\boldsymbol{\theta}$) and on spatial dependencies, e.g., those associated with $\boldsymbol{Z}_s$. We therefore extend $dN(\boldsymbol{\lambda} \mid \boldsymbol{Z}_s)_M$ to include $\boldsymbol{Z}_s$, e.g., we consider now $dN(\boldsymbol{Z} = \boldsymbol{\lambda} \times \boldsymbol{Z}_s)_M$ and the associated spatial average:

$$\langle dN_1\rangle_s = \sum_{k=0}^{\infty} \langle dN_1^{(k)}(\boldsymbol{Z}_1)\rangle_s. \tag{3.8}$$

Unlike (3.2), (3.2a) for the (re-) radiative events associated with the scattering elements in $(\boldsymbol{\lambda}_1, \boldsymbol{\lambda}_1 + d\boldsymbol{\lambda}_1)$, we can have $\langle dN_1^{(k)}\rangle_s \neq 0$ for *all* $k \geqslant 0$, since $\langle dN_1^{(k)}\rangle_s$ is simply the average number of scatterers (per unit domain $d\boldsymbol{Z}_s = d\boldsymbol{r} \times d\boldsymbol{\rho} \times d\boldsymbol{\theta}$) *available* for class k scatter, but not necessarily so acting. Specifically, we write

$$\langle dN_1^{(k)}\rangle_s = dN_1^{(k)}(\boldsymbol{\lambda})_{s1} d\boldsymbol{Z}_{s1} = dN_1^{(k)}(\boldsymbol{\lambda})_{s1} w_1(\boldsymbol{r}_1, \boldsymbol{\rho}_1, \boldsymbol{\theta}_1 \mid \boldsymbol{\lambda}_1) d\boldsymbol{r}_1 \ldots d\boldsymbol{\theta}_1, \tag{3.9}$$

where $dN_{s1}^{(k)}$ is the (spatially, average *random*) number of radiators active in $(\boldsymbol{\lambda}_1, \boldsymbol{\lambda}_1 + d\boldsymbol{\lambda}_1)$ with the spatial properties associated with $\boldsymbol{\theta}_1$ and location $(\boldsymbol{r}_1, \boldsymbol{\rho}_1)$. An important special case occurs when it is reasonable to assume that the spatial properties of the scattering elements in Λ_M are solely deterministic. Then we have $w_1(\boldsymbol{r}, \boldsymbol{\rho}) = \delta(\boldsymbol{r} - c_0\boldsymbol{\lambda})\delta(\boldsymbol{\rho} - c_0\boldsymbol{\lambda})$ etc., for $w_2(\boldsymbol{r}_1, \boldsymbol{\rho}_1, \boldsymbol{r}_2, \boldsymbol{\rho}_2)$ etc., with a consequent simplification in the various moments of the scattered field and received scattered waves [cf. [1], Part II, Section I, et seq.].

Combining radiation and spatial averages, (3.2)–(3.9) above, we obtain the following total averages, as designated by $\langle\ \rangle_{RS} \equiv \langle\ \rangle$:

($n = 1$): *the mean*:

$$\langle dN_1\rangle = \langle dN_1^{(0)}\rangle_{RS} = R_M^{(0)}(\boldsymbol{\lambda}_1)w_1(\boldsymbol{r}_1, \boldsymbol{\rho}_1, \boldsymbol{\theta}_1 \mid \boldsymbol{\lambda}_1)d\boldsymbol{\lambda}_1 \ldots d\boldsymbol{\theta}_1, \tag{3.10}$$

for coherent reradiation ($k = 0$). For the second-order moments one has from (3.7c) directly

($n = 2$); *second-order*:

$$\langle dN_1 dN_2\rangle = \{R_M^{(0)}(\boldsymbol{\lambda}_1)R_M^{(0)}(\boldsymbol{\lambda}_2) + R_M^{(1)}(\boldsymbol{\lambda}_1)\delta(\boldsymbol{\lambda}_2 - \boldsymbol{\lambda}_1) + R_M^{(2)}(\boldsymbol{\lambda}_1, \boldsymbol{\lambda}_1)\}w_2(\boldsymbol{r}_1, \boldsymbol{\rho}_1; \boldsymbol{r}_2, \boldsymbol{\rho}_2; \boldsymbol{\theta}_1, \boldsymbol{\theta}_2)d\boldsymbol{\lambda}_1 \ldots d\boldsymbol{\theta}_2, \tag{3.11}$$

and similarly for the third and higher-order moments, by obvious extensions of (3.7d), (3.9). In general, from (3.6) one has

($n \geq 1$); *nth-order*:

$$\langle dN_1 \ldots dN_n\rangle = \sum_{k=0}^{n} \langle dN_1^{(k)} \ldots dN_n^{(k)}\rangle_{RS}, \tag{3.12}$$

where the detailed "anatomy" of the components (k) is obtained by the "rules" (1)–(5) above.

The *local physics* of the scattering process associated with $d\Lambda_M$ is embodied now in the *statistics* $R_M^{(0)}$, $R_M^{(1)}$, $R_M^{(2)}$, etc., which also includes the effects of the boundary conditions, since by definition there is an actual physical scattering process taking place, for which this is the interactive model.

Here, specifically, we have $R_M^{(0)} = \sigma_M^{(0)}(\boldsymbol{\lambda}_1)J_{M1}(\boldsymbol{\lambda}_1)$, where $\sigma_M^{(0)}$ is the physical density (per unit region Λ_M) of scattering elements in Λ_M which can coöperate to produce coherent, or specular reradiation, and where J_{M1} is the jacobian appropriate to the coördinate system used. Thus, $R_M^{(0)}$ is the average number of scatterers in $(\boldsymbol{\lambda}_1, \boldsymbol{\lambda}_1 + d\boldsymbol{\lambda}_1)$ for class ($k = 0$) inhomogeneities, producing coherent (re-) radiation. [For surfaces ($M = S, B$), or volumes ($M = V$), we have $J_M = c^2\boldsymbol{\lambda}_1, \ = c^3\boldsymbol{\lambda}_1^2 \sin\theta_1$, cf. fig. 2.1, respectively.]

Similarly, we have $R_M^{(1)} = \sigma_M^{(1)}(\boldsymbol{\lambda}_1)J_{m1}(\boldsymbol{\lambda}_1)$, the average number of scatterers producing purely independent scatter ($k = 1$) from elements in $(\boldsymbol{\lambda}_1, \boldsymbol{\lambda}_1 + d\boldsymbol{\lambda}_1)$, while $R_M^{(2)} = \sigma_M^{(2)}(\boldsymbol{\lambda}_1)\sigma_M^{(2)}(\boldsymbol{\lambda}_2)J_{M1}J_{M2}\rho^{(2)}(\boldsymbol{\lambda}_1, \boldsymbol{\lambda}_2)_M$ is the corresponding average number producing second-order, or "pair-coupled" (re-) radiation, for elements jointly in $(\boldsymbol{\lambda}_1, \boldsymbol{\lambda}_1 + d\boldsymbol{\lambda}_1; \boldsymbol{\lambda}_2, \boldsymbol{\lambda}_2 + d\boldsymbol{\lambda}_2)$ in Λ_M. As before, $\sigma_M^{(2)}$ is the physical density of such scatterers, while $\rho_M^{(2)}$ is a normalized correlation function, connecting the radiation events at $(\boldsymbol{\lambda}_1, \boldsymbol{\lambda}_2)$, such that

$$\lim_{\boldsymbol{\lambda}_1 \leftrightarrow \boldsymbol{\lambda}_2} \rho_M^{(2)}(\boldsymbol{\lambda}_1, \boldsymbol{\lambda}_2) = 1; \qquad \lim_{|\boldsymbol{\lambda}_2 - \boldsymbol{\lambda}_1| \to \infty} \rho_M^{(2)} = 0. \tag{3.13}$$

Furthermore, we must have

$$\rho_M^{(2)}(\boldsymbol{\lambda}_1, \boldsymbol{\lambda}_2) = \rho_M^{(2)}(\boldsymbol{\lambda}_2, \boldsymbol{\lambda}_1); \qquad \rho_M^{(2)} \geq 0, \tag{3.14}$$

the former since there is no preferred designation of location, and the latter, since there cannot be a negative number of scatterers interacting. (This second condition rules out such functions as $\rho_M^{(2)} = \sin a\,|\boldsymbol{\lambda}_1 - \boldsymbol{\lambda}_2|/a\,|\boldsymbol{\lambda}_1 - \boldsymbol{\lambda}_2|$, for example, while $\rho_M^{(2)} = e^{-a|\boldsymbol{\lambda}_2 - \boldsymbol{\lambda}_1|}$ is acceptable.) We may continue in the same way for $k \geq 3$. In fact, along with obvious generalizations of the limits (3.13) we have generally $1 \geq \rho_M^{(k)}(\boldsymbol{\lambda}_1, \ldots, \boldsymbol{\lambda}_k) \geq 0$, $k \geq 1$, with $\rho_M^{(k)}$ interchangeable in its arguments. Because of these properties, we can also regard $\rho_M^{(k)}$, $k \geq 1$, as (proportional to) a *probability density*, $w_k(\boldsymbol{\lambda}_1, \ldots, \boldsymbol{\lambda}_k)$, which measures the probability (density) of the occurrence of kth-order coupled scattering elements in Λ_M.

The fully detailed structure of the $R_M^{(k)}$ depend, of course, on the local physics of the medium ($M = V$) or interfaces ($M = S, B$), as we shall note in Section 5, cf. eqs. 5.4, 5.5. The probability distributions of the point-process dN itself are generally not explicitly obtainable, but it can be shown that $dN^{(1)}$ ($k = 1$), obeys a generalized poisson process, and that under some circumstances dN is conditionally gaussian, with possibly nonvanishing means [cf. Ref. [1], Part I, Section II D]. Also, various models of the coupling, embodied in $\rho_M^{(k)}$, may be constructed, where we may take advantage of the fact that $\rho_M^{(k)}$is proportional to a proper pdf and treat the scattering domain as developable in hierarchies of Markoff processes, obeying conditions like those cited in Section 1.4 of Ref. [6], for example. These higher order statistical questions, however, do not restrict our treatment, as long as we are concerned solely with such physically important quantities as mean values (coherent radiation, in certain cases, cf. (3.2), (3.10), and (4.4)), and mean intensities, covariances and associated spectra, of the scattered fields, i.e. the lower order moments, *when this is permitted by the Langevin equation in force*, cf. (4.5), (4.6a, b). For other governing Langevin equations, however, some structural knowledge of *all* moments of the point process dN is needed, as we shall see presently, cf. (4.6c, d, e).

Finally, when the scattering domains are distinct, e.g., a surface and a volume, their respective developments $dN(\boldsymbol{Z})_M$, $dN(\boldsymbol{Z})_{M'}$, $M \neq M'$ according to (3.1), are also statistically independent, not only within each hierarchy (k: M; l: M'), but between hierarchies, e.g.

$$\langle F(dN_M)\, G(dN_{m'})\rangle = \langle F(dN_M)\rangle\langle G(dN_{M'})\rangle,$$

with

$$\mathrm{d}N_M = \sum_k \mathrm{d}N_M^{(k)}, \qquad \mathrm{d}N_{M'} = \sum_l \mathrm{d}N_{M'}^{(l)}, \tag{3.15}$$

or more generally, symbolically in terms of pdf's of the $\mathrm{d}N$'s, $w_n(\mathrm{d}N_M, \ldots, \mathrm{d}N_{M'} \ldots) = w_n(\mathrm{d}N_M \ldots)w_n(\mathrm{d}N_{M'} \ldots)$, $n \geqslant 1$, as well as $\langle \mathrm{d}N_M^{(k)} \mathrm{d}N_{M'}^{(l)} \rangle = \langle \mathrm{d}N_M^{(k)} \rangle \langle \mathrm{d}N_{M'}^{(l)} \rangle$, $k = l$, $k \neq l$, etc., among the individual component processes.

4. Langevin equations and the scattered field

The ensemble of dynamical equations governing propagation in the medium and scattering from inhomogeneities within it and from its bounding interfaces is called a Langevin equation [3][5]. There are many varieties, depending on the physics involved.

One typical form of Langevin equation for propagation and scattering, which includes many important special cases of practical interest, and which is itself a (linear) special case of the Stokes–Navier equation [7], is given by the following (scalar) wave equation for propagation in linear and otherwise lossless, homogeneous media:[6]

$$\left(\nabla^2 - \frac{1}{c_0^2}\left[1 + \hat{g}_M * \frac{\mathrm{d}^2}{\mathrm{d}t^2}\right]\right) p = -\hat{g}_T * S_{\mathrm{in}} = -G_T\,; \qquad p = p(\boldsymbol{R}, t) \tag{4.1}$$

(in rectangular coördinates). Here $\hat{g}_M *$ is the scattering operator, and G_T is the input signal source function represented respectively by

$$\hat{g}_M * = \int_{Z_M} \mathrm{d}\hat{g}_M * \equiv A_M \int_{Z_M} \mathrm{d}N(\boldsymbol{Z})_M \int h_M(\tau, t\,|\,\boldsymbol{Z})(\ )_{t-\tau}\,\mathrm{d}\tau \tag{4.2a}$$

cf. (2.1); and

$$\begin{aligned} G_T(t, \boldsymbol{\xi} = \boldsymbol{R} - \boldsymbol{r}) &= \int_{-\infty \mathrm{i}+d}^{\infty \mathrm{i}+d} A_T(\boldsymbol{\xi}, s/2\pi\mathrm{i}) S_{\mathrm{in}}(s/2\pi\mathrm{i}, \boldsymbol{\xi}) \mathrm{e}^{st}\,\mathrm{d}s/2\pi\mathrm{i} \\ &= \mathcal{F}_s^{-1}\{G_{0T}\}, \; G_T \in V_T, \end{aligned} \tag{4.2b}$$

with $G_{0T} = [A_T S_{in}]_s$ in which A_T is the (frequency-dependent) aperture weighting of the active signal source, and $S_{\mathrm{in}}(= \mathcal{F}_t\{S_{\mathrm{in}}\})$ is the amplitude

[5] Ref. [6], Section 10.1, for example. Here, because of the spatially distributed character of the medium, these are partial differential equations, rather than ordinary ones.

[6] Operators are indicated here by superscript (^), viz. $\hat{g}$, $\hat{H}$, etc., and (*).

spectrum of the signal applied to the transmitting aperture A_T at location ($\boldsymbol{\xi}$), cf. Fig. (4.1). Here V_T, c_0 are respectively the domain occupied by the signal source ($G_T \notin V_T = 0$) and the mean speed of propagation of the (scalar) wavefield $p(\boldsymbol{R}, t)$; $\{*\}$ denotes temporal convolution, and the source operator $\hat{g}_T *$ is $\mathscr{F}_s^{-1}\{\hat{G}_{0T}\} = \int_{\mathrm{Br}} A_T \cdot (\ )e^{st} ds/2\pi i$, where the paths of integration in (4.2b) and for $\hat{G}_{0T}$ are Bromwich contours [8].

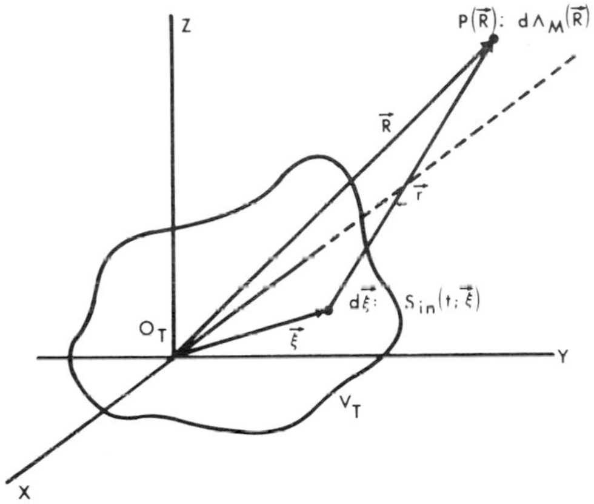

Fig. 4.1. Geometry of source aperture in V_T, input signal applied at $d\vec{\xi}$, at $\vec{\xi}$, and a scattering element $d\Lambda_M$ at $P(\vec{R})$.

Since the medium is assumed to be linear [cf. (4.1)], we may write for the field $p(\boldsymbol{R}, t)$ at point $P(\boldsymbol{R})$, after scattering,

$$p(\boldsymbol{R}, t) = p_{\mathrm{scat}}(\boldsymbol{R}, t) + p_{\mathrm{inc}}(\boldsymbol{R}, t). \tag{4.3}$$

From the relation

$$\left(\nabla^2 - \frac{1}{c_0^2}\frac{d^2}{dt^2}\right) p_{\mathrm{inc}} = -G_T \quad \text{if} \quad G_T \in V_T; \tag{4.4}$$

$$= 0, \qquad \text{if} \quad G_T \notin V_T,$$

and (4.3), we obtain directly from (4.1) the Langevin equation governing p_{scat}:

$$\left(\nabla^2 - \frac{1}{c_0^2}\frac{d^2}{dt^2}\right) p_{\mathrm{scat}-M} = \hat{g}_M * [\ddot{p}_{\mathrm{inc}} + \ddot{p}_{\mathrm{scat}-M}]/c_0^2. \tag{4.5}$$

The scattering operator $\hat{g}_M *$ is, in effect, the dynamic (i.e. time-dependent) *aperture* of the scattering region Λ_M, and includes the random dopplers of the moving scattering elements (e.g., moving "sources"). A condition that the Langevin equation has the forms (4.1), (4.5) is that (*i*), the dopplers are small (vs. c_0); (*ii*), $p_{\text{scat}}(\boldsymbol{R}, t)$ is in the far-field vis-à-vis (most of) the scatterers activated; and (*iii*), the dopplers are at most weakly dependent on location.[7]

We shall briefly consider four special cases of (4.5), in particular à propos of scattering problems in underwater acoustics. These are (for at most a single interface interaction):

I. *Interface scatter*:

$$\left(\nabla^2 - \frac{1}{c_0^2}\frac{d^2}{dt^2}\right)_{\text{scat}-M} = \hat{g}_M * \ddot{p}_{\text{inc}}/c_0^2, \quad M = S(\text{urface}),$$

B(ottom). [*Intermode* scattering (IV. below) is assumed negligible.] (4.6a)

II. *Weak volume scatter*:

Eq. (4.6a), $M \to V$(olume); this is the 1st Born approximation, where $\hat{g}_V * \ddot{p}_{\text{scat}-V}$ is negligible i.e. here there is ignorable coupling between the scattering elements and the scattered field. (4.6b)

III. *General volume scatter*:

Eq. (4.5), $M \to V$; (4.6c)

IV. *Intermode scattering*:

(a) *Interface* ⊗ *volume*: $p_{\text{inc}} \to p_{\text{scat}-M'}$; $M' = S, B$; $M = V$

$$\therefore \text{Eq. (4.4)} \to \left(\nabla^2 - \frac{1}{c_0^2}\frac{d^2}{dt^2}\right) p_{\text{scat}-V} = \hat{g}_V * [\ddot{p}_{\text{scat}-M'} + \ddot{p}_{\text{scat}-V}]/c_0^2 \quad (4.6d)$$

(b) *Volume* ⊗ *interface*: $p_{\text{inc}} \to p_{\text{scat}-M'}$; $M' = V$; $M = S, B$

$$\hat{g}_M * \ddot{p}_{\text{scat}-M} = 0,$$

$$\therefore \text{Eq. (4.4)} \to \left(\nabla^2 - \frac{1}{c_0^2}\frac{d^2}{dt^2}\right) p_{\text{scat}-M} = \hat{g}_M * \ddot{p}_{\text{scat}-V}/c_0^2, \quad M = S, B. \quad (4.6e)$$

Note that reciprocity does not hold here, i.e. (a) $\not\leftrightarrow$ (b), for intermode scattering, because of the different properties and order of interaction among the scattering elements. For (4.6d) we use the solutions of (4.6a) for $\ddot{p}_{\text{scat}-M'}$, whereas for (4.6e) $p_{\text{scat}-V}$ is obtained generally from (4.5), $M = V$,

[7] Ref. [1], Appendix A.2. This restriction can be removed, however.

cf. (4.6c), by the iterative technique outlined in Eqs. (4.15) et seq. below, unless the volume scatter is *weak*, in which case the much simpler results of (4.6b) [or (4.6a), $M \to V$] apply.

Cases I, II: *Noninteracting Scattered Fields, eqs.* (4.6a, b)

Solutions for Cases I, II, cf. (4.6a, b) are readily obtained, since $\hat{g}_M * \ddot{p}_{\text{scat}-M}$ vanishes or is ignorable. We outline the approach: first take the generalized fourier transform (g.f.t) of both sides of (4.6a, b), where we write the differential operator

$$\hat{L}_0(\boldsymbol{R} \mid s) \equiv \nabla^2 - s^2/c^2. \tag{4.7}$$

Next, since we are dealing only with the scattered, e.g. random radiation in (4.6a, b), the "point-" process $\mathrm{d}N_M$ becomes $\mathrm{d}N'_M = \Sigma_{k=1}^{\infty} \mathrm{d}N^{(k)}$: the *coherent component* $\mathrm{d}N_M^{(0)}$ ($k = 0$), cf. Sec. 3, *is omitted.* For the same reason, the filter operator $\hat{h}_M *$ is replaced by $\hat{h}'_M * = \int (h_M - \langle h_M \rangle)(\ )\mathrm{d}\tau$, where $\langle h_M \rangle \neq 0$ if there is coherent radiation [cf. Section I, Part II, Ref. (1)]. We obtain accordingly

$$\hat{L}_0 p_{0-\text{scat}\,M} = \hat{g}'_M \circledast (s^2/c_0^2[p_{\text{inc}}]); \qquad M = S, B, V \tag{4.8}$$

$$\hat{g}'_{0M} \equiv \mathcal{F}_t \left\{ A_M \int_{Z_M} \mathrm{d}N'(\boldsymbol{Z}) \hat{h}'_M * \right\} = \int_{Z_M} \mathrm{d}N'(\boldsymbol{Z})_M \hat{H}_{0M}; \tag{4.8a}$$

where $\hat{g}'_{0M}$ is the (g.f.t.) of $\hat{g}'_M *$, cf. (4.2a), and $\circledast$ denotes the convolution in the (complex angular) frequency s. Specifically, we have

$$\begin{aligned} p_{\text{scat}\ M} &= \mathcal{F}_s^{-1}\{p_{0-\text{scat}-M}\} = \int_{-\infty i+d}^{\infty i+d} p_{0-\text{scat}\,M} e^{st}\,\mathrm{d}s/2\pi i \\ p(\boldsymbol{R}, s)_{0-\text{scat}\,M} &= \mathcal{F}_t\{p_{\text{scat}-M}\} = \int_{-\infty}^{\infty} p_{\text{scat}-M} e^{-st}\,\mathrm{d}t. \end{aligned} \tag{4.9a}$$

In addition, we write

$$\begin{aligned} \hat{H}_{0M} = \hat{H}_{0M}(s \mid \boldsymbol{R}) &= A_M \mathcal{F}_t\{\hat{h}'_M(t \mid \ldots \mid \boldsymbol{R})\} \\ &= A_M \int_{-\infty}^{\infty} e^{-st}\,\mathrm{d}t \int_{-\infty}^{\infty} (\ )_{t-\tau}\,\mathrm{d}\tau e^{-st} h'_M \mathrm{d}\tau, \end{aligned} \tag{4.9b}$$

$$\therefore \hat{g}'_{0M} \circledast \left(\frac{s^2}{c_0^2} p_0\right) = \int_{Z_M} \mathrm{d}N'(\boldsymbol{Z})_M \int_{-\infty i+d}^{\infty i+d} \hat{H}_{0M}(s - s' \mid \boldsymbol{R}) \times \left(\frac{s'}{c_0^2}\right)^2 p_0(\boldsymbol{R}, s')\mathrm{d}s'/2\pi i. \tag{4.9c}$$

Next, we define the integral operator $\hat{M}_0$ by the inverse relation

$$\hat{M}_0 \equiv \hat{L}_0^{-1} = -\int_{\Lambda_M} \frac{e^{-s|\boldsymbol{R}-\boldsymbol{\rho}'|/c_0}}{4\pi|\boldsymbol{R}-\boldsymbol{\rho}'|}(\ )d\boldsymbol{\rho}'; \quad \left(=\int_{V_T} G_0(\boldsymbol{R},\boldsymbol{\xi}\mid s)(\ )d\boldsymbol{\xi}\right). \tag{4.10}$$

This last relation follows, since for the signal *source*, $\Lambda_M \to V_T$, $\boldsymbol{\rho}' \to \boldsymbol{\xi}$, $(M \to V_T)$. Clearly the green's function G_0 is $\hat{M}_0\delta(\boldsymbol{\rho}'-\boldsymbol{\rho}) = G_0(\boldsymbol{R},\boldsymbol{\rho}\mid s) = -\exp\{-s|\boldsymbol{R}-\boldsymbol{\rho}|/c_0\}/4\pi|\boldsymbol{R}-\boldsymbol{\rho}|$. Thus, from (4.2b) and (4.4) we get with the aid of (4.10)

$$\hat{M}_0^{-1}p_{0-\text{inc}} = -G_{0T}, \quad \text{or} \quad p_{0-\text{inc}} = -\hat{M}_0 G_{0T} = \int_{V_T} \frac{e^{-s|\boldsymbol{R}-\boldsymbol{\xi}|/c_0}}{4\pi|\boldsymbol{R}-\boldsymbol{\xi}|}[A_T S_{\text{in}}]d\boldsymbol{\xi}. \tag{4.11}$$

Accordingly, (4.8) becomes

$$\begin{aligned} p_{\text{scat}-M} &= \mathcal{F}_s^{-1}\{p_{0-\text{scat }M}\} = \mathcal{F}_s^{-1}\left\{\hat{M}_0\hat{g}'_{0M} \circledast \left(\frac{s^2}{c_0^2}p_{0-\text{inc}}\right)\right\} \\ &= \mathcal{F}_s^{-1}\left\{\hat{M}_0\hat{g}'_{0M} \circledast \left(\frac{-s^2}{c_0^2}\right)\hat{M}_0 G_{0T}\right\}, \end{aligned} \tag{4.12}$$

which at once specifies $p_{0-\text{scat}-M}$ $(=\mathcal{F}_s^{-1}\{p_{\text{scat}-M}\})$.

The *received* scattered field $X_M(t)$, or waveform following the receiving aperture $A_R(=\tilde{\mathcal{F}}_t\{\alpha_R(\boldsymbol{\eta},t)\})$, is obtained from

$$X_M(t) = \hat{h}_R * p_{\text{scat}-M} = \int_{V_R} d\boldsymbol{\eta}\int_{-\infty}^{\infty} \alpha_R(\boldsymbol{\eta}, t-\tau)p_{\text{scat}}(\boldsymbol{r}(\boldsymbol{\eta}),\tau)_M d\tau, \tag{4.13}$$

cf. eq. (17.6a), Ref. [2], with $\hat{h}_R *$ the space-time operator representing the action of the receiving aperture on the *incoming* field, here $p_{\text{scat}-M}$, eq. (4.12). The detailed geometrical structure of (4.13) applied to (4.12) is described in Ref. [1], Section 1 (Part I), for the general bistatic situation of Fig. 2.1.

From (4.12), and (4.2a) modified to omit the coherent component $dN_M^{(0)}$, it is at once evident that the scattered field and the received waveform [(4.13) in (4.12)] become respectively

$$p_{\text{scat}-M} = \sum_{k=1}^{\infty} p_{\text{scat}-M}^{(k)} = \sum_{k=1}^{\infty} \mathcal{F}_s^{-1}\left\{\hat{M}_0\hat{g}_{0M}^{(k)\prime} \circledast \left(\frac{-s^2}{c_0^2}M_0 G_{0T}\right)\right\}; \tag{4.14a}$$

$$\begin{aligned} \therefore X_M(t) &= \sum_{k=1}^{\infty} X_M^{(k)}(t) = \sum_{k=1}^{\infty} \hat{h}_R * \mathcal{F}_s^{-1}\left\{\hat{M}_0\hat{g}_{0M}^{(k)\prime} \circledast \left(\frac{-s^2}{c_0^2}\hat{M}_0 G_{0T}\right)\right\}; \\ &= \sum_{k=1}^{\infty}\int_{Z_M} U_M(t\mid \boldsymbol{Z})dN_M^{\prime(k)}(\boldsymbol{Z}), \end{aligned} \tag{4.14b}$$

so that the basic received waveform U_M is given by

$$U_M = \hat{h}_R * \mathscr{F}_s^{-1}\left\{\hat{M}_0\mathscr{F}_t(\hat{h}'_M) \circledast \left(\frac{-s^2}{c_0^2}\hat{M}_0 G_{0T}\right)\right\}, \tag{4.14c}$$

cf. eq. (6), Part I, Ref. [1], et seq. for the practically very important geometrical details. The scattered field and received waveform X_M are simple linear sums of independent components, cf. eq. (3.1), representing various orders of (self-) coupled scattering from the independent classes of inhomogeneities which are activated at any given instant, cf. fig. (3.1). Similarly, the moments of $p_{\text{scat}-M}$ and $X_M(t)$ are accordingly *linear functionals* (L) *of the moments of* $dN^{(k)}$ $(k \geq 1)$. Thus, the first-degree, first-order moment, $\langle p_{\text{scat}-M}\rangle = \langle X_M\rangle = L(\Sigma_1^\infty \langle dN^{(k)}\rangle) = 0$, by (3.2a); a typical second-order (second-degree) moment is

$$\langle p_{\text{scat}}(\boldsymbol{R}_1, t_1)_M p_{\text{scat}}(\boldsymbol{R}_2, t_2)_M\rangle = L(\langle dN_1^{(1)} dN_2^{(1)}\rangle + \langle dN_1^{(2)} dN_2^{(2)}\rangle) \quad (\neq 0),$$

cf. (3.7c), and so on.

An important consequence of this *linear* functional relation between the $dN^{(k)}$ and the corresponding component of the field or received waveform, cf. (4.14), is that the degree $(n \geq 1)$ of the moment of X_M, or $p_{\text{scat}-M}$, which is to be determined, contains no higher order components of $dN^{(k)}$ than $k = n$, cf. (3.6). Thus, a second degree (first- or second-order) moment of X_M is determined solely by inhomogeneities of order k (≤ 2), e.g. "specular point" scatterers $(k = 1)$ and (independent) continuum scattering of order $(k = 2)$. For possible contributions from higher order scattering elements $(n \geq 3)$, we need to consider $\langle X_M^n(t)\rangle$, for example, cf. Section VI, Ref. [1] (Part I). The simplicity of these results resides in the fact that there is no interaction between the scattered field and the scatterers in these models, cf. (4.6a, b). At the present level of experimental resolution these models do indeed appear to be valid for underwater environments. Reference [1] provides an extensive analysis of this important area of application. [For the details of the solutions of (4.14) and the associated moments, etc., of $X_M(t)$, see Part I (Ref. [1]). For various specific applications to underwater scattering problems, see Part II, Ref. [1].][9]

Cases III, IV: *Interacting Scattered Field and Scatterers*: *eqs.* (4.6c, d, e):

However, in many other instances we cannot ignore the coupling between the scattering element and the composite scattered field, so that a Langevin equation like (4.5), or (4.6c), must be invoked, instead of (4.6a, b). Using the integral operator (4.10), and (4.2a), with $\hat{g}'_{0M} = \mathscr{F}_t\{\hat{g}'_M\}$, we find that (4.5) becomes

$$\left[1-\hat{M}_0\hat{g}'_{0M}\circledast\frac{s^2}{c_0^2}\right]p_{0-\text{scat}-M}=\hat{M}_0\hat{g}'_{0M}\circledast\left(\frac{s^2}{c_0^2}p_{0-\text{inc}}\right), \tag{4.15}$$

which yields, formally, on multiplying by the inverse operator $[1-\hat{M}_0\hat{g}'_{0M}*s^2/c_0^2]^{-1}$ and expanding the *perturbation-theoretical series* [9], [10]:

$$p_{0-\text{scat}-M}=\sum_{q=0}^{\infty}\left(\hat{M}_0\hat{g}'_{0M}\circledast\frac{s^2}{c_0^2}\right)^q\hat{M}_0\hat{g}'_{0M}\circledast\left(\frac{s^2}{c_0^2}p_{0-\text{inc}}\right), \tag{4.16}$$

(where as before, cf. (4.11), (4.12), the operators act in order successively to the right). [For the 1st-Born approximation, all terms $q\geq 1$ in (4.16) are dropped, and one gets (4.12) again.] Equation (4.16) is a generalization of Tatarskii [9] and others relations [10].

Now $p_{\text{scat}-M}$ is no longer a linear functional of the $\mathrm{d}N'_M$ $(=\Sigma_1^\infty \mathrm{d}N_M^{(k)})$, since $\hat{g}'_{0M}$ appears to all orders $(q\geq 0)$. For $q\geq 1$ we must also omit the $(k=1)$-component, $\mathrm{d}N_M^{(1)}$, since there can be no fully independent scattering elements when interacting with the composite scattered field itself. All orders of moments of the $\mathrm{d}N'$ now appear in even the lowest order moment of $p_{\text{scat}-M}$, e.g. the mean, $\langle p_{\text{scat}-M}\rangle$, which is generally nonvanishing [9], [10]. For example, with the "rules" of Section 3 above, we see that for $\langle p_{\text{scat}-M}\rangle$

$$\begin{aligned}
&q=0: && \langle \mathrm{d}N'\rangle=0, \quad \text{as before;}\\
&q=1: && \langle \mathrm{d}N_1^{(2)}\mathrm{d}N_2^{(2)}\rangle\text{: } k=2 \text{ only;} \quad k=0,1 \text{ terms omitted;}\\
&q=2: && k=3 \text{ only: } \langle \mathrm{d}N_1'^{(3)}\mathrm{d}N_2'^{(3)}\mathrm{d}N_3'^{(3)}\rangle\\
&q=3: && k=(2,4)\text{: } \langle \mathrm{d}N_i'^{(2)}\mathrm{d}N_j'^{(2)}\rangle\langle \mathrm{d}N'^{(2)}\mathrm{d}N'^{(2)}\rangle, \quad \text{all combinations}\\
& && \langle \mathrm{d}N_1'^{(4)}\mathrm{d}N_2'^{(4)}\mathrm{d}N_3'^{(4)}\mathrm{d}N_4'^{(4)}\rangle\\
&q=4: && k=(2,3,4,6), \quad \text{in appropriate combinations, etc.}
\end{aligned} \tag{4.17}$$

Similar expansions for the covariance $\langle p_{\text{scat}-M1}p_{\text{scat}-M2}\rangle$ of the scattered field, and for higher order moments, may be developed.[8] In all instances, these may be expressed in terms of appropriate Feynman diagrams [9, 10], which can often be manipulated and/or approximated so as to permit the summation of the operator series (4.16). When this is possible for the mean, $\langle p_{\text{scat}-M}\rangle$, from (4.16), the resulting relation is known as a form of *Dyson's equation* [9, 10]. The corresponding relation for the covariance

[8] The analysis is currently being carried out, by the author and others, for a variety of statistical models, which include moving scatterers and intermode interactions pertinent to the underwater régime.

$\langle p_{\text{scat}-M1} p_{\text{scat}-M2}\rangle$ is a form of the *Bethe–Salpeter equation* [9], [10]. (The details of these developments are outside the scope of the present paper[8].)

For intermodal interactions (4.6d, e) the various Langevin equations become, respectively

$$[(B,S)\otimes V]:\ p_{0\text{-scat}-V|S\otimes V} = \sum_{q=0}^{\infty}\left(\hat{M}_0\hat{q}'_{0V}\circledast\frac{s^2}{c_0^2}\right)^q \hat{M}_0\hat{g}'_{0M'}\circledast\left(\frac{s^2}{c_0^2}\hat{M}_0\hat{g}'_{0M'}\right)$$
$$\circledast\left(\frac{s^2}{c_0^2}p_{0\text{-inc}}\right),\qquad M' = S, B. \tag{4.18}$$

With *weak* intermodal interactions, we have the $q=0$ term only, e.g.

$$[M'\otimes V]:\ p_{0\text{-scat}-V|M'\otimes V} \doteq \hat{M}_0\hat{g}'_{0V}\circledast\left(\frac{s^2}{c_0^2}\hat{M}_0\hat{g}'_{0M'}\right)\circledast\left(\frac{s^2}{c_0^2}p_{0\text{-inc}}\right). \tag{4.18a}$$

From (4.18) [and (3.15)] it is at once evident that the mean $\langle p_{0\text{-scat}-V|S\otimes V}\rangle$ vanishes, since $\langle(\hat{g}'_{0V})^q\hat{g}'_{0M'}\rangle = \langle(\hat{g}'_{0V})^q\rangle\langle\hat{g}'_{0M'}\rangle = 0$, $\langle\hat{g}'_{0M'}\rangle = 0$. Thus, there is no Dyson's equation here, but second- and higher-order moments of the $S\otimes V$-interaction field do not vanish, so that we may expect a form of the Bethe–Salpeter equation [9], [10] here also, although its structure will necessarily be different from that associated with eq. (4.16). Similarly, we get for (4.6a), the $V\otimes S$ (or B) interaction,

$$[V\otimes M]:\ p_{0\text{-scat}\,M|V\otimes M} = \hat{M}_0\hat{g}'_{0M}\circledast\frac{s^2}{c_0^2}\sum_{q=0}^{\infty}\left(\hat{M}_0\hat{g}'_{0V}\circledast\frac{s^2}{c_0^2}\right)^q\hat{M}_0\hat{g}'_{0V}\circledast\left(\frac{s^2}{c_0^2}p_{0\text{-inc}}\right), \tag{4.19}$$

where we have used (4.16) with $M - V$. The weak-interaction case now becomes

$$p_{0\text{-scat}\,M|V\otimes M} \doteq \hat{M}_0\hat{g}'_{0M}\circledast\frac{s^2}{c_0^2}\left(\hat{M}_0\hat{g}'_{0V}\circledast\left(\frac{s^2}{c_0^2}p_{0\text{-inc}}\right)\right),\qquad M = S, B. \tag{4.19a}$$

Again there is no Dyson equation, either for (4.19) or (4.19a), but higher-order moments in both instances do exist, when once more we apply the "rules" outlined in Section 3 preceeding.[8]

Finally, we remark for these generally strongly interacting Cases (III, IV), that it is no longer strictly possible to resolve the individual, contributing orders of inhomogeneity ($k \geq 1$) in the observed data, as in Cases I, II, cf. comments (4.14c)ff. Indirect methods of inferring the contributions ($k \geq 1$) must therefore be developed, in such instances.

5. An example: acoustic scattering from the random moving ocean surface

An important practical situation here is the application of underwater acoustics to such problems as communication and target detection, where interaction with the random moving surface of the ocean can strongly influence the transmission and reception of the desired signals. Our example is taken from Ref. [1] (Part II): we shall consider briefly the covariance of the received (underwater acoustic) scatter return from the ocean surface with nonmoving source and receiving platforms and an isovelocity ocean ($\nabla c_0 = 0$), as a concise illustration of the application of our general scattering approach [Sections 2–4], which also displays various elements of the techniques of multivariate analysis.

The governing Langevin equation here is (4.6a); [we assume negligible intermode coupling.] The desired received waveform is $X_S(t)$, given by (4.13), with (4.12), and the detailed geometry of the various operators is given in Section **IC**, Part I (Ref. [1]). Application of the BPSM, (2.2), to (4.14), following the results of Section 3 cf. eqs. (47–51), Part I, Ref. [1], we can write for the covariance of the received incoherent surface scatter components ($k = 1, 2$), cf. eq. (25), Part II, Ref. [1], the general relation:

$$K_{X-\langle X\rangle}(t_1, t_2)_S =$$

$$\frac{A_0^2}{2}\,\mathrm{Re}\left\{e^{i\omega_0\tau}\iint_{\Lambda_S^2} \hat{u}_0(t_1 - T_{01})^*\hat{u}_0(t_2 - T_{02})\, g_{S2}^{(2)} e^{i\omega_0(t_{01}-T_{02})}\,\Delta F_{\zeta\dot{\zeta}}\,\hat{K}_{0S}^{(12)}\,d\boldsymbol{\lambda}_1 d\boldsymbol{\lambda}_2\right\},$$

$$\tau = t_2 - t_1, \qquad (5.1)$$

where $K_{X-\langle X\rangle}$ vanishes for all $t_{1,2} - T_{01,2}$ outside ΔT_S, the input signal duration. Here $T_0 = T_0(\boldsymbol{\lambda})$ is the path travel time from $O_T \to P(\boldsymbol{\lambda})$ on $S \to O_R$, cf. Fig. (2.1); $A_0\hat{u}_0 = \hat{S}_0$ = complex signal envelope; ω_0 ($= 2\pi f_0$) = carrier angular frequency; [for these narrowband signals $S_{in} = \mathrm{Re}(\hat{S}_0 e^{i\omega_0 t})$, cf. (4.2b)]. In addition, we have

(i) $g_{S2}^{(2)} = g_{S2}^{(2)}(\boldsymbol{\lambda}_1, \boldsymbol{\lambda}_2 \mid f_0) = (\mathscr{A}_R\mathscr{A}_T)^*_{\boldsymbol{\lambda}_1}(\mathscr{A}_R\mathscr{A}_T)_{\boldsymbol{\lambda}_2}/(4\pi)^4 c^2 \lambda_1\lambda_2 R_R(\boldsymbol{\lambda}_1)\boldsymbol{R}_R(\boldsymbol{\lambda}_2)$, (5.2)

which is the second-order "geometrical" factor involving the complex beam patterns ($\mathscr{A}_T, \mathscr{A}_R$) of source and receiver, as well as spreading loss in the course of propagation from $O_T \to P(\boldsymbol{\lambda}) \to O_R$; (a far-field condition is assumed).

The functional multivariate character of $K_{X-\langle X\rangle}$ appears in the *interaction covariance*

(ii) $\Delta F_{\zeta\dot{\zeta}} \equiv F_{2-\zeta\dot{\zeta}} - (F^*_{1-\zeta\dot{\zeta}})_1(F_{2-\zeta\dot{\zeta}})_2$; (5.3a) here F_1, F_2 are the joint first- and second-order characteristic functions (c.f.'s) of the surface elevation $\zeta(\boldsymbol{r}, t)$ (about the mean, $\langle\zeta\rangle = 0$) and vertical speed $\dot{\zeta}(\boldsymbol{r}, t)$, ($*$ denotes the complex conjugate), specialized to the appropriate surface geometry and interface doppler[9], e.g.

$$F_{1-\zeta\dot{\zeta}} = \langle\exp[-ik_0\hat{b}(\zeta + \dot{\zeta}\Delta t_1)]\rangle_{\zeta,\dot{\zeta}}, \qquad \Delta t_1 \equiv t_1 - T_{01} \tag{5.3b}$$

$$\hat{b} \equiv \cos\theta_T + \cos\theta_R; \qquad k_0 = \omega_0/c_0;$$

$$F_{2-\zeta\dot{\zeta}} = \langle\exp\{ik_0[\hat{b}_1\zeta_1 - \hat{b}_2\zeta_2 + \hat{b}_1\Delta t_1\dot{\zeta}_1 - \hat{b}_2\Delta t_2\dot{\zeta}_2]\}\rangle_{\zeta_1\ldots\dot{\zeta}_2} \tag{5.3c}$$

$$\hat{b}_1 \equiv \cos\theta_{T1} + \cos\theta_{R1}, \qquad \text{etc.}$$

Gaussian statistics provide a good description of the random part of the sea surface in the deep ocean [cf. Appendix A.3, Ref. [1]], so that (5.3b), for example becomes explicitly[9]

$$F_{1-\zeta\dot{\zeta}} = \exp\{-k_0^2\hat{b}^2[\sigma_\zeta^2 + \sigma_{\dot{\zeta}}^2(\Delta t)^2]\}, \qquad \sigma_\zeta^2 = \langle\zeta^2\rangle, \qquad \sigma_{\dot{\zeta}}^2 = \langle\dot{\zeta}^2\rangle. \tag{5.3d}$$

In a similar fashion, $F_{2-\zeta\dot{\zeta}}$ is a four-variable multivariate c.f.-multivariable gaussian in $\zeta(\boldsymbol{r}_1, t_1)$, $\dot{\zeta}(\boldsymbol{r}_1, t_1)$, $\zeta(\boldsymbol{r}_2, t_2)$, $\dot{\zeta}(\boldsymbol{r}_2, t_2)$ (for the random ocean surface model used here, cf. eq. 29, Ref. [1], Part II).[9]

Finally, for the BPSM, eq. (2.2), assumed in the present analysis, $\hat{K}_{0S}^{(12)}(\boldsymbol{\lambda}_1, \boldsymbol{\lambda}_2)$ in (5.1) is the *covariance of the scattering cross-section*, defined by

$$\hat{K}_{0S}^{(12)}(\boldsymbol{\lambda}_1, \boldsymbol{\lambda}_2) = \langle\gamma_0(\boldsymbol{\lambda}_1)\gamma_0(\boldsymbol{\lambda}_2)\rangle_{\gamma_0}\langle \mathrm{d}N(\boldsymbol{Z}_1)_S \mathrm{d}N(\boldsymbol{Z}_2)_S\rangle, \tag{5.4}$$

which is seen to be weighted by a second-order average depending on the number of actual scattering elements ($\sim \mathrm{d}N_0$) in the surface domain A_S, commonly ensonofied, and "viewed" by the receiving beam ($\mathscr{A}_R$). From (3.11) and the subsequent discussion in Section 3, (5.4) becomes

$$\begin{aligned}\hat{K}_{0S}^{(12)}(\boldsymbol{\lambda}_1, \boldsymbol{\lambda}_2) &= \langle\gamma_0(\boldsymbol{\lambda}_1)^2\rangle J_{S1}\sigma_S^{(1)}(\boldsymbol{\lambda}_1)\delta(\boldsymbol{\lambda}_2 - \boldsymbol{\lambda}_1) + \langle\gamma_0(\boldsymbol{\lambda}_1)\gamma_0(\boldsymbol{\lambda}_2)\rangle \times \\ &\quad \times J_{S1}J_{S2}\sigma_S^{(2)}(\boldsymbol{\lambda}_1)\sigma_S^{(2)}(\boldsymbol{\lambda}_2)\rho_S^{(2)}(\boldsymbol{\lambda}_1, \boldsymbol{\lambda}_2) \\ &= \hat{K}_{0S}^{(1)}\delta(\boldsymbol{\lambda}_2 - \boldsymbol{\lambda}_1) + K_{0S}^{(2)}(\boldsymbol{\lambda}_1, \boldsymbol{\lambda}_2),\end{aligned} \tag{5.5}$$

with $J_{S1} = c^2\lambda_1$, etc., the surface jacobian here, cf. Fig. 2.1. The first term of (5.5) embodies the contribution of the "specular-point" [5], independent

[9] Here $\Delta t_{1,2}$ is small vis-à-vis the correlation time of $\zeta(\boldsymbol{r}, t)$. An "exact" treatment replaces $\hat{b}(\zeta + \dot{\zeta}\Delta t)$ by $\hat{b}\zeta(\boldsymbol{\lambda}, t - R_{R/c})$. See Appendix A.2, Ref. 1. Thus, the exact treatment $O(c^{-1})$ Ref. [1] requires $\dot{\zeta}$, $\sigma_{\dot{\zeta}}$, etc. $= O$ in the text (Ref. [1], Part II) generally.

scatterers ($k = 1$), while the second represents the effects of coupled or continuum scattering elements ($k = 2$). The former are the poisson scatterers of the FOM theory,[10] which are essentially independent of frequency and constitute the dominant scattering mechanism at high frequencies [$\doteq 0$ ($f_0 > 5$ kHz)].[11] The latter, which fall off $O(f_0^{-1} - f_0^{-2})$ usually, stem from the relatively smooth continuous component of the random surface [as distinct from the isolated "specular-points" or facets of type ($k = 1$)], and produce the significant scattering at the lower frequencies [$\doteq 0$ ($f_0 < 1$ kHz)].[11] The cross-section γ_0 can be expressed in more detail: $\gamma_0 = \mathcal{R}(\boldsymbol{\lambda}_1 | f_0)\hat{\gamma}_0(\boldsymbol{\lambda}_1)$, where $\mathcal{R}$ (with $|\mathcal{R}| \leq 1$) is the plane-wave reflection coefficient at $\boldsymbol{\lambda}_1$, and is a weak function of frequency f_0. [For water → air, water → rigid bottom, $\mathcal{R} \doteq -1, +1$ respectively; furthermore, $|\gamma_0| \leq 1$, since the scattering elements do not reradiate more energy than they receive.] Shadowing effects, likewise, are embodied in the mean physical densities $\sigma_S^{(1)}$, $\sigma_S^{(2)}$, of the scatterers on the average interacting with the incident radiation.

Many important special cases of the rather general result (5.1) can be derived from it, e.g. "weak-scattering", "strong-scattering"; far-field, plane-wave irradiation of the surface, surface wave spectra, scatter intensities, various geometries, etc. The details are outside the scope of this paper, and are treated specifically in Ref. [1] (Part II).[9]

6. Concluding remarks

In the preceeding sections we have described in moderate detail the principal elements of a new approach to the treatment of ("classical") scattering problems involving random media and their (random) boundaries. The chief feature of the method is the use of a hierarchy of independent random "point-" processes with which to develop the scattering elements $d\Lambda_M$ into various distinct and independent orders ($k \geq 0$) of inhomogeneity [Sections 2,3]. This technique has the advantage, among

[10] FOM ≡ "Faure–Ol'shevskii–Middleton"; after P. Faure, Theoretical models of reverberation noise, *J. Acoust. Soc. Am.* **36,** 259–268, Feb. (1964); V.V. Ol'shevskii, *Characteristics of Sea Reverberation, Consultant's Bureau* (Plenum Press, New York) 1967; original Russian edition, Moscow, 1966; and Middleton, Ref. [2], 1967, 1972.

[11] These frequency estimates are rather loose; further experiments are required to determine the frequency "limits" more precisely, limits which depend in part on the relative "populations" (i.e. sizes of $\hat{K}_{0S}^{(1)}$ vs. $\hat{K}_{0S}^{(2)}$) of ($k = 1$) and ($k = 2$) classes of surface inhomogeneity, a matter which must be experimentally observed, in order to "calibrate" the theory to the particular sea surface under study. See Sections V, VII, VIII, of Ref. [1]. See the comments in Section 6 concerning "phenomenologicality".

others, of enabling us to treat surface (i.e., interface) scattering in terms of "point" or localized differential scattering elements, as in the case of volume scatter.

The result is that the distributed boundary conditions of the classical surface treatments, (as embodied in the usual formulation of the Helmholtz integral equation, for example) are transformed into equivalent relations expressed now as statistics of the point-processes used in the development of the scattering elements themselves. The detailed statistical-physical properties of these point-processes are in turn governed by those of the individual component inhomogeneities associated with each scattering region $d\Lambda_M$. In addition, the pertinent physics of the radiation and reradiation processes are introduced through an appropriate Langevin equation.

Thus, unlike earlier approaches based solely on a poisson model[10, 11] ($k = 1$) (now predicted, and in part experimentally observed to be valid in a strong first approximation at sufficiently high frequencies [Section V, Ref. [1] (Part II)]) we here combine the "classical" treatment (based on the ($k = 2$) class of inhomogeneities for second-order moments and appropriate for the lower frequencies) with the newer poissonian models, *in what is no longer a quasi-phenomenological approach* [Ref. [2], Part I, for example], but one which is capable of validation at the same physical level of modeling as the conventional "classical" approaches, which, in any case, do not yield the poisson component [cf. Ref. [1], Introduction, ***B***, ***C*** (Part I); ***A*** (Part II), also Sections V, VII, Part II].

Several varieties of Langevin equation are considered in Section 4, including cases where the typical scattering element interacts with the (total) scattered field. In such instances the component orders (k) of inhomogeneity are not resolveable in the observed scattered field, unless the interaction between scatterers and scattered field is weak (or effectively nonexistent, as is that between acoustic surface and volume scatter at the present level of measurement). Our present analysis is a generalization, also, in that the doppler structures of the moving scattering elements are included. The point-process formalism also has the further advantage of clarifying the various orders of interaction which appear in the general perturbation-theoretical series for the scattered field [cf. (4.16)], and which form the elements of the Feynman diagrams by which one seeks to sum these series, to obtain the equivalent Dyson's and Bethe-Salpeter equations, respectively for the mean and covariance of the scattered field.

Finally, as an example of the new approach, and as a topic of multivariate analysis, we have discussed an application to underwater acoustic scattering from the moving ocean surface (Section 5). The new approach, in any case, is applicable to scattering problems generally, and is readily extended to vector fields.[8]

References

[1] Middleton, D. (1974). Characterization of active underwater acoustic channels. I. A new formulation for random media and interfaces; II. Scattering in an isovelocity ocean. Technical Report ARL–TR–74–61, December, Applied Research Laboratories, the University of Texas at Austin, P.O.Box 8029, Austin, Texas 78712. See "References", Parts I, II.

[2] Middleton, D. (1967). A statistical theory of reverberation and similar first-order scattered fields, III, IV, *IEEE Trans. Information Theory* (PGIT), IT–18, No. 1, Jan., 1972, pp. 35–90; vide Sec. 18.2, Parts I, II, ibid, IT–**13**, July, pp. 372–414.

[3] Middleton, D. (1970). Multidimensional detection and extraction of signals in random media, *Proc. IEEE* **58**, No. 5, May, pp. 646–706, Section IIIB.

[4] Bartlett, M.S. (1955). *An Introduction to Stochastic Processes*, Camb. Univ. Press, Section 3.42. See also "A note on the mathematical theory of correlated random points", by P.I. Kuznetzov and R.L. Stratonovich, Article No. 6, 101–115 in *Nonlinear Transformations of Stochastic Processes*, Ed. by P.I. Kuznetzov, R.L. Stratonovich, and V.I. Tikhonov (trans. and ed. by J. Wise and D.C. Cooper), Pergamon Press (New York), 1965. Our present treatment is an extension and generalization, with the new feature of the hierarchical development in independent processes, cf. (3.1) et seq. and its specific application in scattering situations.

[5] Barrick, D.E. (1968). Rough surface scattering based on the specular point theory, IEEE Trans. Antennas and Propagation, Vol. *AP–16*, No. 4, July, pp. 449–454. See also Ref. 80, in Ref. [1], Part I.

[6] Middleton, D. (1960). *Introduction to Statistical Communication Theory*. McGraw-Hill, New York.

[7] Neubert, J.A., and Lumley, J.L. (1970). Derivation of the stochastic Helmholtz equation for sound propagation in a turbulent fluid. *J. Acoust. Soc. Am.* **18**, May, pp. 1212–1218.

[8] MacLachlan, N.W. (1939). *Complex Varables and Operational Calculus*, Camb. Univ. Press, Ch. 4; also, Ref. [2], Section 13.

[9] Tatarskii, V.I. (1971). *The Effects of the Turbulent Atmosphere on Wave Propagation*, trans. Israel Program for Scientific Translations; U.S. Department of Commerce, NTIS, Springfield, Va. 22151, Vol. TT–68–50464. See Eq. (5), Chapter 5, et seq.

[10] Dence, D., and Spence, J.E. (1973). *Wave Propagation in Random Anisotropic Media*, pp. 121–181, in: *Probabilistic Methods in Applied Mathematics* Vol. 3, Ed. by Baruchta-Reid, Academic Press (New York).

P.R. Krishnaiah, ed., *Mutivariate Analysis–IV*
© North-Holland Publishing Company (1977) 431–441

SOME SYSTEM APPROACHES TO WATER RESOURCES PROBLEMS III. OPTIMAL CONTROL OF DAM STORAGE

Yu. A. ROZANOV
I.I.A.S.A., Laxenburg, Austria

Some stochastic aspects of dam storage theory are considered in this paper. In particular optimal control based on some reliable lower estimates of unknown (uncertain) system parameters with the corresponding operational program is developed. Also, statistical equilibrium in dam storage (random) processes are analyzed and general conditions for such a phenomenon are established.

1.

A water reservoir operation depends on a proper time period $(t_0, t_0 + T)$ and is usually based on a so-called operational graph. This can be represented by a monotone function $z = z(x)$ which shows the amount of water to be released during the considered period $(t_0, t_0 + T)$, if the total volume of available water will be x (see Fig. 1).

x = AVAILABLE WATER (INITIAL VOLUME + INFLOW);
w = WATER DEMAND;
z = WATER RELEASE;
r = (OPERATIONAL) RESERVOIR CAPACITY.

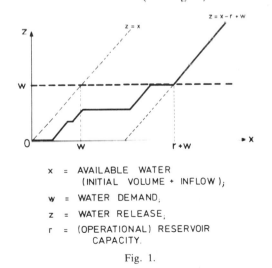

Fig. 1.

Of course, one does not release the corresponding amount of water z all at once; its distribution over time depends, in particular, on water demands per time unit and channel capacity. If these river basin characteristics are constant during the considered time period $(t_0, t_0 \pm T)$, then a local operation policy may be of the following type: The amount of water Δz_t per time unit Δt released with constant discharge at the current time interval $(t, t+\Delta t)$ is

$$\Delta z_t = \min\{c, z(x_t) - z_t\},$$

where

x_t = water available during the period (t_0, t);

z_t = water already released from the reservoir during the period (t_0, t);

c = operational constant limited by channel capacity.

Generally, a reservoir is designed to meet water demands as well as to prevent floods. Thus the operational graph must be chosen according to proper multiobjective decision making. Water demands and flood possibility obviously depend on time, so one has to determine the operational graph $z_k = z_k(x_k)$, for each time period $(t_k, t_k + T)$, $t_k = t_0 + kT$; $k = 0, 1, \ldots$.

Let us set $T = 1$ and let y_k be the reservoir volume at the beginning of k-period and ξ_k be the total inflow

$$\xi_k = \int_{t_k}^{t_k+T} \dot{\xi}_t \, dt,$$

where $\dot{\xi}_t$ is the inflow per time usually identified with the so-called hydrograph.

We have

$$x_k = y_k + \xi_k, \qquad y_{k+1} = x_k - z_k, \qquad k = 0, 1, \ldots. \tag{1}$$

Suppose for each period $(t_k, t_k + T)$ we are given a total water demand w_k, and the loss function $f_k(z_k)$ reflects loss in the case of water deficit $w_k - z_k$ (see Fig. 2). The problem is to determine optimal reservoir

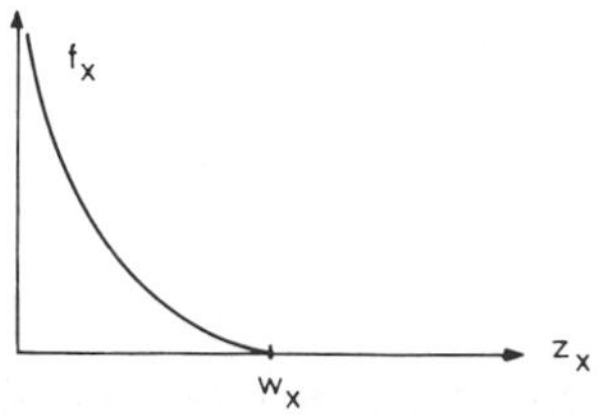

Fig. 2.

operation taking into account not only the current water demands, but also possible future water deficits and floods.

Because of a water channel's limited capacity, a flood is usually connected with a high peak hydrograph on a comparatively small base-time Δ (see Fig. 3). Flood damage seems to be a function of the corresponding high peak hydrograph; but by using approximations shown in Fig. 3, one may estimate the damage by a proper function $f(\eta, \delta)$ of two parameters η and δ.

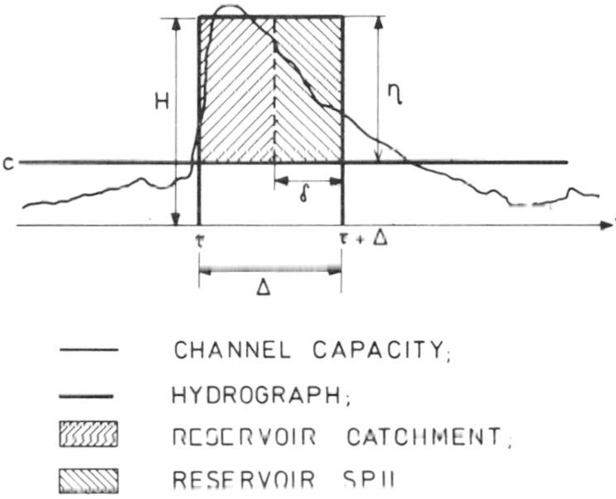

Fig. 3.

Flood damage, as we understand it, is usually incomparably high with respect to the water deficit loss. Thus it seems reasonable to assume that a proper reservoir capacity for a flood catchment can be established disregarding water demands. Let r_k be the corresponding reservoir operational volume during the operational time period $(t_k, t_k + T)$. So if R is the absolute reservoir volume, then the remaining $R - r_k$ represents the "flood catchment" capacity which helps to reduce the damage cost from $f_k(\eta, \Delta)$ to $f_k(\eta, \delta)$, where

$$\eta = H - c, \qquad \delta = \Delta - \frac{R - r}{H - c}$$

(see Fig. 3). The proper operational volume $r = r_k$ may be determined, for example, from the condition that the damage cost may exceed some upper crucial level C only with a small probability ε:

$$\mathbf{P}\{f_k(\eta, \delta) > C\} \leq \varepsilon.$$

With regard to water demands, one can be careful about a water deficit only up to the first "wet" period τ when excess water (not necessarily a flood) enters the reservoir. So the problem is to determine operation functions $z_k = z_k(x_k)$ in such a way so as to minimize the total water deficit loss

$$\sum_{k=0}^{\tau-1} f_k(z_k) \rightarrow \min. \tag{2}$$

One may immediately notice the main difficulty that arises here: the water deficit loss defined above depends on the reservoir inflow $\xi_0, \xi_1, \ldots,$ which is uncertain and must be treated as a random process.

One of the possible approaches to such minimization problems traditionally offered by optimal control theory is minimization of expected loss:

$$\mathbf{E} \sum_{k=0}^{\tau-1} f_k(x_k) \rightarrow \min,$$

over all possible control functions

$$z_k = z_k(x_k, \omega)$$

which in our case depend not only on available water $x = x_k$, but also on river basin data "ω" up to the current time period.

The corresponding optimization technique involves (conditional) probabilities distribution of the random variables $\xi_0, \xi_1, \ldots$; this can hardly be used in practice, as there are usually no reliable data available to ensure that this or that sophisticated probabilistic model fits the reality.

We will next consider a rather simple probabilistic model of the inflow process $\xi_0, \xi_1, \ldots,$ and a proper reservoir control which is optimal in the sense of some kind of minimax principle; this seems to be reliable from a practical point of view.

2.

The reservoir inflow process

$$\xi_0, \xi_1, \ldots,$$

arises as a result of basic river flow, rainfall in different river basin areas, etc.; a mechanism of the random variables $\xi_0, \xi_1, \ldots,$ formation is rather complicated to discuss here in detail.

Let us call a series of the considered periods

$$(t_k, t_k + T); \qquad k = 0, 1, \ldots, \tau - 1$$

up to the first "wet" period τ, $\tau > 0$, regular season, and a series of the inflow variables

$$\xi_k\,; \qquad k = 0, 1, \ldots, \tau - 1,$$

regular process. It seems reasonable to assume that the inflow variables $\xi_0, \ldots, \xi_n$ associated with the regular season are independent (under condition $\tau > n$) on the beginning of the future "wet" period τ; furthermore, probabilities distributions of $\xi_0, \ldots, \xi_n$ with respect to the condition $\tau > n + k$ are the same for all future periods $n + k$; $k = 0, 1, \ldots$.

During the regular season there are comparitively minor random fluctuations in the regular inflow process $\xi_0, \xi_1, \ldots$, mainly due to such random events as rainfall in different river basin areas. Water requires some transient time to flow from an area to the reservoir. We believe that if such a transient time for any area is comparatively small with respect to the chosen reservoir period T, then, for purposes of possible future water deficit estimation and sensible water supply during the regular season, one may treat the regular inflows $\xi_0, \xi_1, \ldots$, as independent (random) variables.

Actually, our minimization problem (see (2)) concerns an optimal water supply during the regular time interval $(t_0, t_0 + T)$; the main difficulty is estimating possible future water deficits up to the wet τ-period.

Let us set

$$P_n = \mathbf{P}\{\tau > n\,|_{0<\tau\leq N}\}; \qquad n = 0, 1, \ldots, .$$

For example, one may assume (not unreasonably) that the "waiting time" for the wet period has a probabilities distribution of the exponential type:

$$P_0 = \mathbf{P}\{\tau > 0\} = 1,$$

$$P_n = \mathbf{P}\{\tau > n\} = q^n/1 - q^N, \qquad n = 1, \ldots, N,$$

where a parameter q may be interpreted as the probability of being regular for each of the considered (independent) time-periods $(t_k, t_k + T)$; a parameter N arises in a case when the wet period certainly occurs during the annual cycle (i.e., the melting of snow certainly occurs before the summer season, etc.).

Let us fix all inflow variables ξ_k in the regular part of the inflow process and consider the expected water deficit loss

$$\Phi(z, \xi) = \mathbf{E}\left\{\sum_{k=0}^{\tau-1} f_k(z_k)\,\bigg|_{\xi_0, \ldots, \xi_{\tau-1};\, 0<\tau\leq N}\right\} \tag{3}$$

as a function of the random process

$$\xi = \{\xi_0, \ldots, \xi_{N-1}\},$$

and the control parameters

$$z = \{z_1, \ldots, z_{N-1}\}.$$

The optimal control parameters

$$z_k^0 = z_k^0(x;\xi); \qquad k = 0, 1, \ldots, N-1$$

which minimize the loss function $\Phi(x, \xi)$:

$$\Phi(z^0, \xi) = \min_z \Phi(z, \xi)$$

can be easily determined by standard dynamic programming, i.e.

$$z_{N-1}^0 \colon f_{N-1}(z) \to \min_{0 \leq z \leq x} f_{N-1}(z) = \Phi_{N-1}(x),$$

$$\ldots\ldots\ldots\ldots\ldots\ldots$$

$$z_k^0 \colon f_k(z) + \frac{P_{k+1}}{P_k}\Phi_{k+1}(x - z + \xi_{k+1}) \to \tag{4}$$

$$\min_{0 \leq z \leq x}\left\{f_k(z) + \frac{P_{k+1}}{P_k}\Phi(x - z + \xi_{k+1})\right\} = \Phi_k(x),$$

$$k = N-2, \qquad N-3, \ldots, 0.$$

At each step of the program (4) one can use the following proposition concerning the minimization procedure

$$\sum_{i=1}^m F_i(z_i) \to \min \quad \text{subject to} \sum_{i=1}^m z_i \leq x \ (z_i \geqq 0)$$

for *convex* functions F_i.[1] Here parameter x may be interpreted as a resource distributed amongst consumers; let us suppose it is distributed in some units Δx.

Lemma. *If $z_i^0(x)$; $i = 1, \ldots, m$ is the optimal distribution with respect to the parameter x, then*

$$z_i^0(x + \Delta x) = z_i^0(x) + \delta_{ij}\Delta x; \qquad i = 1, \ldots, \tag{5}$$

where δ_{ij} is a Kronecker symbol and the corresponding preferable "j" is determined by the condition

[1] At each kth step of the program (4) we have $m = 2$, $z_1 = z$, $z_2 = x - z$, $F_1(z_1) = f_k(z)$, $F_2(z_2) = (P_{k+1}/P_k)\Phi_{k+1}(x - z + \xi_{k+1})$.

$$F_j(z_j^0+\Delta x)-F_j(z_j^0)=\min_{1\le i\le m} F_i(z_i^0+\Delta x)-F_i(z_i^0).$$

The function

$$F(x)=\min\sum_{i=1}^{m} F_i(x)$$

is convex and

$$F(x+\Delta x)-F(x)=F_j(z_j^0+\Delta x)-F_j(z_j^0);$$

the recurrent equation (5), lets us determine $z_i^0(x)$; $x=0,\Delta x, 2\Delta x,\dots,$. Note that this very simple minimization procedure is not valid in a case of non-convex functions. Suppose that we consider the water distribution problem with loss functions $F_1(z_1)$ and $F_2(z_2)$, which has arisen because of current and future water deficits. Further suppose that the future loss can be reduced only by a significant water supply $z_2>\Delta x$, but that the current loss F_1 becomes less even with a minor water supply Δx. Then, according to Equation (5), one must meet current water demands using *all* available water:

$$z_1^0(x)=x, \qquad z_2^0(x)=0$$

for any x. Obviously, this procedure is wrong in a case represented in Fig. 4, where

$$z_1^0(x)=0, \qquad z_2^0(x)=x$$

for $x=2\Delta x$.

It is worth noting that program (4) gives us the optimal control functions $z_k^0=z_k^0(x;\xi)$ which minimize *all* expected values

$$\mathbf{E}\left\{\sum_{j=n}^{\tau-1} f_k(z_k)\,\Bigg|_{x_n=x;\ \xi_{n+1},\dots,\xi_\tau,\ n<\tau\le N}\right\}=$$

$$=\frac{1}{P_n}\sum_{k=n}^{N-1} P_k f_k(z_k).$$

Thus if we set

$$z_k=z_k^0(x_k,\xi)$$

in the water balance Equation (1), then the corresponding reservoir process

$$(x_k, y_k, z_k); \qquad k=0,1,\dots,$$

will be optimal (i.e. water deficit loss will be minimal).

But one can not implement this optimal process because at each current n-period, the future inflow variables $\xi_{n+1}, \ldots,$ remain unknown.

Our suggestion is to substitute uncertain variables $\xi_0, \xi_1, \ldots,$ with some reliable *lower* estimates, $\underline{\xi}_0, \underline{\xi}_1, \ldots,$ such that

$$\mathbf{P}\{\xi_k \geqslant \underline{\xi}_k\} \sim 1 - \alpha, \tag{6}$$

where $1 - \alpha$ is the proper confidence level, $(0 < \alpha < 1)$:

The corresponding control functions $z_k - z_k^0(x, \underline{\xi})$; $k = 0, 1, \ldots,$ appear to be optimal with respect to some kind of minimax criterion when we consider that only $100 \cdot (1 - \alpha)$ per cent of possible inflows ξ_k satisfy the condition (6).

Let us consider an arbitrary reservoir control based on operation graphs:

$$z_k = z_k(x); \qquad k = 0, 1, \ldots .$$

We believe there is no sense in a control for which the water deficit loss $\Phi(z, \xi)$ can increase when the reservoir inflows are increasing. So let us consider the regular control in which the water deficit loss $\Phi(z, \xi)$ is a monotone decreasing function of each inflow variable ξ_k, $k = 0, 1, \ldots$.

The control suggested above, i.e.

$$z_k^0 = z_k^0(x, \xi); \qquad k = 0, 1, \ldots, \tag{7}$$

is regular for any $\xi = \{\xi_k\}$.

Indeed, according to the general Equation (5), all parameters

$$z_k^0(x, \xi), \qquad y_{k+1}^0 = x - z_k^0(x, \xi)$$

are monotone increasing functions of $x = x_k$ and, in a case where the inflow ξ_k is increasing, we deal with the increased variables

$$\begin{array}{lll}
x_k = y_k^0 + \xi_k & z_k^0 \ \text{ and } \ y_{k+1}^0, \\
x_{k+1} = y_{k+1}^0 + \xi_{k+1}, & z_{k+1}^0 \ \text{ and } \ y_{k+2}^0, \\
\vdots & \vdots \\
x_{N-1} = y_{N-1}^0 + \xi_{N-1}, & z_{N-1}^0.
\end{array}$$

Hence the water deficit loss

$$\Phi(z^0, \xi) = \sum_{n=0}^{N-1} P_n f_n(z_n^0)$$

will be reduced because each local loss $f_n(z_n^0)$ is a monotone decreasing function of the water supply z_n^0.

For any regular control $z = \{z_k(x)\}$ we have

$$\Phi^*(z) = \max_{\xi \geq \underline{\xi}} \Phi(z, \xi) = \Phi(z, \underline{\xi}),$$

and the reservoir control $z^0 = \{z_k^0(x, \underline{\xi})\}$ is optimal, in the sense that it gives the minimum of the maximum loss $\Phi^*(z)$ over all reservoir control policies $z = \{z_k(x)\}$,

$$\min_z \Phi^*(z) = \Phi^*(z^0).$$

Moreover, the pair $(z^0, \underline{\xi})$ is a saddle point in our reservoir game against nature with its strategy $\xi = \{\xi_k\}$:

$$\min_z \max_{\xi \geq \underline{\xi}} \Phi(z, \xi) = \max_{\xi \geq \underline{\xi}} \min_z \Phi(z, \xi) = \Phi(z^0, \underline{\xi}). \tag{8}$$

Indeed,

$$\min_z \Phi(z, \xi) \geq \Phi[z^0(\cdot, \xi), \xi]$$

and

$$\max_{\xi \geq \underline{\xi}} \min_z \Phi(z, \xi) \geq \max_{\xi \geq \underline{\xi}} \Phi[z^0(\cdot, \xi), \xi] = \Phi[z^0(\cdot, \underline{\xi})\underline{\xi}] = \Phi(z^0, \underline{\xi}).$$

3.

Most practical applications of stochastic reservoir models are based on the assumption that the random process

$$(x_k, y_k, z_k); \qquad k = 0, 1, \ldots,$$

in the reservoir system (see (1)) eventually reaches a so-called statistical equilibrium; this actually means that the probabilities distribution P_n of

$$(x_{k+n}, y_{k+n}, z_{k+n}); \qquad k = 0, 1, \ldots,$$

tends to some limit:

$$\lim_{N \to \infty} P_n = P, \tag{9}$$

and this limit probability distribution P is invariant with respect to the annual time shift transformation

$$(x_k, y_k, z_k) \to (x_{k+\Delta}, y_{k+\Delta}, z_{k+\Delta}),$$

where Δ means the entire year period; moreover, the frequency of any annual event A during a series of years N also tends toward the corresponding probability $P(A)$:

$$\lim_{N\to\infty} \frac{\nu_N(A)}{N} = P(A), \tag{10}$$

where $\nu_N(A)$ is the number of years in which the event A occurs.

Let us consider the arbitrary water release policy; the only assumption is that the current release $z_k = z_k(x)$ does not exceed the water demands w_k, if there is no water excess:

$$z_k \leqslant w_k \quad \text{if} \quad x_k \leqslant r_k + w_k$$

(remember that r_k is the upper operational reservoir volume — see Fig. 1).

We assume that the current water demands w_k, as well as the river basin inflows ξ_k, do not physically depend on the reservoir existence; further, the random process (ξ_k, w_k); $k = 0, 1, \ldots,$ can be considered as a part of the process (ξ_k, w_k), $-\infty < k < \infty$, which is stationary with respect to the annual time shift transformation. One may treat (ξ_k, w_k) as a component of general climatological process $\omega = \omega_t$, $-\infty < t < \infty$, (in the considered river basin) assuming that the annual time shift transformation does not change the probabilities distribution.

One can treat the operational upper level r_k in the same way, because it depends only on the actual reservoir capacity R and ω_t, $t \leqslant t_k$.

Naturally, we can expect statistical equilibrium (see (10)) only under some ergodicity conditions for the process $\omega = \omega_t$, $-\infty < t < \infty$, and under such conditions, the following result holds true: suppose that during a long range operation, the total reservoir inflow $\Sigma_{k=0}^n \xi_k$ sometimes becomes comparatively high with respect to the total water demands $\Sigma_{k=0}^n w_k$; suppose more precisely that the sequence

$$\eta_n = \sum_{k=0}^{n} (\xi_k - w_k); \qquad n = 0, 1, \ldots,$$

with non-zero probability may exceed the reservoir level R (or at least the operational capacity $r = r_n$; $n = 0, 1, \ldots,$) at some random time period $n = \tau$. Then the statistical equilibrium phenomenon holds true; furthermore, we find the ergodic process (x_k^*, y_k^*, z_k^*); $-\infty < k < \infty$, stationary concerning the annual cycle such that

$$(x_k, y_k, z_k) = (x_k^*, y_k^*, z_k^*), \qquad k \geqslant \tau. \tag{11}$$

The limit in (10) coincides with the probabilities distribution P of the process

$$(x_k^*, y_k^*, z_k^*); \qquad k = 0, 1, \ldots,$$

and

$$\mathrm{Var}\,(P_n - P) \leq 2\mathbf{P}\{\tau > n\}. \tag{12}$$

Note that in most interesting cases the τ distribution is of an exponential type so the convergence rate (accounting to (12)) is very high.

All of the results presented above can be obtained by obvious modification of the "imbedded stationary processes" methods developed in [4], where the specific z-shape reservoir policy was analyzed. The main idea is based on the phenomenon whereby all possible trajectories of the reservoir process will be at the same point (the reservoir will be full) at the considered τ-period, no matter what the initial reservoir conditions.

References

[1] Moran, P.A. (1964). *The Theory of Storage.* London, Metheuen.

[2] Prabhu, N.V. (1964). *Time-Dependent Results in Storage Theory.* London, Metheuen.

[3] Borovokov, (1972). *Probabilistic Processes in Queueing Theory.* Moscow, Nauka (Publishing Co.).

[4] Rozanov, Yu. A. (1974). "Some System Approaches to Water Resources Problems II: Statistical Equilibrium in Processes in Dam Storage." IIASA RR-75-4, Research Report, International Institute for Applied Systems Analysis, Laxenburg, Austria. (To appear in proceedings of the International Conference on Recent Trends in Statistics, Calcutta.)

PART V

CONTINGENCY TABLES, DIRECTIONAL DATA STATISTICS, AND PATTERN RECOGNITION

P.R. Krishnaiah, ed., *Multivariate Analysis–IV*

SOME APPLICATIONS OF A METHOD OF IDENTIFYING AN ELEMENT OF A LARGE MULTIDIMENSIONAL POPULATION

Herman CHERNOFF*
Massachusetts Institute of Technology, Cambridge, Massachusetts

1. Introduction

Galton dealt with the problem of classifying fingerprints so as to facilitate the storage of files which would permit easy identification of criminals through their fingerprints. This problem still poses technical difficulties because of the very large files that are maintained by government agencies such as the Federal Bureau of Investigation (FBI). Nevertheless there seems to be very little in the way of a theoretical framework from which this problem has been discussed. Some recent work on record linkage by Sunter and Felligi [9] represents a notable exception.

Two other problems of a similar nature are that of identifying a chemical compound from its mass spectrogram and that of identifying the medical record of an individual from possibly distorted information available. Of the three problems mentioned I shall discuss the chemical compound problem in most detail. The common elements of these problems are the following:

(1) There is a large file or library of records, each identifying an individual. For example, in the chemical problem the records are mass spectrograms and the individuals are chemical compounds.

(2) Information of a related sort becomes available on an individual *target* who may or may not be represented in the file. In the medical record problem, the information presented may be a name, a social security number, sex, age, height and weight which become available when a prospective patient appears.

(3) It is desired to determine which if any individual in the file corresponds to the information presented.

* This research was supported in part under ONR contract N00014–75–C–0555 (NR–042–331).

(4) Much of the data in the files and on the target are subject to noise or random error.

We shall present an approach to this class of problems which derives from cost considerations. Most of the detail will be applied to the chemical problem on which the author has access to some data. This approach will be elaborated in a later paper [3].

2. General approach

To fix ideas let us think in terms of the fingerprint identification problem. The FBI has a file of the order of magnitude of 10^7 sets of fingerprints. Each day it is required to process 30 000 new sets. Each new set must be compared with those in the files to see if they correspond to an individual whose fingerprints are already present (possibly under a different name). If not, the new set is added to the file.

If it could be assumed that all the data in the files and on the targets were noise free, with no random errors, then the task of filing, and comparing the incoming prints could be handled by hash coding [8] or related methods.

When the possibility of error must be taken into account, then those methods are no longer applicable. The general approach to be presented is based on the following cost considerations.

Assume that the file has N elements identified as $X_1, X_2, \ldots, X_N$ and that the X_i may be regarded as independent observations on a multivariate random variable $X \in \boldsymbol{X}$ with probability density $f_X(x)$. When a target element $Y \in \boldsymbol{Y}$ is presented, it is required to select a subset $\delta(Y)$ of $\boldsymbol{X}$, so that each element $X \in \delta(Y)$ is to be compared carefully with Y. If the cost of each such comparison is c, then the expected cost of comparison, given $Y = y$, will be $cN\mathbf{P}\{X \in \delta(y)\}$. Counterbalancing this cost is that of not realizing that the individual is represented in the file if indeed he is. If π is the probability that a random target Y is represented in the file and k_1 is the cost of "missing" him then the conditional expected cost due to missing, given $Y = y$, is $\pi k_1 \mathbf{P}\{X \in \delta(y) \mid Y = y\}$ where X and Y have the joint distribution f_{XY} associated with the file and target records for the same individual.

If $\boldsymbol{X}$ and $\boldsymbol{Y}$ are the same space, corresponding to measurements of the same variables, then, we would expect the joint distribution of X and Y to be such that X and Y are close to each other and that their marginal distributions would be equal, i.e., $f_X(x) = f_Y(x)$. In general $\boldsymbol{X}$ and $\boldsymbol{Y}$ can be different spaces. For example, X could be more detailed than Y or vice versa.

To return to our costs, a relatively simple analysis shows that the choice of δ which minimizes the overall cost is given by

$$\delta(y) = \left\{x:\ \lambda(x, y) = \frac{f_{XY}(x, y)}{f_X(x) f_Y(y)} \geq k\right\}$$

where

$$k = \frac{cN}{\pi k_1}$$

and the expected cost is given by

$$C = \pi k_1 \left\{ \iint_{\lambda < k} f_{XY}(x, y) \mathrm{d}x\, \mathrm{d}y + k \iint_{\lambda \geq k} f_X(x) f_Y(y) \mathrm{d}x\, \mathrm{d}y \right\}.$$

Not surprisingly, the optimal procedure is equivalent to a likelihood–ratio test. More precisely, let H_1 be the hypothesis that X and Y have the joint distribution $f_{XY}(x, y)$ and let the alternative H_2 be the hypothesis that X and Y are independent with densities $f_X(x)$ and $f_Y(y)$ respectively. Then $\lambda(X, Y)$ is the likelihood-ratio, and our procedure is equivalent to searching each element x of $\boldsymbol{X}$ for which the likelihood-ratio test would lead to the acceptance of hypothesis H_1, that x and y correspond to the same individual versus the alternative H_2, that they correspond to independently selected individuals.

As seems reasonable, the constant k is large when the parameters determining the relative cost of search c/k_1 is large. Indeed

$$C = \pi k_1(\alpha + k\beta)$$

where α and β are the probabilities of type 1 and type 2 error respectively.

It should be noted that the idea of using the likelihood-ratio has appeared in the information retrieval literature, although with the exception of the Sunter–Fellegi paper cited before, it seems to have typically appeared in an ad hoc fashion. See for example [4].

3. The chemical compound problem

In the chemical compound problem, a record of compounds are kept where the individuals are identified by their mass spectrograms. For each of 120 to 400 masses, an intensity is listed in a file. If an unknown compound is to be compared with the elements of the file, these mass spectra are compared. Traditionally these comparisons have been done manually with satisfactory results. Recently files have grown, and a need to automate the search has developed.

One primitive approach consists of identifying each intensity in the spectrogram of a compound as large or small in which cases the mass is coded as 1 or 0, respectively. Thus each spectrum is filed as a high dimensional vector consisting of ones and zeros. The distance between two spectra is measured as the number of masses (components of the vector) on which the spectra disagree. Then the most "likely" compounds in the file to correspond to the unknown target are those for which the distance from the target spectrum to their spectra are smallest.

Grotch [5] observed that by simply counting mismatches, one gives the same weight to the matching of two ones as to the matching of two zeros. Since ones appear much less frequently than zeros in his files (for a reason we shall mention later), he felt that these coincidences should be weighted differently. Accordingly he suggested a modified distance scheme where a mismatch gives rise to a one in the sum leading to the distance but a match of two ones should lead to a contribution of $-w$. A match of two zeros would not contribute anything. He discovered empirically that a value of $w \doteq 2$ seemed to be most effective for his objectives and contributed substantial improvement in performance.

The improved approach can be represented in terms of a two by two table for each mass. This table has weight w_{ij} corresponding to a reading of i on the compound in the file and a reading of j on the (target) compound being identified

Table 1

Weights w_{ij}

file \ target	0	1
0	w_{00}	w_{01}
1	w_{10}	w_{11}

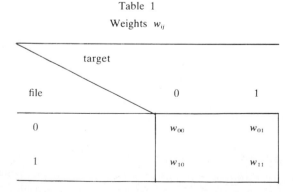

where $w_{00} = 0$, $w_{01} = w_{10} = +1$, and $w_{11} = -w$. The distance is the sum of the weights corresponding to each of the masses.

The table with general w_{ij} represents an empty generalization of the Grotch scheme as long as the same table is used for each mass and $w_{01} = w_{10}$. This is clear since the transformation

$$w^*_{ij} = \frac{w_{ij} - w_{00}}{w_{10} - w_{00}}$$

yields the Grotch method without affecting the order relations among distances being measured. On the other hand if the w_{ij} are allowed to vary from mass to mass, one does obtain more flexibility. Indeed, the likelihood-ratio method reduces exactly to such a generalization. If we assume that the readings for distinct masses are independently distributed then we may write

$$f_{XY}(x, y) = \prod_m f_{(X_m, Y_m)}(x_m, y_m)$$

$$f_X(x) = \prod_m f_{X_m}(x_m), \qquad f_Y(y) = \prod_m f_{Y_m}(y_m)$$

where $X = (X_1, X_2, \ldots, X_m, \ldots)$, $x = (x_1, x_2, \ldots, x_m, \ldots)$ and Y and y are treated similarly. Then

$$\log \lambda(X, Y) = \sum_m \log \lambda_m(X_m, Y_m)$$

where

$$\log \lambda_m(x, y) = \log \frac{f_{X_m Y_m}(x, y)}{f_{X_m}(x) f_{Y_m}(y)}$$

Thus the likelihood-ratio method is equivalent to measuring distance by a weighing scheme where the weights for the m-th mass are computed by $w_{ij}^{(m)} = -\log \lambda_m(i, j)$.

How does one compute the appropriate weights or λ's? By studying a sample of compounds on which duplicate spectra have been obtained, one can estimate $f_{X_m, Y_m}(x, y)$ for $x = 0, 1$ and $y = 0, 1$ and hence the marginal probabilities $f_{X_m}(x) = f_{Y_m}(x)$ for $x = 0, 1$ and these estimates can be used to estimate $w_{ij}^{(m)}$.

Grotch [5] carried out these calculations. He used a library of 15000 mass spectra. From this file 883 redundant pairs of spectra were examined. He observed that the normalized weights $w_{11}^{(m)*} = (w_{11}^{(m)} - w_{00}^{(m)})/$

$(w_{10}^{(m)} - w_{00}^{(m)})$ tend to average about -2 although they vary considerably from mass to mass. More important he evaluated the resulting performance using the $w_{ij}^{(m)}$ (i.e., the likelihood-ratio approach) by comparing several schemes as follows.

In previous work [5] 200 different mass spectra were tested as "unknowns" using a variety of matching procedures. Of these spectra 15 were found to be particularly difficult to identify by most automated matching procedures. To illustrate the efficacy of a few competing procedures, Grotch applied these to the fifteen difficult spectra, using in every case the same library which consists of 6880 mass spectra (a subset of the 15000 mass spectra referred to previously).

Table 2
Confusion in identifying 15 unknowns

	Methods		
Compound	$w = 2$	likelihood-ratio	$1/\sigma^2$
1 Levaline acid	5	2	1
2 Allyl acetate	2	1	4
3 Phenyl isocyanate	4	2	6
4 Acetamide	7	1	2
5 2-Amino pridine	6	0	0
6 4-methyl-cyclohexamone	10+	1	10+
7 Benzil	10+	4	10+
8 Tetrahydrofuran	3	0	3
9 Vanllin	10+	7	10+
10 Piperidine	10+	4	10+
11 2-Ethyl hexyl amine	10+	4	0
12 Dimethyl aniline	10	1	1
13 Methyl isopropyl ketone	10	5	5
14 Benzyl bromide	10+	3	0
15 Allyl iodide	10	5	10+

The efficacy of the matching procedure was given, and is partly reproduced here in Table 2, in terms of the *confusion table* where the entry represents the number of compounds in the library which agree with the unknown as well as or better than the true answer. The ideal matching procedure would always give a *confusion* of zero. Note that 10+ represents a number greater than 10 and the method denoted by $1/\sigma^2$ used $w_{00}^{(m)} = 0$, $w_{10}^{(m)} = w_{01}^{(m)} = 1$ and $w_{11}^{(m)} = -1/\sigma_m^2$ where $\sigma_m^2 = f_{X_m}(0) f_{X_m}(1)$ is the variance of the dichotomous variable X_m.

The likelihood-ratio technique assumes that the sample of compounds giving rise to the estimated weights is a random sample from the library of compounds. It also assumes that the readings for successive masses are independent, i.e., the pairs (X_m, Y_m) are independent. The latter assumption is definitely *not* valid. What we neglected to point out before is the method used to determine whether $X_m = 1$ or 0. An earlier method used a threshold level a on each mass to determine a value of one if the intensity X_m^* at mass m exceeds a. In the interest of data compactification, another approach is currently used to generate binary valued spectra. The masses ranging from 20 to 243 are subdivided into groups of 14 successive masses, and the mass in each group which has the highest intensity level is given a reading of $X_m = 1$. The others are given readings of 0. Thus the X_m are clearly dependent.

This procedure has the advantage that the data of fourteen intensities can be summarized by one four bit integer representing one of fourteen locations instead of by 14 one bit digits. It has the disadvantage of losing some information when compared to using the actual intensities. Furthermore, if the assumption of independence for the intensities of different masses had ever been plausible, it is clear that the X_m are no longer independent. Nevertheless the statistics derived on the assumption of independence, is a well defined statistic whose performance can be studied empirically even though that assumption is not valid.

To amplify a bit, the choice of groups of 14 masses was based in part on the fact that the autocorrelation of the intensities of spectra of organic compounds reveal a cyclic effect with period 14 which is not surprising from a theoretical point of view and which indicates another presence of dependence. Also the early primitive approach based on the binary valued spectra using a threshold was not as effective as the same approach based on the current binary valued spectra when these approaches were compared empirically. We shall comment further on this point later.

4. The use of r levels of intensity

The relative importance of data compression depends on the available hardware and the magnitude of the library. If enough memory capacity is readily available, one may consider using the Grotch technique or its likelihood-ratio generalization as a computationally convenient first step in selecting a substantial number of potential matches to a target. On these, one could carry out more detailed computation to see which if any are the likely candidates. In principle this more detailed computation could be based on the original spectral intensities X_m^*, Y_m^* at mass m which would carry more information than is contained in the 0, 1 levels into which the intensities have been transformed. If there were a suitable model which approximated $f_{X_m^* Y_m^*}$ adequately then one may apply the likelihood-ratio directly. However, empirical fits by simple models seem to be difficult to derive for even the marginal distributions of X_m^* seem to consist of a mixture of log normals after excluding the substantial portion of the data where $X_m^* = 0$.

Alternatively, with some loss of information, the original intensities X_m^* can be grouped directly into r intervals which are labeled $X_m = i$, $1 \leq i \leq r$. Several such groupings will be described later. Then, using a sample of pairs of spectra, one can estimate the joint densities $f_{X_m, Y_m}(x, y)$ for $1 \leq x \leq r$, $1 \leq y \leq r$ from the frequency table of (X_m, Y_m). From these follow the marginal frequencies $f_{X_m}(x)$ and $f_{Y_m}(y) = f_{X_m}(y)$ and

$$w_m(x, y) = -\log [f_{X_m, Y_m}(x, y)/f_{X_m}(x) f_{Y_m}(y)]$$

can be used to compute

$$D(X, Y) = \sum_m w_m(X_m, Y_m) = -\log \lambda(X, Y)$$

a generalized distance from X to Y whose values can be either negative or positive.

Under the assumption of independence, the D measure makes use of more of the available information, as the grouping which converts X_m^* to X gets finer and finer. However, as r increases greater demands are put on the memory capacity of the computer which must store the w_m tables or some suitable approximation and one must ask how much is gained in return for the increased information.

One way to approach the question is to ask the following. Suppose that the decision between H_1 and H_2 must be made on the basis of (X_1, Y_1), $(X_2, Y_2), \ldots, (X_n, Y_n)$ which are n i.i.d random variables with the distribution of (X, Y). How does the cost $C = \pi k_1(\alpha + k\beta)$ depend on n when one uses the appropriate likelihood-ratio test? It will be shown in [3] that if k is large, i.e.

$$\frac{1}{n} \log k = b$$

then $\alpha + k\beta$ is roughly of the order of magnitude of e^{-nI_b} where

$$I_b = -\inf_{0<t<1} \log \int\int f_{XY}^{1-t}(x, y) f_X^t(x) f_Y^t(y) e^{bt} \, dx \, dy$$

and the above integral is represented by a sum when X and Y are discrete random variables.

In this context each pair (X_i, Y_i) can be regarded as contributing a factor of $\exp(-I_b)$ to reducing the cost of our procedure. In this sense I_b represents a measure of information. Related to this measure are two others

$$\hat{I} = \int\int f_{XY}(x, y)[\log f_{XY}(x, y)] dx \, dy - \int f_X(x)[\log f_X(x)] dx - \int f_X(y)[\log f_Y(y)] dy$$

which always exceeds I_b. $\hat{I}$ is the Shannon mutual information and overestimates the potential value of (X_i, Y_i). At the same time I_b exceeds

$$\tilde{I}_b = -\log\left[\int\int [f_{XY}(x, y) f_X(x) f_Y(y) e^b]^{1/2} dx \, dy\right] = -0.5b + \tilde{I}_0$$

which is the value at $t = \frac{1}{2}$ of the integral whose minimum determines I_b. $\tilde{I}_0$ has been studied by Bhattacharya [1], Matusita [7], and I_0 has been studied by Chernoff [2].

5. Illustration

A collection of 5000 mass spectra was examined over mass ranges 31 to 150 and the intensities X_m^* for mass m were normalized subject to $\Sigma_m X_m^* = 1$ for each spectrum. When $X_m^* = 0$, X_m was defined to be one. Then μ_m and σ_m were calculated to be the conditional mean and standard deviation of $\log X_m^*$ given that $X_m^* > 0$. Finally X_m was set equal to i if

$$a_{i-1} < (\log X_m^* - \mu_m)/\sigma_m \leq a_i, \qquad i = 2, \ldots, 10$$

where $a_1 = -\infty$, $a_i = -3.6 + 0.6i$ for $2 \leq i \leq 9$ and $a_{10} = \infty$. This rule assigns the intensity X_m^*, for a given compound and mass, to one of 10 distinct levels X_m corresponding to whether $X_m^* = 0$ or to how many standard deviations away from μ_m is $\log X_m^*$. Indeed for $2 \leq i \leq 9$, each level corresponds to an interval of 0.6 standard deviations.

A set of 102 pairs of spectra, each pair corresponding to one of the compounds listed in the collection of 5000 spectra, was used to obtain pairs (X_m^*, Y_m^*), both components of which were converted to levels by the grouping described above to give (X_m, Y_m). Then the frequencies $g_{ij}^{(m)}$, that is the number of pairs where $(X_m, Y_m) = (i, j)$ were counted and used with the aid of some smoothing to estimate the joint probabilities $f_{X_m, Y_m}(x, y)$ for $1 \leq x \leq 10$, and $1 \leq y \leq 10$ and the marginal probabilities $f_{X_m}(x) = f_{Y_m}(x)$ for $1 \leq x \leq 10$.

From these, the weights $w_{ij}^{(m)} = -\log [f_{X_m Y_m}(i, j)/f_{X_m}(i) f_{Y_m}(j)]$ were computed. The smoothing referred to above had, as one object, the avoidance of unduly large values of $|w_{ij}^{(m)}|$ which could arise if some $g_{ij}^{(m)}$ was zero. Then I_0, $I_{0.2}$ and $\hat{I}$ was calculated for each mass. The 120 values of I_0 averaged 0.115 giving a total of 13.84. The values of $\hat{I}$ tended to be about four times as large as the I_0 values and hence constitute a substantial overestimate. The I_b values are obtained by minimizing an integral with respect to a parameter t between 0 and 1. Generally the minimizing value of t in the computation of I_0 is close to 0.5 and, related to this, the I_b values tend to be close to but a little larger than $I_0 - b/2$. If $N \doteq 5000$, $\pi \doteq 1$, $c/k_1 \doteq 0.1$, and $n \doteq 120$ then $k \doteq 500$ and $b \doteq 0.05$.

For the time being let us concentrate on I_0 as though k were approximately one. The information numbers I_0 are not strictly additive, but the total 13.84 indicates error probabilities of the order of $\exp(-13.84) \approx 10^{-6}$. Assuming that this quantity is quite small, one could afford to discard some information, at least in a preliminary stage. One way of doing this is by using a coarser grouping. Several other groupings based on the original 10 levels were compared. We mention two of these. The original 10 levels will be called grouping I. Grouping II was formed by using 4 levels. Level 1 corresponds to $X = 1$ or 2, level 2 to $X = 3, 4$, or 5, level 3 to $X = 6$ or 7 and level 4 to $X = 8, 9$ or 10. Grouping III was formed by using 2 levels. Level 1 corresponds to $X = 1$, 2 or 3 and level 2 to $X > 3$.

Note that a grouping with r levels requires a memory which retains $r(r+1)/2$ weights for each mass. Thus $r = 10$ requires 55 weights per mass compared to 10 for $r = 4$ and only 3 for $r = 2$. For both groupings II and III the I_0 values tended to be one quarter those of $\hat{I}$ values. For grouping II the average value of I_0 was 0.0945 while for grouping III it was only 0.042. Generally the relative efficiencies of these groupings as measured by the ratios of I_0 values were roughly 1, 4/5 and 1/3 for groups I, II, and III respectively.

For group II, the sum of the I_0 is 11.34 and for group III it is 5.09. Since $\exp(-11.34) = 1.2 \cdot 10^{-5}$ and $\exp(-5.09) = 6.2 \cdot 10^{-3}$, it seems that group II is considerably better than III and not much worse than I.

Moreover the I_0 values for group II have a standard deviation of 0.047 which suggest that if the most informative masses are used, considerable reduction in error probabilities are attainable with the use of data for relatively few masses (assuming that violations of independence are not very harmful to our analysis). Indeed the 50 most informative masses provide a total I_0 of 7.0 under grouping II and $\exp(-7) = 9 \cdot 10^{-4}$.

A rough and ready way to introduce the effect of k and b is to subtract $120b/2$ from the total informations accumulated above. An alternative view which leads to the same algebra is to note that to optimize $\alpha + k\beta$ we tend to reduce β by a factor of $\sqrt{k}$ and increase α, and consequently $\alpha + k\beta$, by the same factor which is, in our illustration, $\sqrt{500} \doteq 22$.

These calculations yield rough estimates of the efficacy of the likelihood-ratio approach using various groupings of the data. These estimates are based on a theoretical framework which assumes independence and may not be reliable. Thus empirical evaluations of these procedures should be, but have not yet been, carried out.

The two level grouping used is equivalent to a binary valued spectrum based on a threshold, and hence probably not as effective as the binary spectra currently used by Grotch. Unfortunately our lack of empirical evaluations now prevent us from directly comparing the efficacy of that type of spectra to those based on the various groupings above.

6. Remarks

We have developed an approach whose performance requires empirical evaluation. What constitutes an adequate evaluation? The confusion table (Table 2) is a limited evaluation of performance based on 15 difficult compounds. Such a table should be supplemented. Having developed an approach, one can select a sample of compounds at random from those in the file and see how well the replicated spectra on these compounds are evaluated. How large are the likelihood-ratios when the targets are matched against the correct compound? How large are the likelihood-ratios when the targets are matched against randomly selected compounds in the file? How large are the 5 or 10 largest likelihood-ratios for each target? In the sampling one may select the targets so that each compound in the file is equally likely. Alternatively one may select the targets according to a sampling scheme which more nearly resembles the distribution of targets that appear in practice (assuming such a distribution is available).

The presence of a small amount of noise has a surprisingly large effect in degrading the information in a variable. In many applications the economics of computation time and data storage demand the evaluation and use of summary statistics that lose little information for the task of identifying an individual. The improvement found by Grotch by using the current binary spectra in place of those derived from a threshold suggest that when dealing with many independent and similarly distributed random variables, the label of the largest or the labels of the two or three largest may carry a substantial part of the available information, especially if these are supplemented by a few extra bits describing the intensities of these largest peaks.

More generally the efficiency issue arises in other applications where feasibility depends on the ability to point out statistics or *features* which are high in information content but low in data storage requirements. For example in fingerprint identification the type of fingerprint (whirl or loop etc.) is low in noise, and high in useful information per bit of storage space. In the medical record case, sex is low in noise. However names as conventionally spelled may be subject to considerable error. For this reason allowance is often made for transpositions of letters, or surnames may be converted to a "soundex" form.

In summary, we have introduced a formulation of the identification problem which indicates the relevance of the likelihood-ratio and of related information numbers which should be helpful in determining how effective certain features will be.

Bibliography

[1] Bhattacharya, A. (1943). On a measure of divergence between two statistical populations, *Bull. Calcutta Math. Soc.* **35**, pp. 99–109.

[2] Chernoff, H. (1972). *Sequential Analysis and Optimal Design*, Soc. for Indust. and Appl. Math., Phila. Pa.

[3] Chernoff, H. The identification of an element of a large population in the presence of noise (to be completed).

[4] Giuliano, V.E. (1965). The interpretation of word association, *Proc. of sympt. on Statistical Associative methods for mechanized documentation*, (Ed. by Stevens, M.E. et al), Nat. Bur. Stds, Misc. Publ. 269, pp. 25–32.

[5] Grotch, S.L. (1971). Computer techniques for identifying low resolution mass spectra, *Anal. Chem.*, **43**, pp. 1362–1370.

[6] Grotch, S.L. (1975). Automatic identification of chemical spectra. A goodness of fit measure derived from hypothesis testing. *Anal. Chem.*, **47**, p. 1285–1289.

[7] Matusita, K. Distance and decision rules, *Ann. of Inst. of Statist. Math.*, **16**, pp. 305–316.

[8] Minsky, M. and Papert, S. (1969). *Perceptrons*, MIT Press, Cambridge, Mass.

[9] Sunter, A.B. and I.P. Fellegi, (1967). An optimal theory of record linkage, *Proc. of 36th Session of the International Statistical Institute*, Sydney, pp. 809–835.

P.R. Krishnaiah, ed., *Multivariate Analysis–IV*
© North-Holland Publishing Company (1977) 457–471

SOME PROBLEMS IN STATISTICAL PATTERN RECOGNITION

Somesh DAS GUPTA*
University of Minnesota, Minneapolis, Minn. U.S.A.

1. Introduction

In this paper we shall review some of the recent work on nonparametric and sequential rules in statistical pattern recognition along with criticisms, and indicate some new results in these areas. Summary reviews of the literature are given by Das Gupta [10] and Kanal [22].

The basic problem in statistical pattern recognition can be formulated as follows. Let ω be a point in a sample space with an associated σ-field of events and a probability measure. Let $X(\omega)$ denote a real-valued vector of measurements on ω and let $I(\omega)$ denote the pattern-class of ω which takes values in $(1, 2, \ldots, k)$. The problem is to predict $I(\omega)$ from the knowledge of $X(\omega)$. Denote the pattern-class probabilities by $\xi = (\xi_1, \ldots, \xi_k)$, where $\xi_i = \mathbf{P}(I = i)$, and the class distributions by $F = (F_1, \ldots, F_k)$, where F_i is the conditional distribution of X, given $I = i$; let F_i admit a density f_i with respect to a σ-finite measure μ.

The above problem can arise in different forms due to different situations and structures of available knowledge as discussed below. The problem may be to classify a single unit or more than one unit which may occur in a single batch or in a sequence. Moreover, the units to be classified may belong to the same pattern-class (i.e., when units are sampled from the space conditioned on some value of I), or to different pattern-classes. In almost all the problems, F and ξ are not completely known although it may be known that $F_1 \times \ldots \times F_k$ belong to a given set Ω. In order to get more information on F and ξ, data in the name of a "training sample" are collected in one of the following ways.

(a) Separate samples from k pattern-class populations, denoted by $\Pi_1, \ldots, \Pi_k$. In this case ξ is interpreted as a prior probability vector.

* This work was supported by U.S. Army Research Grant DAAG–29–76–G–0038 at the University of Minnesota.

(b) Sample from the mixed population (denoted by Π) which is a mixture of $\Pi_1, \ldots, \Pi_k$ in the proportion $\xi_1, \ldots, \xi_k$.

Sometimes when the units to be classified occur in a sequence, observations on the first i units are taken as a training sample to predict the pattern-class of the $(i+1)$th unit.

A training sample may be identified, i.e., the observations are available on both X and I, or unidentified when the observations are available on X.

The so-called nonparametric methods arise when F_i's are not given explicitly in simple parametric forms. It is well known that there is no distribution-free rule for this problem. So the performance of a rule cannot be evaluated (except some asymptotic results and broad bounds) without additional knowledge of the underlying Ω. A Bayes rule can be easily derived when ξ and f_i's are known. The major bulk of the literature is devoted on plug-in versions of a Bayes rule, i.e., when the unknowns are replaced by their respective estimates (derived from the training sample) in the form of the Bayes rule. For this nonparametric problem, generally estimates of f_i's and ξ are used; asymptotic properties of such rules are then easily derived from the asymptotic properties of the estimates used. Rules based on tolerance regions and nearest-neighbors are also discussed in the literature; some of these rules indirectly use estimates of density functions. Another class of rules are suggested following the nonparametric methods for the two-sample problem; these rules are based on general U-statistics, estimates of c.d.f.'s, ranks, and permutation-invariance. A class of rules, called "empirical best-of-class rules" is also under study; these rules are optimum in some sense when they are applied on the identified training sample.

Sequential rules also arise in different situations. There can be rules based on sequential experimentation on the components of $X(\omega)$, although there is not any result in this area worth mentioning (except possibly some heuristic methods). Next, a training sample may be obtained sequentially and rules may be devised based on such sequential experiments. Furthermore, when the units to be classified belong to the same pattern-class, they may also be observed through a sequential experiment. It may be noted that when the units to be classified occur in a sequence from Π, a sequential rule may be devised, although a sequential experimentation in such a case is not meaningful. All the papers in this area deal with direct applications of sequential two-sample tests. Unfortunately, very little has been done so far.

Although there are some papers on Monte-Carlo studies on performances of some rules, the studies on robustness and relative efficiency have to be done much more intensively and carefully. Asymptotic results are of theoretical interest; however, good bounds on error-probabilities and studies on errors of approximation will be more valuable.

2. Notations and preliminaries

Let us first restrict our attention to the case $k = 2$, and the situation where the pattern class of one unit is to be predicted. A decision rule, not depending on the training sample, is given by $\delta = (\delta_1, \delta_2)$, where $\delta_i(x)$ is the probability of deciding i as the correct pattern class, given the observation $X = x$. The probabilities of error for the rule δ are given by

$$\alpha_1(\delta; f) = \int \delta_2 f_1 \mathrm{d}\mu, \qquad \alpha_2(\delta; f) = \int \delta_1 f_2 \mathrm{d}\mu$$

where $f = (f_1, f_2)$. Then the risk with the prior distribution $\xi = (\xi_1, \xi_2)$ and 0 1 loss function (or, the total probability of error) is given by

$$R(\delta; \xi, f) = \xi_1 \alpha_1(\delta; f) + \xi_2 \alpha_2(\delta; f).$$

Given ξ and f, a Bayes rule $\delta^*(\cdot\,; \xi, f)$ which minimizes $R(\delta; \xi, f)$ is given by

$$\delta_1^*(x; \xi, f) = \begin{cases} 1 & \text{if} \quad \xi_1 f_1(x) > \xi_2 f_2(x), \\ 0 & \text{if} \quad \xi_1 f_1(x) < \xi_2 f_2(x). \end{cases}$$

Let $R^*(\xi, f) \equiv R(\delta^*(\cdot; \xi, f); \xi, f)$.

Let θ stand for (ξ, f) or (ξ, F). When θ is known a "good" rule generally depends on θ, and it is denoted by $\delta(\cdot; \theta)$; δ^* is such a rule. We shall drop ξ from θ when the population is not mixed and ξ is not given. When δ explicitly depends on θ, we shall write $\alpha_i(\delta; f)$ as $\alpha_i(\delta; \theta)$.

Now, consider the problem when θ is not completely known. Information on θ is based on a training sample S_N; in case the sampling is done separately from Π_i's, N will stand for the vector of sample sizes. A decision rule, in that case, is denoted by $\delta_N(\cdot, S_N) = (\delta_{N1}(\cdot, S_N), \delta_{N2}(\cdot, S_N))$. In particular, such a rule may be a plug-in version of $\delta(\cdot; \theta)$, given by $\delta(\cdot; \hat{\theta}_N) = \delta_N(\cdot; \hat{\theta}_N)$, where $\hat{\theta}_N$ is generally chosen to be a consistent estimate of θ; we shall often write $\delta_N(\cdot; \hat{\theta}_N)$ by $\hat{\delta}_N = (\hat{\delta}_{N1}, \hat{\delta}_{N2})$.

The conditional probabilities of error and the conditional risk of $\delta_N(\cdot, S_N)$, given S_N, are given by

$$\alpha_{ic}(\delta_N; S_N, f) \equiv \int \delta_{Nj}(x; S_N) f_i(x) \mathrm{d}\mu(x) \quad (i \neq j),$$

and

$$R_c(\delta_N; S_N, \theta) \equiv \sum_{i=1}^{2} \xi_i \alpha_{ic}(\delta_N(\cdot, S_N); f).$$

The unconditional probabilities of error and the risk of δ_N are given by

$$\alpha_i(\delta_N; \theta) = \mathbf{E}_N[\alpha_{ic}(\delta_N; S_N, f)]$$

$$R(\delta_N; \theta) = \mathbf{E}_N[R_c(\delta_N; S_N, \theta)]$$

where $\mathbf{E}_N$ denote the expectation over S_N.

When there are more than one unit to be classified, and it is known that they come from the same population, the above development can easily be extended. However, when the units occur in a sequence one may adopt Samuel's approach, although no results are available in the literature when the densities are not known. When the units to be classified arise from the mixed population one may use the compound decision approach as suggested by Robbins [34] and later developed by Hannan and Robbins [18], Samuel [36, 37] and Yao [53]; however, all these papers assume that the class-densities are known. See Van Ryzin [43] for a similar development when the distributions are unknown. When the units to be classified arise in a sequence from Π and for classifying the ith unit the observations on all the previous identified (by a "teacher") units are used as a training sample, a completely separate theoretical development would be called for. However, all the papers dealing with this problem put emphasis only on the prediction of the class of ith unit using the standard theory discussed above.

3. Nonparametric rules

3.1. *A simple approach*

Most of the papers deal with asymptotic properties of rules based on a training sample S_N, as $N \to \infty$. In particular, the convergences (and their rates) of conditional, as well as unconditional, probabilities of error and risk are dealt with. Special emphasis is given to the rule $\hat{\delta}_N^*$, the plug-in version of a Bayes rule δ^*. Bounds on probabilities of correct classification for some heuristic rules are also available in some papers. The above convergences as $N \to \infty$ and the number of units to be classified tend to ∞ are also discussed in some special cases.

We present the following asymptotic results which can be proved under very general conditions as the size (or sizes) N of the training sample tend to ∞.

Suppose $\delta(x;\hat{\theta}_N) \to \delta(x;\theta)$ in probability (a.s.) for almost all (μ) x. Then

(I) $\alpha_{ic}(\hat{\delta}_N; S_N, f) \to \alpha_i(\delta;\theta)$, in prob. (a.s.),

(II) $\alpha_i(\hat{\delta}_N;\theta) \to \alpha_i(\delta;\theta)$,

(III) $R_c(\hat{\delta}_N; S_N, \theta) \to R(\delta;\theta)$, in prob. (a.s.),

(IV) $R(\delta_N;\theta) \to R(\delta;\theta)$.

In the above result for convergences of risks, the condition "for almost all x" may be relaxed by the condition "for almost all x in the set $\{x: \xi_1 f_1(x) \neq \xi_2 f_2(x)\}$."

In particular suppose δ_1 $(x,\theta) = 1$, if $D(x,\theta) > 0$, where $D(x;\theta)$ is a function of x when θ is known. Moreover, assume that $D(x,\theta)$ equals zero on a null set (although this condition can be slightly relaxed). If $D(x,\hat{\theta}_N) \to D(x,\theta)$ in probability (a.s.) for almost all x, the above results on convergences hold. The primary requirements for the above conditions to hold are that $D(x,\theta)$ is continuous in θ for almost all x, and $\hat{f}_i(x) \to f(x)$ in prob. (a.s.), where $\hat{f}_i$ is an estimate of f_i. The detailed proof of these results will be given elsewhere. For similar but weaker results, see Johns [21] and Glick [16]. It may be noted that the above results do not require δ to be a Bayes rule and $\hat{f}_i(x)$ integrable; these assumptions are used in almost all the papers in this area. The above results can be used to simplify the proofs of many known results.

The problem may also be handled from decision-theoretic viewpoint with provisions for "withholding decision" or "doubtful regions." See Rao [32] for this approach when the distributions are known and Partick's book [29] for some heuristic developments when the distributions are unknown.

3.2. Rules based on estimates of density functions

All the important papers deal with asymptotic properties of δ_N^* with various estimates of θ. Recall that one may define δ^* in either of the following ways:

A. $\delta_1^*(x;\theta)=1$, iff $D(x,\theta)\equiv \xi_1 f_1(x)-\xi_2 f_2(x)\geq 0$.

B. $\delta_1^*(x;\theta)=1$, iff $D_1(x;\theta)\equiv \xi_1 f_1(x)/[\xi_1 f_1(x)+\xi_2 f_2(x)]\geq \frac{1}{2}$.

The following methods for estimating f_i's are mostly used.

(i) Fix–Hodges' [12] method; later, modified by Loftsgaarden and Quesenberry [25].

(ii) Aizerman–Braverman–Rozonoer's method [1].

(iii) Čencov's method [6].

(iv) Parzen–Cacoullos' method [28, 5].

(v) Wolverton–Wagner–Yamato's recursive method [50, 52].

The known results on $\hat{f}_i$ are then used to derive asymptotics for $\hat{\delta}_N^*$.

Fix and Hodges [12] essentially proved (II) with their suggested estimators of f_i's, when the training sample is separately drawn from Π_i's and $\xi=(\frac{1}{2},\frac{1}{2})$. Johns [21] proved the same result with a minor modification of Fix–Hodges' estimates when the training sample is identified and drawn from the mixed population. However, Johns considered the problem with a general loss structure and more general space of values for the pattern-class indicator variable I. Van Ryzin [44] proved the result III (in probability) with estimates (ii), (iii) and (iv), when the training sample is separately drawn from Π_i's. He also studied the rates of these convergences. Van Ryzin [42] proved the result IV with estimates similar to (iv) when the training sample is an identified sample from the mixed population; he also obtained bounds on $R(\hat{\delta}_N^*;\theta)-R(\delta^*;\theta)$ under additional conditions on θ.

Another approach used in the literature is to estimate $D(x;\theta)$ or $D_1(x,\theta)$ directly. Recursive decision rules are considered (for easier updating), especially when the units to be identified occur in a sequence from Π and the correct pattern-class of a unit is known after its prediction. For classifying the ith unit, all the observations on the previous units constitute a training sample. Suppose $\hat{\delta}_N^*$ depends on $D_{N1}(x;S_N)$ or $D_N(x;S_N)$ as δ^* depends on $D_1(x;\theta)$ or $D(x;\theta)$, respectively; that is, D_{N1} and D_N are respective estimates of D_1 and D. It is shown by Van Ryzin [45] that if

$$\text{(V)} \qquad \int [D_N(x;S_N)-D_1(x,\theta)]^2 f_\xi(x)\mathrm{d}x \to 0$$

in probability, where $f_\xi(x)=\xi_1 f_1(x)+\xi_2 f_2(x)$, then the result III (in probability) holds. Van Ryzin [45] suggested a stochastic and recursive algorithm for estimating $D_1(x;\theta)$ for which (V) holds under fairly general conditions. His algorithm essentially involves window-kernels for estimating the density functions. Van Ryzin's work was inspired by the work of Aizerman et al [1] who proved (V) with recursive estimates using essentially method (i) along with an additional assumption that $D_1(x;\theta)$ is a finite linear combination of some known orthonormal functions in L_2. For a generalization of the above work, see Gyorfi [17]. Wolverton and Wagner [51] proved that

$$\text{(VI)} \qquad \int [D_N(x;S_N)-D(x,\theta)]^2 \mathrm{d}x \to 0$$

in prob./a.s. implies the result III in prob./a.s., when f_i's are uniformly continuous (on $\mathbf{R}^m$). They suggested recursive estimates of $D(x,\theta)$ for which VI holds in probability when f_i's are uniformly continuous and in a.s. when, in addition, f_i's satisfy uniform Lipschitz condition. They also studied the rate of convergence when, specifically, f_i's have bounded supports. Similar result on rate of convergence in probability was obtained by Rejtö and Revesz [33]. Watanabe [49, 50] proved (VI) in prob. and a.s. along with their rates using recursive estimates for $D(x,\theta)$ following the method (v). Similar problem was studied by Tanaka [41] when the training sample constitutes dependent observations. Pelto [30] suggested to take the width of the window-kernel for estimating densities as the value which minimizes the deleted-counting estimate of the risk; however, his paper does not give any algorithm and his proofs are all based on heuristics.

It may be remarked that the rates of convergences derived in the papers cited above only reflect the performances of the suggested rules for future prediction, ignoring their performances in the past.

In passing, it may be noted that the estimate of density function by method (i) is not integrable; this estimate is further studied by Moore and Henrichson [27] and Wagner [47]. See also a recent paper by Wahba [48].

3.3. Use of the two-sample test procedures

Let $(X_1,\ldots,X_{n_1})$ and $(Y_1,\ldots,Y_{n_2})$ be the observations on random samples from Π_1 and Π_2 respectively. Let $(Z_1,\ldots,Z_n)$ be the observations on the n units to be classified. These units form a random sample either from Π_1 or from Π_2. Let F_0 be the common c.d.f. of Z_i's. Then the problem is to test $F_0=F_1$ vs. $F_0=F_2$.

Two-sample test statistics are often used to devise rules for the above problem. The basic heuristic ideas can be described as follows.

(a) Use Z's and X's to test $F_0 = F_1$ vs. $F_0 = F_2$; let T_1 be a test statistic such that large values of T_1 lead to the rejection of $F_0 = F_1$. Similarly use Z's and Y's to test $F_0 = F_2$ vs. $F_0 = F_1$; let T_2 be a test statistic such that large values of T_2 lead to the rejection of $F_0 = F_2$. Then define a rule which accepts $F_0 = F_1$ if $T_1 < T_2$, and accepts $F_0 = F_2$ if $T_2 < T_1$. One may also compare the critical levels of T_1 and T_2 instead of comparing T_1 and T_2 directly.

(b) Assume $F_0 = F_1$ and treat Z's and X's as i.i.d. observations. Get an estimate of the divergence between F_1 and F_2 by using a test statistic for testing $F_1 = F_2$ vs. $F_1 \neq F_2$. Similarly assume Z's and Y's as i.i.d. observations and determine the corresponding estimates of the divergence. A rule now can be devised by comparing these two estimates of divergence.

It is well-known that a distribution-free rule cannot be derived for the pattern recognition problem posed above. The performance of a rule can only be judged in specific situations except for some broad asymptotic results.

Das Gupta [9] used the idea (a) and suggested a rule based on Wilcoxon-statistic; he showed that such a rule is consistent (i.e., the error probabilities tend to 0 as $n, n_1, n_2 \to \infty$). Hudimoto [20] also used Wilcoxon-statistic when $F_1(x) \geq F_2(x)$ for all x, and derived some bounds for the probabilities of error. Kinderman [23] used more or less the idea (b) in deriving rules based on rank-scores when $F_1(x) = F_2(x + \theta)$, $\theta > 0$, and studied the asymptotic efficiencies of those rules in Pitman's sense. Chandra and Lee [7] suggested a rule which first uses Wilcoxon-test for $F_1(x) > F_2(x)$ vs. $F_1(x) < F_2(x)$ based on X's and Y's, and then tests $F_0 = F_1$ vs. $F_0 = F_2$ using another Wilcoxon-type statistic and the result of the first test. They studied the asymptotic properties of this rule as n_1 and n_2 tend to ∞. It may be noted that this rule is asymptotically equivalent to Das Gupta's rule [9]. See Chatterjee (*J. Multiv. Anal.*, 1973, **3**, 26–56) for a related work.

A minimum distance rule can be defined following (a) above with T_1 and T_2 as distances between the respective empirical c.d.f.'s. Matusita [26] obtained bounds on the probabilities of error of such a rule for the discrete case with Matusita-distance. Das Gupta [9] proved the consistency (as $n, n_1, n_2 \to \infty$) of such rules and derived bounds for the probabilities of error using Kolmogorov-distance. No detailed studies on these rules are yet available.

For simplicity, consider the minimum distance rule with Kolmogorov-distance for $n = 1$. It can be shown that such a rule decides $F_0 = F_1$ if

$|r_1/n_1 - \frac{1}{2}| < |r_2/n_2 - \frac{1}{2}|$, and $F_0 = F_2$ if $|r_1/n_1 - \frac{1}{2}| > |r_2/n_2 - \frac{1}{2}|$, where $r_1 = \#$ of X_i's $< Z_1$ and $r_2 = \#$ of Y_i's $< Z_1$. We have studied the probabilities of error of this rule, and the results will be reported elsewhere. In particular, suppose $F_1(x) = G(x - \theta_1)$, $F_2(x) = G(x - \theta_2)$, where G is a continuous distribution, symmetric about 0. Then both the conditional probabilities of error tend to $G(-|\theta_1 - \theta_2|/2)$ with probability 1 as $n_1, n_2 \to \infty$.

3.4. *Use of tolerance regions*

The use of tolerance regions (and statistically equivalent blocks) in classification was suggested by Anderson [3]. The basic idea is quite related to the problem of estimating a density function, and in that form it appears in the work of Fix and Hodges [12]. Quesenberry and Gessaman [31] suggested a method for constructing a rule based on tolerance regions which is asymptotically optimal; however, their idea is not very useful since the construction of such a rule depends on some inherent known structures of the distributions. Later, Anderson and Benning [2] and Beakley and Tuteur [4] suggested some other heuristic models. So far no theoretical results are available on these rules. This is due to the fact that very little is known on the performance of a tolerance region under a different distribution. Gessaman and Gessaman [14] studied some of these rules by Monte Carlo method.

3.5. *Empirical best-of-class rules*

The basic idea can be described as follows: Consider a class Δ of rules, and let $C(\delta)$ be the proportions of correct identifications when $\delta \in \Delta$ is applied to identify the observations in the training sample. Let $\delta_N \in \Delta$ be a rule which maximizes $C(\delta)$ in Δ. Then δ_N is called an empirical best Δ-class rule. Let $\delta^* \in \Delta$ be a rule which maximizes the probability of correct classification in the class Δ.

Stoller [40] proved the convergence (in prob.) of the conditional risk of δ_N to the risk of δ^* in the univariate case when Δ is the class of rules determined by single cutoff points. Glick studied the convergence (a.s.) of $C(\delta_N)$ to the probability of correct classification of δ^* with special emphasis on the class of linear rules. The other papers in this area can be found in Duda and Hart (*Pattern Classification and Scene Analysis*, 1973, Wiley), although these are not of much statistical interest.

General results regarding the existence of δ_N and algorithm to determine δ_N are not yet available (except in the case considered by Stoller). General asymptotic results easily follow from the known asymptotic properties of empirical c.d.f.'s.

We suggest the following rule in the multivariate case, which is easy to apply. First treat the problem separately for each variate. However this will lead to some inconsistent decisions or indecision zones. Each of these zones can then be treated successively and separately by each variate to successively reduce the number of such indecision zones. It is believed that this rule will be asymptotically optimal for a large number of classes.

3.6. Rank-distance rules

The basic idea is to find distances of the observations in the training sample from the observation X to be classified and construct rules based on the ranks of these distances and the corresponding pattern-class numbers.

Fix and Hodges [12] suggested the following rule δ_N, termed as 1-NN rule: Classify X to the pattern-class of the nearest neighbor (NN) of X. It can be shown (possibly given in [12]) that under mild conditions, the probability of misclassification

$$\alpha_1(\delta_N; f) \to \int \frac{p_2 f_2(x) f_1(x) \mathrm{d}x}{p_1 f_1(x) + p_2 f_1(x)},$$

where $p_i = \lim n_i/(n_1 + n_2)$, as n_1, n_2 (the sample sizes from Π_1 and Π_2) tend to ∞. For the mixed population case, Cover and Hart [8] have shown that

$$R(\delta_N; \theta) \to \int_{\mathbf{R}^m} \frac{2\xi_1 \xi_2 f_i(x) f_2(x) \mathrm{d}x}{\xi_1 f_1(x) + \xi_2 f_2(x)} \equiv R_0$$

as $N \to \infty$, when the sample space of each X_i is $\mathbf{R}^m$ (or slightly more general). Wagner [46] has shown that under very mild conditions the conditional risk of δ_N tends to R_0 in probability (in a.s. under additional restrictions).

Fix and Hodges [12] also suggested the K_N-NN rule which can be described as follows. Let M_{Ni} be the number of observations in the training sample with the pattern class i that belong to the K_N nearest neighbors of X. Then the K_N-NN rule decides the pattern-class of X as 1, if $M_{N1}/n_1 > M_{N2}/n_2$, where n_i is the number of observations in the training sample with the pattern-class i. One may also consider a rule by comparing M_{N1} and M_{N2}. Johns [21] stated that under mild conditions the risk of the latter rule tends to the risk of δ^* (Bayes), when $K_N \to \infty$, $K_N/N \to 0$. The convergence of the conditional risk of this rule to the risk of δ^* (in different modes) can be obtained from the results in 3.1. Some other theoretical results on K_N-NN rule are claimed in the literature, although they were not proved with rigor.

It may be noted that NN rules are also related to the rules based on estimates of density functions. All the papers in this area deal only with the problem of classifying one unit.

3.7. Conclusion

Many nonparametric rules are suggested in the literature from heuristic viewpoints. Asymptotic properties of most of these rules are not difficult to obtain, although good studies on the rates of convergence and asymptotic expansions of risks would be more useful. The usefulness of a rule is determined by its simplicity, as well as by its robustness. Studies on robustness and small-sample behavior of these rules are quite limited. The relative comparisons (finite-sample or asymptotic) of some of the popular rules in specific situations are also called for.

4. Sequential rules

The sequential pattern recognition may involve sequential experimentation and sequential decision rules. A sequential experiment may arise in any combination of the following three situations. (a) Selection of components of the measurement vector on each unit in the training sample, as well as in the sample of units to be classified. (b) Selection of the sample size of the training sample. (c) Selection of the sample size of the units to be classified when all the units are known to belong to one population. The basic object for using a sequential rule is to attain prescribed probabilities of errors and to reduce the average sample size; in Bayes' formulation (involving probabilities of errors and cost of observations) the object is to reduce Bayes' risk. When the pattern-class densities are known, a Bayes rule can easily be derived following Wald's formulation; however, when the densities are unknown the problem is too involved and no satisfactory results are yet available.

Following the ideas of Hoeffding and Wolfowitz [19], Das Gupta and Kinderman [11] introduced an important notion termed as "classifiability" for the situations (b) and (c) above. Consider three independent random vectors (of the same dimension) X_0, X_1 and X_2, where the c.d.f. of X_i is F_i. It is known that F is either equal to F_1 or F_2, and $F_1 \times F_2$ lies in a given set

Ω. The problem is to decide $F_0 = F_1$ or $F_0 = F_2$ based on a sequence of observations on (X_0, X_1, X_2). This problem is said to be sequentially (finitely) classifiable, if for every $\alpha\,(0 < \alpha < 1)$ there exists a sequential rule (finite sample size rule) which terminates with probability 1 such that the probabilities of that rule are uniformly (in Ω) less than α. Necessary and sufficient conditions for sequential and finite classifiability are given by Das Gupta and Kinderman [11]. With the object of controlling the error probabilities uniformly and arbitrarily it is also important to find out whether it is necessary and sufficient to get observations only on X_1 (or on X_2) or on both X_1 and X_2. This problem is analysed also in Das Gupta and Kinderman [11]. In particular, suppose $F_1 = N_p(\mu_1, \Sigma)$, $F_2 = N_p(\mu_2, \Sigma)$. Then the problem is sequentially classifiable based on observations on (X_0, X_1, X_2) if, and only if, $\mu_1 \neq \mu_2$. The problem is sequentially classifiable or finitely classifiable based on observations on (X_0, X_1) if $\inf_\Omega \|\mu_1 - \mu_2\| > 0$, or $\inf_\Omega (\mu_1 - \mu_2)'\Sigma^{-1}(\mu_1 - \mu_2) > 0$, respectively.

Following Hoeffding and Wolfowitz [19] we present a "minimum distance" sequential rule. Consider n observations on (X_0, X_1, X_2), and let $F_i^{(n)}$ be the empirical c.d.f. based on n observations on X_i $(i = 0, 1, 2)$. Let $\Omega \subset \mathcal{F} \times \mathcal{F}$, where $\mathcal{F}$ is a class of distributions on the space of X_i, and let d be a uniform consistent distance function defined on $\mathcal{F} \times \mathcal{F}$. (We also assume that d is defined on empirical c.d.f.'s.) Assume that

$$\Omega_d = \{F_1 \times F_2 \in \Omega:\ d(F_1, F_2) = 0\}$$

is null. Let $\{C_i\}$ be a sequence of positive constants decreasing to 0, and $\{N_i\}$ be a sequence of strictly increasing positive integers. Now we define the rule as follows. (See [23].)

Take sample of sizes n_1, $n_2 - n_1, \ldots,$ until $\Delta_i \geq C_i$, where $\Delta_i = \max\{d(F_0^{(ni)}, F_1^{(ni)}),\ d(F_0^{(ni)}, F_2^{(ni)})\}$. Setting $N = n_i$ make the terminal decision as follows. Decide $F_0 = F_1$ iff $d(F_0^{(N)}, F_1^{(N)}) < d(F_0^{(N)}, F_2^{(N)})$.

Given α, the sequences $\{n_i\}$ and $\{C_i\}$ can be chosen such that $\mathbf{P}(N < \infty) = 1$ and the probabilities of errors are less than α. If d is Kolmogorov-distance, then $\mathbf{E}\,N < \infty$ besides the above. However, the distribution of N needs further study. For the two-population problem, Kurz and Woinsky [24] suggested a nonparametric sequential rule based on Wilcoxon statistic considering the situations (b) and (c), when $\int F_2(x)\mathrm{d}F_1(x) < \frac{1}{2}$ or $F_2(x) = F_1(x - \theta)$ with known θ. They considered asymptotic properties of their rules as the maximum of the error probabilities (denoted by α) tend to 0. Following the technique of Chow–Robbins they proved the asymptotic efficiency of the sample size, although the result is not very meaningful.

Moreover, they proved that the difference between the maximum of the error probabilities of their rule and α tends to 0 as α tends 0; such a result is almost trivial and throws no light at all on the performance of their rule. One should consider the limit of the ratio of the two above instead of their difference.

Srivastava [39] proposed sequential rules when $F_1 = N_p(\mu_1, \Sigma)$ and $F_2 = N_p(\mu_2, \Sigma)$ in the following two cases: (i) $\mu_1 - \mu_2$ known but Σ unknown; (ii) Both $\mu_1 - \mu_2$ and Σ are unknown. Given α, he constructed a sequential rule for the case (i) such that the error probabilities of the rule are less than α and its sample size is asymptotically efficient (in comparison to the sample size when the parameters are known) as $(\mu_1 - \mu_2)' \Sigma^{-1}(\mu_1 - \mu_2) \to 0$. For the case (ii), he showed that the limits of the error probabilities of his rule are $\leq \alpha$. Srivastava followed the ideas of Chow–Robbins and Simons [37]; however, his proof for the case (i) is incomplete and it is wrong for the case (ii). The main error lies in the fact that the notion of a.s. convergence as $(\mu_1 - \mu_2)' \Sigma^{-1}(\mu_1 - \mu_2) \to 0$ is not well defined.

For the situation (a) described above, there are many apparently good results available in the literature, especially in Fu's book [13]. Unfortunately, most of the results are blind copies of the two-sample sequential rules and they lack sufficient rigor, as well as meaningful formulation. Some heuristic rules, as in Smith and Yau [38], may be studied further with proper rigor.

Practically very few interesting results are available in the study of sequential rules. The problem requires first a meaningful formulation and a useful definition for asymptotic efficiency. For example in Srivastava's case (ii) no asymptotically efficient sequential rule would exist and one may then focus on the loss of efficiency. It seems that the problem may well be studied from Chernoff's viewpoint after introducing sampling cost.

References

[1] Aizerman, M.A., Braverman, E.M. and Rozonoer, L.T. (1964). The probability problem of pattern recognition learning and the method of potential function. *Automat. and Remote Control*, **26,** 1175–1190.

[2] Anderson, M.W. and Benning, R.D. (1970). A distribution-free discrimination procedure based on clustering. *I.E.E.E Trans. Inform. Theory*, **IT-16**, 341–348.

[3] Anderson, T.W. (1966). Some nonparametric multivariate procedures based on statistically equivalent blocks. *Proc. lst Internat. Symp. Multiv. Anal.*, Ed. P.R. Krishnaiah. Academic Press, New York. pp. 5–27.

[4] Beakley, G.W. and Tuteur, F.B. (1972). Distribution-free pattern verification using statistically equivalent blocks. *I.E.E.E Trans. Computers*, C–**21**, 1337–1347.

[5] Cacoullos, T. (1966). Estimation of a multivariate density. *Ann. Inst. Statist. Math.*

[6] Čencov, N.N. (1962). Evaluation of an unknown distribution density from observations. *Soviet Mathematics*, **3**, 1559–1562.

[7] Chanda, K. C. and Lee, J.C. (1975). A class of nonparametric classification rules. To appear in *Some Statistical Methods Useful in Oil Exploration*. Ed. D.B. Owen.

[8] Cover T.M. and Hart, P.E. (1967). Nearest neighbor pattern classification. *I.E.E.E. Trans. Inform. Theory*, IT–**16**, 26–31.

[9] Das Gupta, S. (1964). Nonparametric classification rules. *Sankhya*, A, **26**, 25–30.

[10] Das Gupta, S. (1973). Theories and methods in classification: A review. *Discriminant Analysis and Prediction*. Ed. T. Cacoullos. Academic Press, New York. pp. 77–137.

[11] Das Gupta, S. and Kinderman, A. (1974). Classifiability and designs for sampling. *Sankhya*, A, **36**, 237–250.

[12] Fix, E. and Hodges, J.L. (1951). Nomparametric discrimination: Consistency properties. *U.S. Air Force, School of Aviation Medicine*, Report No. 4. Randolph Field, Texas.

[13] Fu, K.S. (1968). *Sequential methods in pattern recognition and machine learning*. Academic Press, New York.

[14] Gessaman, M.P. and Gessaman, P.H. (1972). A comparison of some multivariate discrimination procedures. *Jour. Amer. Statist. Assoc.*, **67**, 468, 472.

[15] Glick, N. (1969). Estimating unconditional probabilities of correct classification. *Tech. Report No.* 3. Stanford Univ., Department of Statistics, Stanford.

[16] Glick, N. (1972). Sample-based classification procedures derived from density estimators. *Jour. Amer. Statist. Assoc.*, **67**, 116–122.

[17] Gyórfi, L. (1972). Estimation of a probability density and optimal decision functions in reproducing kernel Hilbert space. *Progress in Statistics*, Vol. I. Ed. J. Gani et. al. North-Holland Publishing Co., London.

[18] Hannan, J.F. and Robbins, H. (1955). Asymptotic solution of the compound decision problem for two completely specified distributions. *Ann. Math. Statist.*, **26**, 37–51.

[19] Hoeffding, W. and Wolfowitz, J. (1958). Distinguishability of probability measures. *Ann. Math. Statist.*, **29**, 700–718.

[20] Hudimoto, H. (1964). On a distribution-free two-way classification. *Ann. Inst. Statist. Math.*, **16**, 247–253.

[21] Johns, M.V. (1967). An empirical Bayes approach to nonparametric two-way classification. Studies in Item Analysis and Prediction. Ed. H. Solomon. Stanford University Press, Stanford.

[22] Kanal, L. (1974). Patterns in Pattern Recognition: 1968–1974. *I.E.E.E. Trans. Inform. Theory*, IT–**20**, No. 6, 697–722.

[23] Kinderman, A. (1972). On some properties of classification: classifiability, asymptotic relative efficiency and a complete class theorem. *Tech. Report No. 178*, Dept. of Statistics, University of Minnesota, Minneapolis.

[24] Kurz, L. and Woinsky, M.M. (1969). Sequential nonparametric two-way classification with prescribed maximum asymptotic error. *Ann. Math. Statist.*, **40**, 445–455.

[25] Loftsgaarden, D.O. and Quesenberry, C.P. (1965). A nonparametric estimate of a multivariate density estimation. *Ann. Math. Statist.*, **36**, 1049–1051.

[26] Matusita, K. (1956). Decision rule, based on the distance, for the classification problem. *Ann. Inst. Statist. Math.*, **8**, 67–77.

[27] Moore, D.S. and Henrichson, E.G. (1969). Uniform consistency of some estimates of a density function. *Ann. Math. Statist.*, **40**, 1499–1502.

[28] Parzen, E. (1962). On estimation of a probability density function and mode. *Ann. Math. Statist.*, **33**, 1065–1076.

[29] Patrick, E.A. (1972). *Fundamentals of Pattern Recognition.* Prentice-Hall, Englewood Cliffs, New Jersey.

[30] Pelto, C.R. (1969). Adaptive nonparametric classification. *Technometrics,* **11,** 775–792.

[31] Quesenberry, C.P. and Gessaman, M.P. (1968). Nonparametric discrimination using tolerance regions. *Ann. Math. Statist.,* **39,** 664–673.

[32] Rao, C.R. (1952). *Advanced Statistical Methods in Biometric Research.* Wiley, New York.

[33] Rejtö, L. and Revesz. (1973). Density estimation and pattern classification. *Problems of Control and Inform. Theory,* **2,** 67–80.

[34] Robbins, H. (1964). The empirical Bayes approach to statistical decision function. *Ann. Math. Statist.,* **35,** 1–20.

[35] Samuel, E. (1963). Asymptotic solutions of the sequential compound decision problem. *Ann. Math. Statist.,* **34,** 1079–1094.

[36] Samuel, E. (1963). Note on a sequential compound decision problem. *Ann. Math. Statist.,* **34,** 1095–1097.

[37] Simons, G. (1968). On the cost of not knowing the variance when making a fixed-width confidence interval for the mean. *Ann. Math. Statist.,* **39,** 1946–1952.

[38] Smith, S.E. and Yau, S.S. (1972). Linear sequential pattern classification. *I.E.E.E. Trans. Inf. Theory,* 673–678.

[39] Srivastava, M.S. (1973). A sequential approach to classification: Cost of not knowing the covariance matrix. *Jour. Multiv. Anal.,* **3,** 173–183.

[40] Stoller, D.C. (1954). Univariate two-population distribution-free discrimination. *Jour. Math. Statist. Assoc.,* **49,** 770–777.

[41] Tanaka, K. (1970). On the pattern classification problems by learning (I and II). *Bull. Math. Statist.,* **14,** 31–49 and 61–73.

[42] Van Ryzin, J. (1965). Nonparametric Bayesian decision procedure for (pattern) classification with stochastic learning. *Proc. IV Prague Conf. on Inf. Theory, Statist. Dec. Functions, and Random Processes.* 479–494.

[43] Van Ryzin, J. (1966). Repetitive play on finite statistical games with unknown distributions. *Ann. Math. Statist.,* **37,** 976–994.

[44] Van Ryzin, J. (1966). Bayes risk consistency of classification procedures using density estimation. *Sankhya,* A, **28,** 201–270.

[45] Van Ryzin, J. (1967). A stochastic a posteriori updating algorithm for pattern recognition. *Math. Anal. Appl.,* **20,** 359–379.

[46] Wagner, T. J. (1971). Convergence of the nearest neighbor rule. *I.E.E.E. Trans. Inf. Theory,* IT–**17,** No. 5, 566–571.

[47] Wagner, T.J. (1973). Strong consistency of a nonparametric estimate of a density function. *I.E.E.E. Trans. Systems Man Cybernetics,* SMC–**3.**

[48] Wahba, G. (1975). Optimal convergence properties of variable knot, kernel, and orthogonal series methods for density estimation. *Ann Statist.,* **3,** 15–29.

[49] Watanabe, M. (1972). On asymptotically optimal algorithms for pattern classification problems. *Bull. Math. Statist.,* **15,** 31–48.

[50] Watanabe, M. (1974). On convergences of asymptotically optimal discriminant function for pattern classification problems. *Bull. Math. Statist.,* **16,** 23–34.

[51] Wolverton, C.T. and Wagner, T.J. (1969). Asymptotically optimal discriminant functions for pattern classification. *I.E.E.E. Trans. Inf. Theory,* IT–**16,** No. 2, 258–265.

[52] Yamato, H. (1971). Sequential estimation of a continuous probability density function. *Bull. Math. Statist.,* **14,** 1–12.

[53] Yao, J. (1972). A sequential classification into one of several populations. *Bull. Math. Statist.,* **15,** 19–28.

P.R. Krishnaiah, ed., *Multivariate Analysis–IV*

MULTIVARIATE STATISTICAL PROBLEMS IN METEOROLOGY*

RICHARD H. JONES**
Department of Information and Computer Sciences
University of Hawaii, Honolulu, Hawaii

Brief historical reviews are given of some of the multivariate statistical problems in meteorology with suggestions for improving the current state of the art. Topics covered are statistical aids to local forecasts, time series analysis, statistical-dynamical forecasts, and probability forecasts using logistic models. It is shown that techniques for fitting multivariate mixed models to time series can be modified and applied to statistical-dynamical forecasts.

1. Introduction

Meteorology is one of the *most* multivariate disciplines. The data collected and stored daily on a world wide basis can be measured in tons of magnetic tapes. Data are collected from over a thousand surface stations, hundreds of stations collecting upper air data, weather ships, commercial and military ships and aircraft, satellites, constant level balloons, and other sources. One of the major uses of this vast collection of data is the production of numerical forecasts (Thompson, 1961). Twice daily the raw data are interpolated to grid points at various levels in the atmosphere producing initial value maps. The non-linear equations of motion of the atmosphere are numerically integrated forward in time producing forecasts at the grid points of temperatures, heights of constant pressure surfaces, moisture, and vertical motion. Twelve, twenty-four and thirty-six hour forecasts are quite accurate and useful for such purposes as planning aircraft routes. However, since the grid spacing is on the order of several hundred kilometers, local forecasts present a problem. The major multivariate statistical problems in meteorology are in the areas of statistical-

* Research sponsored by the Air Force Office of Scientific Research, Air Force Systems Command, USAF, under Grant No. AFOSR-72-2405. The United States Government is authorized to reproduce and distribute reprints for Governmental purposes notwithstanding any copyright notation hereon.

** Present address: Department of Biometrics, University of Colorado Medical Center, Denver, Colorado, 80220.

dynamical prediction and the production of local forecasts from current observations and numerical forecasts.

2. Statistical aids to local forecasts

A significant input of multivariate techniques into the field of meteorology was provided by Robert G. Miller's Ph.D. dissertation at Harvard written under the direction of Cochran and Dempster and published as a Travelers Research Center Technical Memo and as an American Meteorological Society Monograph (Miller, 1962). In addition to the dissertation, the computer programs that were produced were used in subsequent meteorological research at Travelers Research Center. The emphasis of this work was on stepwise regression and discriminant analysis procedures. The discriminant analysis programs were used not only for forecasting the occurrence of events such as rain or a closed airfield, but also for estimating the probability of occurrence of these events, which lead to the introduction of probability forecasts.

The extension of this work to operational forecasts was undertaken by the Techniques Development Laboratory, National Weather Service, NOAA, Silver Spring, Maryland. Operational forecasting equations for meteorological variables have been developed for local stations, and the forecasts are sent out by facsimile several times a day. Most of the forecasts are developed by stepwise regression methods choosing predictors from locally available surface and upper air data, and from numerical forecasts.

The discriminant analysis approach to probability forecasts gave way to a new program developed at Travelers Research Center called REEP (Regression Estimation of Event Probabilities) (Miller, 1964). This is simply linear regression of a dichotomous response variable on dichotomized independent variables. The method is computationally very efficient, and used with care can produce useful probability forecasts. It is now being used operationally to produce probability forecasts of cloud amount, ceiling, visibility, precipitation occurrence, precipitation amount, thunderstorms, type of precipitation, and severe local weather.

Recently there has been interest in the use of the logistic function for probability estimation which has the same functional form as the probability in linear discriminant analysis,

$$\mathbf{P} = \left(1 + \exp\left(\beta_0 + \sum_{k=1}^{P} \beta_k x_k\right)\right)^{-1}, \tag{1}$$

where x_k are the independent variables, and the parameters, β_k, are estimated by maximum likelihood (Brelsford and Jones, 1967; Bryan, 1968; Jones, 1968; Press, 1972; Jones, 1975). Probability forecasts based on the logistic function are now being used operationally to forecast the conditional probability of frozen precipitation given that there is precipitation (Glahn and Bocchieri, 1975).

Techniques Development Laboratory has, since 1968, shown that multivariate statistics has a place in operational forecasting. Much of this success has been due to the use of something called model output statistics (MOS) (Klein and Glahn, 1974). The output variables of numerical dynamical forecasts are used as independent variables for statistical forecasts using such methods as stepwise regression for forecasts of dependent variables of interest, and REEP or the logistic function for probability forecasts.

3. Time series analysis

The operational use of time series analysis in meteorology is confined mainly to the inclusion of time lagged values of the independent variables in the regression equations, i.e., autoregression. The stepwise multivariate autoregression computational algorithm developed by Whittle (1963) has been used experimentally for statistical forecasts (Jones, 1964), but since the atmosphere is a statistical-dynamical system, purely statistical methods do not seem to be the proper approach. As pointed out in a recent article by Ramage (1976), the atmospheric scientists often take the other extreme position and treat the atmosphere as a purely dynamical system — given the initial conditions, the state of the atmosphere can be forecast perfectly — any distance into the future. The inability to do this is attributed to the inability to observe the initial state exactly. Ramage introduced the concept of unpredictable turbulent bursts in the atmosphere on all scales, which limit the predictability. This is similar to the innovations in time series analysis — the unpredictable random input of each time point. Ramage also noted that numerical dynamical forecasts are highly contaminated by statistics in the form of smoothing, to prevent certain types of motion from dominating, and empirical corrections introduced at all stages of the forecast. If we conclude that purely statistical forecasts are unsatisfactory because the atmosphere is a dynamical system, and the numerical dynamical forecasts are permeated with empirical statistics, we see that the real problem is one of statistical-dynamical forecasting of many variables.

4. Statistical-dynamical forecasts

There are several approaches to statistical-dynamical forecasting. One is based on Kalman (1960) optimum estimation, where attempts are made at combining observational data with the most recent forecast of the field at the observational time in an optimal manner to produce better initial conditions for use in the dynamical forecasting programs (Jones, 1965; Peterson, 1968). Epstein (1969) introduced the idea of stochastic dynamic prediction where the uncertainty in the initial conditions are represented as probability distributions, and attempts are made to propogate the distributions through the dynamical equations. Although this in general is very difficult, some success has been made in propogating second moment statistics (Peterson, 1973).

Another area of recent interest is sometimes referred to as four dimensional assimilation — updating initial value maps as data becomes available in time. Much of the useful data, such as from aircraft and satellites, is available at times other than the twice daily sampling times (Peterson, 1968; Eddy, 1971).

Recently C.E. Leith (National Center for Atmospheric Research, Boulder, Colorado) and E.N. Lorenz (Massachusetts Institute of Technology) have carried out some interesting numerical experiments (Leith, 1974). The data base consisted of ten years of twice daily analyses of 500 mb height fields for the Northern Hemisphere. A 500 mb height field is the height of a constant pressure surface (500 mb) which varies between 17 000 and 20 000 feet. This level is often used since it represents the "average" atmosphere — approximately half the air being above and half below this level (surface pressure being near 1000 mb). Each field was expanded in a series of spherical harmonics using wave numbers up to eighteen giving 190 coefficients (symmetry across the equator was assumed). Forecasts were made using a very efficient spectral barotropic forecast model devised by Lorenz which produces a 24 hour forecast in about one second on a CDC 7600. In the statistical-dynamical study by Leith (1974), spherical harmonics up to wave number twelve were used giving 91 coefficients and reducing the computer time by a factor of three. Also only the winter months of December, January, and February were used.

The dynamical forecasts were improved by the use of regression, using both initial time coefficients and dynamical forecast coefficients for two different initial times as independent variables. This gives a possible total of 273 independent variables that could be used for each of the 91 dependent variables (the actual coefficient at forecast time). Each spherical harmonic for wave numbers greater than zero have a sine and a cosine coefficient, and for wave number zero, only one coefficient. To reduce the dimensionality of the problem, the coefficients of each harmonic were forecast separately giving a maximum of six independent variables for each dependent variable.

There is a basic difference between dynamical forecasts and statistical forecasts as the forecast interval increases. Dynamical forecasts are constrained so they do not diverge. As the forecast interval increases, a weather map is forecast which may have no correlation with the actual map at the forecast time, but it appears to be a reasonable weather map. In a sense, the dynamical forecast converges to a random weather map as the forecast interval increases, while a statistical forecast converges to the mean weather map (climatology). The error variance of a statistical forecast will converge to the variance of the process as the forecast interval increases, while the error variance of a dynamical forecast approaches twice the process variance.

Leith compared the barotropic dynamical forecast with purely statistical forecasts (using the coefficients of the initial value maps as independent variables), and various combinations of statistical-dynamical forecasts. The dynamical forecast had lower variance for forecasts up to two days. For forecasts of more than two days the statistical forecast was better. Since the dynamical forecast starts off better for short forecast times, it is obvious that there must be a crossing point since for long forecast times the dynamical forecast approaches twice the variance of the statistical forecast.

An improvement over both statistical forecasts and dynamical forecasts was obtained by using the dynamically forecast coefficients as the independent variables in a statistical forecast. This allows the non-linear dynamics to be incorporated in the model, and the use of regression avoids having the estimate approach a random map as the forecast period increases.

5. Possible alternatives

Recent advances in applied regression, model identification, and time series analysis should enable multivariate statistical techniques to be used more effectively in meteorology.

Ridge regression (Hoerl and Kennard, 1970; Marquardt and Snee, 1975) may be better than the stepwise regression techniques now used as aids to local forecasts since the possible independent variables are highly intercorrelated. Rather than throw 50 or 100 variables into the pot and select six or seven by stepwise regression, a smaller number of variables could be selected which should naturally be included based on meteorological considerations, and ridge regression used to estimate the coefficients.

Since probability estimation using the logistic function involves maximum likelihood estimation, for large sample sizes, which are common in meteorology, asymptotic tests such as tests based on minus two log likelihood can be used for model selection. To aid in this selection, Akaike's Information Criterion (AIC) is quite useful (Akaike, 1973). For each subset of independent variables

$$\text{AIC} = -2 \ln \text{likelihood} + 2p$$

is calculated, where p is the number of estimated parameters, and the model for which AIC is a minimum is selected. The term $2p$ is the penalty for estimating p parameters and was derived using decision theoretic and information theoretic arguments.

For numerical weather forecasting, it must be recognized that not only are the initial conditions not known exactly, but also that the dynamical model is not perfect. Kalman's theory assumes that the dynamical model is exact and concentrates on finding the best estimate of the initial conditions. Since the model is not exact, this suggests trying to find good initial estimates *and* using the output from the dynamical forecast as independent variables in a regression forecast. The techniques involved are similar to the methods for fitting mixed autoregressive, moving average models to multivariate time series. The simplest, first order, case is,

$$X_t = AX_{t-1} + B\varepsilon_{t-1} + \varepsilon_t, \tag{2}$$

where X_t is an m dimensional column vector of the time series data collected at time t, A is an m by m matrix of autoregressive coefficients representing the dependence of the data on the previous time point, ε_t is an m vector of the unpredictable innovations (or turbulent bursts) at time t, and B is an m by m matrix of moving average coefficients representing the dependence of the data on the prediction error at the previous time. If A and B are known, or have been estimated, the prediction of the process at time t, using data at time $t-1$ is

$$\hat{X}_t = AX_{t-1} + B\varepsilon_{t-1}. \tag{3}$$

The prediction error at time t is

$$\varepsilon_t = X_t - \hat{X}_t. \tag{4}$$

When the observations at time t, X_t, are available, this can be calculated and the prediction for time $t+1$ made,

$$\hat{X}_{t+1} = AX_t + B\varepsilon_t. \tag{5}$$

This results in a recursive prediction scheme.

The estimation problem is more difficult. Given data at n consecutive time points, $X_1, X_2, \ldots X_n$, initial conditions must be assumed (after subtraction of means) such as $X_0 = 0$, $\varepsilon_0 = 0$. The non-linear estimation problem is to choose A and B such that the determinant of the residual sum of squares and cross product matrix is a minimum,

$$\left| \sum_{t=1}^{n} \varepsilon_t \varepsilon'_t \right| = \min, \tag{6}$$

where ε_t is calculated recursively using equations (3) and (4).

If the time series involves non-lincar dynamics, the model could be written

$$X_t = f(X_{t-1}) + \varepsilon_t. \tag{7}$$

If the dynamics are exact, ε_t again represents the unpredictable innovations, and are uncorrelated in time. If the dynamics are approximate, ε_t will be serially correlated indicating that statistical improvement is possible. If, in addition, X_t can not be observed exactly, additional improvement is possible. Let Y_t be the observed data at time t and X_t the true values of the observed quantities. A recursive procedure assumes that at time t there is a "best" estimate of the true values, $\hat{X}_{t|t}$. This estimate is a combination of the observations at time t, Y_t, and the prediction from the previous time point, $\hat{X}_{t|t-1}$ (Kalman, 1960),

$$\hat{X}_{t|t} = \hat{X}_{t|t-1} + B(Y_t - \hat{X}_{t|t-1}). \tag{8}$$

The forecasts at future times are then calculated using the dynamical equations. For example, a one step forecast is

$$\hat{X}_{t+1|t} = f(\hat{X}_{t|t}). \tag{9}$$

Assuming imperfect dynamics, this can then be improved by regression, i.e.

$$\hat{X}_{t+1|t} = A\,f(\hat{X}_{t|t}). \tag{10}$$

In this equation the dynamic operator $f(\ )$ produces a vector of forecast variables which are then used as independent variables in regression. The final recursive model is

$$\hat{X}_{t+1|t} = A\, f(\hat{X}_{t|t})$$

where (11)

$$\hat{X}_{t|t} = \hat{X}_{t|t-1} + B(Y_t - \hat{X}_{t|t-1}).$$

A and B are estimated such that

$$\left| \sum_{t=1}^{n} (Y_t - \hat{X}_{t\,|\,t-1})(Y_t - \hat{X}_{t\,|\,t-1})' \right| = \min. \tag{12}$$

This involves non-linear estimation methods as with mixed models, but also incorporates the dynamics.

This type of estimation would be feasible in a spectral (spherical harmonic) forecast on individual harmonics or on a grid point model on individual grid points. The interaction among grid points would occur in the dynamical portion of the forecasts.

6. Conclusion

The predictability of the atmosphere is limited. The point has been reached where more effort should be devoted to the combination of statistical and dynamical methods, rather than devoting most of the effort to refining the dynamic portion of the forecast where the improvement for each advance is becoming smaller and smaller.

Acknowledgment

The author is grateful to Harry R. Glahn of the Techniques Development Laboratory for helpful comments on the first draft of this paper. The final version was prepared while the author was visiting the National Center for Atmospheric Research, Boulder, Colorado, which is sponsored by the National Science Foundation.

References

[1] Akaike, H. (1973). Information theory and an extension of the maximum likelihood principle. *Second International Symposium on Information Theory* (B.N. Petrov and F. Csaki, Eds.), Akademiai Kiado, Budapest, 267–281.

[2] Brelsford, W.M. and Jones, R.H. (1967). Estimating probabilities. *Monthly Weather Review*, **95**, 570–576.

[3] Bryan, J.G. (1968). Development of a computational technique for multi-group probability forecasting using a linear exponential-quotient (LEQ) model. *Proceedings of the First Statistical Meteorological Conference*, Hartford, Connecticut, May 27–29, 1968, American Meteorological Society, Boston, 61–75.

[4] Eddy, A. (1971). Four-dimensional interpolation from sets of atmospheric soundings. *Preprints: International Symposium on Probability and Statistics in the Atmospheric Sciences*, Honolulu, Hawaii, June 1–4, 1971, American Meteorological Society, Boston, 116–119.

[5] Epstein, E.S. (1969). Stochastic dynamic prediction. *Tellus*, **21**, 739–759.

[6] Glahn, H.R. and Bocchieri, J.R. (1975). Objective estimation of the conditional probability of frozen precipitation. *Mon. Wea. Rev.*, **103**, 3–15.

[7] Hoerl, A.E. and Kennard, R.W. (1970). Ridge regression: biased estimation for non-orthogonal problems. *Technometrics*, **12**, 55–67.

[8] Jones, R.H. (1964). Prediction of multivariate time series. *Journal of Applied Meteorology*, **3**, 285–289.

[9] Jones, R.H. (1965). Optimal estimation of initial conditions for numerical prediction. *J. Atm. Sci.*, **22**, 658–663.

[10] Jones, R.H. (1968). A non-linear model for estimating probabilities of k events. *Monthly Weather Review*, **96**, 383–384.

[11] Jones, R.H. (1975). Probability estimation using a multinomial logistic function. *Journal of Statistical Computation and Simulation*, **3**, 315–329.

[12] Kalman, R.E. (1960). A new approach to linear filtering and prediction problems. ASME Transactions, Part D (*Journal of Basic Engineering*), **82**, 35–45.

[13] Klein, W.H. and Glahn, H.R. (1974). Forecasting local weather by means of model output statistics. *Bull. Amer. Meteor. Soc.*, **55**, 1217–1227.

[14] Leith, C.E. (1974). Spectral statistical dynamical forecast experiments. Presented at the International Symposium on Spectral Methods in Numerical Weather Prediction, Copenhagen, August 12–16, 1974.

[15] Marquardt, D.W. and Snee, R.D. (1975). Ridge regression in practice. *The American Statistician*, **29**, 3–20.

[16] Miller, R.G. (1962). Statistical prediction by discriminant analysis. *Meteorological Monographs*, **4**, No. 25, Amer. Met. Soc., Boston.

[17] Miller, R.G. (1964). Regression estimation of event probabilities. Travelers Research Center Technical Report No. 1. U.S. Weather Bureau Contract Cwb – 10704, April 1964.

[18] Petersen, D.P. (1968). On the concept and implementation of sequential analysis for linear random fields. *Tellus*, **20**, 673–686.

[19] Petersen, D.P. (1973). Propagation of second-moment statistics of estimation error in a numerical advective scheme. *Preprints: Third Conference on Probability and Statistics in Atmospheric Science*, Boulder, Colorado, June 19–22, 1973. Amer. Met. Soc., Boston, 193–197.

[20] Press, S.J. (1972). *Applied Multivariate Analysis*. Holt, Rinehart and Winston, New York.

[21] Ramage, C.S. (1976). Prognosis for weather forecasting. *Bull. Am. Meteor. Soc.*, **57**, 4–10.

[22] Thompson, P.D. (1961). *Numerical Weather Analysis and Prediction*. MacMillan, New York.

[23] Whittle, P. (1963). On the fitting of multivariate autoregressions, and the approximate canonical factorization of a spectral density matrix. *Biometrika*, **50**, 129–134.

P.R. Krishnaiah, ed., *Multivariate Analysis–IV*

MULTIVARIATE CONTINGENCY TABLES AND SOME FURTHER PROBLEMS IN MULTIVARIATE ANALYSIS

Sir Maurice KENDALL
International Statistical Institute, London, U.K.

The paper discusses a number of outstanding problems in Multivariate Analysis with particular reference to multivariate contingency tables. These included: the problem of dealing with empty cells; Simpson's Paradox arising from the amalgamation of sections of a multivariate table; some problems in the interpretation of chi squared; measures of dependence and difficulties associated with the large number of hypotheses which can be set up for such tables. Reference is also made to non-probabilistic methods, functional relationship and the costs of machine programs.

1.

Multivariate analysis is one of the great and growing areas of statistics but is in some danger of growing in the wrong direction, too much towards theoretical development and too little towards the solution of problems confronting the practicing statistician in his daily work. In this paper I want to describe what I regard as the major unsolved problems in the treatment of multivariate relationship, with particular reference to contingency tables. There are ten groups of them. This is far from being a complete list of all the things I don't know about multivariate analysis. It is, however, a list of those I would most like to know in my statistical work.

2. Empty cells

Table 1 shows, for a recent survey in Fiji, the distribution of 6,779 women by age and number of children. In this particular case one could, of course, compute a product-moment correlation coefficient, but I adduce it to illustrate a different problem. In the north-east corner of the table there are a number of zeros for the simple reason that there are relatively few women in the upper age groups. In the south-west corner there are a number of zeros for a different reason, that very young nubile women have had no time to generate large families.

Table 1

Total number of children	Age groups 15 – 19	20 – 24	25 – 29	30 – 34	35 – 39	40 – 44	45 – 49	50 – 54	55 – 59	60 – 64	65 – 69	70 – 74	75 – 79	80 – 84	85 – 89	90 – 94	95 – 99
0	154	235	97	54	35	27	32	35	24	23	8	4	2	1	0	0	5
1	79	344	165	50	44	25	20	30	21	17	9	5	2	1	0	2	5
2	17	256	275	98	53	36	26	32	17	18	7	5	0	0	0	0	3
3	4	121	250	165	84	49	29	36	20	15	12	6	1	2	0	3	1
4	0	39	188	177	80	45	26	48	33	23	13	6	2	2	1	0	1
5	0	10	87	190	119	73	38	59	36	15	13	16	3	4	0	0	1
6	0	1	44	122	100	76	49	50	39	29	14	10	1	3	0	0	2
7	0	0	10	69	117	84	50	54	34	27	13	9	4	1	2	2	2
8	0	0	5	35	65	67	62	66	26	19	24	14	3	1	1	1	1
9	0	0	1	16	36	60	48	44	28	16	7	13	0	3	1	2	1
10	0	0	0	9	37	98	88	139	82	58	15	13	3	3	3	3	0

3.

If certain cells in a contingency table are theoretically zero no problem arises—there is no special reason why such a table should be rectangular. But in the south-west corner it is difficult to say what is biologically impossible, at least to the extent of ruling out of consideration all the zero cells which arise in a moderate sized sample. Even small frequencies, which may well occur, provide difficulties because they impair tests of independence. The problem is not to be solved by amalgamating adjacent rows or columns. For one thing, in an un-ordered table rows and columns are not adjacent in any real sense. But even if they are, as in Table 1, amalgamation in the south-west creates a heavy loss of information in the north-west.

4. Simpson's paradox

In any case, the amalgamation of a multivariate table into marginals may result in seriously misleading results, something pointed out by Yule (1912) and more recently emphasized by Simpson (1951). Consider the simple $2 \times 2 \times 2$ table (Table 2).

Table 2

A = sex of patient with B (= treatment for a disease), C = cure

Males (A thousands)				Females (not A thousands)				Males and Females together, thousands			
	B	not B			B	not B			B	not B	
C	10	100	110	C	100	50	150	C	110	150	260
not C	100	730	830	not C	50	20	70	not C	150	750	900
	110	830	940		150	70	220		260	900	1160

In the A group association between B and C is negative; so also in the not-A group. But in the two together association is positive. I consider the data again below.

5. Shortcomings of chi-squared

It is standard practice to test the indepencence of variables in a contingency table by working out the frequencies (based on margins) which would appear should independence exist and to test in the Type III χ^2 distribution. The justification for this procedure rests on the facts (1) that *asymptotically* the distribution of the statistics concerned tends to the Type III form and (2) that the sample is large enough (or the cell-frequencies are large enough) for asymptotic theory to apply.

There are two forms of statistic in common use, the wider applications of the second having had to wait the general availability of computers.

Pearson's form

$$\chi^2_{\mathrm{P}} = \sum \frac{(F-T)^2}{T} \tag{1}$$

where F is the observed frequency, T the theoretical frequency and summation takes place over all the cells of the table.

Maximum likelihood

$$\chi^2_{\mathrm{ML}} = 2 \sum F \log \frac{F}{T}\,. \tag{2}$$

Both tend to the Type III form and are therefore usually regarded as equivalent within acceptable arithmetic limits. It comes as a shock, then to find that in practice they may give widely differing results. For Table 2, taking fixed bivariate margins and leaving, therefore, 7 degrees of freedom

$$\chi^2_{\mathrm{P}} = 2760, \qquad \chi^2_{\mathrm{ML}} = 1791.$$

The reason is that the χ^2_{P} approximation breaks down if $F - T \not< T$ *for any one cell.* In fact, for the log likelihood

$$\begin{aligned} -2 \log L &= 2 \sum F \log \frac{F}{T} \\ &= -2 \sum \{T + F - T\} \log \left\{1 + \frac{F-T}{T}\right\} \qquad (3) \\ &= \sum \frac{(F-T)^2}{T} + \mathrm{O}\left\{\frac{(F-T)^3}{T^2}\right\} \end{aligned}$$

and the last term cannot be neglected.

6.

Now it may be argued that both results are so highly significant that the difference does not matter. But it clearly does, in at least two respects:

(1) It is important to consider the deviations $F - T$ individually and not to rely on χ^2 alone. The situation is similar to that of regression, where it is equally important to look at the residuals individually and not to confine attention to their sum of squares.

(2) When independence is rejected it is sometimes desirable to set up a measure of dependence, and the coefficients usually prepared for the purpose (Karl Pearson's, with modifications by Tchuprow and Cramér) depend critically on χ^2_P.

7. Measure of dependence

For the past thirty or forty years statisticians have been mainly concerned to test for independence and let it go at that, without paying much attention to the measurement of dependence. I think we must reconsider this matter. How does one set up a measure of dependence in Table 1, for example? There clearly is a relationship, but one would like to measure its intensity for comparison with other years, other countries or even within the country, as for instance if the table is broken down into the three ethnic groups of Fiji, Fijian, Indian and Chinese.

8.

It may well be that for more complicated types of relationship there is no single summary statistic expression of the dependence. We are apt to rely too heavily on the correlation coefficient even in continuous variation. The question then arises whether we need to categorise the nature of the contingency table and to devise different measures according to its general pattern. The problem is not easy even in two dimensions. For four or five it becomes quite formidable. A precise formulation of the concept of 'shape' even for ordered tables, is a subject to which the mathematicians might well give some attention. Analogous problems arise in many other multivariate contents such as pattern recognition, cluster analysis and component analysis. Another way of stating the same point is that we may require to set up a model to fit to the data. Failing that we may be thrown back on the complexity inherent in the original table.

9. Multiplicity of hypotheses

It has been emphasized by many writers, notably Plackett and Goodman, that the number of hypotheses which can be set up for a table in more than two dimensions is very large. For a three-way table there are 17, for a four-way table 166 and for a five-way table several thousands. In the past tables of more than three dimensions have been rare, or at least have rarely appeared in the literature. Nowadays we can generate them quite easily and in fact need to do so if causal relations are really to be brought to light. It is then necessary to set about hypothesis testing in a systematic way. Consider, for example, a four-way table. We have four one-way (univariate) marginal totals, 6 two-way margins, 4 three-way margins and the four-way table itself. Unless one of the variables is marked out as a regressand ('dependent' or 'explained' variable) and we are concerned with its dependence on the others—analogous to the regression situation — we are primarily interested in the inter-dependence of the others and the problem is to know in what order to go about the exploration.

10.

Suppose that in a four-way table the variables are denoted by A, B, C, D, with respectively r, s, t, u arrays, one obvious way to begin is to take the univariate margins as fixed, calculate the 'independent' frequencies and test chi-squared with $(rstu - r - s - t - u + 3)$ degrees of freedom. If this is not significant we can regard the variables as independent. In practice, such an event would be rare. If independence is rejected we could proceed by testing all the r three-way tables, all the s three-way tables etc., obtained by considering the $r, s, \ldots$ values of $A, B, \ldots$ separately. We could then go on to test all the possible two-way tables, for example the rs tables obtained for separate pairs of values of A, B. Perhaps this is what we should do, but the interpretation of the results becomes more and more complicated and clear conclusions are hard to draw.

11.

For three-way tables the situation is tractable. Machine programs now exist to work out the χ^2 values corresponding to the fixation of various types of margin. For example, with the data of Table 2 we have Table 3.

Table 3

Marginal Totals Fitted[a]	χ^2_{ML}	Degress of freedom
	1791	7
A	1310	6
B	1418	6
C	1418	6
$A+B$	936.3	5
$B+C$	1044	5
$C+A$	936.3	5
$A+B+C$	562.7	4
AB	655.7	4
BC	974.7	4
CA	655.7	4
$AB+C$	282.0	3
$BC+A$	493.5	3
$CA+B$	282.0	3
$AB+BC$	212.8	2
$AC+BC$	212.8	2
$AB+AC$	1.378	2
$AB+AC+BC$	0.038	1

[a] If the two-way marginal AB is fitted then the assumption is that the one-way marginals A and B are also fitted.

The most parsimonious acceptable fit is $AB+CA$. It appears that the apparent positive relation between treatment and cure in the total table (both sexes together) is due to association between sex and treatment and between sex and cure.

12.

For tables in further dimensions the number of possibilities increases rapidly. An exhaustive enumerator for more than four is practically impossible. There then arises the interesting question (rather like that of retaining non-redundant variables in regression) whether one can find a systematic way through the multiplicity of possibilities which enables us to discard a number of them. Certainly one can do this by rule of thumb. But we may still be faced with the fact that several different models of equivalent parsimony fit the data equally well.

13. Dependency relationships

In an ordinary regression relationship of a regressand on a set of regressors it is customary to find some dependence among the regressor variables. This creates difficulty in interpretation, especially if there are near-collinearities among them and their covariance matrix approaches degeneracy. Consider the analogous problem when one of the variables in a multivariate contingency array is to be related to the others: for example, the dependence of birth rate y on martital status, education, ethnic grouping and proportion using contraceptives. There are prior grounds for supposing these regressor-type variables to be inter-correlated — if they were independent then the relationship of y could be considered on each separately.

14.

One familiar method sometimes advocated in this context is to substitute numerical dummy values for the categories. E.G. an ethnic division into two groups can be denoted by $(1,0)$; an ordered classification into three classes by $(-1,0,1)$; an unordered classification by $(1,0,0)$, $(0,1,0)$, $(0,0,1)$. An ordinary regression can then be carried through. However this approach does not solve the problem and indeed obscures it. The effect of writing in dummy values is merely to average relationships which should be kept distinct.

15.

I cannot see any general satisfactory solution to this problem. Some ordered classifications can, perhaps, be scaled. Others, like ethnic classification, should be kept distinct and a separate study carried out on each, melding only being permitted when it is shown that the relationship under study is the same for them all. But these are expedients, not really a fundamental solution. It may be that in ordered categorisation we are setting an impossible task if we wish to exhibit a relationship with the simplicity of a regression model. The problem emphasizes the need for further study of relationships and partial relationships among categorized variables. Until this problem is solved (if it can be solved) we have severe difficulty in assigning to a dependent variable the relative contributions of the individual regressor-type variables. Perhaps a multiplicative model with log y as the regressand, might be suitable in some cases but the interpretation of the interaction terms is not always easy.

16. The selection of subsets of variables

In field surveys, market research and medicine the number of variables may be large. The use of regressions in such cases is common but in my opinion is to be condemned for reasons I have set out in Multivariate Analysis (1975). Where interrelationship rather than relationship is concerned some of the resources available for continuous variables are not available, for example principal component analysis. On the other hand, even if the data were available in abundance, the prospect of analysing a 20-dimensional contingency table is horrifying. The problem then arises whether we can (a) simplify the situation by transforming to orthogonal 'variables' (b) develop a rejection routine for eliminating variables which are not contributory to the purpose of the analysis in hand or (c) selecting sub-sets for analysis and if so which sub-sets. It would be useful to have a systematic search routine to deal with such cases but I find it hard to think of one. In practice, it seems one has to proceed by a good deal of trial and error or appeal to extraneous non-statistical reasons for the selection.

17.

In the extreme case the number of variables may exceed the number of individuals on which they are observed. This may happen, for example, in a medical context where a large number of symptoms are observed on relatively few patients. Even if the variables are continuous the problem of classification or discrimination is bedevilled by the fact that the covariance matrix of the observations is degenerate; or, to put it another way, on the sample of this size there are a number of collinearities in the variables, though such may not exist in the full population. The difficulty is enhanced for classificatory variables.

18.

It seems to me useless to say that such data present impossible problems just because they fall outside the methods developed by statisticians. The problem has to be tackled de novo in the light of what is the object of the study. In discrimination, for example, it would seem fruitful to follow the distribution-free line suggested in Kendall and Stuart, vol 3 (1975) and to take the variables one by one, ignoring those which will be found to add nothing to the discriminatory process. In cluster analysis the problem is more difficult. One can work on the basis of similarity indices and although there are some drawbacks in using them as the complement of measures of distance, this seems to be the best method now available; but there may well be better methods.

19. Non-probabilistic methods

The theory of statistics has been dominated so long by probabililistic ideas that there is some danger of trying to apply them in certain areas where they are quite inappropriate. We may thave a set of objects which form the complete population, e.g. in a study of the towns in a given country, or the countries of the world. Even if the subset available for scrutiny is only part of the population it may well not be a random sample, e.g. patients arriving for treatment at a hospital or the visible stars in the sky. Such aggregates give rise to problems of classification, of dependence and interdependence relations and even of hypothesis testing. It seems to me that the logic of reasoning about such aggregates may need extensive study. That we have to reason about them is certain; but how we put confidence in our conclusions or measure the strength of our convictions is very much an open question.

20.

As an example of the way in which existing probabilistic ideas can be taken over in a misleading way, consider the application of the

Mahalanobis distance in cluster analysis. The attraction of the Mahalanobis distance is that it is scale-invariant, whereas most measures of distance in cluster analysis are not. But in cluster analysis we are concerned with setting up a measure of distance between points, and the Mahalanobis distance between a pair of points does not exist. In fact, it depends on the inverse of the covariance matrix of the pair, and that matrix is degenerate. To use the covariance matrix of a whole set of points to measure the distance between pairs seems to me to beg the question. The whole point of cluster analysis is that we only undertake it if we suspect that the observations do not form a single homogeneous group.

21. Functional relationship and variables subject to error

The theory of regression is so easy, compared with the fitting of functional relations where the variables are all subject to experimental error, that for decades statisticians have concentrated attention on the former. It is known that the latter leads to unidentifiable models, but that identifiability can be achieved by making supplementary assumptions. But little has been done about the estimation problem in multivariate situations other than the bivariate case. And yet one suspects that in many statistical situations where functional relations are the real concern the attack is conducted by a regression model in which the regressors are wrongly assumed to be observed without error.

22.

A typical example occurs in econometrics. The so-called errors-in-equation model is of the form

$$By + \Gamma z = u$$

where B is an $m \times m$ matrix of coefficients, y is an m-rowed column vector of endogenous variables, Γ is a $m \times k$ matrix of coefficients, z a -row column vector of pre-determined variables and u is a column vector of random components. It hardly has to be stated that in economic practice y and z are subject to considerable error but in nearly all the econometric work I have seen this is assumed away by ignoration. The assumptions required to make the model identifiable, and the estimation procedures required to estimate B and Γ when it is, await detailed study.

23. Machine programs

All kinds of machine programs exist for carrying out the more complicated routines of multivariate analysis and I am glad that various societies, including the American Statistical Association, have set up a committee (the latter under the chairmanship of Dr. Ivor Francis) to evaluate package programmes purporting to deal with statistical problems.[1] I should like to see some studies on the relative effectiveness of available programs and in particular the relative costs of running them. (I know this depends on the extent of the data to be analysed, but some specimen figures could be given.) After all, a computer company asked to quote on a particular job has to estimate the cost, presumably on the basis of number of instructions, size of the data and the output required, but no figures seem to be available for a given job of the relative *costs* of different programmes on different machines.

24.

I hope that perhaps some of the delegates to this symposium may have some suggestions to make about methods of tackling the above problems or may be stimulated to study them. Until they are solved, however empirically, we shall be severely handicapped in the practical application of multivariate methods.

References

Kendall, M.G. and Stuart, A. (1975). *The Advanced Theory of Statistics*, vol 3, Third edition. London, Charles Griffin & Co.

Kendall, M.G. (1975). *Multivariate Analysis*, London, Charles Griffin & Co.

Simpson, C.H. (1951). The interpretation of interactions in contingency tables, *J.R. Statist. Soc., B*, **13**, 238.

Yule, G.U. (1912). On the methods of measuring association between two attributes. *Jour. Roy. Stat. Soc.*, **75**, 579.

[1] A similar study, funded by the Social Science Research Council is being carried out in the UK. It will prepare a register of statistical packages available for survey analysis and list their features.

P.R. Krishnaiah, ed., *Multivariate Analysis–IV*
© North-Holland Publishing Company (1977) 495–511

MAHALANOBIS DISTANCES AND ANGLES*

K.V. MARDIA
University of Leeds, Leeds, U.K.

Mahalanobis distance has become one of the most fundamental concepts in Multivariate Analysis. In certain studies, Mahalanobis "angles" have also been appearing. This paper examines these concepts when the common covariance matrix may be singular and investigates their main properties. Their roles in various techniques such as cluster analysis, factor analysis, genetics, growth curves, discrimination analysis, multinormality tests, etc. are outlined and certain new results are provided. A cophenetic of practical importance is also investigated.

1. Introduction

Mahalanobis distance has become one of the most fundamental concepts in Multivariate Analysis since its introduction in 1930 by Mahalanobis. (For an early history, see the inaugural address to the Symposium by Professor R.C. Bose.) Let $\boldsymbol{x}_1$ and $\boldsymbol{x}_2$ be two independent random vectors (r.v.'s) having p components with

$$E(\boldsymbol{x}_i) = \boldsymbol{\mu}_i, \qquad \operatorname{cov}(\boldsymbol{x}_i) = \boldsymbol{\Sigma}, \qquad i = 1, 2.$$

If $\boldsymbol{\Sigma}$ is a p.d. matrix, the Mahalanobis distance (squared) between the two populations with r.v.'s $\boldsymbol{x}_1$ and $\boldsymbol{x}_2$ is defined as

$$D^2(\boldsymbol{\mu}_1, \boldsymbol{\mu}_2, \boldsymbol{\Sigma}) = (\boldsymbol{\mu}_1 - \boldsymbol{\mu}_2)'\boldsymbol{\Sigma}^{-1}(\boldsymbol{\mu}_1 - \boldsymbol{\mu}_2), \tag{1.1}$$

which will be called the Mahalanobis distance between $\boldsymbol{\mu}_1$ and $\boldsymbol{\mu}_2$. In the same spirit, we can define the Mahalanobis angle $\theta(\boldsymbol{\mu}_1, \boldsymbol{\mu}_2, \boldsymbol{\Sigma})$ between $\boldsymbol{\mu}_1$ and $\boldsymbol{\mu}_2$ subtended at the origin by

$$\cos\theta(\boldsymbol{\mu}_1, \boldsymbol{\mu}_2, \boldsymbol{\Sigma}) = \boldsymbol{\mu}_1'\boldsymbol{\Sigma}^{-1}\boldsymbol{\mu}_2/\{D(\boldsymbol{\mu}_1, \boldsymbol{0}, \boldsymbol{\Sigma})D(\boldsymbol{\mu}_2, \boldsymbol{0}, \boldsymbol{\Sigma})\}. \tag{1.2}$$

In fact, the spaces generated by

$$\boldsymbol{y}_i = \boldsymbol{\Sigma}^{-\frac{1}{2}}\boldsymbol{x}_i, \qquad i = 1, 2 \tag{1.3}$$

* Presented at the 4th Int. Symp. on Multivariate Analysis.

can be called Mahalanobis spaces, where $\boldsymbol{\Sigma}^{\frac{1}{2}}$ is a symmetric square root of $\boldsymbol{\Sigma}$. Then the definitions (1.1) and (1.2) are obvious.

Section 2 examines the concept when $\boldsymbol{\Sigma}$ may be singular. Section 3 gives the main properties of Mahalanobis distances and angles. There are various multivariate techniques which use the concepts and Section 4 briefly describes the salient points. Section 5 deals with certain cophenetics.

2. Singular case

Let us now assume that $\boldsymbol{\Sigma}$ is of rank r $(r < p)$. There exists a matrix $\boldsymbol{N}$ $(p \times \overline{p - r})$ of rank $p - r$ such that

$$\boldsymbol{N}'\boldsymbol{\Sigma} = \boldsymbol{0}, \qquad \boldsymbol{N}'\boldsymbol{N} = \boldsymbol{I}. \tag{2.1}$$

The $p - r$ columns of $\boldsymbol{N}$ may be the $p - r$ eigenvectors corresponding to the zero eigenvalues of $\boldsymbol{\Sigma}$.

Definition 1. If

$$\boldsymbol{N}'(\boldsymbol{\mu}_1 - \boldsymbol{\mu}_2) = \boldsymbol{0}, \tag{2.2}$$

we define the *Mahalanobis distance* between $\boldsymbol{\mu}_1$ and $\boldsymbol{\mu}_2$ as

$$D^2(\boldsymbol{\mu}_1, \boldsymbol{\mu}_2, \boldsymbol{\Sigma}) = (\boldsymbol{\mu}_1 - \boldsymbol{\mu}_2)'\boldsymbol{\Sigma}^-(\boldsymbol{\mu}_1 - \boldsymbol{\mu}_2), \tag{2.3}$$

where $\boldsymbol{\Sigma}^-$ is any g-inverse of $\boldsymbol{\Sigma}$, i.e. $\boldsymbol{\Sigma}\boldsymbol{\Sigma}^-\boldsymbol{\Sigma} = \boldsymbol{\Sigma}$.

It may be noted that (2.2) implies that $(\boldsymbol{\mu}_1 - \boldsymbol{\mu}_2)$ lies in the column space of $\boldsymbol{\Sigma}$, i.e. there exists a vector $\boldsymbol{d}$ such that

$$(\boldsymbol{\mu}_1 - \boldsymbol{\mu}_2) = \boldsymbol{\Sigma}\boldsymbol{d}. \tag{2.4}$$

In sequel, we allow any of the three quantities $\boldsymbol{\mu}_1, \boldsymbol{\mu}_2, \boldsymbol{\Sigma}$ to be constant or stochastic. For example, with a r.v. $\boldsymbol{x}$ with $E(\boldsymbol{x}) = \boldsymbol{\mu}$, $\operatorname{cov}(\boldsymbol{x}) = \boldsymbol{\Sigma}$, we have

$$D^2(\boldsymbol{x}, \boldsymbol{\mu}, \boldsymbol{\Sigma}) = (\boldsymbol{x} - \boldsymbol{\mu})'\boldsymbol{\Sigma}^-(\boldsymbol{x} - \boldsymbol{\mu}). \tag{2.5}$$

as the "Mahalanobis distance" between $\boldsymbol{x}$ and $\boldsymbol{\mu}$. In this case, we always have

$$\boldsymbol{N}'(\boldsymbol{x} - \boldsymbol{\mu}) = \boldsymbol{0} \tag{2.6}$$

with probability 1 since $\boldsymbol{x} - \boldsymbol{\mu}$ lies in the column space of $\boldsymbol{\Sigma}$. Hence the condition given by (2.2) is automatically satisfied in probability.

Definition 1 has implicitly appeared in literature in various different forms (see, e.g. Gower, 1966a; Khatri, 1968; Rao and Mitra, 1971).

We now give certain main properties.

Theorem 1. *D^2 remains the same for any $\boldsymbol{\Sigma}^-$.*

Proof. From Theorem 2.4.1 of Rao and Mitra (1971, p. 26), the most general form of g-inverse of $\boldsymbol{\Sigma}$ for given $\boldsymbol{\Sigma}^-$ is

$$\boldsymbol{V} = \boldsymbol{\Sigma}^- + \boldsymbol{N}\boldsymbol{U}_1 + \boldsymbol{U}_2\boldsymbol{N}'$$

where $\boldsymbol{U}_1$ and $\boldsymbol{U}_2$ are arbitrary. Using this in (2.3) and (2.5) in conjuction with (2.2) and (2.6), we get the result. For the non-singular case, the result was first formulated by Bose and Roy (1938).

Theorem 2. *D^2 is invariant under those linear transformations which preserve the rank of $\boldsymbol{\Sigma}$.*

Proof. Consider the transformations

$$\boldsymbol{y}_i = \boldsymbol{B}\boldsymbol{x}_i + \boldsymbol{c} \tag{2.7}$$

where

$$\operatorname{rank}(\boldsymbol{B}\boldsymbol{\Sigma}\boldsymbol{B}') = \operatorname{rank}(\boldsymbol{\Sigma}), \tag{2.8}$$

and $\operatorname{rank}(\boldsymbol{\Sigma})$ denotes the rank of $\boldsymbol{\Sigma}$.

The new distance is found to be

$$(\boldsymbol{\mu}_1 - \boldsymbol{\mu}_2)'\boldsymbol{A}(\boldsymbol{\mu}_1 - \boldsymbol{\mu}_2)$$

where

$$\boldsymbol{A} = \boldsymbol{B}'(\boldsymbol{B}\boldsymbol{\Sigma}\boldsymbol{B}')^-\boldsymbol{B}.$$

From Mitra (1968), $\boldsymbol{A}$ is a g-inverse of $\boldsymbol{\Sigma}$ whenever (2.8) holds. (Also see Rao and Mitra, 1971, Lemma 4.1.1 and Khatri, 1971, p. 98). Hence the result follows on using Theorem 1.

It may be noted that for $D^2(\boldsymbol{x}, \boldsymbol{\mu}, \boldsymbol{\Sigma})$, the invariance with respect to $\boldsymbol{\Sigma}^-$ holds with probability 1.

Consider an application of Theorem 2. We can always select $\boldsymbol{B}$ of the order $r \times p$ such that $\boldsymbol{B}\boldsymbol{\Sigma}\boldsymbol{B}'$ is non-singular. Then the distance defined by (2.3) becomes the usual Mahalanobis distance with respect to the random vectors

$$\boldsymbol{y}_i = \boldsymbol{B}\boldsymbol{x}_i, \qquad i = 1, 2,$$

having a non-singular covariance matrix of the order $r \times r$. Thus, we can view (2.3) as the usual Mahalanobis distance (with non-singular $\boldsymbol{\Sigma}$) in the lower space.

Obviously, we can select $\boldsymbol{B}$ with its columns as the eigenvectors corresponding to the non-zero eigenvalues of $\boldsymbol{\Sigma}$. Also, if $\operatorname{rank}(\boldsymbol{\Sigma}) = \operatorname{rank}(\boldsymbol{\Sigma}_{11}) = r$, we can take $\boldsymbol{B} = (\boldsymbol{I}_r, \boldsymbol{0})$ where $\boldsymbol{\Sigma}_{11} = (\sigma_{ij});\ i, j = 1, \ldots, r$.

We now note various important points.

(a) *Multi-Populations.* In obvious notation, let there be k-populations denoted by $\Pi(\boldsymbol{\mu}_i, \boldsymbol{\Sigma})$ $i = 1, \ldots, k$. We define the Mahalanobis distance of these populations from a population $\Pi(\boldsymbol{\mu}, \boldsymbol{\Sigma})$ as

$$D_k^2(\boldsymbol{\mu}, \boldsymbol{\Sigma}) = \frac{1}{k} \sum_{i=1}^{k} D^2(\boldsymbol{\mu}_i, \boldsymbol{\mu}, \boldsymbol{\Sigma}^-) \tag{2.9}$$

provided

$$\boldsymbol{N}'(\boldsymbol{\mu}_i - \boldsymbol{\mu}) = \boldsymbol{0}, \qquad i = 1, \ldots, k. \tag{2.10}$$

Various properties of (2.9) can be proved as above. For $\boldsymbol{\mu} = \Sigma \boldsymbol{\mu}_i / k$, (2.9) leads to the population version of Hotelling's T_0^2.

(b) *Sample Counterparts.* Let $\boldsymbol{x}_{ij},\ j = 1, \ldots, n_i,\ i = 1, \ldots, k$ be the k-sample. Suppose the *total* (corrected) matrix of sum of squares and products is $\boldsymbol{W}$, i.e.

$$\boldsymbol{W} = \sum_{i=1}^{k} \sum_{j=1}^{n_i} (\boldsymbol{x}_{ij} - \bar{\boldsymbol{x}})(\boldsymbol{x}_{ij} - \bar{\boldsymbol{x}})'.$$

Then (2.9) is equivalent to

$$D_k^2(\bar{\boldsymbol{x}}, \boldsymbol{S}) \propto \sum_{i=1}^{k} n_i D^2(\bar{\boldsymbol{x}}_i, \bar{\boldsymbol{x}}, \boldsymbol{W}^-) = V, \qquad \text{say}, \tag{2.11}$$

which is valid under (2.10). Similarly, various different cases can be treated. There are sometimes reasons to omit the factor n_i (see Gower, 1966b). $\boldsymbol{W}^-$ is usually the within matrix.

(c) *Individual Distances.* Let $\boldsymbol{x}_1, \ldots, \boldsymbol{x}_n$ be a random sample from $\Pi(\boldsymbol{\mu}, \boldsymbol{\Sigma})$. Suppose that $\boldsymbol{S}$ is the sample covariance matrix. The individual distances between $\boldsymbol{x}_i$ and $\boldsymbol{x}_j$ are given by

$$D_{ij}^2 = (\boldsymbol{x}_i - \boldsymbol{x}_j)'\boldsymbol{S}^-(\boldsymbol{x}_i - \boldsymbol{x}_j). \tag{2.12}$$

These are related to

$$g_{ij} = (\boldsymbol{x}_i - \bar{\boldsymbol{x}})'\boldsymbol{S}^-(\boldsymbol{x}_j - \bar{\boldsymbol{x}}) \tag{2.13}$$

where $g_{ij}/(g_{ii}g_{jj})^{\frac{1}{2}}$ is the Mahalanobis angle subtended at $\bar{\boldsymbol{x}}$ by $\boldsymbol{x}_i$ and $\boldsymbol{x}_j$ and g_{ii} is the Mahalanobis distance between $\boldsymbol{x}_i$ and $\bar{\boldsymbol{x}}$. For the non-singular case, the usual definition involves a factor in n.

(d) *Square-root transformation.* The transformation (1.3) has no meaning when $\boldsymbol{\Sigma}$ is singular. Let $\boldsymbol{\Sigma}^{+}$ be the Moore–Penrose inverse of $\boldsymbol{\Sigma}$. We can then replace (1.3) by

$$\boldsymbol{y}_i = (\boldsymbol{\Sigma}^{+})^{\frac{1}{2}}(\boldsymbol{x}_i - \boldsymbol{\mu}_i) \tag{2.14}$$

to define the Mahalanobis space where $\boldsymbol{\Sigma}^{\frac{1}{2}}$ is a symmetric square root of $\boldsymbol{\Sigma}$ and $(\boldsymbol{\Sigma}^{+})^{\frac{1}{2}} = (\boldsymbol{\Sigma}^{\frac{1}{2}})^{+}$. That is, if $\boldsymbol{\Sigma} = \boldsymbol{P}\boldsymbol{D}_{\lambda}\boldsymbol{P}'$, we have, $\boldsymbol{\Sigma}^{\frac{1}{2}} = \boldsymbol{P}\boldsymbol{D}_{\lambda}^{\frac{1}{2}}\boldsymbol{P}'$, where $\boldsymbol{P}$ is an orthogonal matrix, $\boldsymbol{D}_{\lambda}$ = the diagonal matrix of the eigenvalues of $\boldsymbol{\Sigma}$. For the individual distances, we can use

$$\boldsymbol{y}_i = (\boldsymbol{S}^{+})^{\frac{1}{2}}(\boldsymbol{x}_i - \bar{\boldsymbol{x}}), \qquad i = 1, \ldots, n.$$

(e) *Mahalanobis Angle.* We can define the Mahalanobis angle θ subtended by $\boldsymbol{\mu}_1$ and $\boldsymbol{\mu}_2$ at $\boldsymbol{0}$ as

$$\cos\theta = \boldsymbol{\mu}_1'\boldsymbol{\Sigma}^{-}\boldsymbol{\mu}_2/\{D(\boldsymbol{\mu}_1, \boldsymbol{0}, \boldsymbol{\Sigma})D(\boldsymbol{\mu}_2, \boldsymbol{0}, \boldsymbol{\Sigma})\}$$

provided

$$\boldsymbol{N}'\boldsymbol{\mu}_1 = \boldsymbol{N}'\boldsymbol{\mu}_2 = \boldsymbol{0}.$$

Fisher (1938) seems to be the first to use this concept. Theorems 1 and 2 obviously hold for θ after some straightforward modifications.

(f) *Semi-Norm.* Let $\boldsymbol{A}$ be a p.s.d. matrix. The semi-normed vector space V is defined by

$$\|\boldsymbol{x}\|_{\boldsymbol{A}}^2 = \boldsymbol{x}'\boldsymbol{A}\boldsymbol{x} \quad \text{for all} \quad \boldsymbol{x} \in V.$$

Let $\boldsymbol{x}$ be a random vector such that $\mathbf{E}(\boldsymbol{x}) = \boldsymbol{a}$ and $\operatorname{cov}(\boldsymbol{x}) = \boldsymbol{\Sigma}$. Let V be a space generated by the random variable $\{\boldsymbol{x} - \boldsymbol{a}\}$. Then a distance between any points of V can be defined by a semi-norm $\|\boldsymbol{x} - \boldsymbol{a}\|_{\boldsymbol{\Sigma}^-}$ with respect to $\boldsymbol{\Sigma}$. If $\boldsymbol{a}$ is not $\mathbf{E}(\boldsymbol{x})$, we define a vector space V_1 of the points generated by $\boldsymbol{x} - \boldsymbol{a}$ such that

$$\boldsymbol{x} - \boldsymbol{a} = \boldsymbol{\Sigma}\boldsymbol{d} \quad \text{for some} \quad \boldsymbol{d} \in R_p.$$

That is, the earlier discussion can be put into an equivalent framework of a semi-normed vector space.

3. Properties

3.1 *Distance properties*

Mahalanobis distances (1.1) and (2.3) become Euclidean distances for the Mahalanobis spaces defined by (1.3) and (2.14) for the non-singular and singular cases respectively. Hence, the following properties hold. Let $D(P, Q)$ denote the distance between two points P and Q in the Mahalanobis space. Then

(i) *Symmetry.* $D(P, Q) = D(Q, P)$.

(ii) *Non-negativity.* $D(P, Q) \geq 0$. We have

$$D(P, Q) = 0 \quad \text{if and only if} \quad P = Q.$$

(iii) *Triangular Inequality.* $D(P, Q) \leq D(P, R) + D(R, Q)$.

We have

$$D_p \leq D_{p+q}, \tag{3.1}$$

where D_p is based on only p variables out of $p+q$ variables in D_{p+q}.

There are some other requirements for distances in classification problems such as stability. D_p^2 is stable if $\lim_{p\to\infty} D_p^2$ is finite (Mahalanobis, 1936; Rao, 1954). For example, consider the case when

$$\boldsymbol{\mu}_1' = (\delta, \mathbf{0}'), \quad \boldsymbol{\mu}_2 = \mathbf{0}, \quad \sigma_{ij} = \rho, \quad i \neq j, \quad \sigma_{ii} = 1.$$

Then

$$D(\boldsymbol{\mu}_1, \boldsymbol{\mu}_2; \boldsymbol{\Sigma}) = |\,\delta\,|\,(\{1 + (p-2)\rho\}/[(1-\rho)\{1 + (p-1)\rho\}])^{\frac{1}{2}}.$$

We have

$$\lim_{p\to\infty} D(\boldsymbol{\mu}_1, \boldsymbol{\mu}_2; \boldsymbol{\Sigma}) = |\,\delta\,|/(1-\rho)^{\frac{1}{2}}.$$

Therefore D^2 reaches a *stable* value as $p \to \infty$ so that there is no point in increasing the number of measurements indefinitely. However, if $\lim_{p\to\infty} D_p^2 = \infty$ then the situation is different. For example, if $\boldsymbol{\Sigma} = \boldsymbol{I}$, $\boldsymbol{\mu}_1 - \boldsymbol{\mu}_2 = \mathbf{1}$, $D_p^2 \to \infty$.

3.2. *Sampling moments*

We now give some sampling moments for certain Mahalanobis distances from any underlying population.

Theorem 1. *Let* g_{ij} *be defined by* (2.12) *where* $\boldsymbol{\Sigma}$ *is non-singular and* $n - 1 > p$.

(a) $E(g_{ii}) = p, \quad E(g_{ij}) = -p/(n-1), \quad i \neq j.$

(b) $E(g_{ii}^2) = c, \quad E(g_{ij}^2) = \dfrac{1}{n-1}(np - c),$

$$E(g_{ii}g_{jj}) = (np^2 - c)/(n-1), \quad E(g_{ii}g_{ij}) = -c/(n-1),$$
$$E(g_{ii'}g_{ij'}) = (2c - np)/(n-1)^{(2)}, \quad E(g_{ii}g_{jj'}) = (2c - np^2)/(n-1)^{(2)},$$
$$E(g_{ii'}g_{jj'}) = \{np(p+2) - 6c\}/(n-1)^{(3)},$$

where

$$n^{(r)} = n(n-1)\dots(n-r+1), \quad c = E(b_{2,p}), \quad b_{2,p} = \frac{1}{n}\Sigma g_{ii}^2, \tag{3.2}$$

$b_{2,p}$ being the multivariate measure of kurtosis. The suffixes are unequal for different symbols.

Proof. Using the results

$$\sum_{i=1}^{n} (\boldsymbol{x}_i - \bar{\boldsymbol{x}})'\boldsymbol{S}^{-1}(\boldsymbol{x}_i - \bar{\boldsymbol{x}}) = np, \quad \sum_{i=1}^{n} (\boldsymbol{x}_i - \bar{\boldsymbol{x}})'\boldsymbol{S}^{-1}(\boldsymbol{x}_j - \bar{\boldsymbol{x}}) = 0, \quad i \neq j,$$

in conjuction with the fact that the g_{ii}'s (or g_{ij}'s $i \neq j$) are identical variables, the above results can be obtained. For example,

$$E(g_{ii}) = \frac{1}{n} E\left(\sum_{i=1}^{n} g_{ii}\right) = p.$$

Corollary 1. *As $n \to \infty$, $\operatorname{corr}(g_{ii'}, g_{jj'}) = 0$ except when $i = i' = j = j'$ provided $c - p^2 = O(1/n^i)$ for some $i \geq 0$.*

These results can be used to obtain the moments of the functions of $\boldsymbol{G} = (g_{ij})$. It may be noted that if $\boldsymbol{D} = (D_{ij}^2)$ where D_{ij}^2 is defined at (2.12), then

$$\boldsymbol{H}_p\boldsymbol{D}\boldsymbol{H}_p = -2\boldsymbol{G}, \quad \boldsymbol{H}_p = \boldsymbol{I} - n^{-1}\boldsymbol{1}\boldsymbol{1}', \tag{3.3}$$

where $\boldsymbol{1}$ is a column vector of ones. Hence from the individual distances, we can obtain the Mahalanobis distances and angles of points with the origin at $\bar{\boldsymbol{x}}$.

Theorem 2. *Let V be defined by (2.11) where $\boldsymbol{\Sigma}$ is non-singular and $n - 1 > p$, $n = \Sigma n_i$. If the observations are from $\Pi_i(\boldsymbol{\mu}, \boldsymbol{\Sigma})$ $i = 1, \dots, k$, we have*

$$\mathrm{E}(V) = p(k-1)/(n-1), \tag{3.4}$$

$$\operatorname{var}(V) = \operatorname{var}_N(V)\left\{1 + \frac{(n-3)}{2n(n-1)} C_X E(C_Y)\right\} \tag{3.5}$$

where

$$\operatorname{var}_N(V) = 2p(k-1)(n-k)(n-p-1)/\{(n+1)(n-1)^2(n-2)\}, \tag{3.6}$$

$$C_X = (n-1)\left[n(n+1)\left\{\sum_{j=1}^{k} n_j^{-1} - k^2/n\right\} - 2(k-1)(n-k)\right] \tag{3.7}$$

and

$$C_Y = (n-1)\{(n+1)b_{2,p} - (n-1)p(p+2)\}/\{p(n-3)(n-p-1)\} \tag{3.8}$$

with

$$b^*_{2,p} = n\sum_{i=1}^{k}\sum_{j=1}^{n_i}[(\boldsymbol{x}_{ij}-\bar{\boldsymbol{x}})'\boldsymbol{W}^{-1}(\boldsymbol{x}_{ij}-\bar{\boldsymbol{x}})]^2.$$

Proof. See Mardia (1971).

There are various important particular cases of interest. With the help of moments, we can obtain an approximation to the distribution of V.

We note in passing that

$$\mathbf{P}\{D^2(\boldsymbol{x}, \boldsymbol{\mu}, \boldsymbol{\Sigma}) < \varepsilon\} \geq 1 - \beta_{2,p}/\varepsilon^2, \qquad \varepsilon > 0, \tag{3.9}$$

where $\beta_{2,p} = \mathbf{E}\{D^4(\boldsymbol{x}, \boldsymbol{\mu}, \boldsymbol{\Sigma})\}$ is the population counterpart of $b_{2,p}$.

If $\boldsymbol{\Sigma}$ is of rank r, p is to be replaced by r in the above Theorems.

3.3. *Multinormal theory*

In the underlying populations in Theorem 1 and Theorem 2 are normal, the expressions can be simplified. For example,

$$E(b_{2,p}) = p(p+2)(n-1)/(n+1).$$

In fact, the distribution theory is well developed. Let $\boldsymbol{G}_r$ be a principal submatrix of order r from $\boldsymbol{G}$. Then

$$n^{-1}\boldsymbol{H}_r^{-\frac{1}{2}}\boldsymbol{G}_r\boldsymbol{H}_r^{-\frac{1}{2}} \sim B_r(\tfrac{1}{2}p, \tfrac{1}{2}(n-p-1)), \quad p \geq r, \quad n \geq p+r+1, \tag{3.10}$$

where B_r has a matrix $(r \times r)$ beta distribution of Type I. In particular, $g_{11}/(n-1)$ is distributed as univariate beta of Type I. This fact can be used to obtain higher moments (Khatri, 1959; Mardia, 1974).

If $\boldsymbol{\Sigma}$ is singular of rank r, we can use Khatri's representation of singular multi-normal density (Khatri, 1968). It is found that

(i) $(\boldsymbol{x}-\boldsymbol{\mu})'\boldsymbol{\Sigma}^-(\boldsymbol{x}-\boldsymbol{\mu}) \sim \chi^2_r$ (e.g. see Rao and Mitra, 1971, p. 173).

(ii) Let $\rho = \int\{f(\boldsymbol{x})g(\boldsymbol{x})\}^{\frac{1}{2}}d\boldsymbol{x}$ be Bhattacharya's measure of divergence between two densities f and g. For $N_p(\boldsymbol{\mu}_i, \boldsymbol{\Sigma})$, $i = 1, 2$, we have in the notation of (2.3),

$$D^2(\boldsymbol{\mu}_1, \boldsymbol{\mu}_2, \boldsymbol{\Sigma}) = -8\log_e\rho \quad \text{if} \quad \boldsymbol{N}'(\boldsymbol{\mu}_1 - \boldsymbol{\mu}_2) = 0.$$

4. Techniques involving D^2

We give various multivariate techniques where D^2 plays an important role. We will not go into any detail where the uses of D^2 are well-known such as for testing equality of two mean vectors, etc.

Factor Analysis. Let

$$x = Lf + \varepsilon \tag{4.1}$$

be the usual factor analysis model where L is a $p \times k$ matrix of the factor loadings having rank k, f and ε are independent and

$$\operatorname{cov}(f) = I, \quad \operatorname{cov}(x) = \Sigma, \quad \operatorname{cov}(\varepsilon) = \Psi,$$

Ψ being a non-singular diagonal matrix. For two individuals 1 and 2, let us write $x_i = Lf_i + \varepsilon_i$, $i = 1, 2$ and $\delta = x_1 - x_2$. For Bartlett's method and Thomson's method of estimating factors (Lawley and Maxwell, 1963, pp. 88–91), the corresponding distances between the estimated factors for the two individuals are found to be

$$D_B^2 = \delta'\hat{\Psi}^{-\frac{1}{2}}(\hat{\Psi}^{-\frac{1}{2}}\hat{L}\hat{L}'\hat{\Psi}^{-\frac{1}{2}})^{+}\hat{\Psi}^{-\frac{1}{2}}\delta - \delta'\hat{\Psi}^{-1}\hat{L}(\hat{L}'\hat{\Psi}^{-1}\hat{L})^{-2}\hat{L}'\hat{\Psi}^{-1}\delta \tag{4.2}$$

and

$$D_T^2 = \delta'\hat{\Sigma}^{-1}\hat{L}\hat{L}'\hat{\Sigma}^{-1}\delta = \delta'\hat{\Psi}^{-1}\hat{L}(I + \hat{L}'\hat{\Psi}^{-1}\hat{L})^{-2}\hat{L}'\hat{\Psi}^{-1}\delta \tag{4.3}$$

where $\hat{\Sigma}$, $\hat{L}$, $\hat{\Psi}$ are the m.l.e.'s of Σ, L, Ψ respectively. The right-hand side of (4.3) has been obtained on using

$$\hat{\Sigma}^{-1}\hat{L} = \hat{\Psi}^{-1}\hat{L}(I + \hat{L}'\hat{\Psi}^{-1}\hat{L})^{-1}.$$

Gower (1966a) gives certain expressions in the form of a g-inverse but his final equations (22) and (26) seem to require some modification if we follow the usual definition of a g-inverse. The expressions (22) and (26) for uniqueness with respect to g-inverses, imply $x_1 - x_2 \in \mathcal{M}(L)$ which cannot be true except when $e_1 = e_2$ in probability since $(x_1 - e_1) - (x_2 - e_2) \in \mathcal{M}(L)$, where $\mathcal{M}(L)$ denotes the space generated by the columns of L.

We now show that

$$D_B^2 \geq D_T^2 \tag{4.4}$$

and the equality holds if and only if $D_B = D_T = 0$.

Using the right hand sides of (4.2) and (4.3), it can be shown on noting the positive definiteness of the quadratic forms in δ that there exists a non-singular transformation from δ to ω such that

$$D_B^2 = \sum_{i=1}^{k} \omega_i^2/\lambda_i^2, \qquad D_T^2 = \sum_{i=1}^{k} \omega_i^2/(1 + \lambda_i)^2, \qquad \lambda_i > 0,$$

where λ_i's are the eigenvalues of $\hat{L}'\hat{\Psi}^{-1}\hat{L}$. Hence (4.4) follows. The implication of this result is that Bartlett's scores are better in separating out individuals.

The above discussion applies to any p.d. $\boldsymbol{\Psi}$.

Genetic Differences. Following Rao (1954), let us take the model of observable characters $\boldsymbol{x}$ as (4.1) where $\boldsymbol{f}$ denotes the genetic factors, $\boldsymbol{\varepsilon}$ the environmental factors, $E(\boldsymbol{x})=\boldsymbol{\mu}$, $E(\boldsymbol{f})=\boldsymbol{\nu}$, $E(\boldsymbol{\varepsilon})=\boldsymbol{0}$, $\text{cov}(\boldsymbol{f})=\boldsymbol{\Gamma}$ and $\text{cov}(\boldsymbol{\varepsilon})=\boldsymbol{\Psi}$, $\boldsymbol{\Psi}$ is any p.s.d. matrix. If $\boldsymbol{\delta}$ denotes the difference between $\boldsymbol{\nu}$ vectors for two groups then the genetic distance D_0^2 and the distance between the two $\boldsymbol{\mu}$'s are defined by

$$D_0^2=\boldsymbol{\delta}'\boldsymbol{\Gamma}^-\boldsymbol{\delta}, \qquad D^2=\boldsymbol{\delta}'\boldsymbol{L}'\boldsymbol{\Sigma}^-\boldsymbol{L}\boldsymbol{\delta}, \tag{4.5}$$

provided

(i) $\boldsymbol{\delta}$ lies in the column space of $\boldsymbol{\Gamma}$ and (ii) $\boldsymbol{L}\boldsymbol{\delta}$ lies in the column space of $\boldsymbol{\Sigma}$ where $\boldsymbol{\Sigma}=\boldsymbol{L}\boldsymbol{\Gamma}\boldsymbol{L}'+\boldsymbol{\Psi}$ is the covariance matrix of $\boldsymbol{x}$. We also allow $\boldsymbol{\Gamma}$ to be singular; for a genetic example, see Rao (1954). We prove the following extended version of Rao's inequality.

$$D^2\leqslant D_0^2 \tag{4.6}$$

where the equality holds if and only if

$$\boldsymbol{\delta}=\boldsymbol{\Gamma}\boldsymbol{L}'\boldsymbol{\Sigma}^+\boldsymbol{L}\boldsymbol{\delta}. \tag{4.7}$$

From (i), there exists $\boldsymbol{d}$ such that $\boldsymbol{\delta}=\boldsymbol{\Gamma}\boldsymbol{d}$. Define

$$\boldsymbol{e}=\boldsymbol{\Gamma}^{\frac{1}{2}}\boldsymbol{d}, \qquad \boldsymbol{A}'=\boldsymbol{\Gamma}^{\frac{1}{2}}\boldsymbol{L}, \qquad \boldsymbol{B}=\boldsymbol{A}'(\boldsymbol{A}\boldsymbol{A}'+\boldsymbol{\Psi})^+\boldsymbol{A}. \tag{4.8}$$

Then under condition (ii), (4.6) can be written as

$$\boldsymbol{e}'\boldsymbol{B}\boldsymbol{e}\leqslant\boldsymbol{e}'\boldsymbol{e}. \tag{4.9}$$

This is true if all the eigenvalues of $\boldsymbol{B}$ are less than or equal to 1. To prove this result, let $r=\text{rank}(\boldsymbol{A}\boldsymbol{A}'+\boldsymbol{\Psi})$. Since $\boldsymbol{\Psi}$ and $\boldsymbol{A}\boldsymbol{A}'$ are p.s.d., there exists a matrix $\boldsymbol{T}$ $(p\times r)$ of rank r such that (see Mitra and Rao, 1968; Khatri, 1971, p. 153);

$$\boldsymbol{\Psi}=\boldsymbol{T}\boldsymbol{D}_1\boldsymbol{T}', \quad \boldsymbol{A}\boldsymbol{A}'=\boldsymbol{T}\boldsymbol{D}_2\boldsymbol{T}'$$

where $\boldsymbol{D}=\boldsymbol{D}_1+\boldsymbol{D}_2$ is a non-singular diagonal matrix, $\boldsymbol{D}_1$ and $\boldsymbol{D}_2$ are diagonal matrices such that $\boldsymbol{D}_2$ has the first s elements unity and rest zero, $s=\text{rank}(\boldsymbol{A}\boldsymbol{A}')$. Since $(\boldsymbol{\Psi}+\boldsymbol{A}\boldsymbol{A}')^+=(\boldsymbol{T}^+)'\boldsymbol{D}^{-1}\boldsymbol{T}^+$, we have

$$\text{e.v.}(\boldsymbol{B})=\text{e.v.}[\boldsymbol{A}\boldsymbol{A}'(\boldsymbol{\Psi}+\boldsymbol{A}\boldsymbol{A}')^+]=\text{e.v.}(\boldsymbol{D}^{-1}\boldsymbol{D}_2)\leqslant 1. \tag{4.10}$$

where e.v. $(\boldsymbol{B})$ denotes the non-zero eigenvalues of $\boldsymbol{B}$. Hence (4.9) is established. The equality in (4.9) holds if and only if $\boldsymbol{B}\boldsymbol{e}=\boldsymbol{e}$ which leads to (4.7) after using (4.8). If $\boldsymbol{\Gamma}$ is non-singular and $\text{rank}(\boldsymbol{L})=k$ $(\leqslant p)$ then (4.7) is always true provided $\boldsymbol{\Psi}=\boldsymbol{0}$. Hence, in this case as observed by Rao

(1954) it is possible to transform $\boldsymbol{x}$ in such a way that the random component is zero for at least k linearly independent new variables.

If $\boldsymbol{\Psi} = \boldsymbol{0}$, (ii) is automatically satisfied since (i) implies (ii). Writing D_p in place of D, we have from (4.7),

$$\lim_{p\to\infty}\left\{\sup_p D_p^2\right\} \leq D_0^2.$$

A judicious choice of characters $x_1, x_2, \ldots$ would give adequate information about D_0^2.

Growth Curves. Let $\boldsymbol{W}$ be the within covariance matrix based on samples from k p-variate populations. Suppose that the desired comparisons among populations are partially confounded with k 'growth' effects which can be represented as the columns of a $p \times k$ matrix $\boldsymbol{G}$ (see Rao, 1966a; Burnaby, 1966). If $\boldsymbol{\delta}$ denotes the differences between the means of two populations, then Rao (1966a) and Burnaby (1966) have shown that

$$D^2 = \boldsymbol{\delta}'\boldsymbol{W}^{-1}\boldsymbol{\delta} = D_Q^2 + D_M^2$$

where D_Q^2 is the constrained distance given by

$$D_Q^2 = \boldsymbol{\delta}'\boldsymbol{A}\boldsymbol{\delta}, \qquad \boldsymbol{A} = \boldsymbol{W}^{-1} - \boldsymbol{W}^{-1}\boldsymbol{G}(\boldsymbol{G}'\boldsymbol{W}^{-1}\boldsymbol{G})^{-1}\boldsymbol{G}'\boldsymbol{W}^{-1}.$$

We remove the requirements of $\boldsymbol{W}$ and $\boldsymbol{G}'\boldsymbol{G}$ to be non-singular and show that when (i) $\boldsymbol{\delta}$ lies in the column space of $\boldsymbol{W}$

$$D^2 = \boldsymbol{\delta}'\boldsymbol{W}^{-}\boldsymbol{\delta} = D_Q^2 + D_M^2, \tag{4.11}$$

where

$$D_Q^2 = \boldsymbol{\delta}'\boldsymbol{Q}(\boldsymbol{QWQ})^{-}\boldsymbol{Q}\boldsymbol{\delta}, \qquad \boldsymbol{Q} = \boldsymbol{I} - \boldsymbol{G}(\boldsymbol{G}'\boldsymbol{G})^{-}\boldsymbol{G}'. \tag{4.12}$$

This problem as in Burnaby (1966) can be looked upon as maximizing $(\boldsymbol{l}'\boldsymbol{\delta})^2/(\boldsymbol{l}'\boldsymbol{W}\boldsymbol{l})$ with respect to $\boldsymbol{l}$ when $\boldsymbol{l}'\boldsymbol{W}\boldsymbol{l} \neq 0$ and $\boldsymbol{l}'\boldsymbol{G} = \boldsymbol{0}$. The latter restriction implies a general solution $\boldsymbol{l} = \boldsymbol{Q}\boldsymbol{z}$ where $\boldsymbol{z}$ is an arbitrary vector. This problem reduces to maximizing $(\boldsymbol{z}'\boldsymbol{Q}\boldsymbol{\delta})^2/\boldsymbol{z}'(\boldsymbol{QWQ})\boldsymbol{z}$ subject to $\boldsymbol{z}'(\boldsymbol{QWQ})\boldsymbol{z} \neq 0$ for which the answer is given by Rao (1966a) and Khatri (1968), because on account of condition (i) $\boldsymbol{Q}\boldsymbol{\delta}$ lies in the column space of $\boldsymbol{QWQ}$. It may be noted that if $\boldsymbol{W}$ and $\boldsymbol{G}'\boldsymbol{G}$ are non-singular then (see also, Gower, 1975)

$$\boldsymbol{Q}(\boldsymbol{QWQ})^{-}\boldsymbol{Q} = \boldsymbol{A}, \qquad \boldsymbol{A}^{+} = \boldsymbol{QWQ}.$$

Note that D_Q^2 denotes the distance in the space orthogonal to the space where the growth effects are confounded. Gower (1975) gives some new estimation procedures.

Discrimination Analysis. Various appearances of D^2 in discriminatory analysis are well-known (for example, see Kshirsagar, 1972). One of the most important significant results is that the probability of misclassifying an individual $\boldsymbol{x}$ to one of the populations $N(\boldsymbol{\mu}_i, \boldsymbol{\Sigma})$, $i = 1, 2$ is $\Phi(-\frac{1}{2}D)$ where $\Phi(\cdot)$ is the distribution function of $N(0,1)$. Using Khatri's representation for the singular case, this result can be proved. Rao and Mitra (1971, p. 205) give the appropriate discriminant function. Analogues of the Mahalanobis distance also appear in the discrimination problem for Gaussian Processes (see Rao and Vardarajan, 1963). Cacoullos (1973) considers $k (k \geq 2)$ diffusion processes $\Pi_i : N(t\boldsymbol{\mu}_i, t\boldsymbol{\Sigma})$ $i = 1, \ldots, k$ and the nearest Π_i to $\Pi : N(t\boldsymbol{\mu}, t\boldsymbol{\Sigma})$ is selected in the sense $D_i^2 = \min_{1 \leq j \leq k} D_j^2$ where $D_j^2 = (\boldsymbol{\mu}_j - \boldsymbol{\mu})'\boldsymbol{\Sigma}^{-1}(\boldsymbol{\mu}_j - \boldsymbol{\mu})$.

Outliers. Let $\boldsymbol{x}_1, \ldots, \boldsymbol{x}_r$ be a random sample. Let $\boldsymbol{P} = n\boldsymbol{S}$ and let $\boldsymbol{P}_{-i}$ denote $\boldsymbol{P}$ after deleting the ith observation. Then Wilks' Criterion (1963) for testing a single extreme outlier is $\min_{i=1\ldots n} R_i$, where

$$R_i = |\boldsymbol{P}_{-i}|/|\boldsymbol{P}| = 1 - \frac{1}{n-1} D_i^2.$$

That is, the criterion reduces to $U = \max_{i=1\ldots n} D_i^2$. On an intuitive basis this criterion has been recommended independently by Cox (1968) and Healy (1968). From Section 2, various properties of U can be investigated. For a detailed survey, see Andrews et al. (1973) and Gnanandesikan and Kettenring (1972) where a measure with $\boldsymbol{\Sigma}^{-1}$ replaced by $\boldsymbol{\Sigma}^{-b}$ in the distance is also considered. For outliers influencing the mean vector as well as the covariance matrix, one should consider $\max_i (\boldsymbol{x}_i - \bar{\boldsymbol{x}})'\boldsymbol{S}_{-i}^{-1}(\boldsymbol{x}_i - \bar{\boldsymbol{x}})$ where the ith observation is removed in the sample covariance matrix $\boldsymbol{S}_{-i}$.

A Test for Multinormality. Mardia (1970, 1974) has given a test of multinormality based on the following measures of multivariate skewness and kurtosis,

$$b_{1,p} = \frac{1}{n^2} \sum_{i=1}^{n} \sum_{j=1}^{n} g_{ij}^3, \quad b_{2,p} = \frac{1}{n} \sum_{i=1}^{n} g_{ii}^2. \tag{4.13}$$

These measures have appeared naturally from various analytical robustness studies. The population counterparts of $b_{1,p}$ and $b_{2,p}$ are given by

$$\beta_{1,p} = \mathbf{E}\{(\boldsymbol{x} - \boldsymbol{\mu})'\boldsymbol{\Sigma}^{-1}(\boldsymbol{y} - \boldsymbol{\mu})\}^3, \qquad \beta_{2,p} = \mathbf{E}\{(\boldsymbol{x} - \boldsymbol{\mu})'\boldsymbol{\Sigma}^{-1}(\boldsymbol{x} - \boldsymbol{\mu})\}^2 \tag{4.14}$$

respectively where the random vectors $\boldsymbol{x}$ and $\boldsymbol{y}$ are distributed identically and independently; $E(\boldsymbol{x}) = \boldsymbol{\mu}$, $\operatorname{cov}(\boldsymbol{x}) = \boldsymbol{\Sigma}$. For the multi-normal case $\beta_{1,p} = 0$ and $\beta_{2,p} = p(p+1)$. Their definitions extend to the singular case, when p is to be replaced by $r = \text{rank}(\boldsymbol{\Sigma})$ in the results of Mardia (1970, 1974). Mardia and Zemroch (1975) give an algorithm for calculation of $b_{1,p}$ and $b_{2,p}$.

Robustness. The following conclusions are arrived at in Mardia (1971, 1974, 1975). (i) Hotelling's T^2 is robust but it is more sensitive to $\beta_{1,p}$ than $\beta_{2,p}$. (ii) Two-sample T^2 is more sensitive to $\beta_{1,p}$ if $n_1 \neq n_2$. (iii) The size of the normal theory tests on covariance matrices would be seriously influenced by $\beta_{2,p}$ but the effect of $\beta_{1,p}$ may not be so serious.

The Effect of Increasing p. Rao (1966b) studies the effect of increasing the number of characters on tests of significance based on Hotelling's T^2 or D^2. It is found that for small samples, the power can decrease in the number of characters from q to $p+q$ unless the increase in the corresponding true Mahalanobis distance is of certain order.

Principal Co-ordinate Analysis of D^2. Let D_{ij}^2 denote the Mahalanobis distance between the means of the ith and jth population, $i, j = 1 \ldots k$. Rao (1952) shows that the first t canonical variates maximize the total D^2 in t dimensions. Gower (1966a) proves a closely related result that the configuration using the space defined by some only of the principal co-ordinate axes is the same as that provided by the canonical variates for the same dimensions where the principal co-ordinate technique gives a set of k points referred to rectangular principal axes with inter-point distances D_{ij}^2. For a recent application of this technique in cluster analysis, we refer to Rao (1971).

5. D^2 and the Karl Pearson distance

Various empirical studies have appeared to explore the relation between D^2 and Karl Pearson's coefficient (K^2) of racial likeness. (See Penrose, 1954; Gower, 1972.) Such correlations are commonly known as cophenetics. In the examples considered, the two measures are found to be highly correlated and therefore recommendations are made that the practical worker may as well use K^2 since it is simpler to calculate. The possibility that these observed correlations are spurious needs investigation.

We investigate the correlation between D^2 and K^2. If $\boldsymbol{x}$ is a random vector with mean $\boldsymbol{\mu}$ and covariance matrix $\boldsymbol{\Sigma}$, we define the Karl Pearson distance (squared) as

$$K^2 = (\boldsymbol{x} - \boldsymbol{\mu})'\boldsymbol{\Sigma}_0^{-1}(\boldsymbol{x} - \boldsymbol{\mu}), \qquad \boldsymbol{\Sigma}_0 = \text{diag}(\boldsymbol{\Sigma}). \tag{5.1}$$

Lemma 1. *Let* $\mathbf{x} \sim N_p(\boldsymbol{\mu}, \boldsymbol{\Sigma})$ *with* $\sigma_{ii} = 1$. *Suppose* $D_1^2 = \mathbf{x}'\boldsymbol{\Sigma}^{-1}\mathbf{x}$, $K_1^2 = \mathbf{x}'\mathbf{x}$. *Then*

$$\mathrm{E}(K_1^2) = \Delta_1^2 + p, \qquad \mathrm{E}(K_1^4) = \Delta_1^4 + 4\Delta_2^2 + 2p\Delta_1^2 + p^2 + 2\,\mathrm{tr}\,\boldsymbol{\Sigma}^2,$$

$$\mathrm{E}(D_1^2) = \Delta_0^2 + p, \qquad \mathrm{E}(D_1^4) = \Delta_0^4 + (p+2)(p+2\Delta_0),$$

$$\mathrm{E}(D_1^2K_1^2) = \Delta_0^2\Delta_1^2 + (p+4)\Delta_1^2 + p\Delta_0^2 + p(p+2),$$

where

$$\Delta_0^2 = \boldsymbol{\mu}'\boldsymbol{\Sigma}^{-1}\boldsymbol{\mu}, \qquad \Delta_1^2 = \boldsymbol{\mu}'\boldsymbol{\mu}, \qquad \Delta_2^2 = \boldsymbol{\mu}'\boldsymbol{\Sigma}\boldsymbol{\mu}.$$

Proof. Use the transformation $\mathbf{x} = \boldsymbol{\mu} + \boldsymbol{\Sigma}^{\frac{1}{2}}\mathbf{y}$ and simplify.

Theorem 1. *If* $\mathbf{x} \sim N_p(\boldsymbol{\mu}, \boldsymbol{\Sigma})$ *and* $\rho_{ij} = \mathrm{corr}(x_i, x_j)$, *then*

$$\rho(D^2, K^2) = \left\{1 + p^{-1}\sum_{i \neq j}\sum \rho_{ij}^2\right\}^{-\frac{1}{2}}. \tag{5.2}$$

Proof. Since D^2 and K^2 are invariant under scale changes, we can take $\sigma_{ii} = 1$ and then apply Lemma 1.

Corollary. *If* $\rho_{ij} = \rho$, *we have*

$$\rho(D^2, K^2) = \{1 + (p-1)\rho^2\}^{-\frac{1}{2}}. \tag{5.3}$$

From (5.2), we note that $\rho(D^2, K^2)$ is always non-negative. As expected, $\rho(D^2, K^2) = 1$ if and only if $\rho_{ij} = 0$ for all i and j. For fixed ρ, from (5.3), as $p \to \infty$, $\rho(D^2, K^2) \to 0$, which affirms that the two measures would be different for the equi-correlation case.

For Rao's data (1948) as used by Penrose (1954), we have

$$p = 8, \qquad \sum_{i \neq j}\sum \rho_{ij}^2 = 3.4848.$$

Hence from (5.2), $\rho(D^2, K^2) = 0.83$. Penrose obtains a value of 0.96 empirically. Our method assumes that the differences in the mean vectors between populations are zero which could explain the discrepancy.

We can write the modified value of (5.2) from Lemma 1 when K^2 and D^2 are not taken from $\boldsymbol{\mu}$ but the conclusions are similar.

The above is the asymptotic value of the following sample correlation. For k-sample problem, let

$$D_{ij}^2 = w_{ij}(\bar{\mathbf{x}}_i - \bar{\mathbf{x}}_j)'\boldsymbol{\Sigma}^{-1}(\bar{\mathbf{x}}_i - \bar{\mathbf{x}}_j), \qquad K_{ij}^2 = w_{ij}(\bar{\mathbf{x}}_i - \bar{\mathbf{x}}_j)'\boldsymbol{\Sigma}_0^{-1}(\bar{\mathbf{x}}_i - \bar{\mathbf{x}}_j),$$

$$w_{ij} = n_in_j/(n_i + n_j) \quad \text{for} \quad i, j = 1, \ldots, k.$$

Let r_1 be the correlation between the pairs (D^2_{ij}, K^2_{ij}).. Assuming that $\bar{x}_i \sim N_p(\boldsymbol{\mu}_i, \boldsymbol{\Sigma})$, $i = 1,\ldots,k$ and $n_1 = \ldots = n_k = n$ (for simplicity), we have for large k, $E(r_1) = \rho_1$ where

$$\rho_1 = (2p + 4\bar{\Delta}_1 + \overline{AB})/\{(2p + 4\bar{\Delta}_0 + \overline{AA})(2\,\mathrm{tr}\,\boldsymbol{\Sigma}^2 + 4\bar{\Delta}_2 + \overline{BB})\}^{\frac{1}{2}} \quad (5.4)$$

with

$$\overline{rAB} - \sum_{i<j} \Delta^{*2}_{ij}\Delta^2_{ij} - r\bar{\Delta}_0\bar{\Delta}_1,$$

$$\overline{rAA} = \sum_{i<j} \Delta^4_{ij} - r\bar{\Delta}^2_0, \qquad \overline{rBB} = \sum_{i<j} \Delta^{*4}_{ij} - r\bar{\Delta}^2_1,$$

$$\Delta^2_{ij} = \tfrac{1}{2}n(\boldsymbol{\mu}_i - \boldsymbol{\mu}_j)'\boldsymbol{\Sigma}^{-1}(\boldsymbol{\mu}_i - \boldsymbol{\mu}_j), \qquad \Delta^{*2}_{ij} = \tfrac{1}{2}n(\boldsymbol{\mu}_i - \boldsymbol{\mu}_j)'(\boldsymbol{\mu}_i - \boldsymbol{\mu}_j),$$

$$\bar{\Delta}_0 = \frac{1}{r}\sum_{i<j} \Delta^2_{ij}, \qquad \bar{\Delta}_1 = \frac{1}{r}\sum_{i<j} \Delta^{*2}_{ij}, \qquad r = \tfrac{1}{2}k(k-1),$$

$$\bar{\Delta}_2 = \frac{n}{2r}\sum_{i<j} (\boldsymbol{\mu}_i - \boldsymbol{\mu}_j)'\boldsymbol{\Sigma}(\boldsymbol{\mu}_i - \boldsymbol{\mu}_j).$$

For large n, it is found that $\rho_1 \to \overline{AB}/(\overline{AA}\,\overline{BB})^{\frac{1}{2}}$.

We can prove (5.4) using the results in Lemma 1. Its extension to the case for unequal sample sizes follows from (5.4) on replacing $(\frac{1}{2}n)^{\frac{1}{2}}(\boldsymbol{\mu}_i - \boldsymbol{\mu}_j)$ by $w_{ij}^{-\frac{1}{2}}(\boldsymbol{\mu}_i - \boldsymbol{\mu}_j)$. Again, it can be seen that (5.4) is positive. For $\boldsymbol{\mu}_j = \mathbf{0}$, $j \geq 2$, $\boldsymbol{\mu}_1 \neq \mathbf{0}$, we have

$$k\bar{\Delta}_0 = 2a, \quad k\bar{\Delta}_1 = 2b, \quad k\bar{\Delta}_2 = 2c, \quad a = \Delta^2_{11}, \quad b = \Delta^{*2}_{11}, \quad c = \boldsymbol{\mu}_1'\boldsymbol{\Sigma}\boldsymbol{\mu}_1$$

$$k^2\overline{AA} = 2a^2(k-2), \qquad k^2\overline{BB} = 2b^2(k-2), \qquad k^2\overline{AB} = 2ab(k-2)$$

so that if a, b, c are finite, $\rho_1 \to \rho(D^2, K^2)$. Thus for moderately large k, it seems there is no spurious correlation in spite of one of the populations being different from the other; Gower (1971) argues that for small k there would be high correlation in such circumstances. When $\boldsymbol{\Sigma}$ is replaced by its estimators, the above type of correlations are studied in Khatri and Mardia (1975).

In general, we should distinguish between the following two aspects of distances. If two sets of characters contain the same information then one set can be ignored for discrimination purposes since D^2 eliminates correlated (redundant) variables. However if groups are not established as in Cluster Analysis, K^2 should be used since different groups can be distinguished because of correlated characteristics within the postulated groups (Gower, 1972). That is, separate populations must be recognizable when D^2 is used. Exceptions are, of course, problems of outliers, testing multinormality, etc.

Acknowledgement

I wish to express my gratitude to Professor C.G. Khatri for many helpful comments and also for the shorter proofs of some results.

References

Andrews, D.F., Gnanadesikan, R. and Warner, J.L. (1973). Methods for assessing multivariate normality. *Multivariate Analysis.* (Ed. P.R. Krishnaiah), **3**, 95–115. Academic Press, New York.

Bose, R.C. and Roy, S.N. (1938). The distribution of the studentised D^2-statistic. *Sankhyā*, **4**, 19–38.

Burnaby, T.P. (1966). Growth-invariant discriminant functions and generalized distances. *Biometrics*, **22**, 96–110.

Cacoullos, T. (1973). Distance, discrimination and error. *Discriminant Analysis and Applications* (Ed. T. Cacoullos), 61–75. Academic Press, New York.

Cox, D.R. (1968). Notes on some aspects of regression analysis. *J. Roy. Statist. Soc.*, A, **131**, 265–279.

Fisher, R.A. (1938). The statistical utilization of multiple measurements. *Ann. Eug.*, **8**, 376–386.

Gnanadesikan, R. and Kettenring, J.R. (1972). Robust estimates, residuals, and outlier detection with multiresponse data. *Biometrics*, **28**, 81–124.

Gower, J.C. (1966a). Some distance properties of latent root and vector methods used in multivariate analysis. *Biometrika*, **53**, 325–338.

Gower, J.C. (1966b). A Q-technique for the calculation of canonical variates. *Biometrika*, **53**, 588–590.

Gower, J.C. (1971). Statistical methods of comparing different multivariate analyses of the same data. *Mathematics in the Archaeological and Historical Sciences*, 138–149. (Eds. Hodson, F.R., Kendall, D.G. & Tăutu, P.). Edinburgh University Press, Edinburgh.

Gower, J.C. (1972). Measures of taxonomic distance and their analysis. *The Assessment of Population Affinities in Man.* (Eds. Weiner, J.S. and Huizinga, J.), 1–24. Clarendon Press, Oxford.

Gower, J.C. (1975). Growth free canonical varietes and generalized inverses. (Unpublished).

Healy, M.J.R. (1968). Multivariate normal plotting. *J. Roy. Statist. Soc.*, C, **17**, 157–161.

Khatri, C.G. (1959). On the mutual independence of certain statistics. *Ann. Math. Statist.*, **30**, 1258–1262.

Khatri, C.G. (1966). A note on a MANOVA model applied to problems in growth curve. *Ann. Inst. Statist. Math.*, **18**, 75–86.

Khatri, C.G. (1968). Some results for the singular normal multivariate regression models. *Sankhyā*, **30**, A, 267–280.

Khatri, C.G. (1971). *Mathematics of Matrices* (In Gujarati). Gujarat University, Ahmedabad.

Khatri, C.G. and Mardia, K.V. (1975). Correlation between Mahalanobis and Karl Pearson distances for small samples. (Unpublished).

Kshirsagar, A.M. (1972). *Multivariate Analysis.* M. Decker, New York.

Lawley, D.N. and Maxwell, A.E. (1963). *Factor Analysis as a Statistical Method.* Butterworths, London.

Mahalanobis, P.C. (1930). On tests and measures of group divergence. *J. Asiat. Soc.* (Bengal), **26**, 541–588.

Mahalanobis, P.C. (1936). On the generalized distance in statistics. *Proc. Nat. Inst. Sci.*, **2**, 49–55.

Mardia, K.V. (1970). Measures of multivariate skewness and kurtosis with applications. *Biometrika*, **57**, 519–530.

Mardia, K.V. (1971). The effect of nonnormality on some multivariate tests and robustness to nonnormality in the linear model. *Biometrika*, **58**, 105–121.

Mardia, K.V. (1974). Applications of some measures of multivariate skewness and kurtosis in testing normality and robust studies. *Sankhyā*, **36**, B, 115–128.

Mardia, K.V. (1975). Assessment of multinormality and the robustness of Hotelling's T^2 test. *J. Roy. Stat. Soc.*, C, **24**, 163–171.

Mardia, K.V. and Zemroch, P.J. (1975). Measures of multivariate skewness and kurtosis. *J. Roy. Stat. Soc.*, C, Algorithm AS84, **24**, 262–265.

Mitra, S.K. (1968). On a generalized inverse of a matrix and applications. *Sankhyā*, A, **30**, 323–330.

Mitra, S.K. and Rao, C.R. (1968). Simultaneous reduction of a pair of quadratic forms. *Sankhyā*, A, 313–332.

Penrose, L.E. (1954). Distance, size and shape. *Ann. Eugen.*, **18**, 337–343.

Rao, C.R. (1948). The utilization of multiple measurements in problems of biological classification (with discussion). *J. Roy. Stat. Soc.*, B, **10**, 159–203.

Rao, C.R. (1952). *Advanced Statistical Methods in Biometrical Research.* Wiley, New York.

Rao, C.R. (1954). On the use and interpretation of distance functions in statistics. *Bull. Int. Stat. Inst.*, **34**, 90–97.

Rao, C.R. (1966a). Discriminant function between hypotheses and related problems. *Biometrika*, **53**, 339–345.

Rao, C.R. (1966b). Covariance adjustment and related problems in multivariate analysis. *Multivariate Analysis.* (Ed. P.R. Krishnaiah). Vol. I, 87–103, Academic Press, New York.

Rao, C.R. (1971). Taxonomy in anthropology. *Mathematics in the Archaeological and Historical Sciences.* (Eds. Hodson, F.R., Kendall, D.G. and Tăutu, P.), 19–29. Edinburgh university Press, Edinburgh.

Rao, C.R. and Mitra, S.K. (1971). *Generalized Inverse of Matrices and its Applications.* Wiley, New York.

Rao, C.R. and Vardarajan, V.S. (1963). Discrimination of Gaussian processes. *Sankhyā*, A, **25**, 303–330.

Wilks, S.S. (1963). Multivariate statistical outliers. *Sankhyā*, A, **25**, 674–693.

P.R. Krishnaiah, ed., *Multivariate Analysis–IV*
© North-Holland Publishing Company (1977) 513–522

PROBLEMS OF ASSOCIATION FOR BIVARIATE CIRCULAR DATA AND A NEW TEST OF INDEPENDENCE

Madan L. PURI* and J.S. RAO**
Indiana University, Bloomington, Ind., U.S.A.

1. Introduction and summary

Statistical data where the observations are directions in either two or three dimensions occur naturally in many diverse fields such as geology, biology, astronomy and medicine among others. These directions may be represented as unit vectors, that is, as points on the circumference of the unit circle if two dimensional or on the unit sphere in case they are three dimensional. These directions may also be represented in terms of angles with respect to a fixed "zero direction". It is natural to require that statistical techniques for such directional data have to be independent of this arbitrary chosen zero direction, as well as the sense of rotation, that is, whether one takes clockwise or anticlockwise as the positive direction. These natural restrictions rule out the application of most of the standard statistical methods for directional data. For instance, even the usual "Arithmetic Mean" and the "Standard Deviation" fail to be meaningful measures of location and dispersion. This novel area of statistics has been receiving increasing attention only recently and most of the statistical developments in this area have occurred during the past two decades, especially after the appearance of a paper by Fisher (1953). For a general survey of this field, the reader is referred to Mardia (1972) and Batschelet (1965).

In this paper we shall restrict our attention to circular data, that is, data on two dimensional directions. In many instances, we may measure more than one direction corresponding to each "individual" in the population either because we wish to gain more information than what a single

* Work supported by the Air Force Office of Scientific Research, AFSC, USAF, under Grant No. AFOSR 76–2927. Reproduction in whole or in part permitted for any purpose of the U.S. Government. Part of this research was done while the author was a senior Research Fellow of the SRC at the University of Leeds, England.

** Present address: Dept of Statistics, University of Wisconsin, Madison, Wisconsin, 53706.

measurement could give or because we wish to study the interrelationships, if any (for example, correlation and regression), between such variates. Also, one might be interested in regression problems like predicting the paleocurrent direction on the basis of cross-bedding dip directions or pebble orientations, or in problems of correlation like measuring the association between the flight directions of pigeons and the prevalent wind direction. This important area of multivariate directional data analysis has not so far received much attention, with the result that the research work (on multivariate situations) has been very limited. To acquaint the reader with some background material, we give in section two, a brief review of some of the literature on association and independence for bivariate circular data. In section three, we introduce a new test of independence for measurements on a torus, that is, bivariate circular data. The asymptotic distribution theory of the proposed test statistic is derived in the last section.

2. Problems of association and independence

As stated in section one, we give here a brief survey of the papers of Downs (1974), Mardia (1975), Gould (1969) and Rothman (1971), all of which deal with problems of association for angular variables. While the first two of these papers attempt to define a measure of correlation for angular variates, the third discusses problems of regression, and the last introduces a test of independence. In what follows, $(X_1, Y_1), \ldots, (X_n, Y_n)$ will denote a random sample of observations on a torus, that is, X_i as well as Y_i are angles in $[0, 2\pi)$.

Downs (1974) defines a measure of "rotational correlation" for circular data which is in many ways analogous to the product moment correlation on the line. Let

$$\bar{C}_x = n^{-1} \sum_{i=1}^{n} \cos X_i, \qquad \bar{S}_x = n^{-1} \sum_{i=1}^{n} \sin X_i,$$

$$\bar{C}_y = n^{-1} \sum_{i=1}^{n} \cos Y_i, \qquad \bar{S}_y = n^{-1} \sum_{i=1}^{n} \sin Y_i,$$

be the arithmetic means of cosine and sine components of X and Y. Let

$$T = \frac{1}{n} \sum_{i=1}^{n} \begin{pmatrix} \cos Y_i - \bar{C}_y \\ \sin Y_i - \bar{S}_y \end{pmatrix} \begin{pmatrix} \cos X_i - \bar{C}_x \\ \sin X_i - \bar{S}_x \end{pmatrix}.$$

Set

$$S_x^2 = 1 - (\bar{C}_x^2 + \bar{S}_x^2), \qquad S_y^2 = 1 - (\bar{C}_y^2 + \bar{S}_y^2),$$

and

$$S_{xy} = |T(T'T)^{-\frac{1}{2}}| \cdot \mathrm{tr}(T'T)^{\frac{1}{2}}$$

where $(T'T)^{\frac{1}{2}}$ is the unique symmetric positive definite matrix whose square is $(T'T)$ and $(T'T)^{-\frac{1}{2}}$ is its inverse. Note that the determinant $|T(T'T)^{-\frac{1}{2}}|$ is ± 1 because of the orthogonality of the matrix. Also if R_x^2 and R_y^2 denote the squared lengths of the resultants corresponding to the directions $(X_1, \ldots, X_n)$ and $(Y_1, \ldots, Y_n)$ respectively, then it is easily seen that

$$S_x^2 = 1 - R_{x/n^2}^2 \quad \text{and} \quad S_y^2 = 1 - R_{y/n^2}^2$$

which are the commonly used measures of variation. A justification for the rather complex definition of S_{xy} is not hard to find and the reader is referred to Downs (1974). Downs defines the circular rotational correlation $\gamma_c = S_{xy}/S_x \cdot S_y$. It can be seen that γ_c lies between -1 and $+1$ attaining one of the extreme values only when the Y-deviation (from its resultant direction) is a constant multiple of an orthogonal transformation (that is, a rotation) of the X-deviation for every pair (X_i, Y_i) in the sample. One has to keep in mind however that this may not be an appropriate measure of association when the correlation is not strictly rotational. γ_c remains invariant if the origins are changed for the X and Y measurements. The sampling distribution of γ_c has not been investigated so far and this limits the use of γ_c for statistical inference.

Mardia's (1975) correlation coefficient for circular data is based on the ranks of the observations, and is defined as follows: In analogy with the linear situation, a perfect correlation is said to exist between X and Y if the whole probability mass is concentrated on

$$(lX \pm mY + Z) \bmod 2\pi = 0$$

for some positive integers l, m and a fixed angular quantity Z. Define $X_i^* = lX_i \pmod{2\pi}$, $Y_i^* = mY_i \pmod{2\pi}$, and let the linear ranks of $X_1^*, \ldots, X_n^*$ be $1, \ldots, n$ respectively and those of $Y_1^*, \ldots, Y_n^*$ be $r_1, \ldots, r_n$ respectively. The angles (X_i^*, Y_i^*) are then replaced by the uniform scores $(2\pi i/n, 2\pi r_i/n)$. Now let R_1^2 and R_2^2 denote the lengths of the resultants corresponding to the directions $2\pi(i - r_i)/n$, $i = 1, \ldots, n$ and $2\pi(i + r_i)/n$, $i - 1, \ldots, n$ respectively Mardia (1975) defines

$$\gamma_0 = \max(R_1^2/n^2, R_2^2/n^2)$$

as the circular rank correlation coefficient. It is clear that γ_0 lies between zero and one, and that it remains invariant under changes of zero-directions of X and Y. We have $R_1^2/n^2 = 1$ for perfect "positive" dependence, and $R_2^2/n^2 = 1$ for perfect "negative" dependence. γ_0 will be close to zero if X and Y are uncorrelated. Mardia (1975) also discusses the asymptotic null distribution of γ_0 and gives a table of critical values.

Gould (1969), on the other hand, considers an analogue of the normal theory linear regression for circular variables. Let (x_i, Y_i), $i = 1, \ldots, n$ be observations on directions such that $Y_i, i = 1, \ldots, n$ are independently distributed as circular normal random variables $\mathrm{CN}(\alpha + \beta x_i, \mathcal{K})$, that is, with density

$$f(y_i) = \frac{1}{2\pi I_0(\mathcal{K})} \exp(\mathcal{K} \cos(y_i - \alpha - \beta x_i))$$

where $x_1, \ldots, x_n$ are known numbers while α, β and $\mathcal{K}$ are unknown parameters. Since the logarithm of the likelihood function is

$$- n \log 2\pi - n \log I_0(\kappa) + \mathcal{K} \sum_{i=1}^{n} \cos(Y_i - \alpha - \beta x_i),$$

the MLE's (Maximum Likelihood Estimates) $\hat{\alpha}$ and $\hat{\beta}$ of α and β are the solutions of the equations

$$\sum_{i=1}^{n} \sin(Y_i - \hat{\alpha} - \beta x_i) = 0,$$

$$\sum_{i=1}^{n} x_i \sin(Y_i - \hat{\alpha} - \hat{\beta} x_i) = 0.$$

These solutions can be obtained by an iterative procedure discussed in Gould (1969). After solving for $\hat{\alpha}$ and $\hat{\beta}$, the MLE $\hat{\mathcal{K}}$ of $\mathcal{K}$ is obtained from the equation

$$\frac{I_1(\hat{\mathcal{K}})}{I_0(\hat{\mathcal{K}})} = \sum_{i=1}^{n} \cos(Y_i - \hat{\alpha} - \hat{\beta} x_i)/n.$$

Finally, for testing independence, Rothman (1971) assumes that the observations $(X_1, Y_1), \ldots, (X_n, Y_n)$ are from a continuous d.f. (distribution function) $F(x, y)$ whose marginal d.f.'s are $F_1(x)$ and $F_2(y)$. The problem is to test the hypothesis $H_0 : F(x, y) = F_1(x) \cdot F_2(y) \forall (x, y)$. With respect to the given origins for the two variates, let

$$F_{n1}(x) = n^{-1} \sum_{i=1}^{n} I(X_i, x),$$

$$F_{n2}(y) = n^{-1} \sum_{i=1}^{n} I(Y_i; y),$$

$$F_n(x, y) = n^{-1} \sum_{i=1}^{n} I(X_i; x) I(Y_i; y),$$

where

$$I(s; t) = \begin{cases} 1 & \text{if} \quad s \leq t, \\ 0 & \text{if} \quad s > t. \end{cases}$$

Thus $F_{n1}(x)$, and $F_{n2}(y)$ and $F_n(x, y)$ are the empirical d.f.'s of the X's, Y's and (X, Y)'s respectively. Distribution free tests of the form

$$\int\int T_n^2(x, y) \mathrm{d}F_n(x, y), \quad \text{where} \quad T_n(x, y) = [F_n(x, y) - F_{n1}(x) F_{n2}(y)]$$

were considered earlier by Blum, Kiefer and Rosenblatt (1961) and since they are not invariant under different choices of the origins for X and Y, they are not applicable to the circular case. To circumvent this problem, Rothman (1971) suggested the modified statistic

$$C_n = n \int_0^{2\pi} \int_0^{2\pi} Z_n^2(x, y) \mathrm{d}F_n(x, y),$$

where

$$Z_n(x, y) = \Big[T_n(x, y) + \int\int T_n(x, y) \mathrm{d}F_n(x, y) - \int T_n(x, y) \mathrm{d}F_{n1}(x) - \int T_n(x, y) \mathrm{d}F_{n2}(y) \Big].$$

The statistic C_n has the desired invariance property and its asymptotic distribution theory under the null hypothesis has been derived by Rothman (1971).

3. A new test for co-ordinate independence of circular data

Let $\mathscr{F}$ denote the family of probability distributions on the circumference $[0, 2\pi)$ of the circle with the property $F(\alpha + 2\pi) = F(\alpha) + \frac{1}{2}$ for all α. For instance circular distributions with axial symmetry would be in this class. In this section we propose a test which is applicable to testing independence when the marginals F_1 and F_2 belong to this class $\mathscr{F}$. When dealing with axial data, each observed axis will be represented by both its antipodal points for the purposes of this test. The proposed test may also be applied for testing independence in the non-axial case. But in this case, corresponding to every observed direction, we should add its antipodal point also to the data thus doubling the original sample size. The effect such "doubling" would have on the power of the test procedure, is presently being investigated. Thus from now on, the random sample $(X_1, Y_1), \ldots, (X_n, Y_n)$ referred to in Sections 3 and 4 corresponds to the axial data with both ends represented as two distinct sample points or the "doubled" sample in the general non-axial case. This ensures that the marginal distributions F_1 and F_2 belong to the class $\mathscr{F}$.

As before, let $(X_1, Y_1), \ldots, (X_n, Y_n)$ denote a random sample of angular variates on the basis of which we wish to test the null hypothesis of independence. For any fixed x in $[0, 2\pi)$, let $N_1(x)$ denote the number of X_i's that fall in the half-circle $[x, x+\pi)$. Similarly, for any fixed y in $[0, 2\pi)$, let $N_2(y)$ denote the number of Y_i's which fall in the half circle $[y, y+\pi)$. Also, let $N(x, y)$ denote the number of observations (X_i, Y_i) that fall in the quadrant $[x, x+\pi)\times[y, y+\pi)$. Now defining the indicator variables

$$I_i(x) = \begin{cases} 1 & \text{if} \quad X_i \in [x, x+\pi), \\ 0 & \text{otherwise} \end{cases} \tag{3.1}$$

and

$$\bar{I}_i(y) = \begin{cases} 1 & \text{if} \quad Y_i \in [y, y+\pi], \\ 0 & \text{otherwise} \end{cases} \tag{3.2}$$

we obtain

$$\begin{aligned} N_1(x) &= \sum_{i=1}^{n} I_i(x), \qquad N_2(y) = \sum_{i=1}^{n} \bar{I}_i(y), \\ N(x, y) &= \sum_{i=1}^{n} I_i(x)\,\bar{I}_i(y). \end{aligned} \tag{3.3}$$

If the hypothesis of independence holds, then we should have

$$N(x, y) = N_1(x)\cdot N_2(y)/n$$

by the usual arguments. Thus we define

$$D_n(x, y) = n^{-\frac{1}{2}}[N(x, y) - N_1(x)N_2(y)/n] \tag{3.4}$$

as a measure of discrepancy between the observed and expected (under the hypothesis of independence) frequencies. Since $T_n(x, y)$ depends specifically on the choices of x and y, we suggest the (invariant) statistic

$$\begin{aligned} T_n &= \int_0^{2\pi}\int_0^{2\pi} D_n^2(x, y)\,\frac{dx}{2\pi}\,\frac{dy}{2\pi} \\ &= \frac{1}{4\pi^2 n}\int_0^{2\pi}\int_0^{2\pi}\left[N(x, y) - \frac{N_1(x)N_2(y)}{n}\right]^2 dx\,dy \end{aligned} \tag{3.5}$$

for testing independence. The integrand $D_n^2(x, y)$ is much like the usual chi-square test for independence from a 2×2 table. We now derive a computional form for T_n in terms of the X and Y spacings.

In view of (3.1), (3.2) and (3.3), we have

$$\begin{aligned} nD_n^2(x,y) &= \left[\left(1-\frac{1}{n}\right)\left(\sum_{i=1}^{n} I_i(x)\bar{I}_i(y) - \frac{1}{n}\left(\sum_{i\neq j} I_i(x)\bar{I}_j(y)\right)\right]^2 \\ &= \left(1-\frac{1}{n}\right)^2\left\{\sum I_i\bar{I}_i + \sum I_{ij}\bar{I}_{ij}\right\} \\ &\quad -\frac{2}{n}\left(1-\frac{1}{n}\right)\left\{\sum I_i\bar{I}_{ij} + \sum I_{ij}\bar{I}_i + \sum I_{ij}\bar{I}_{ik}\right\} \\ &\quad +\frac{1}{n^2}\left\{\sum I_i\bar{I}_j + \sum I_{ij}\bar{I}_k + \sum I_i\bar{I}_{jk} + \sum I_{ij}\bar{I}_{k1}\right\} \end{aligned}$$

where $I_{ij} = I_i(x)I_j(x)$, $\bar{I}_{ij} = \bar{I}_i(y)\bar{I}_j(y)$, and summations are over all distinct subscripts. It is easy to check that

$$\int_0^{2\pi} I_i(x)\mathrm{d}x = \int_0^{2\pi} \bar{I}_i(y)\mathrm{d}y = \pi,$$

$$\int_0^{2\pi} I_i(x)I_j(x)\mathrm{d}x = \pi - D_{ij},$$

$$\int_0^{2\pi} \bar{I}_i(y)\bar{I}_j(y)\mathrm{d}y = \pi - \bar{D}_{ij},$$

where D_{ij} and $\bar{D}_{ij}$ are the "circular" distances between (X_i, X_j) and (Y_i, Y_j) respectively. Omitting the routine computations, we obtain

$$\begin{aligned} T_n = \frac{1}{4\pi^2 n}\Bigg[&(n-1)\pi^2 - \frac{\pi}{n}\left\{\sum(\pi - D_{ij}) + \sum(\pi - \bar{D}_{ij})\right\} \\ &+\left(1-\frac{1}{n}\right)^2\left\{\sum(\pi - D_{ij})(\pi - \bar{D}_{ik})\right\} \\ &-\frac{2}{n}\left(1-\frac{1}{n}\right)\left\{\sum(\pi - D_{ij})(\pi - \bar{D}_{ik})\right\} \\ &+\frac{1}{n^2}\left\{\sum(\pi - D_{ij})(\pi - \bar{D}_{k1})\right\}\Bigg], \end{aligned}$$

again the summations run over all the distinct subscripts. The statistic T_n being a function of the circular distances $\{D_{ij}\}$ and $\{\bar{D}_{ij}\}$, is clearly invariant under rotations of either coordinate axis.

4. The asymptotic null distribution

To derive the asymptotic null distribution of T_n, we will utilize the methods of Fourier analysis similar to those in Rao (1972) and Rothman (1971). Since $D_n(x, y)$ is a doubly periodic function (in both x and y) we may find the Fourier expansion

$$D_n(x,y) = \sqrt{n} \sum_{k=-\infty}^{\infty} \sum_{m=-\infty}^{\infty} Z_{km}\, e^{-ikx}\, e^{-imy}, \tag{4.1}$$

where

$$Z_{km} = \frac{1}{4\pi^2} \int_0^{2\pi} \int_0^{2\pi} \left[\frac{N(x,y)}{n} - \frac{N_1(x)}{n} \cdot \frac{N_2(y)}{n} \right] e^{ikx}\, e^{imy}\, dx\, dy \tag{4.2}$$

$$= \frac{1}{n} \sum_{j=1}^{n} \left(\int_0^{2\pi} I_j(x) \frac{e^{ikx}}{2\pi}\, dx \right) \left(\int_0^{2\pi} \bar{I}_j(y) \frac{e^{imy}}{2\pi}\, dy \right)$$

$$- \frac{\sum_{j=1}^{n} \left(\int_0^{2\pi} I_j(x) \frac{e^{ikx}}{2\pi}\, dy \right)}{n} \times \frac{\sum_{j=1}^{n} \left(\int_0^{2\pi} \bar{I}_j(y) \frac{e^{imy}}{2\pi}\, dy \right)}{n}.$$

From the definitions of $I_j(x)$ and $\bar{I}_j(y)$, it follows that

$$Z_{km} = \begin{cases} 0, & \text{if } k \text{ or } m \text{ is even,} \\ \dfrac{1}{\pi^2 mk} \left[\left(\dfrac{\sum_{j=1}^{n} e^{ikX_j}}{n} \right) \left(\dfrac{\sum_{j=1}^{n} e^{imY_j}}{n} \right) - \left(\dfrac{\sum_{j=1}^{n} e^{ikX_j + imY_j}}{n} \right) \right], & \text{if both } k \text{ and } m \text{ are odd.} \end{cases}$$

Thus from (3.5), (4.1) and an application of Parseval's theorem, we have

$$T_n = n \sum_k \sum_m |Z_{km}|^2.$$

If can be verified that under the null hypothesis of independence

$$\mathbf{E} Z_{km} = 0, \qquad \forall k, m$$

$$\mathbf{E} Z_{km} Z_{k'm'} = \begin{cases} \dfrac{\delta_{kk'} \delta_{mm'}}{\pi^4 k^2 m^2} \left(\dfrac{n-1}{n^2} \right), & \text{if both } k \text{ and } m \text{ are odd,} \\ 0, & \text{otherwise} \end{cases}$$

where δ_{jk} is the usual Kronecker delta. Thus, the random Fourier coefficients Z_{km} have zero means and are uncorrelated for distinct pairs

(k, m). It may be remarked here that the above expectations may be calculated under the assumption that the X_i's and Y_i's have a uniform distribution, which under the null hypothesis are further independent, in view of the fact we could use a probability integral transformation as in Blum, Kiefer and Rosenblatt (1961). This transformation does not change the statistic T_n since the numbers $N(x, y)$, $N_1(x)$, $N_2(y)$ remain unchanged under monotone transformations.

Now by arguments similar to those in Rao (1972) or Rothman (1971), we have the result that T_n has asymptotically the same distribution as the sum of the squares of independent normal variables X_{km} with zero means and variances $(\pi^4 k^2 m^2)^{-1}$ for odd k and m. Thus the asymptotic characteristic function of T_n is given by

$$\psi(t) = \prod_{k \text{ odd}} \prod_{m \text{ odd}} \left(1 - \frac{2it}{\pi^4 k^2 m^2}\right)^{-1/2}$$

$$= \prod_{k=1}^{\infty} \prod_{m=1}^{\infty} \left(1 - \frac{2it}{\pi^4 (2k-1)^2 (2m-1)^2}\right)^{-2}.$$

This characteristic function can be formally inverted as in Rao (1972). On the other hand by a result of Zolotarev (1961), if $F(x)$ denotes the asymptotic d.f. of T_n, then the upper tail probabilities relating to T_n may be approximated as follows:

$$\lim_{x \to \infty} \frac{1 - F(x)}{\mathbf{P}[\chi_4^2 > \pi^4 x]} = \prod_{k=1}^{\infty} \prod_{m=1}^{\infty} \left(1 - \frac{1}{(2k-1)^2 (2m-1)^2}\right)^{-2}$$

$$(k, m) \neq (1, 1)$$

where χ_4^2 denotes a random variable having the chi-square distribution with 4 degrees of freedom.

Acknowledgment

The authors wish to thank Professor Christopher Bingham for pointing out the somewhat restrictive nature of the test proposed here.

References

[1] Batschelet, E. (1965). Statistical Methods for the Analysis of Problems in Animal Orientation and Certain Biological Rhythms, Amer. Inst. Biol. Sciences, Washington.

[2] Blum, J.R., Kiefer, J. and Rosenblatt, M. (1961). Distribution free test of independence based on the sample distribution function, *Ann. Math. Statist.* **32**, 485–492.

[3] Downs, T.D. (1974). Rotational Angular Correlations, in *Biorhythms and Human Reproduction*, (M. Ferin et al. Eds.), John Wiley, New York: 97–104.
[4] Fisher, R.A. (1953). Dispersion on a sphere, *Proc. Roy. Soc.* London A**217**, 295–305.
[5] Gould, A.L. (1969). A regression technique for angular variates, *Biometrics*, **25**, 683–700.
[6] Mardia, K.V. (1972). *Statistics of Directional Data*, Academic Press, New York.
[7] Mardia, K.V. (1975). Statistics of Directional Data, to appear in *Jour. Roy. Statist. Soc.*, Series B.
[8] Rao, J. S. (1972). Some variants of chi-square for testing uniformity on the circle, *Z. Wahrscheinlichkeitstheorie Verw. Geb.* **22**, 33–44.
[9] Rothman, E.D. (1971). Tests of coordinate independence for a bivariate sample on torus, *Ann. Math. Statist.*, **42**, 1962–1969.
[10] Zolotarev, V.M. (1961). Concerning a certain probability problem, *Theor. Probability Appl*, 201–204.

P.R. Krishnaiah, ed., *Multivariate Analysis–IV*
© North-Holland Publishing Company (1977) 523–545

ASYMPTOTIC EXPANSIONS FOR ERROR RATES AND COMPARISON OF THE *W*-PROCEDURE AND THE *Z*-PROCEDURE IN DISCRIMINANT ANALYSIS

Minoru SIOTANI* and Ruey-Hwa WANG
Kansas State University, Manhattan, Kans., U.S.A.

Two discriminant procedures based on Anderson's W-criterion and John–Kudo's Z-criterion are compared with respect to error rates or probabilities of misdiscrimination. To make an extensive comparison in the case of moderate sample sizes, the cubic terms are added to the asymptotic expansion formulae for the error rates obtained by Okamoto [8, 9] and Memon–Okamoto [7] up to the quadratic terms. Tables of values of the coefficients contained in the cubic terms and tables which give us some information on the comparison of two procedures are prepared. The discussion on the values of parameters validating the expansion formulae is made.

1. Introduction

Recent rapid development for the general distribution in the multivariate analysis gives us the formulae based on which we can investigate comparisons of a number of procedures proposed for the same problem. As an example, we are concerned in this paper with comparison of two discriminant procedures for the two p-variate normal populations π_1: $N_p(\mu_1, \Sigma)$ and π_2: $N_p(\mu_2, \Sigma)$, where μ_1, μ_2, and Σ are unknown. Let x be a new vector observation to be assigned to either π_1 or π_2, $\bar{x}_i$, $i = 1, 2$ be sample mean vectors based on samples of sizes N_i taken from π_i and $S: p \times p$ be an unbiased estimator of Σ such that nS has the Wishart distribution with n degrees of freedom (d.f.) and covariance matrix Σ and is independent of $\bar{x}_i$, $i = 1, 2$. Then Anderson's criterion [1] is defined by

$$W = [x - \tfrac{1}{2}(\bar{x}_1 + \bar{x}_2)]' S^{-1} (\bar{x}_1 - \bar{x}_2) \tag{1.1}$$

and the discrimination is in many applications performed in the following way:

* Work of this author was supported in part by the Faculty Research Award Committee, Kansas State University.

$$\begin{aligned}&\text{Assign } x \text{ to the population } \pi_1 \text{ if } W \geq 0, \text{ and}\\&\text{assign } x \text{ to the population } \pi_2 \text{ if } W < 0.\end{aligned} \tag{1.2}$$

We call this the *W-procedure.*.

John [4] and Kudo [5] proposed the criterion

$$Z = \frac{N_1}{N_1+1}(x-\bar{x}_1)'S^{-1}(x-\bar{x}_1) - \frac{N_2}{N_2+1}(x-\bar{x}_2)'S^{-1}(x-\bar{x}_2) \tag{1.3}$$

as a competitor to Anderson's W. The discrimination procedure is

$$\begin{aligned}&\text{Assign } x \text{ to the population } \pi_1 \text{ if } Z \leq 0, \text{ and}\\&\text{assign } x \text{ to the population } \pi_2 \text{ if } Z > 0.\end{aligned} \tag{1.4}$$

We call this the *Z-procedure*, which is also the maximum likelihood procedure. Das Gupta [3] proved that the Z-procedure is unbiased, admissible and minimax in the class of invariant discrimination procedures.

We want to investigate the comparison of the W-procedure and the Z-procedure based on the error rates or the probabilities of misdiscrimination, which are defined and denoted by

$$P_1 = \mathbf{P}\{W < 0 \mid x \in \pi_1\} \text{ and } P_2 = \mathbf{P}\{W \geq 0 \mid x \in \pi_2\} \text{ for the } W\text{-procedure,}$$

$$P_1^* = \mathbf{P}\{Z > 0 \mid x \in \pi_1\} \text{ and } P_2^* = \mathbf{P}\{Z \leq 0 \mid x \in \pi_2\} \text{ for the } Z\text{-procedure.}$$

When $N_1 = N_2 = N$, we have $Z = -[2N/(N+1)]W$, so that the W and Z procedures are equivalent. If both N_1 and N_2 (and n) are sufficiently large, then $N_i/(N_i+1) \approx 1$, $i = 1, 2$ and hence both procedures are also equivalent. But since the criterion Z takes account of sample sizes explicitly and the criterion W does not, some differences are expected between error rates of the two procedures when N_1 and N_2 are different and at least one of them is of moderate magnitude.

Okamoto [8] gave an expansion formula for P_i $(i = 1,2)$ and Memon and Okamoto [7] obtained an asymptotic expansion for P_i^* $(i = 1,2)$ both up to terms of the second order with respect to $(N_1^{-1}, N_2^{-1}, n^{-1})$. Unfortunately it was found by numerical examination that considerably large sample sizes, N_i, $i = 1,2$ are needed to obtain good estimates of P_i and P_i^* from their expansion formulae as the dimensionality p becomes large, in which case the both procedures are equivalent for practical applications.

In Section 2, we shall give the cubic term to make Okamoto's and Memon–Okamoto's formulae more accurate so as to cover the cases of moderate sample sizes. After the calculation process is outlined (Section

2.2), tables of coefficients of $1/N_iN_jN_k$ ($N_3 = n$) in the cubic term are given (Section 2.3). In Section 3, we discuss on the validity of the formulae under a given criterion and try to summarize the information for the validity in the form of charts showing the lower boundaries of regions of valid points (N_1, N_2) at which the asymptotic formulae give us good approximations to P_i and P_i^* for fixed p and D, the Mahalanobis distance between two populations π_1 and π_2.

In the investigation of methods for estimating error rates when the W-procedure is used, Lachenbruch and Mickey [6] and also Cochran [2] stated that the "OS-method", the method based on Okamoto's formula for P_i, gives good estimates of error rates in the discriminant analysis. So our extended formulae will serve as a useful basis for the estimation of the error rates.

The comparison of the W-procedure and the Z-procedure is given in Section 4. We use as a criterion for the comparison

$$\Delta = (P_1 + P_2) - (P_1^* + P_2^*) \tag{1.5}$$

and we say that the W-procedure (the Z-procedure) is *better* than the Z-procedure (the W-procedure) if $\Delta < 0$ (if $\Delta > 0$) and the both procedures are *equivalent* when $\Delta = 0$. The superiority of one procedure to the other is not consistent and depends on the dimensionality p, the Mahalanobis distance D and sample sizes N_1, N_2. We considered in this paper the cases when $p = 1(1)11, 13, 15, 17, 20, 25$; $D = 0.6, 1.0, 1.4, 1.6, 2.0, 2.4, 3.0, 3.6, 4.0, 5.0$, and $N_1, N_2 \geq 30$. In many cases the Z-procedure has been found to be better than the W-procedure except the cases of smaller p and larger D. But for fixed D, the superiority becomes changeable depending on values of N_1 and N_2 as p increases. Details of the comparison are summarized in Table 4.2.

2. Expansion formulae for P_i and P_i^*

2.1. Final formulae

$$P_1 = \Phi(-\tfrac{1}{2}D) + \left[\frac{a_1}{N_1} + \frac{a_2}{N_2} + \frac{a_3}{n}\right] + \left[\frac{b_{11}}{N_1^2} + \frac{b_{12}}{N_1N_2} + \frac{b_{22}}{N_2^2} + \frac{b_{13}}{N_1n} + \frac{b_{23}}{N_2n} + \frac{b_{33}}{n^2}\right]$$
$$+ \left[\frac{c_{111}}{N_1^3} + \frac{c_{112}}{N_1^2N_2} + \frac{c_{122}}{N_1N_2^2} + \frac{c_{222}}{N_2^3} + \frac{c_{113}}{N_1^2n} + \frac{c_{123}}{N_1N_2n} + \frac{c_{223}}{N_2^2n} + \frac{c_{133}}{N_1n^2} + \frac{c_{233}}{N_2n^2} + \frac{c_{333}}{n^3}\right]$$
$$+ O_4, \tag{2.1}$$

$$P_2 = \text{the expression obtained by interchanging } N_1 \text{ and } N_2 \text{ in } P_1, \tag{2.2}$$

where $D^2 = (\mu_1 - \mu_2)'\Sigma^{-1}(\mu_1 - \mu_2)$, square of the Mahalanobis distance, $\Phi(c)$ = the c.d.f. of $N(0,1)$, a_i's and b_{ij}'s are given by Okamoto [8] in Corollary 2 with the correction of b_{33} in [9], and

$$\begin{aligned}
c_{111} &= (48D^6)^{-1}[d_0^{12} + 9(p+4)d_0^{10} + 3(p+4)(9p+34)d_0^8 \\
&\quad + 3(p+2)(p+4)(9p+68)d_0^6 + 60(p+2)(p+4)(3p+4)d_0^4 \\
&\quad + 168p(p+2)(p+4)d_0^2], \\
c_{112} &= (16D^6)^{-1}[d_0^{12} + 5(p+8)d_0^{10} + (3p^2+154p+488)d_0^8 \\
&\quad - (p+2)(9p^2-124p-1040)d_0^6 \\
&\quad - 4(p+2)(23p^2-228p-320)d_0^4 - 120p(p+2)(p-8)d_0^2], \\
c_{122} &= (16D^6)^{-1}[d_0^{12} + (p+44)d_0^{10} - 5(p^2-10p-120)d_0^8 \\
&\quad + (3p^3-122p^2+704p+2880)d_0^6 \\
&\quad + 12(3p^3-62p^2+264p+320)d_0^4 + 72p(p-6)(p-8)d_0^2], \\
c_{222} &= (48D^6)^{-1}[d_0^{12} - 3(p-16)d_0^{10} + 3(p^2-34p+248)d_0^8 \\
&\quad - (p-8)(p-12)(p-46)d_0^6 - 12(p-6)(p-8)(p-16)d_0^4 \\
&\quad - 24(p-4)(p-6)(p-8)d_0^2], \\
c_{113} &= (16D^4)^{-1}(p-1)[d_0^{10} + 6(p+6)d_0^8 + (9p^2+118p+376)d_0^6 \\
&\quad + 8(7p^2+72p+156)d_0^4 + 40(p+2)(p+12)d_0^2], \\
c_{123} &= (8D^4)^{-1}\ (p-1)[d_0^{10} + 2(p+16)d_0^8 - (3p^2-34p-296)d_0^6 \\
&\quad - 24(p+4)(p-9)d_0^4 - 24(p+4)(p-6)d_0^2], \\
c_{223} &= (16D^4)^{-1}(p-1)[d_0^{10} - 2(p-14)d_0^8 + (p^2-34p+216)d_0^6 \\
&\quad + 8(p-6)(p-10)d_0^4 + 8(p-4)(p-6)d_0^2], \\
c_{133} &= (16D^2)^{-1}(p-1)[(p+1)d_0^8 + (3p^2+31p+24)d_0^6 \\
&\quad + 12(2p^2+15p+10)d_0^4 + 24(p^2+8p+4)d_0^2], \\
c_{233} &= (16D^2)^{-1}(p-1)[(p+1)d_0^8 - (p^2-15p-12)d_0^6 \\
&\quad - 4(2p^2-13p-6)d_0^4 - 8p(p-4)d_0^2], \\
c_{333} &= \tfrac{1}{48}(p-1)[(p+1)(p+3)d_0^6 + 12(p+1)^2 d_0^4 + 24p^2 d_0^2].
\end{aligned}$$

and d_0^i are constants defined by

$$d_0^i = (d^i/dc^i)\Phi(c)|_{c=-\frac{1}{2}D} \qquad (i = 1, 2, \ldots). \tag{2.3}$$

$$\begin{aligned} P_1^* = \Phi(-\tfrac{1}{2}D) &+ \left[\frac{a_1^*}{N_1} + \frac{a_2^*}{N_2} + \frac{a_3^*}{n}\right] \\ &+ \left[\frac{b_{11}^*}{N_1^2} + \frac{b_{12}^*}{N_1N_2} + \frac{b_{22}^*}{N_2^2} + \frac{b_{13}^*}{N_1n} + \frac{b_{23}^*}{N_2n} + \frac{b_{33}^*}{n^2}\right] \\ &+ \left[\frac{c_{111}^*}{N_1^3} + \frac{c_{112}^*}{N_1^2N_2} + \frac{c_{122}^*}{N_1N_2^2} + \frac{c_{222}^*}{N_2^3} + \frac{c_{113}^*}{N_1^2n} + \frac{c_{123}^*}{N_1N_2n} + \frac{c_{223}^*}{N_2^2n} \right. \\ &\qquad \left. + \frac{c_{133}^*}{N_1n^2} + \frac{c_{233}^*}{N_2n^2} + \frac{c_{333}^*}{n^3}\right] \\ &+ O_4, \end{aligned} \tag{2.4}$$

P_2^* = the expression obtained by interchanging N_1 and N_2 in P_1^*, (2.5)

where a_i^* and b_{ij}^* are the same coefficients as those given by Memon and Okamoto [7] in the corollary of the main theorem, and

$$\begin{aligned} c_{111}^* = (48D^6)^{-1}[&-d_0^{12} + 3(p-28)d_0^{10} - 3(p^2-54p+744)d_0^8 \\ &+ (p^3-78p^2+2408p-22176)d_0^6 \\ &- 48(7p^2-218p+1576)d_0^4 - 288(p^2-34p+208)d_0^2], \end{aligned}$$

$$\begin{aligned} c_{112}^* = (16D^6)^{-1}[&3d_0^{12} - (5p-184)d_0^{10} + (p^2-190p+3624)d_0^8 \\ &+ (p^3+14p^2-2112p+27360)d_0^6 \\ &+ 80(p^2-94p+912)d_0^4 - 5760(p-8)d_0^2], \end{aligned}$$

$$\begin{aligned} c_{122}^* = (16D^6)^{-1}[&-9d_0^{12} + (3p-428)d_0^{10} + (5p^2+98p-6552)d_0^8 \\ &+ (p^3+90p^2+968p-38208)d_0^6 \\ &+ 192(2p^2+17p-404)d_0^4 + 192(p-8)(p+24)d_0^2], \end{aligned}$$

$$\begin{aligned} c_{222}^* = (48D^6)^{-1}[&27d_0^{12} + 9(3p+128)d_0^{10} + 3(3p^2+278p+5144)d_0^8 \\ &+ (p^3+150p^2+7424p+75648)d_0^6 \\ &+ 576(p^2+36p+208)d_0^4 \\ &+ 288(p^2+46p+128)d_0^2], \end{aligned}$$

$$\begin{aligned} c_{113}^* = (16D^4)^{-1}(p-1)[&d_0^{10} - 2(p-20)d_0^8 + (p^2-42p+448)d_0^6 \\ &+ 4(p^2-50p+384)d_0^4 - 192(p-6)d_0^2], \end{aligned}$$

$$c^*_{123} = (8D^4)^{-1}\ (p-1)[-3d_0^{10} + 2(p-46)d_0^8 + (p^2+42p-800)d_0^6 + 4(p^2+52p-528)d_0^4 + 192(p-6)d_0^2],$$

$$c^*_{223} = (16D^4)^{-1}(p-1)[9d_0^{10} + 2(3p+136)d_0^8 + (p^2+110p+2336)d_0^6 + 4(p^2+118p+1536)d_0^4 + 384(p+9)d_0^2],$$

$$c^*_{133} = (16D^2)^{-1}(p-1)[-(p+1)d_0^8 + (p^2-15p-12)d_0^6 + 4(2p^2-13p-6)d_0^4 + 8p(p-4)d_0^2],$$

$$c^*_{233} = (16D^2)^{-1}(p-1)[3(p+1)d_0^8 + (p^2+61p+48)d_0^6 + 4(2p^2+71p+42)d_0^4 + 8(p^2+32p+12)d_0^2],$$

$$c^*_{333} = \tfrac{1}{48}(p-1)[(p+1)(p+3)d_0^6 + 12(p+1)^2 d_0^4 + 24p^2 d_0^2].$$

2.2. Note on the derivation

The derivation of the formulae was performed in the similar way as Okamoto's [8] for P_i and Memon–Okamoto's [7] for P^*_i. It is enough here to give a short note.

Let us first consider the characteristic functions $\phi(t)$ of $(W - \frac{1}{2}D^2)/D$ for P_1 and $\psi(t)$ of $(Z + D^2)/2D$ for P^*_1 when π_1 is the true population of x. Then, according to the method based on the Taylor expansion for deriving an asymptotic expansion in the multivariate analysis (general exposition on this is seen in Siotani [10]), $\phi(t)$ and $\psi(t)$ can be expanded respectively in the following form:

$$\phi(t) = \Theta \cdot e^{A} \big|_{\mu_1=0,\ \mu_2=\mu_0,\ \Sigma=I} \tag{2.6}$$

$$\psi(t) = \Theta \cdot e^{A^*} \big|_{\mu_1=0,\ \mu_2=\mu_0,\ \Sigma=I} \tag{2.7}$$

where $\mu_0' = (D, 0, \ldots, 0)$,

$$A = \tfrac{1}{2}(D\theta - 2D^{-1}\theta\mu + D^{-2}\theta^2\sigma^2), \tag{2.8}$$

$$\begin{aligned} A^* = &-\tfrac{1}{2}D\theta + \tfrac{1}{2}D^{-1}\theta\alpha(\alpha+1)\beta[(\mu_1-\mu_2)'\Sigma^{-1}(\mu_1-\mu_2)] \\ &-\tfrac{1}{2}[(\alpha+1)\mu_1 - \alpha\mu_2]'[I-(I+D^{-1}\theta\beta\Sigma^{-1})^{-1}][(\alpha+1)\mu_1-\alpha\mu_2] \\ &-\tfrac{1}{2}\log|I+D^{-1}\theta\beta\Sigma^{-1}| \end{aligned} \tag{2.9}$$

with the notations

$$\mu = -\tfrac{1}{2}(\mu_1+\mu_2)'\Sigma^{-1}(\mu_1-\mu_2), \qquad \sigma^2 = (\mu_1-\mu_2)'\Sigma^{-2}(\mu_1-\mu_2),$$

$$\theta = -it\ (i=\sqrt{-1}),\ \alpha = N_2(N_1+1)/(N_1-N_2),\ \beta = (N_1-N_2)/(N_1+1)(N_2+1)$$

and Θ is a differential operator defined by

$$\Theta = \exp\left[\frac{1}{2N_1}\,\partial_1'\partial_1 + \frac{1}{2N_2}\,\partial_2'\partial_2 - \mathrm{tr}(\partial) - \frac{n}{2}\log\left|I - \frac{2}{n}\,\partial\right|\right] \tag{2.10}$$

where

$$\partial_k' = (\partial/\partial\mu_{k1}, \partial/\partial\mu_{k2}, \ldots, \partial/\partial\mu_{kp}), \qquad k = 1, 2$$

$$\partial = (\partial_{ij}): p \times p,\ \partial_{ij} = \partial_{ji} = \tfrac{1}{2}(1 + \delta_{ij})\partial/\partial\sigma_{ij},\ \delta_{ij} = 1 \text{ for } i = j,\ = 0 \text{ for } i \neq j.$$

In the actual calculation, the right-hand side of (2.10) is expanded with respect to N_1^{-1}, N_2^{-1} and n^{-1}. In (2.6) and (2.7), D contained in A and A^* is a constant and the differentiation is performed at $\mu_1 = 0$, $\mu_2 = \mu_0$ and $\Sigma = I$. Since A^* involves N_1 and N_2 explicitly, we need to evaluate each derivative of $\exp(A^*)$ in (2.7) up to the necessary order so as to obtain the final expansion up to the third order with respect to $(N_1^{-1}, N_2^{-1}, n^{-1})$.

After quite heavy computations, the characteristic functions $\phi(t)$ and $\psi(t)$ are evaluated in the following forms:

$$\phi(t) = [1 + L(\theta; D) + Q(\theta; D) + C(\theta; D)]\exp(\tfrac{1}{2}\theta^2) + O_4, \tag{2.11}$$

$$\psi(t) = [1 + L^*(\theta; D) + Q^*(\theta; D) + C^*(\theta; D)]\exp(\tfrac{1}{2}\theta^2) + O_4, \tag{2.12}$$

where $L(\theta; D)$, $L^*(\theta; D)$ are the first order terms and $Q(\theta; D)$, $Q^*(\theta; D)$ are the second order terms in the expansions for $\phi(t)$ and $\psi(t)$, all of which are given in [7] and [8]. $C(\theta; D)$ and $C^*(\theta; D)$ are the third order terms and given in Siotani and Wang [11]. It is noted that those terms are polynomials in θ and

$$\int_{-\infty}^{\infty} e^{-\theta c}\, d\Phi^{(r)}(c) = \theta^r \exp(\theta^2/2). \tag{2.13}$$

Hence the application of the inverse formula of the Fourier transformation to $\phi(t)$ and $\psi(t)$ gives

$$\mathbf{P}\{W < \tfrac{1}{2}D^2 + cD \mid \pi_1\} = [1 + L(d; D) + Q(d; D) + C(d; D)]\Phi(c) + O_4 \tag{2.14}$$

$$\mathbf{P}\{Z \leq 2Dc - D^2 \mid \pi_1\} = [1 + L^*(d; D) + Q^*(d; D) + C^*(d; D)]\Phi(c) + O_4 \tag{2.15}$$

since $\theta^r \exp(\theta^2/2)$ corresponds to $d^r\Phi(c)$, where $d = d/dc$. P_1 is now obtained from (2.14) by putting $c = -\tfrac{1}{2}D$ and the coefficients c_{ijk} of the cubic terms are given by

$$c_{ijk} = N_iN_jN_k\,[\varepsilon_{ijk}L_i(d;D)L_j(d;D)L_k(d;D)$$
$$+\varepsilon_{ijk}\{(1+\delta_{jk})L_i(d;D)Q_{jk}(d;D)$$
$$+(1+\delta_{ij})L_k(d;D)Q_{ij}(d;D)+(1+\delta_{ik})L_j(d;D)Q_{ik}(d;D)\} \tag{2.16}$$
$$+C_{ijk}(d;D)]\Phi(c)|_{c=-\frac{1}{2}D} \qquad (i \leqslant j \leqslant k = 1,2,3)$$

where δ_{ij} are the Kronecker symbols, $\varepsilon_{ijk} = 1/6$ for $i = j = k$; $= 1/2$ for $i = j \neq k$ or $i = k \neq j$ or $i \neq j = k$; $= 1$ for $i \neq j \neq k \neq i$, $N_3 = n$, L_i's, Q_{ij}'s are the same functions as in [8] and C_{ijk}'s are given in [11]. Similarly we obtain from (2.15) with $c = \frac{1}{2}D$

$$P_1^* = 1-[1+L^*(d;D)+Q^*(d;D)+C^*(d;D)]\Phi(c)|_{c=\frac{1}{2}D}+O_4$$
$$=[1+L^*(-d;D)+Q^*(-d;D)+C^*(-d;D)]\Phi(c)|_{c=-\frac{1}{2}D}+O_4 \tag{2.17}$$

and the coefficients c^*_{ijk} are written as

$$c^*_{ijk} = N_iN_jN_k\,[\varepsilon_{ijk}L^*_i(-d;D)L^*_j(-d;D)L^*_k(-d;D)$$
$$+\varepsilon_{ijk}\{(1+\delta_{jk})L^*_i(-d;D)Q^*_{jk}(-d;D)$$
$$+(1+\delta_{ik})L^*_j(-d;D)Q^*_{ik}(-d;D) \tag{2.18}$$
$$+(1+\delta_{ij})L^*_k(-d;D)Q^*_{ij}(-d;D)\}$$
$$+C^*_{ijk}(-d;D)]\Phi(c)|_{c=-\frac{1}{2}D} \qquad (i \leqslant j \leqslant k = 1,2,3).$$

Functions L^*_i's, Q^*_{ij}'s are given in [7] and C^*_{ijk} in [11].

Final forms of c_{ijk} and c^*_{ijk} in Section 2.1 can be obtained by eliminating D appeared in (2.16) and (2.18) by repeated applications of the recurrence relation

$$Dd_0^i = 2d_0^{i+1}+2(i-1)d_0^{i-1}, \qquad i = 1,2,\ldots \tag{2.19}$$

2.3. Tables of coefficients c_{ijk} and c^*_{ijk}

We calculated extensive tables of (a_i, b_{ij}, c_{ijk}) and $(a^*_i, b^*_{ij}, c^*_{ijk})$ for $p = 1(1)20, 25, 30, 40, 50$ and $D = 0.4(0.2)2.0(0.4)4.0, 3.0, 5.0(1.0)8.0$. (See Tables 2.1 and 2.2.) But we present here only the tables of c_{ijk} and c^*_{ijk} for $p = 1, 2, 3, 5, 7, 10, 20, 50$ and $D = 1, 2, 3, 4, 6, 8$ which are supplementary portions to Table I of a_i, b_{ij} in [8] and Table I of a^*_i, b^*_{ij} in [7], respectively.

3. The validity of the expansion formulae

It is helpful to have the information about the values of parameters for which the expansion formulae in Section 2.1 give us good approximations to the values of error rates. To do so, we define that the formulae (2.1) and (2.2) for the W-procedure (the formulae (2.4) and (2.5) for the Z-procedure) are *valid* if

$$\max_{i=1,2}\left\{\frac{\text{Absolute value of the third order term in the expansion for } P_i,\ (P_i^*)}{\Phi(-\frac{1}{2}D)}\right\} < 0.003 \quad (3.1)$$

holds. The number 0.003 was chosen as a compromising value after we examined the values of (N_1, N_2) satisfying the inequalities with 0.005 and 0.001 for several values of p and D. In the case of $n = N_1 + N_2 - 2$, the validity depends on N_1, N_2, p and D. We call a pair (N_1, N_2) satisfying (3.1) for fixed p and D a *valid point* of the formulae (2.1) and (2.2) for the error rates of the W-procedure ((2.4) and (2.5) for the error rates of the Z-procedure) and denote the set of all valid points for given p and D by $R_{p,D}$ $(R^*_{p,D})$. Let

$$U_D = 0.003 \cdot \Phi(-\tfrac{1}{2}D). \quad (3.2)$$

This is an upper bound for the absolute value of the third order term satisfying the inequality criterion for a fixed D and any p. Table 3.1 gives the values of U_D. Consulting with these values, we find the *points of the lower boundaries* of $R_{p,D}$ and $R^*_{p,D}$. For the extreme case where $N_1 = \infty$ or $N_2 = \infty$, the inequality (3.1) becomes

$$\text{Max}\left\{\frac{|c_{111}|}{U_D}, \frac{|c_{222}|}{U_D}\right\} < N_1^3 \ (\text{or } N_2^3),\quad \text{Max}\left\{\frac{|c^*_{111}|}{U_D}, \frac{|c^*_{222}|}{U_D}\right\} < N_1^3 \ (\text{or } N_2^3). \quad (3.3)$$

Table 3.1

Values of U_D

D	0.4	0.6	0.8	1.0	1.2	1.4	1.6	1.8
U_D	0.0^2126	0.0^2115	0.0^2103	0.0^3926	0.0^3823	0.0^3726	0.0^3636	0.0^3552
D	2.0	2.4	2.8	3.0	3.6	4.0	5.0	6.0
U_D	0.0^3476	0.0^3345	0.0^3242	0.0^3200	0.0^3108	0.0^4683	0.0^4186	0.0^5405

Let us denote the minimum value of N_1 (or N_2) satisfying (3.3) by $N_{1,\infty}^{(p,D)}$ (or $N_{2,\infty}^{(p,D)}$) for P_i and by $N_{1,\infty}^{*(p,D)}$ (or $N_2^{*(p,D)}$) for P_i^*. Obviously $N_{1,\infty}^{(p,D)} = N_{2,\infty}^{(p,D)}$ and $N_1^{*(p,D)} = N_{2,\infty}^{*(p,D)}$, and the boundaries of $R_{p,D}$ and $R^*_{p,D}$ are symmetric about the line $N_1 = N_2$ in the (N_1, N_2)-plane. Figure 3.1 is a diagram

Table 2.1

Values of c_{ijk}

p	D	c_{111}	c_{222}	c_{112}	c_{122}	c_{113}	c_{223}	c_{123}	c_{133}	c_{233}	c_{333}
1	1	0.0007	0.0007	0.0022	0.0022	0	0	0	0	0	0
	2	0.0005	0.0005	0.0014	0.0014	0	0	0	0	0	0
	3	−0.0002	−0.0002	−0.0005	−0.0005	0	0	0	0	0	0
	4	−0.0003	−0.0003	−0.0009	−0.0009	0	0	0	0	0	0
	6	0.0000	0.0000	0.0001	0.0001	0	0	0	0	0	0
	8	0.0000	0.0000	0.0001	0.0001	0	0	0	0	0	0
2	1	0.4073	−0.0356	0.7792	0.3362	−0.2716	0.0159	−0.0577	−0.1872	0.0576	−0.0461
	2	0.0113	−0.0000	0.0227	0.0113	−0.0492	0.0113	−0.0378	−0.1815	0.0302	−0.1512
	3	0.0007	0.0001	0.0014	0.0009	0.0113	0.0019	−0.0273	0.0190	−0.0319	−0.0873
	4	0.0001	−0.0003	−0.0001	−0.0004	0.0264	−0.0029	−0.0069	0.1683	−0.0359	0.1552
	6	0.0002	−0.0000	0.0005	0.0002	0.0103	−0.0001	0.0058	0.0975	0.0102	0.2310
	8	0.0000	0.0000	0.0001	0.0001	0.0010	0.0002	0.0010	0.0104	0.0033	0.0353
3	1	−0.0094	−0.0015	−0.0036	0.0041	−1.1284	−0.0171	0.1747	−0.7083	0.2159	−0.0770
	2	−0.0024	0.0014	−0.0033	0.0005	−0.2760	0.0265	−0.0681	−0.6049	0.1210	−0.2420
	3	0.0005	0.0008	−0.0017	−0.0014	0.0042	0.0110	−0.0819	0.0683	−0.0774	0.1093
	4	0.0014	−0.0001	0.0003	−0.0012	0.0742	−0.0034	−0.0304	0.5399	−0.1080	0.8639
	6	0.0007	−0.0001	0.0009	0.0002	0.0284	−0.0008	0.0110	0.2991	0.0199	0.8376
	8	0.0001	0.0000	0.0001	0.0001	0.0024	0.0003	0.0020	0.0308	0.0083	0.1199
5	1	−2.2408	−0.0351	−0.9851	0.1646	−4.8448	−0.6091	3.7878	−3.8907	1.2143	0.2200
	2	−0.1602	−0.0014	−0.1404	0.0184	−1.6711	0.0227	0.1663	−2.7222	0.6654	0.4839
	3	−0.0073	0.0015	−0.0345	−0.0016	−0.1512	0.0492	−0.2639	0.2150	−0.1951	2.5742
	4	0.0082	0.0006	−0.0059	−0.0033	0.2438	0.0076	−0.1535	2.1056	−0.4319	5.3991
	6	0.0022	−0.0001	0.0015	−0.0001	0.0946	−0.0030	0.0141	1.1179	0.0041	3.8557
	8	0.0002	0.0000	0.0002	0.0001	0.0073	0.0003	0.0041	0.1101	0.0203	0.5149

Table 2.1 (continued)

Values of c_{jk}

7	1	−4.5480	−0.7277	7.5878	6.1276	−11.645	−2.5350	15.394	−11.825	3.7544	1.5183
	2	−0.5865	−0.0307	−0.2963	0.2595	−5.0020	−0.1021	1.2476	−7.4407	1.9963	3.6295
	3	−0.0322	−0.0011	−0.[illegible]221	0.0289	−0.6386	0.1115	−0.5271	0.3552	−0.3248	9.8352
	4	0.0227	0.0009	−0.0297	−0.0009	0.5277	0.0418	−0.4226	5.1224	−1.1136	16.521
	6	0.0052	−0.0000	0.0013	−0.0006	0.2119	−0.0049	−0.0024	2.6624	−0.1163	10.410
	8	0.0003	−0.0000	0.0003	0.0000	0.0153	0.0001	0.0058	0.2549	0.0304	1.3407
10	1	−3.4579	−4.3735	6[illegible].393	38.296	−28.708	−9.5858	57.754	−37.616	12.116	6.7146
	2	−1.9774	−0.1739	−0.3176	1.4858	−15.4[illegible]4	−0.7146	5.6485	−21.777	6.2610	16.061
	3	−0.1207	−0.0190	−0.4070	0.1983	−2.3688	0.2433	−1.0328	0.3672	−0.4798	35.059
	4	0.0651	−0.0009	−0.1212	0.0254	1.2093	0.1559	−1.2162	13.298	−3.1167	51.933
	6	0.0131	0.0001	−0.0017	−0.0009	0.5125	−0.0027	−0.0885	6.7906	−0.6432	29.840
	8	0.0006	−0.0000	0.0003	−0.0000	0.0345	−0.0003	0.0063	0.6312	0.0297	3.7279
20	1	126.77	−64.596	1023.3	576.11	−165.20	−99.999	600.24	−336.49	110.27	77.671
	2	−17.734	−2.4329	6.0623	21.363	−130.24	−10.129	66.520	−176.71	54.882	183.32
	3	−1.2151	−0.3347	−3.6321	3.0586	−23.793	0.8865	−2.9119	−4.7640	0.4970	342.49
	4	0.4979	−0.0548	−1.3311	0.5687	6.4700	1.4411	−9.3998	89.463	−24.051	450.72
	6	0.0854	−0.0005	−0.0619	0.0113	3.1325	0.1249	−1.3739	44.623	−8.1906	232.88
	8	0.0032	0.0000	−0.0006	−0.0000	0.1909	0.0002	−0.0305	3.9481	−0.3367	27.966
50	1	3992.3	−1363.1	22452.	12231.	−1841.6	−1804.9	10828.	−5646.7	1869.6	1459.8
	2	−289.58	−50.043	206.68	446.22	−2092.1	−203.97	1260.9	−2801.1	908.51	3407.3
	3	−21.276	−7.2437	−[illegible]9.286	65.220	−417.74	1.3153	9.9673	−159.26	47.671	5950.2
	4	7.4607	−1.4244	−24.450	13.293	70.243	23.171	−142.21	1226.5	−370.81	7372.8
	6	1.1593	−0.0420	−[illegible].3693	0.4526	40.155	3.6699	−26.752	601.85	−158.14	3576.2
	8	0.0363	−0.0005	−0.0283	0.0066	2.2523	0.1146	−1.0597	50.977	−10.812	418.39

Table 2.2

Values of c^*_{ijk}

p	D	c^*_{111}	c^*_{222}	c^*_{112}	c^*_{122}	c^*_{113}	c^*_{223}	c^*_{123}	c^*_{133}	c^*_{233}	c^*_{333}
1	1	−0.0084	0.0256	−0.0073	−0.0054	0	0	0	0	0	0
	2	−0.0137	0.0222	−0.0052	0.0005	0	0	0	0	0	0
	3	−0.0133	0.0006	0.0015	0.0100	0	0	0	0	0	0
	4	−0.0094	−0.0081	0.0069	0.0081	0	0	0	0	0	0
	6	−0.0020	−0.0000	0.0043	−0.0021	0	0	0	0	0	0
	8	−0.0001	0.0004	0.0005	−0.0006	0	0	0	0	0	0
2	1	0.1192	0.2220	0.5100	0.6353	−0.0619	−0.0852	−0.1666	−0.0576	−0.0721	−0.0461
	2	−0.0038	0.0227	−0.0000	0.0265	−0.0189	−0.0492	−0.0076	−0.0302	−0.1210	−0.1512
	3	−0.0078	−0.0017	−0.0029	0.0154	−0.0069	−0.0411	0.0339	0.0319	−0.0448	−0.0873
	4	−0.0066	−0.0081	0.0035	0.0105	0.0016	−0.0094	0.0243	0.0359	0.0966	0.1552
	6	−0.0016	0.0006	0.0036	−0.0016	0.0036	0.0190	−0.0066	−0.0102	0.1178	0.2310
	8	−0.0001	0.0005	0.0005	−0.0006	0.0005	0.0035	−0.0019	−0.0033	0.0170	0.0353
3	1	0.0390	0.0033	−0.0466	−0.0056	−0.1615	−0.2714	−0.5382	−0.2159	−0.2765	−0.0770
	2	0.0080	0.0061	−0.0232	0.0052	−0.0643	−0.1550	−0.0983	−0.1210	−0.3630	−0.2420
	3	−0.0019	−0.0062	−0.0096	0.0159	−0.0256	−0.0964	0.0553	0.0774	−0.0865	0.1093
	4	−0.0040	−0.0082	−0.0001	0.0126	−0.0034	−0.0135	0.0574	0.1080	0.3239	0.8639
	6	−0.0013	0.0012	0.0029	−0.0011	0.0057	0.0421	−0.0091	−0.0199	0.3390	0.8376
	8	−0.0001	0.0006	0.0004	−0.0006	0.0009	0.0073	−0.0034	−0.0083	0.0474	0.1199
5	1	0.3601	−0.5785	−1.0639	−1.8135	−0.0618	−0.5332	−1.0709	−1.2143	−1.4621	0.2200
	2	0.0487	−0.0666	−0.1205	−0.1451	−0.2798	−0.5823	−0.6201	−0.6654	−1.3914	0.4839
	3	0.0105	−0.0220	−0.0314	0.0010	−0.1181	−0.2691	0.0212	0.1951	−0.1752	2.5742
	4	0.0009	−0.0090	−0.0081	0.0158	−0.0363	−0.0093	0.1434	0.4319	1.2418	5.3991
	6	−0.0007	0.0025	0.0015	0.0002	0.0056	0.1012	−0.0011	−0.0041	1.1260	3.8557
	8	−0.0001	0.0007	0.0003	−0.0005	0.0014	0.0160	−0.0057	−0.0203	0.1507	0.5149

Table 2.2 (continued)

Values of c^*_{ijk}

7	1	3.1663	0.4949	3.5751	1.2007	1.1245	0.0373	0.0522	−3.7544	−4.3158	1.5183
	2	0.1309	−0.1734	−0.2292	−0.3323	−0.7373	−1.3725	−1.7467	−1.9963	−3.4482	3.6295
	3	0.0233	−0.0475	−0.0657	−0.0366	−0.3167	−0.5572	−0.1803	0.3248	−0.2944	9.8352
	4	0.0049	−0.0109	−0.0179	0.0169	−0.1050	0.0063	0.2455	1.1136	2.8953	16.521
	6	−0.0003	0.0041	0.0003	0.0018	0.0005	0.1783	0.0259	0.1163	2.4297	10.410
	8	−0.0001	0.0008	0.0002	−0.0004	0.0015	0.0262	−0.0065	−0.0304	0.3156	1.4507
10	1	16.873	10.046	35.652	29.287	6.7461	4.2617	8.4515	−12.116	−13.385	6.7146
	2	0.3819	−0.3478	−0.2647	−0.7524	−2.0757	−3.4368	−4.9680	−6.2610	−9.2555	16.061
	3	0.0418	−0.1056	−0.1449	−0.1398	−0.3989	−1.2916	−0.9679	0.4798	−0.5925	35.059
	4	0.0089	−0.0161	−0.0376	0.0132	−0.2892	0.0372	0.4010	3.1167	7.0648	51.933
	6	0.0002	0.0067	−0.0014	0.0050	−0.0143	0.3297	0.1059	0.6432	5.5043	29.840
	8	−0.0000	0.0010	0.0001	−0.0002	0.0010	0.0445	−0.0049	−0.0297	0.6907	3.7279
20	1	234.71	200.53	629.78	595.61	92.304	81.045	161.68	−110.27	−115.96	77.671
	2	3.2458	−0.1305	3.4378	0.6560	−15.876	−21.623	−36.349	−54.882	−66.950	183.32
	3	0.0710	−0.4959	−0.7296	−0.9686	−6.8935	−7.9469	−10.978	−0.4970	−3.7701	342.49
	4	−0.0052	−0.0645	−0.1753	−0.0744	−1.8628	−0.0516	0.4258	24.051	41.362	450.72
	6	0.0002	0.0182	−0.0057	0.0215	−0.0887	1.1813	0.7909	8.1906	28.242	232.88
	8	0.0000	0.0019	−0.0001	0.0007	−0.0027	0.1332	0.0300	0.3367	3.2748	27.966
50	1	4854.5	4613.7	14041.	13803.	1353.8	1777.3	3551.5	−1869.6	−1907.5	1459.8
	2	54.939	32.280	123.63	102.43	−241.02	−278.08	−516.13	−908.51	−984.10	3407.3
	3	−0.8400	−4.1831	−7.5365	−10.026	−104.36	−108.81	−193.29	−47.671	−63.920	5950.2
	4	−0.5556	−0.7494	−2.0179	−1.7983	−22.243	−8.8913	−17.659	370.81	484.90	7372.8
	6	−0.0132	0.0809	0.0009	0.1320	0.4498	7.9327	8.6901	158.14	285.58	3576.2
	8	−0.0001	0.0065	0.0009	0.0075	0.0334	0.7267	0.5470	10.812	29.354	418.39

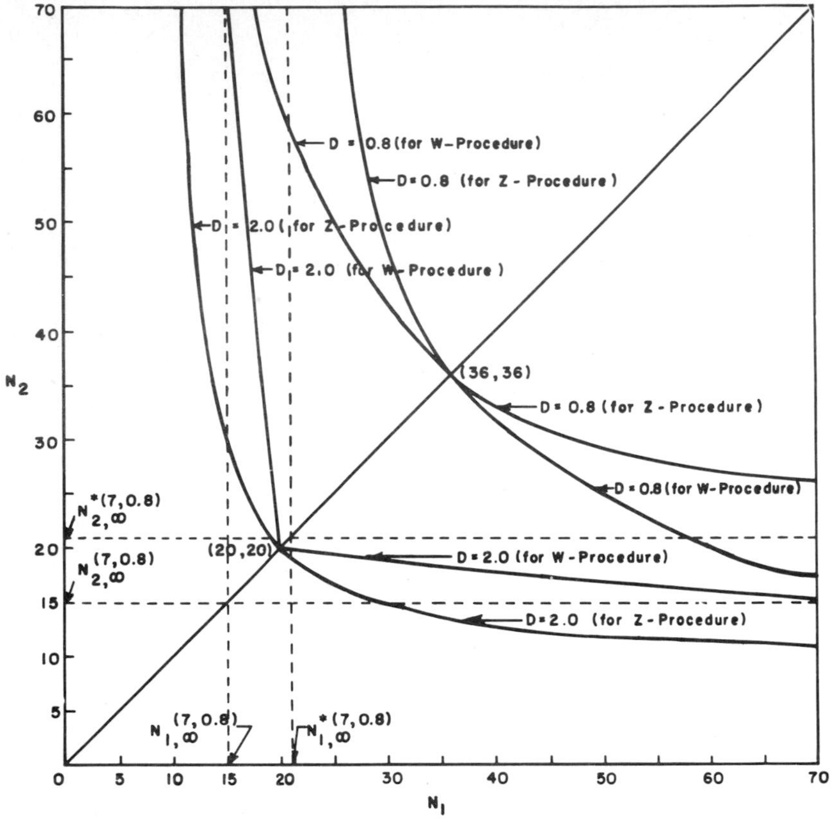

Fig. 3.1. Boundaries of $R_{7,D}$ and $R^*_{7,D}$ when $D = 0.8$, 2.0.

showing the boundaries of $R_{p,D}$ and $R^*_{p,D}$ when $p = 7$ and $D = 0.8, 2.0$. Charts of this kind have been prepared in [11] for selected values of p and D.

4. Comparison

We now consider the comparison of the W-procedure and the Z-procedure with respect to error rates calculated by the expansion formulae given in Section 2.1. Since, for $N_1 = N_2$, both procedures are equivalent and hence $P_i = P^*_i$, $i = 1, 2$ and since coefficients a_i, b_{ij}, c_{ijk} and a^*_i, b^*_{ij}, c^*_{ijk} do not depend on N_i and n, we have the following equalities:

$$a_1 + a_2 = a_1^* + a_2^*, \qquad a_3 = a_3^*, \tag{4.1}$$

$$b_{11} + b_{12} + b_{22} = b_{11}^* + b_{12}^* + b_{22}^*, \quad b_{13} + b_{23} = b_{13}^* + b_{23}^*, \quad b_{33} = b_{33}^*, \tag{4.2}$$

$$\begin{aligned} &c_{111} + c_{112} + c_{122} + c_{222} = c_{111}^* + c_{112}^* + c_{122}^* + c_{222}^*, \qquad c_{333} = c_{333}^*, \\ &c_{113} + c_{123} + c_{223} = c_{113}^* + c_{123}^* + c_{223}^*, \qquad c_{133} + c_{233} = c_{133}^* + c_{233}^*. \end{aligned} \tag{4.3}$$

(These relations were used for checking our formulae in Section 2.1.) Using the above relations, the difference Δ defined by (1.5) can be written as

$$\Delta = \left(\frac{1}{N_1} - \frac{1}{N_2}\right)^2 \left[\Delta_b + \left(\frac{1}{N_1} + \frac{1}{N_2}\right) \Delta_{c(1)} + \frac{1}{n} \Delta_{c(2)}\right] + O_4 \tag{4.4}$$

where

$$\begin{aligned} \Delta_b = b_{12}^* - b_{12}, \qquad &\Delta_{c(1)} = (c_{112}^* + c_{122}^*) - (c_{112} + c_{122}), \\ &\Delta_{c(2)} = c_{123}^* - c_{123}. \end{aligned} \tag{4.5}$$

It can be proved that each term of higher order with respect to $(1/N_1, 1/N_2, 1/n)$ in the difference Δ has the factor $(N_1^{-1} - N_2^{-1})^2$, so that $\Delta = 0$ when $N_1 = N_2$. When N_1 and N_2 are different but both are sufficiently large, Δ is negligibly small. Our numerical comparison is based on the expansion formula (4.4) for Δ; that is, based on the sign of the expression in brackets, assuming that the absolute value of the term O_4 is small enough. For this, values of Δ_b, $\Delta_{c(1)}$ and $\Delta_{c(2)}$ are given in Table 4.1 for $p = 1(1)11, 13, 15, 17, 20, 25$; $D = 0.6, 1.0, 1.4, 1.6, 2.0, 2.4, 3.0, 3.6, 4.0, 5.0$.

From Table 4.1, we can obtain some information on the comparison without further computation. It is seen that for smaller p, the sign of Δ is determined by the leading term Δ_b for almost all sample sizes if they are not so small, since in those cases the absolute values of Δ_b have magnitude of the same degree as those of $\Delta_{c(1)}$ and $\Delta_{c(2)}$, which are multiplied by $(N_1^{-1} + N_2^{-1})$ and n^{-1}, respectively. But this dominant role by Δ_b becomes ambiguous as p increases when D is small, where the values of Δ_b are positive while the both $\Delta_{c(1)}$ and $\Delta_{c(2)}$ are negative and absolute values of $\Delta_{c(1)}$ and $\Delta_{c(2)}$ divided by N_i and n become comparable with Δ_b when at least one of N_1 and N_2 is not large. For example, let us consider the case when $p = 10$ and $D = 0.6$. Then, we have from Table 4.1

$$\Delta = \left(\frac{1}{N_1} - \frac{1}{N_2}\right)^2 \left[6.44958 + \left(\frac{1}{N_1} + \frac{1}{N_2}\right)(-182.9997) + \frac{1}{n}(-95.9132)\right] + O_4$$

whose sign is changeable according to values of N_1 and N_2. Δ is negative at point $(N_1, N_2) = (40,90)$ or $(90, 40)$ in which case the W-procedure is better

Table 4.1

Values of Δ_b, $\Delta_{c(1)}$ and $\Delta_{c(2)}$

p	D	Δ_b	$\Delta_{c(1)}$	$\Delta_{c(2)}$	p	D	Δ_b	$\Delta_{c(1)}$	$\Delta_{c(2)}$
1	0.6	0.01366	−0.01327	0	5	0.6	1.28495	−10.5104	−9.67775
	0.8	0.01694	−0.01605	0		0.8	0.93761	−4.25189	−6.72509
	1.0	0.01925	−0.01762	0		1.0	0.72338	−2.05988	−4.85878
	1.4	0.02063	−0.01686	0		1.4	0.46670	−0.64094	−2.57928
	1.6	0.01970	−0.01465	0		1.6	0.38181	−0.38479	−1.83113
	2.0	0.01512	−0.00757	0		2.0	0.25709	−0.14367	−0.78640
	2.4	0.00816	−0.00118	0		2.4	0.16997	−0.04706	−0.15491
	3.0	−0.00303	0.01242	0		3.0	0.08331	0.00574	0.28517
	3.6	−0.01101	0.01735	0		3.6	0.03285	0.01775	0.35437
	4.0	−0.01350	0.01687	0		4.0	0.01349	0.01686	0.29695
	5.0	−0.01164	0.00896	0		5.0	−0.00463	0.00686	0.08735
2	0.6	0.09311	0.27977	−0.25149	6	0.6	2.00006	−24.4195	−18.0226
	0.8	0.07448	0.09216	−0.16561		0.8	1.45549	−9.80958	−12.5667
	1.0	0.06327	0.02879	−0.10899		1.0	1.11946	−4.70896	−9.12654
	1.4	0.04851	−0.00687	−0.03467		1.4	0.71762	−1.43065	−4.94080
	1.6	0.04233	−0.01024	−0.00831		1.6	0.58551	−0.84754	−3.57215
	2.0	0.03025	−0.00755	0.03025		2.0	0.39320	−0.31002	−1.66355
	2.4	0.01827	−0.00023	0.05291		2.4	0.26100	−0.10428	−0.50413
	3.0	0.00236	0.01025	0.06117		3.0	0.13188	0.00041	0.33286
	3.6	−0.00827	0.01491	0.04682		3.6	0.05752	0.02258	0.51915
	2.0	−0.01181	0.01449	0.03122		4.0	0.02869	0.02189	0.45977
	5.0	−0.01120	0.00731	−0.00264		5.0	−0.00069	0.00875	0.16424

Table 4.1 (continued)

Values of Δ_b, $\Delta_{c(1)}$ and $\Delta_{c(2)}$

3	0.6	0.33148	−0.08752	−1.48584	7	0.6	2.87407	−46.6962	−30.1249
	0.8	0.24711	−0.06706	−1.01482		0.8	2.08846	−18.6988	−21.0529
	1.0	0.19529	−0.05406	−0.71307		1.0	1.60355	−8.93959	−15.3417
	1.4	0.13215	−0.03617	−0.33615		1.4	1.02430	−2.68806	−8.40887
	1.6	0.11022	−0.02872	−0.20944		1.6	0.83446	−1.58356	−6.14695
	2.0	0.07564	−0.01512	−0.03025		2.0	0.55956	−0.57468	−2.99439
	2.4	0.04861	−0.00326	0.07628		2.4	0.37225	−0.19599	−1.07221
	3.0	0.01855	0.00930	0.15719		3.0	0.19124	−0.00907	0.34681
	3.6	−0.00004	0.01407	0.12017		3.6	0.08768	0.02960	0.71841
	4.0	−0.00675	0.01350	0.08774		4.0	0.04724	0.02953	0.66814
	5.0	−0.00989	0.00630	0.00818		5.0	0.00413	0.01198	0.27169
4	0.6	0.72876	−3.04196	−4.39672	8	0.6	3.90699	−79.2673	−46.6784
	0.8	0.53482	−1.25973	−3.03789		0.8	2.83651	−31.6856	−32.6740
	1.0	0.41532	−0.62927	−2.17531		1.0	2.17566	−15.1148	−23.8672
	1.4	0.27155	−0.21181	−1.11438		1.4	1.38673	−4.52028	−13.1934
	1.6	0.22339	−0.13272	−0.76364		1.6	1.12868	−2.65544	−9.71575
	2.0	0.15124	−0.05293	−0.27222		2.0	0.75616	−0.96033	−4.86966
	2.4	0.09918	−0.01614	0.02279		2.4	0.50373	−0.33042	−1.90649
	3.0	0.04554	0.00826	0.21590		3.0	0.26140	−0.02407	0.31490
	3.6	0.01366	0.01495	0.22203		3.6	0.12331	0.03897	0.95405
	4.0	0.00169	0.01418	0.17463		4.0	0.06918	0.04012	0.92712
	5.0	−0.00769	0.00609	0.03676		5.0	0.00984	0.01668	0.41398

Table 4.1 (continued)

Values of Δ_b, $\Delta_{c(1)}$ and $\Delta_{c(2)}$

p	D	Δ_b	$\Delta_{c(1)}$	$\Delta_{c(2)}$	p	D	Δ_b	$\Delta_{c(1)}$	$\Delta_{c(2)}$
9	0.6	5.09883	−124.0595	−68.3766	15	0.6	15.5870	−757.358	−345.450
	0.8	3.69964	−49.5291	−47.9203		0.8	11.2952	−301.672	−242.981
	1.0	2.83578	−23.5977	−35.0663		1.0	8.64485	−143.288	−178.736
	1.4	1.80493	−7.03442	−19.5044		1.4	5.48507	−42.4376	−101.181
	1.6	1.46815	−4.12577	−14.4388		1.6	4.45561	−24.8204	−75.9893
	2.0	0.98301	−1.48963	−7.38011		2.0	2.97926	−8.95292	−40.8628
	2.4	0.65544	−0.51578	−3.05430		2.4	1.99047	−3.15768	−19.1759
	3.0	0.34235	−0.04593	0.22497		3.0	1.05469	−0.39661	−2.21285
	3.6	0.16443	0.05084	1.23807		3.6	0.52629	0.18309	3.78865
	4.0	0.09448	0.05399	1.24179		4.0	0.31720	0.22356	4.58249
	5.0	0.01640	0.02305	0.59539		5.0	0.07425	0.10563	2.74436
10	0.6	6.44958	−182.9997	−95.9132	17	0.6	20.3543	−1153.58	−510.411
	0.8	4.67786	−73.0161	−67.2819		0.8	14.7477	−459.363	−359.227
	1.0	3.58391	−34.7514	−49.3019		1.0	11.2853	−218.106	−264.473
	1.4	2.27889	−10.3376	−27.5517		1.4	7.15786	−64.6596	−150.136
	1.6	1.85290	−6.05714	−20.4763		1.6	5.81353	−37.7439	−113.004
	2.0	1.24010	−2.18530	−10.6165		2.0	3.88666	−13.6184	−61.2186
	2.4	0.82738	−0.76029	−4.56298		2.4	2.59731	−4.81829	−29.2038
	3.0	0.43409	−0.07598	0.06489		3.0	1.37849	−0.62954	−4.04001
	3.6	0.21103	0.06536	1.54244		3.6	0.69077	0.25406	5.03869
	4.0	0.12317	0.07145	1.61720		4.0	0.41843	0.32056	6.37094
	5.0	0.02385	0.03125	0.82017		5.0	0.10054	0.15415	3.96725

Table 4.1 (continued)

Values of Δ_b, $\Delta_{c(1)}$ and $\Delta_{c(2)}$

11	0.6	7.95924	−258.0146	−129.9818	20	0.6	28.6972	−1974.44	−844.995
	0.8	5.77116	−102.8918	−91.2493		0.8	20.7897	−785.974	−595.104
	1.0	4.42007	−48.9389	−66.9371		1.0	15.9062	−373.054	−438.544
	1.4	2.80861	−14.5369	−37.5453		1.4	10.0852	−110.330	−249.710
	1.6	2.28292	−8.51217	−27.9885		1.6	8.18990	−64.4987	−188.399
	2.0	1.52744	−3.07001	−14.6695		2.0	5.47459	−23.2821	−102.868
	2.4	1.01954	−1.07217	−6.47985		2.4	3.65926	−8.26642	−49.9104
	3.0	0.53652	−0.11561	−0.17750		3.0	1.94512	−1.12481	−8.06581
	3.6	0.26312	0.08268	1.89913		3.6	0.97860	0.38994	7.33974
	4.0	0.15523	0.09280	2.05840		4.0	0.59559	0.51276	9.82552
	5.0	0.03218	0.04144	1.09261		5.0	0.14655	0.25135	6.39400
13	0.6	11.45529	−463.9761	−220.4899	25	0.6	45.7802	−4075.08	−1681.25
	0.8	8.30302	−184.8944	−154.9621		0.8	33.1613	−1621.68	−1184.93
	1.0	6.35643	−87.8675	−113.8588		1.0	25.3680	−769.432	−874.109
	1.4	4.03532	−26.0523	−64.2112		1.4	16.0794	−227.403	−499.380
	1.6	3.27873	−15.2435	−48.0781		1.6	13.0558	−132.916	−377.736
	2.0	2.19286	−5.49727	−25.5884		2.0	8.72607	−48.0085	−207.974
	2.4	1.46455	−1.93087	−11.7275		2.4	5.83374	−17.1137	−102.683
	3.0	0.77407	−0.22891	−0.95772		3.0	3.10538	−2.42918	−19.0118
	3.6	0.38374	0.12635	2.74740		3.6	1.56799	0.70498	13.4424
	4.0	0.22946	0.14843	3.15847		4.0	0.95834	0.97859	17.9790
	5.0	0.05146	0.06851	1.79754		5.0	0.24076	0.48998	12.3031

than the Z-procedure, while it is positive at point $(N_1, N_2) = (40, 160)$ or $(160, 40)$ in which case the Z-procedure is better than the W-procedure. According to our numerical computation, the similar behaviour of Δ is appeared when D is in a small neighbourhood of zero even if p is small. But this neighbourhood is not an interesting region for practical applications, because the discriminant analysis itself is not significant.

To have a security for the smallness of absolute value of the term O_4, and to avoid the complexity, we consider the points (N_1, N_2) and values of D restricted by

$$N_1, N_2 \geqslant 30 \quad \text{and} \quad 5.0 \geqslant D \geqslant 0.6 \tag{4.6}$$

which may cover important cases for practical applications.

The comparison is summarized in Tables 4.2–4.4. Table 4.2 gives the summary of the comparsion with the following entries:

Z (W): the Z- (W-) procedure is always better than the W- (Z-) procedure at all points (N_1, N_2) where $N_1, N_2 \geqslant 30$ and $N_1 \neq N_2$.

W or Z: the conclusion for the superiority is changeable depending on values of N_1 and N_2.

Details for the case "W or Z" are given in Table 4.3 and Table 4.4. Let N_1^* denote either N_1 or N_2 and N_2^* denote the other. Table 4.3 gives two critical values N_0^* and M_0^*;

N_0^*: critical value of N_2^* such that for a given value of N_1^*, the W-procedure is always bettter than the Z-procedure at all points (N_1^*, N_2^*) or (N_2^*, N_1^*) with $N_2^* \geqslant N_0^*$, otherwise the superiority is reversed.

M_0^*: when $N_1^*, N_2^* \geqslant M_0^*$, the W-procedure is always better than the Z-procedure at all points $(N_1^*, N_2{}^*)$ or (N_2^*, N_1^*) and the superiority is reversed when $M_0^* > N_1^*$, $N_2^* \geqslant 30$.

Entries N_0 and M_0 in Table 4.4 are similar critical values, i.e.,

N_0: critical value of N_2^* such that for a given value of N_1^*, the Z-procedure is better than the W-procedure whenever $N_2^* \geqslant N_0$, otherwise the W-procedure is better.

M_0: whenever $N_1^*, N_2^* \geqslant M_0$, the Z-procedure is always better than the W-procedure and if $N_1^*, N_2^* < M_0$, the superiority is reversed.

We hope that Tables 4.2–4.4 give pretty good information on the choice between the W and the Z procedures in the practical use. Finally it is noted that at points (N_1^*, N_2^*) around the critical points, both procedures are practically equivalent, and for almost valid points defined in Section 3, the superiority of one procedure to the other is determined by the sign of Δ_b.

Table 4.2

Comparison of the W-procedure and the Z-procedure. Summary

D \ p	1	2	3	4	5	6	7	8	9	10	11	13	15	17	20	25
0.6	Z	Z	Z	Z	Z	Z	W or Z	W or Z	W or Z	W or Z	W or Z	W or Z	W or Z	W or Z	W or Z	W or Z
0.8	Z	Z	Z	Z	Z	Z	Z	Z	W or Z	W or Z	W or Z	W or Z	W or Z	W or Z	E or Z	W or Z
1.0	Z	Z	Z	Z	Z	Z	Z	Z	Z	Z	Z	W or Z	W or Z	W or Z	W or Z	W or Z
1.4	Z	Z	Z	Z	Z	Z	Z	Z	Z	Z	Z	Z	Z	Z	W or Z	W or Z
1.6	Z	Z	Z	Z	Z	Z	Z	Z	Z	Z	Z	Z	Z	Z	Z	W or Z
2.0	Z	Z	Z	Z	Z	Z	Z	Z	Z	Z	Z	Z	Z	Z	Z	Z
2.4	Z	Z	Z	Z	Z	Z	Z	Z	Z	Z	Z	Z	Z	Z	Z	Z
3.0	W	Z	Z	Z	Z	Z	Z	Z	Z	Z	Z	Z	Z	Z	Z	Z
3.6	W	W	Z	Z	Z	Z	Z	Z	Z	Z	Z	Z	Z	Z	Z	Z
4.0	W	W	W	Z	Z	Z	Z	Z	Z	Z	Z	Z	Z	Z	Z	Z
5.0	W	W	W	W	W	W or Z	Z	Z	Z	Z	Z	Z	Z	Z	Z	Z

Table 4.3

Comparison of the W-procedure and the Z-procedure; Details for the entry "W or Z" (a): Values of N_0^* and M_0^*

p	D \ N_1^*	30	35	40	45	50	55	60	65	70	75	80	85	90	95	100	110	120	130	140	M_0^*
6	5.0	399	356	327	303	285	271	259	248	238	229	221	2[illegible]3	206	199	193	180	168	157	147	144

Table 4.4

Comparison of the W-procedure and the Z-procedure; Details for the entry "W or Z" (b): Values of N_0 and M_0

p	D \ N_1^*	30	32	34	35	36	38	40	42	44	45	46	48	50	55	M_0
7	0.6	50	46	43	42	40										38
8	0.6	91	80	71	68	65	60	56	53	50	49					47
9	0.6	191	149	123	114	107	95	86	79	74	72	70	66	63	57	56
	0.8	38	36													34

Table 4.4 (continued)

Comparison of the W-procedure and the Z-procedure; Details for the entry "W or Z" (b): Values of N_0 and M_0

p	D \ N_1^*	30	32	34	35	36	38	40	45	50	55	60	65	70	75	80	85	90	95	100	M_0
10	0.6	789	372	252	219	194	150	138	106	88	77	70									65
	0.8	53	48	45	43	42	40														39
11	0.6	∞	∞	1037	650	479	321	246	163	126	106	93	84	77							74
	0.8	73	65	58	57	55	52	49	46												44
13	0.6	∞	∞	∞	∞	∞	∞	∞	585	301	213	169	143	126	114	105	98	92			91
	0.8	148	123	107	100	95	86	79	67	59											55
	1.0	47	43	41	39	38															37
15	0.6	∞	∞	∞	∞	∞	∞	∞	∞	2493	592	356	264	214	183	162	147	135	126	119	109
	0.8	428	279	212	191	174	149	132	104	88	77	70									65
	1.0	71	64	59	57	55	51	49													44
17	0.6	∞	∞	∞	∞	∞	∞	∞	∞	∞	∞	1458	621	411	316	261	226	201	182	168	
	0.8	∞	2072	649	491	398	294	236	164	130	110	97	87	80							75
	1.0	107	94	85	81	77	71	67	58	52											51
20	0.6	∞	∞	∞	∞	∞	∞	∞	∞	∞	∞	∞	∞	5722	1168	681	494	395	333	292	
	0.8	∞	∞	∞	∞	∞	12995	1196	399	254	193	159	138	123	112	103	96	91			91
	1.0	220	177	149	139	131	117	106	88	77	68	62									61
	1.4	41	38	36																	35
25	0.6	∞	∞	∞	∞	∞	∞	∞	∞	∞	∞	∞	∞	∞	∞	∞	∞	11453	1970	1118	
	0.8	∞	∞	∞	∞	∞	∞	∞	∞	3837	742	435	318	256	218	192	172	158	146	137	
	1.0	∞	1227	584	469	395	303	250	179	143	122	107	97	88	82						79
	1.4	69	64	59	57	55	52	49													45
	1.6	42	39	37	36																36

p	D \ N_1^*	105	110	115	120	125	130	135	140	145	150	160	170	180	190	195	M_0
17	0.6	156	147	140	133	128											126
20	0.6	261	238	220	206	194	183	175	167	161	155						153
25	0.6	799	631	528	458	408	369	339	315	295	278	252	232	216	204	198	197
	0.8	129	123	117													116

References

[1] Anderson, T.W. (1951). Classification by multivariate analysis. *Psychometrika* **16**, 31–50.

[2] Cochran, W.G. (1968). Commentary on "Estimation of error rates in discriminant analysis". *Technometrics* **10**, 204–205.

[3] Das Gupta, S. (1965). Optimum classification rules for classification into two multivariate normal populations. *Ann. Math. Statist.* **36**, 1174–1184.

[4] John, S. (1960). On some classification problems — I. *Sankhya* **22**, 301–308.

[5] Kudo, A. (1959). The classificatory problem viewed as a two-decision problem. *Memoires of the Faculty of Science, Kyushu Univ., Ser. A.* **13**, 96–125.

[6] Lachenbruch, P.A. and Mickey, M.R. (1968). Estimation of error rates in discriminant analysis. *Technometrics* **10**, 1–11.

[7] Memon, A.Z. and Okamoto, M. (1971). Asymptotic expansion of the distribution of the Z statistic in discriminant analyis. *J. Multivariate Analysis* **1**, 294–307.

[8] Okamoto, M. (1963). An asymptotic expansion for the distribution of the linear discriminant function. *Ann. Math. Statist.* **34**, 1286–1301.

[9] Okamoto, M. (1968). Correction to "An asymptotic expansion for the distribution of the linear discriminant function". *Ann. Math. Statist.* **39**, 1358–1359.

[10] Siotani, M. (1974). *Recent Development in Asymptotic Expansions for the Nonnull Distributions for the Multivariate Test Statistics.* Technical Report No. 32, Department of Statistics, Kansas State Univ., Manhattan, Kansas. (A portion of this paper has been published in *A Modern Course on Statistical Distributions in Scientific Work*, edited by Patil, Kotz and Ord; Volume 1, 299–317 (1975)).

[11] Siotani, M. and Wang, R.H. (1975). *Further Expansion Formulae for Error Rates and Comparison of the W- and the Z-Procedures in Discriminant Analysis.* Technical Report No. 33, Department of Statistics, Kansas State Univ., Manhattan, Kansas.

LIST OF CONTRIBUTED PAPERS

C. ALF (Bowling Green State University): Results on Laws of Large Numbers for Weighted Sums of Banach-Space Valued Random Variables.

F.B. ALT and P. ZUNDE (Georgia Institute of Technology): Method of Mixtures Clustering Using Multivariate Uniform Distributions.

P.M. BENTLER (University of California, Los Angeles): General Model for Multivariate Structural Analysis.

V.P. BHAPKAR and G.W. SOMES (University of Kentucky): Selection of Largest Proportion for Matched Samples.

L. BREIMAN and W. MEISEL (Technology Service Corporation): The Use of Interpoint Distance in Multidimensional Data Analysis and Pattern Recognition.

M. CAPOBIANCO (St. Johns University): Some Multivariate Aspects of Stagraphics.

R.L. COCCARI and F. LEES (Cleveland State University): A Model of U.S. Investment of Western Europe.

V. COHEN and J. OBADIA (University of Paris, France): Comparison of Samples and Multivariate Analysis*.

D.J. DEWAAL (University of the Orange Free State, South Africa): On Asymptotic Distributions of the J-th ESF's of the Characteristic Roots of the Non Central Wishart Matrix.

J.H. FRIEDMAN (Standard University): Some Nonparametric Multivariate Procedures Based on Adaptive Multivariate Decision Trees.

W. GERSCH and J. YONEMOTO (University of Hawaii): Toward Automatic Classification of Multivariate EEG's Using Parametric Time Series Models.

T. GERSTENKORN (Lodz University, Poland): Multidimensional Truncated Polya Distribution*.

B. GYIRES (Kossuth Lajos University, Hungary): On an Extension of a Theorem of Linnik and Zinger*.

J.J. HARTON (Sperry Univac, Vandenberg AFB, California): Projections, Generalized Inverses, Linear and Non-Linear Programming: A Practical Approach to Computer (Trajectory) Optimization.

R.J. HOFMANN (Miami University): A Generalized Regression Estimate Formula for Factor Scores.

S.T. HUANG and S. CAMBANIS (University of North Carolina): Homogeneous Chaos and Path Properties of Spherically Invariant Processes.

N.L. JOHNSON (University of North Carolina) and S. KOTZ (Temple University): On a Multivariate Generalized Occupancy Problem.

* Presented by title

D. KAZAKOS and T. PAPANTONI-KAZAKOS (Rice University): Feature Extraction from Multivariate Gaussian Observations in Certain Estimation and Detection Problems.

F. KAZIM (Princeton University): The Distribution of a Criterion for Testing Temporal Independence in Partially Homogeneous Gaussian Random Fields.

B.K. KIM (University of Michigan) and J. VAN RYZIN (University of Wisconsin): A Histogram Type Estimator of a Bivariate Density.

T. J. VAN W. KOTZE (South African Medical Research Council, South Africa): A Multivariate Multinomial Model.

W.E. LARIMORE (Analytic Sciences Corporation): Statistical Inference on Vector Random Fields.

J. LEE (Aerospace Research Laboratories): Isolating Constants of Motion for the Homogeneous Turbulence.

K.L. MEHRA and M. SUDHAKARA RAO (University of Alberta, Canada): Weak Convergence in d_q-Metrics of Multidimensional Empirical Processes.

M. PAVEL (Université René Descartes, France): Pattern Recognition Categories*.

P.J. PHILLIP (American Hospital Association): A Taxonomy of Community Hospitals.

L.D. PITT (University of Virginia): Inequalities for the Gaussian Measure of Symmetric Convex Sets.

E. PRICE and V.D. VANDELINDE (Johns Hopkins University): Densities with Minimum Fisher Information and Location.

G.S. SADHU (State University of New York at Buffalo) and H.L. WEINERT (Johns Hopkins University): Recursive Computation of L-Splines Based on Stochastic Least-Squares Estimation.

B.K. SINHA and N. GIRI (University of Montreal, Canada): On the Optimality and Non-Optimality of Some Multivariate Normal Test Procedures.

M. SOBEL (University of Minnesota) and V.R.R. UPPULURI (Oak Ridge National Laboratories): Generating Functions of Generalized Incomplete Dirichlet Integrals.

K.P. SOHORGHUBER (Vereinigten Fettwarenindustrie Ges, Austria): Factorial Analysis of Regions with Economic Variables per Head Produce False Results*.

J.N. SRIVASTAVA and D.W. MALLENBY (Colorado State University): On the Probability of Correct Search in Search Linear Models.

G. STOCKMAN and L. KANAL (University of Maryland): A System for the Rapid Extraction of Features from Waveform Data.

H. TALPAZ (Texas A&M University): Estimating Intrinsically Nonlinearizable Multivariate Models by an Efficient Search Optimization Method.

H. YASSAEE (Arya–Mehr University of Technology, Iran and University College, London, U.K.): On Comparison of Estimators in Multidimensional Contingency Tables: Logit-Linear Model.

* Presented by title.

AUTHOR INDEX

Balakrishnan, A.V., 341–349
Barlow, R.E., 227–237
Beutler, F.J., 351–370
Bose, R.C., 3–22

Chang, T.C., 105–118
Chernoff, H., 445–456

Das Gupta, S., 457–471
Dvoretzky, A.V., 23–34

Fraser, D.A.S., 35–53
Fujikoshi, Y., 55–71

Helstrom, C.W., 371–385
Hida, T., 239–245
Hitsuda, M., 247–265

James, A.T., 73–77
Jones, R.H., 473–481

Kailath, T., 387–406
Kallianpur, G., 267–281
Kendall, M.G., 483–494
Khatri, C.G., 79–94
Kiefer, J., 143–158
Krishnaiah, P.R., 95–103; 105–118
Kshirsagar, A.M., 129–139

Lee, J.C., 95–103; 105–118
Lewis, P.A.W., 159–174
Liu, L.H., 159–174
Ljung, L., 387–406
Lukacs, E., 119–128

Mardia, K.V., 495–511
McHenry, C., 129–139
Melamed, D., 351–370
Middleton, D., 407–430

Ng, K.W., 35–53

Olkin, I., 175–191

Parzen, E., 283–295
Puri, M.L., 513–522
Proschan, F., 227–237

Rajput, B., 297–309
Rao, C.R., 193–208
Rao, M.M., 311–324
Rao, J.S., 513–522
Robinson, D.W., 159–174
Rosenblatt, M., 159–174
Rozanov, Yu.A., 431–441

Statulevičius, V.A., 325–337
Siotani, M., 523–545
Sylvan, M., 175–191

Vakhania, N.N., 297–309

Wang, R.-H., 523–545
Watanabe, H., 247–265

Zachs, S., 209–223
Zeigler, B.P., 351–370